Guidebook to
Cytokines and Their Receptors

Other books by Sambrook & Tooze Publications

Guidebook to the Cytoskeletal and Motor Proteins
Edited by Thomas Kreis and Ronald Vale

Guidebook to the Extracellular Matrix and Adhesion Proteins
Edited by Thomas Kreis and Ronald Vale

Guidebook to the Homeobox Genes
Edited by Denis Duboule

Guidebook to the Secretory Pathway
Edited by Jonathan Rothblatt, Peter Novick, and Tom Stevens

Guidebook to Cytokines and Their Receptors
Edited by Nicos A. Nicola

Guidebook to the Small GTPases
Edited by Marino Zerial and Lukas A. Huber

Guidebook to the Calcium-Binding Proteins
Edited by Marco Celio; co-edited by Thomas Pauls and Beat Schwaller

Guidebook to Protein Toxins and Their Use in Cell Biology
Edited by Rino Rappuoli and Cesare Montecucco

Guidebook to Molecular Chaperones and Protein-folding Catalysts
Edited by Mary-Jane Gething

Guidebook to
Cytokines and Their Receptors

Edited by
Nicos A. Nicola

*Walter and Eliza Hall
Institute of Medical Research,
and Cooperative Research Centre
for Cellular Growth Factors,
PO, Royal Melbourne Hospital,
Victoria, Australia*

A SAMBROOK AND TOOZE PUBLICATION
at
OXFORD UNIVERSITY PRESS
OXFORD   NEW YORK   TOKYO

Oxford University Press, Great Clarendon Street, Oxford OX2 6DP

Oxford New York

Athens Auckland Bangkok Bogato Bombay Buenos Aires
Calcutta Cape Town Dar es Salaam Delhi Florence Hong Kong
Istanbul Karachi Kuala Lumpur Madras Madrid Melbourne
Mexico City Nairobi Paris Singapore Taipei Tokyo Toronto
and associated companies in
Berlin Ibadan

Oxford is a trade mark of Oxford University Press

Published in the United States
by Oxford University Press Inc., New York

A catalogue record for this book is available from the British Library

Library of Congress Cataloging in Publication Data
Guidebook to cytokines and their receptors /
edited by Nicos A. Nicola. – 1st ed.
(Guidebooks)
'A Sambrook and Tooze publication.'
Includes bibliographical references.
1. Cytokines. 2. Cytokines—Receptors. I. Nicola, Nicos, 1950–
II. Series: Guidebook series (Oxford, England)
[DNLM: 1. Cytokines. 2. Immunity, Cellular.
3. Receptors, Cytokine. QW 568 G946 1994]
QR185.8.C95G85   1994   616.07'9—dc20   94–19802
ISBN 0 19 859947 1 (Hbk)
ISBN 0 19 859946 3 (Pbk)

Printed in Great Britain by The Bath Press Ltd.

Contents

Contributors

Bharat B. Aggarwal Cytokine Research Section, University of Texas, M.D. Anderson Cancer Center, 1515 Holcombe Blvd., Houston, Texas 77030, USA. Tel. (713) 792–3503, Fax (713) 794–1613.

Michael Aguet current address May 1994: Department of Molecular Biology, Genentech, Inc., 460 Point San Bruno, South San Francisco, CA 94080, USA. Tel. (415) 225–1000, Fax 225–6000.

Kari Alitalo Molecular/Cancer Biology Laboratory, Department of Pathology, P. Box 21 (Haartmaninkatu 3), 00014 University of Helsinki, Finland. Tel. 358 0 434 6434, Fax 358 0 434 6448.

Julian L. Ambrus, Jr. Jewish Hospital of St. Louis, Washington University School of Medicine, Rheumatology Division, Box 8045, 660 S. Euclid St., St. Louis, MO 63110, USA. Tel. (314) 362–8602, Fax (314) 362–0419.

R.J. Armitage Department of Cellular Immunology, Immunex Research and Development Corporation, 51 University Street, Seattle, WA 98101, USA. Tel. (206) 587–0430, Fax (206) 233–9733.

Y.-A. Barde Department of Neurobiochemistry, Max-Planck Institute for Psychiatry, 82152 Planegg-Martinsried, Am Klopferspitz 18A, Germany. Tel. 49 89 8578 3614, Fax 49 89 8578 3749, e-mail barde@vms.biochem.mpg.de.

Philip A. Barker Department of Neurobiology, Stanford University School of Medicine, Room D225—Sherman Fairchild Science Building, Stanford, CA 94305–5401, USA. Tel. (415) 723–5811, Fax (415) 725–0388, e-mail PBarker@cmgm.stanford.edu.

Ora Bernard The Walter and Eliza Hall Institute of Medical Research, Post Office, Royal Melbourne Hospital, Victoria 3050, Australia. Tel. 61 3 345 2494, Fax 61 3 347 0852.

Alan Bernstein Division of Molecular and Developmental Biology, Samuel Lunenfeld Research Institute, Mount Sinai Hospital, 600 University Avenue, Toronto, Canada M5G IX5. Tel. (416) 586–8273, Fax (416) 586–8588.

David E. Blumenthal Washington University School of Medicine, Rheumatology Division, Box 8045, 660 S. Euclid St., St. Louis, MO 63110, USA. Tel. (314) 362–8602, Fax (314) 362–0419.

A. Gregory Bruce Bristol-Myers Squibb, Pharmaceutical Research Institute, 3005 First Avenue, Seattle, WA 98121, USA. Tel. (206) 727–3649, Fax (206) 727–3602.

Anthony Burgess Ludwig Institute for Cancer Research, Post Office, Royal Melbourne Hospital, Victoria 3050, Australia. Tel. 61 3 347 3155, Fax 61 3 347 1938.

Graham Carpenter Department of Biochemistry, School of Medicine, Vanderbilt University Nashville, TN 37232–0146, USA. Tel. (615) 322–6678, Fax (615) 322–2931.

Tim Clackson Department of Protein Engineering, Genentech, Inc., 460 Point San Bruno Blvd., South San Francisco, CA 94080–4990, USA. Tel. (415) 225–6193, Fax (415) 225–3734.

Steven C. Clark Genetics Institute, 87 Cambridge Park Drive, Cambridge, MA 02140–2387, USA. Tel. (617) 876–1170, Fax (617) 576–2979.

Lena Claesson-Welsh Ludwig Institute for Cancer Research, Box 595, Uppsala Branch, S-751 24 Uppsala, Sweden. Tel. 46–18–17 41 46, Fax 46–18–50 68 67.

Paolo M. Comoglio Department of Biomedical Sciences and Oncology, University of Torino Medical School, Corso Massimo, D'Azeglio 52, Torino, Italy I-10126. Tel. 39–11–670–7739, Fax 39–11–650–9105.

Rik Derynck Department of Growth and Development and Anatomy, Programs in Cell Biology and Developmental Biology, University of California at San Francisco, 521 Parnassus Avenue, San Francisco, CA 94143–0640, USA. Tel. (415) 476–7322, Fax (415) 476–1499.

Robert J. Donnelly Department of Molecular Genetics and Microbiology, Robert Wood Johnson Medical School, University of Medicine and Dentistry of New Jersey, 675 Hoes Lane, Piscataway, NJ 08854–5635, USA. Tel. (908) 235–4826, Fax (908) 235–5223.

Steven K. Dower Immunex Research and Development Corporation, 51 University Street, Seattle, WA 98101–2977, USA. Tel. (206) 587–0430, Fax (206) 233–9733.

Napoleone Ferrara Department of Cardiovascular Research, Genentech, Inc., 460 Point San Bruno Blvd., South San Francisco, CA 94080, USA. Tel. (415) 225–2968, Fax (415) 225–6327.

Richard J. Ford Department of Molecular Pathology, Box 89, M.D. Anderson Cancer Center, 1515 Holcombe Blvd., Houston, TX 77030–4095, USA. Tel. (713) 792–3121, Fax (713) 794–4672.

David P. Gearing Department of Molecular Biology, Systemix, Inc., 3155 Porter Drive, Palo Alto, CA 94304, USA. Tel. (415) 813–6605, Fax (415) 856–4919.

Stephen E. Gitelman Department of Pediatrics, University of California at San Francisco, 521 Parnassus Avenue, San Francisco, CA 94143–0136, USA. Tel. (415) 476–3748, Fax (415) 476–1343.

Mark A. Goldsmith Gladstone Institute of Virology and Immunology, San Francisco General Hospital, Departments of Medicine and Microbiology and Immunology, University of California at San Francisco, P.O. Box 419100, San Francisco, CA 94141–9100, USA. Tel. (415) 695–3800, Fax (415) 826–1817.

Eugene Goldwasser Department of Biochemistry and Molecular Biology, The University of Chicago, 920 East 58th Street, Chicago, IL 60637–1432, USA. Tel. (312) 702–1348, Fax (312) 702–0439.

Raymond G. Goodwin Immunex Research and Development Corporation, 51 University Street, Seattle, WA 98101, USA. Tel. (206) 587–0430, Fax (206) 233–9733.

Nicholas M. Gough The Walter and Eliza Hall Institute of Medical Research, Post Office, Royal Melbourne Hospital, Parkville, Victoria 3050, Australia, Tel. 61–3–345 2500, Fax

61–3–347 0852, (Present address: AMRAD Operations Pty Ltd, 17–27 Cotham Road, Kew, Victoria 3101, Australia. Tel. 61–3–8530022, Fax 61–3–8530202).

Patrick Gray Department of Leukocyte Biochemistry, ICOS Corporation, 22021 20th Ave. SE, Bothell, WA 98021–4406, USA. Tel. (206) 485–1900, Fax (206) 485–1961.

Andrea Graziani Department of Biomedical Sciences and Oncology, University of Torino Medical School, Corso Massimo, D'Azeglio 52, Torino, Italy I-10126. Tel. 39–11–670–7739, Fax 39–11–650–9105.

Warner C. Greene please address correspondence to Mark A. Goldsmith in this listing.

Madhu Gupta Department of Biochemistry and Molecular Biology, The University of Chicago, 920 East 58th Street, Chicago, IL 60637–1432, USA. Tel. (312) 702–1348, Fax (312) 702–0439.

Nobuyuki Harada DNAX Research Institute of Molecular and Cellular Biology, Inc., 901 California Avenue, Palo Alto, CA 94304–1104, USA. Tel. (415) 852–9196, Fax (415) 496–1200.

Carl-Henrik Heldin Ludwig Institute for Cancer Research, Box 595, Uppsala Branch, S-751 24 Uppsala, Sweden. Tel. 46–18–17 41 46, Fax 46–18–50 68 67.

Douglas J. Hilton The Walter and Eliza Hall Institute of Medical Research, PO Royal Melbourne Hospital, Parkville Vic 3050, Australia. Tel. 61–3–345–2559, Fax 61–3–345–2616.

Richard Horuk The Department of Protein Chemistry, Genentech, Inc., 560 Point San Bruno Blvd., South San Francisco, CA 94080, USA. Tel. (415) 225–1158, Fax (415) 225–5945.

Maureen Howard DNAX Research Institute of Molecular and Cellular Biology, Inc., 901 California Avenue, Palo Alto, CA 94304–1104, USA. Tel. (415) 852–9196, Fax (415) 496–1200.

Tadamitsu Kishimoto Department of Medicine III, Osaka University Medical School 2-2, Yamada-oka, Suita, Osaka, 565, Japan. Tel. 81–6–879–3830, Fax 81–6–879–3839.

Petra I. Knaus Whitehead Institute for Biomedical Research, 9 Cambridge Center, Cambridge, MA 02142, USA. Tel. (617) 258–5226, Fax (617) 258–9872.

Derek LeRoith NIH-Diabetes and Digestive and Kidney Diseases, Building 10, Room 8S239, 9000 Rockville Pike, Bethesda, MD 20892, USA. Tel. (301) 496–8090, Fax (301) 480–4386.

Edward J. Leonard Immunopathology Section, Laboratory of Immunobiology, NIH, National Cancer Institute-Frederick Cancer Research and Development Center, P.O. Box B, Frederick, MD 21702–1201, USA. Tel. (301) 846–1560, Fax (301) 846–6145.

Rachel Ben Levy Department of Chemical Immunology, The Weizmann Institute of Science, Rehovot 76100, Israel. Tel. 972 8 343026, Fax 972 8 344141.

Harvey F. Lodish Department of Biology, Massachusetts Institute of Technology, Cambridge, MA 02142–1479, USA. Tel. (617) 258–5216, Fax (617) 258–9872.

Thomas F. Lüscher Cardiology, Cardiovascular Research, University Hospital, CH-3010 Bern, Switzerland. Tel. 41 31 632 96 53, Fax 41 31 382 10 69.

G. Lutfalla Institut de Génétique Moléculaire de Montpellier, CNRS—1919 route de Mende—UMR 9942, BP.5051.34033, Montpellier Cedex 1, France. Tel. 33 67 61 36 76/78, Fax 33 67 04 02 45/31.

Paul Matthew Centre for Animal Biotechnology, School of Veterinary Science, The University of Melbourne, Australia; please use telephone and fax numbers provided for Ora Bernard in this listing.

Andrew N.J. McKenzie Laboratory of Molecular Biology, Hills Road, Cambridge CB2 2QH, UK. Tel. 0223–402286, Fax 0223–412178.

Ian McNiece Developmental Hematology Group, Amgen Inc., 1840 DeHavilland Drive, Thousand Oaks, California 91320–1789, USA. Tel. (805) 447–3204, Fax (805) 499–7506.

Isabelle Millet Department of Epidemiology and Public Health, Yale University School of Medicine, 60 College Street, P.O. Box 3333, New Haven, CT 06510, USA. Tel. (203) 785–2915, Fax (203) 785–6130.

Atsushi Miyajima DNAX Research Institute of Molecular and Cellular Biology, 901 California Avenue, Palo Alto, CA 94304–1104, USA. Tel. (415) 852–9196, Fax (415) 496–1200.

K.E. Mogensen Institut de Génétique Moléculaire de Montpellier, CNRS—1919 route de Mende—UMR 9942, BP.5051.34033, Montpellier Cedex 1, France. Tel. 33 67 61 36 76/78, Fax 33 67 04 02 45/31.

Benny Motro Division of Molecular and Developmental Biology, Samuel Lunenfeld Research Institute, Mount Sinai Hospital, 600 University Avenue, Toronto, Canada M5G IX5. Tel. (416) 586–8273, Fax (416) 586–8588.

Tuija Mustonen Molecular/Cancer Biology Laboratory, Department of Pathology, P. Box 21 (Haartmaninkatu 3), 00014 University of Helsinki, Finland. Tel. 358 0 434 61, Fax 358 0 434 6448.

Shigekazu Nagata Department of Molecular Biology, Osaka Bioscience Institute, 6–2–4 Furuedai, Suita, Osaka 565, Japan. Tel. 81–6–872–4814, Fax 81–6–871–6686.

Anthony E. Namen Immunex Corporation, 51 University St., Seattle, WA 98101, USA. Tel. (206) 587–0430, Fax (206) 233–9733.

Masashi Narazaki Institute for Molecular and Cellular Biology, Osaka University 1–3, Yamada-oka, Suita, Osaka 565, Japan. Tel. 81 6 877 5802, Fax 81 6 877 1955.

Elliott Nickbarg Genetics Institute, 87 CambridgePark Drive, Cambridge, MA 02140, USA. Tel. (617) 876–1210 x 8134, Fax (617) 576–2979.

Nicos A. Nicola The Walter and Eliza Hall Institute of Medical Research, PO Royal Melbourne Hospital, Victoria 3050, Australia. Tel. 61 3 345 2526, Fax 61 3 345 2616.

Toshiya Ogorochi (deceased) DNAX Research Institute of Molecular and Cellular Biology, 901 California Avenue, Palo Alto, CA 94304–1104, USA. Tel. (415) 852–9196, Fax (415) 496–1200.

Joost J. Oppenheim Laboratory of Molecular Immunoregulation, Biological Response Modifier Program, NCI-Frederick Cancer Research and Development Center, P.O. Box B, Frederick, MD 21702–1201, USA. Tel. (301) 698–1551, Fax (301) 846–1673.

John E. Park Department of Cardiovascular Research, Genentech, Inc., 460 Point San Bruno Blvd., South San Francisco, CA 94080, USA. Tel. (415) 225–2968, Fax (415) 225–6327.

Linda S. Park Immunex Research and Development Corporation, 51 University Street, Seattle, WA 98101, USA. Tel. (206) 587–0430, Fax (206) 233–9733.

Steven J. Raker The Walter and Eliza Hall Institute of Medical Research, Post Office, Royal Melbourne Hospital, Parkville, Victoria 3050, Australia, Tel. 61–3–345 2500, Fax 61–3–347 0852, (Present address: AMRAD Operations Pty Ltd, 17–27 Cotham Road, Kew, Victoria 3101, Australia. Tel. 61–3–8530022, Fax 61–3–8530202).

Shrikanth Reddy Cytokine Research Section, Department of Clinical Immunology and Biological Therapy, University of Texas, M.D. Anderson Cancer Center, 1515 Holcombe Blvd., Houston, Texas 77030, USA. Tel. (713) 792–3503, Fax (713) 794–1613.

Jean-Christophe Renauld Ludwig Institute for Cancer Research (Brussels Branch), University of Louvain, Brussels, Belgium. Tel. 32 2 764 74 59, Fax 32 2 762 94 05.

Richard J. Robb Onco Therapeutics, Inc., 1002 Eastpark Blvd., Cranbury, NJ 08512, USA. Tel. (609) 655–5300, Fax (609) 655–1755.

Charles T. Roberts, Jr. NIH-Diabetes and Digestive and Kidney Diseases, Building 10, Room 8S239, 9000 Rockville Pike, Bethesda, MD 20892, USA. Tel. (301) 496–8090, Fax (301) 480–4386.

Larry R. Rohrschneider Fred Hutchinson Cancer Research Center, 1124 Columbia Street, Seattle, WA 98104–2092, USA. Tel. (206) 667–4441, Fax (206) 667–6522.

Timothy M. Rose Fred Hutchinson Cancer Research Center, 1124 Columbia St. EP-310-N, Seattle, WA 98109, USA. Tel. (206) 467–8100 x 3385, Fax (206) 282–5065.

Nancy H. Ruddle Department of Epidemiology and Public Health, Yale University School of Medicine, 60 College Street, P.O. Box 3333, New Haven, CT 06510, USA. Tel. (203) 785–2915, Fax (203) 785–6130.

Colin J. Sanderson The Western Australian Research Institute for Child Health, P.O Box D184, Perth, WA 6001, Australia. Tel. 61 9 340–8504, Fax 61 9 388–3414.

John W. Schrader The Biomedical Research Centre, University of British Columbia, 2222 Health Sciences Mall, Vancouver, British Columbia, Canada V6T 1Z3. Tel. (604) 822–7810, Fax (604) 822–7815.

Robert D. Schreiber Department of Pathology, Washington University School of Medicine, Box 8118, 660 S. Euclid Avenue, St. Louis, MO 63110–1093, USA. Tel. (314) 362–8747, Fax (314) 362–8888.

Surendra Sharma Section of Experimental Pathology, Department of Pathology, Roger Williams Medical Center-Brown University, Providence, RI 02908, USA. Tel. (401) 456–6562, Fax (401) 456–6569.

Jae-Hung Shieh Developmental Hematology Group, Amgen, Inc., 1840 DeHavilland Drive, Thousand Oaks, CA 91320–1789, USA. Tel. (805) 447–3204, Fax (805) 499–7506.

Richard J. Simpson Ludwig Institute for Cancer Research, Melbourne Tumor Biology Branch, PO Royal Melbourne Hospital, Parkville, Victoria 3050, Australia. Tel. 61 3 347–3155, Fax 61 3 347–1938.

John E. Sims Department of Molecular Genetics, Immunex Research and Development Corporation, 51 University Street, Seattle, WA 98101, USA. Tel. (206) 587–0430, Fax (206) 233–9733.

Concepció Soler Department of Biochemistry, School of Medicine, Vanderbilt University, Nashville, TN 37232–0146, USA. Tel. (615) 322–3315, Fax (615) 322–4349.

E. Richard Stanley Albert Einstein College of Medicine, 1300 Morris Park Avenue, Bronx, NY 10461, USA. Tel. (718) 430–2344, Fax (718) 823–5877.

Dennis D. Taub Clinical Services Program, Program Resourcs, Inc./DynCorp, NCI-Frederick Cancer Research and Development Center, P.O. Box B, Frederick, MD 21702–1201, USA. Tel. (301) 846–1491, Fax (301) 846–6022.

Jan Tavernier Roche Research Gent, Josef Plateaustraat 22, B-9000 Gent, Belgium. Tel. 32 9 225 76 98, Fax 32 9 233 11 19.

Gilles Uzé Institut de Génétique Moléculaire de Montpellier, CNRS—1919 route de Mende—UMR 9942, BP.5051.34033, Montpellier Cedex 1, France. Tel. 33 67 61 36 76/78, Fax 33 67 04 02 45/31, e-mail UZE@arthur.citi2.fr.

Satu Vainikka Molecular/Cancer Biology Laboratory, Department of Pathology, P. Box 21 (Haartmaninkatu 3), 00014 University of Helsinki, Finland. Tel. 358 0 434 61, Fax 358 0 434 6448.

Stephan D. Voss Immunology/Biologic Therapy Program, University of Wisconsin Medical School, K4/448, 600 Highland Avenue, Madison, WI 53792–0001, USA. Tel. (608) 263–9069, Fax (608) 263–4226.

Francesca Walker Ludwig Institute for Cancer Research, Post Office, Royal Melbourne Hospital, Victoria 3050, Australia. Tel. 61 3 347 3155, Fax 61 3 347 1938.

Stephanie S. Watowich The Whitehead Institute for Biomedical Research, 9 Cambridge Centre, Cambridge, MA 02142, USA, Tel. (617) 258–5227.

Dunzhi Wen Amgen Center, 1840 DeHavilland Drive, Thousand Oaks, CA 91320, USA. Tel. (805) 447–3864, Fax (805) 499–7464.

Bengt Westermark Department of Pathology, University Hospital, S-751 85 Uppsala, Sweden, please contact through Carl-Henrik Heldin in this listing.

Stanley F. Wolf Genetics Institute, 87 CambridgePark Drive, Cambridge, MD 02140, USA. Tel. (617) 876–1210 x 8134, Fax (617) 576–2979.

George D. Yancopoulos Regeneron Pharmaceuticals, Inc., 777 Old Saw Mill River Road, Tarrytown, NY 10591–6707, USA. Tel. (914) 347–7000, Fax (914) 345–7650.

Yosef Yarden Department of Chemical Immunology, The Weizmann Institute of Science, Rehovot 76100, Israel. Tel. 972 8 343974, Fax 972 8 344141.

Teizo Yoshimura Immunopathology Section, Laboratory of Immunobiology, NIH, National Cancer Institute-Frederick Cancer Research and Development Center, P.O. Box B, Frederick, MD 21702–1201, USA. Tel. (301) 846–1560, Fax (301) 846–6145.

Gerard Zurawski Department of Molecular Biology, DNAX Research Institute, 901 California Avenue, Palo Alto, CA 94304–1104, USA. Tel. (415) 852–9196, Fax (415) 496–1214.

Abbreviations

AIGF	androgen-induced growth factor	KDR	kinase insert domain-containing receptor
ARIA	acetylcholine receptor-inducing activity	K-FGF	Kaposi's sarcoma-derived fibroblast growth factor
BCGF	B cell growth factor		
BDNF	brain-derived neurotrophic factor	KGF	keratinocyte growth factor
BFU-E	burst forming unit-erythroid	KL	Kit ligand (see stem cell factor)
BMP	bone morphogenetic protein	LIF	leukaemia inhibitory factor
CFU-E	colony-forming unit-erythroid	M6P	mannose-6-phosphate
CNTF	ciliary neurotrophic factor	MCP-1	monocyte chemoattractant protein
CSF	colony-stimulating factor	M-CSF	macrophage-CSF (CSF-1)
CTAP	connective tissue activating protein	MGSA	melanoma growth-stimulating activity
EGF	epidermal growth factor	MIP	macrophage inflammatory peptide
ENA-78	epithelial cell-derived neutrophil activator	NAP	neutrophil activating peptide
		NDF	neu differentiation factor
EPO	erythropoietin	NGF	nerve growth factor
ET	endothelin	NK	natural killer
FGF	fibroblast growth factor	NRG	neuregulins
FLK-1	fetal liver kinase-1	NT	neurotrophin
FLT-2	fms-like tyrosine kinase	OSM	oncostatin-M
GAF	glia activating factor	PA	plasminogen activator
GCP	granulocyte chemotactic peptide	PDGF	platelet-derived growth factor
G-CSF	granulocyte-CSF	PF	platelet factor
GGF	glia growth factor	PL	placental lactogen
GH	growth hormone	PlGF	placental growth factor
GM-CSF	granulocyte-macrophage CSF	PRL	prolactin
GPA	growth promoting activity	PVGF	pox virus EGF-like protein
GPI	glycosyl-phosphatidylinositol	SCF	stem cell factor
GRB2	growth factor receptor bound protein-2	SF	scatter factor (hepatocyte growth factor)
GRO	growth-related cytokine	SH2,SH3	src homology domains 2 or 3
HB-EGF	heparin-binding EGF	SLF	steel factor (stem cell factor)
HGF	hepatocyte growth factor	SOS	son of sevenless
HTLV	human T-cell leukaemia virus	TG	thromboglobulin
IFN	interferon	TGF	transforming growth factor
IGF	insulin-like growth factor	TK	tyrosine kinase
IL	interleukin	TNF	tumour necrosis factor
IP-10	interferon-inducible protein-10	TRK	tyrosine receptor kinase (neurotrophins)
IRS-1	insulin receptor substrate-1	VEGF	vascular endothelial growth factor
ISGF	interferon-stimulated gene (transcription) factor	VPF	vascular permeability factor
		VVGF	vaccinia virus growth factor

■ Alternative names and abbreviations

apidogenesis inhibitory factor	interleukin-11
BAF	(B-cell activating factor), interleukin-1
BCDF	(B-cell differentiation factor), interleukin-6
BCGFI	(B-cell growth factor I), interleukin-4
BCGFII	(B-cell growth factor II), interleukin-5
BPA	(burst-promoting activity), interleukin-3
BSF-1	(B-cell stimulating factor 1), interleukin-4

BSF-2	(B-cell stimulating factor-2), interleukin-6
cachectin	TNFα
CLMF	(cytotoxic lymphocyte maturation factor) interleukin-12
CNDF	(cholinergic neuronal differentiation factor), leukaemia inhibitory factor
CSFα	GM-CSF
CSF-2	GM-CSF
CSF-3	G-CSF
CSIF	(cytokine synthesis inhibitory factor), interleukin-10
cytotoxic factor	TNFβ
cytotoxin	TNFα
D-factor	leukaemia inhibitory factor
DIF	(differentiation inducing factor), TNFα
DIF	(differentiation inhibiting factor), leukaemia inhibitory factor
DRF	(differentiation retarding factor), leukaemia inhibitory factor
EDF	(eosinophil differentiation factor), interleukin-5
Eo-CSF	(eosinophil colony-stimulating factor), interleukin-5
EP	(endogenous pyrogen), interleukin-1
ETAF	(epidermal cell-derived thymocyte-activating factor), interleukin-1
FAF	(fibroblast activating factor), interleukin-1
GM-CSFβ	G-CSF
H-1	(haemopoietin-1), interleukin-1
haemopoietin-2	interleukin-3
HCGF	(haemopoietic cell growth factor), interleukin-3
hepatopoietin A	hepatocyte growth factor
hepatotrophin	hepatocyte growth factor
HILDA	(human interleukin for Dal cells), leukaemia inhibitory factor
HMW-BCGF	(high molecular weight BCGF), interleukin-14
HRG	heregulin
HPCSA	(histamine-producing cell stimulating activity), interleukin-3
HRG	heregulin
HSF	(hepatocyte stimulating factor), interleukin-6, leukaemia inhibitory factor, and interleukin-1
hybridoma growth factor	interleukin-6
immune interferon	interferon-γ
interferon-β2	interleukin-6
LAF	(lymphocyte-activating factor), interleukin-1
LEM	(leucocyte endogenous mediator), interleukin-1
LMW-BCGF	(low molecular weight B-cell growth factor), BCGF-12kDa
LP-1	(lymphopoietin-1), interleukin-7
lymphotoxin	TNFβ
macrophage cytotoxin	TNFα
MAF	(macrophage activating factor), interferon-γ
MCAF	(monocyte chemotactic and activating factor), MCP-1
MCGF	(mast cell growth factor), interleukin-3 and interleukin-10
MCF	(macrophage cytotoxic factor), TNFα
MCGFII	(mast cell growth factor II), interleukin-4
MEA	(mast cell growth enhancing activity), interleukin-9
MGF	(macrophage growth factor), CSF-1
MGI-1G	(macrophage and granulocyte inducer-1, granulocyte), G-CSF
MGI-1GM	(macrophage and granulocyte inducer-1, granulocyte and macrophage), GM-CSF
MGI-1M	(macrophage and granulocyte inducer-1, macrophage), CSF-1
MLPLI	(melanocyte-derived lipoprotein lipase inhibitor), leukaemia inhibitory factor
MULTI-CSF	(multipotential colony-stimulating factor), interleukin-3
necrosin	TNFα

NKSF	(NK cell stimulating factor), interleukin-12
OAF	(osteoclast activating factor), interleukin-1
PBGF	(pre-B-cell growth factor), interleukin-7
PIF	(proteolysis inducing factor), interleukin-1
pluripoietin	G-CSF
pluripoietin-a	GM-CSF
PSF	(P-cell stimulating factor), interleukin-3
PSH	(pan-specific haemopoietin), interleukin-3
SA	(synergistic activity), interleukin-3
SCAF	(stem cell activating factor), interleukin-3
somatomedin	insulin-like growth factors
TCGF	(T-cell growth factor), interleukin-2
TCGFII	(T-cell growth factor II), interleukin-4
TCGFIII	(T-cell growth factor III), interleukin-9
TGPF	(thymocyte growth promoting factor), interleukin-10
TRF	(T-cell replacing factor), interleukin-5
TSA	(Thy-1 stimulating activity), interleukin-3
TSF	(T-cell stimulating factor), interleukin-12
tumour cytotoxic factor	hepatocyte growth factor

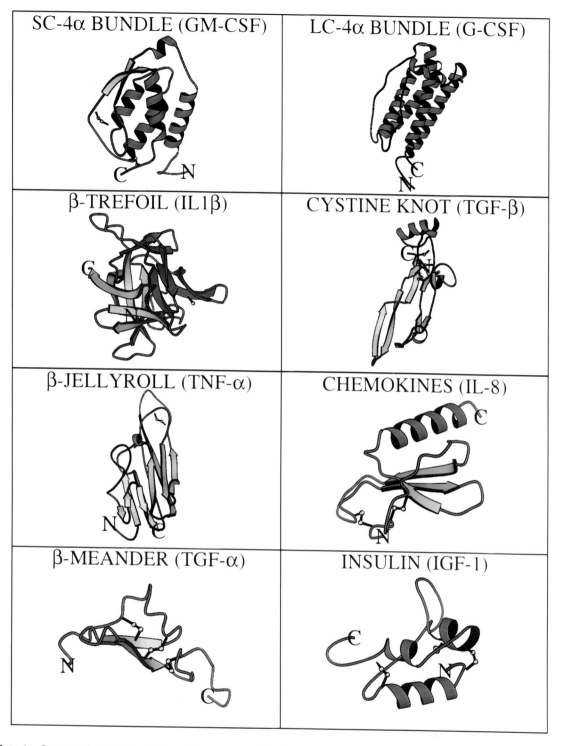

Plate 1. Structural motifs of cytokines. The structural family name is indicated with the specific cytokine example used shown in brackets (see text). Structures were drawn using MOLSCRIPT (Kraulis 1991). α-helices are shown in blue, β-strands in yellow or red, and loops are in green. Disulphide bonds are shown as ball-and-stick models and the N-terminal (N) and C-terminal (C) of each structure is shown. For the β-trefoil structure the β-strands forming the β-barrel are shown in yellow and the strands forming the triangular arrays are shown in red.

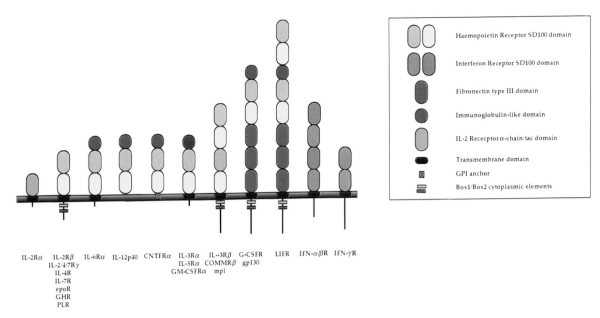

Plate 2. Schematic view of the haemopoietin/interferon receptor family.
Abbreviations: IL = interleukin, epo = erythropoietin, GH = growth hormone, PL = prolactin, CNTF = ciliary neurotrophic factor, G-CSF = granulocyte colony-stimulating factor, GM-CSF = granulocyte-macrophage colony-stimulating factor, COMM = common receptor chain to IL-3, IL-5, and GM-CSF, LIF = leukaemia inhibitory factor, IFN = interferon, R = receptor.
(This figure was adapted from Figure 1 of Gearing and Ziegler 1993.)

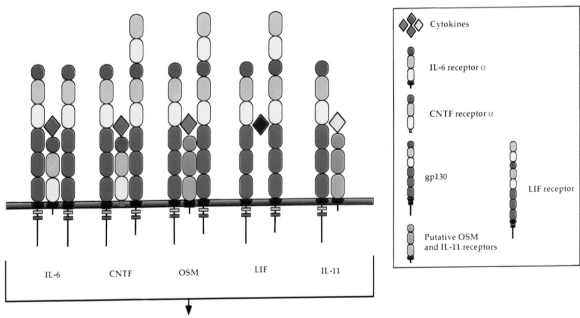

Plate 3. Shared receptor subunits may contribute to cytokine pleiotropy.
Abbreviations: IL = interleukin, OSM = oncostatin M, CNTF = ciliary neurotrophic factor, LIF = leukaemia inhibitory factor.

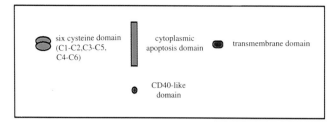

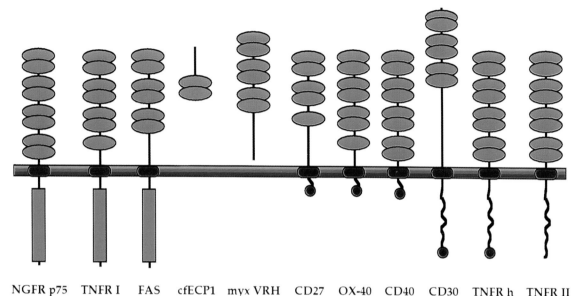

Plate 4. Schematic view of the nerve growth factor/tumour necrosis factor receptor family.
Abbreviations: NGF = nerve growth factor, TNF = tumour necrosis factor, TNFRh = tumour necrosis factor receptor homologue, cfECP1 = EFCP1 protein from *Cladosporium fulvum*, SalF19R, protein from the Shope fibroma virus, R = receptor.
(This figure was adapted from Figure 2 of Bazan 1993).

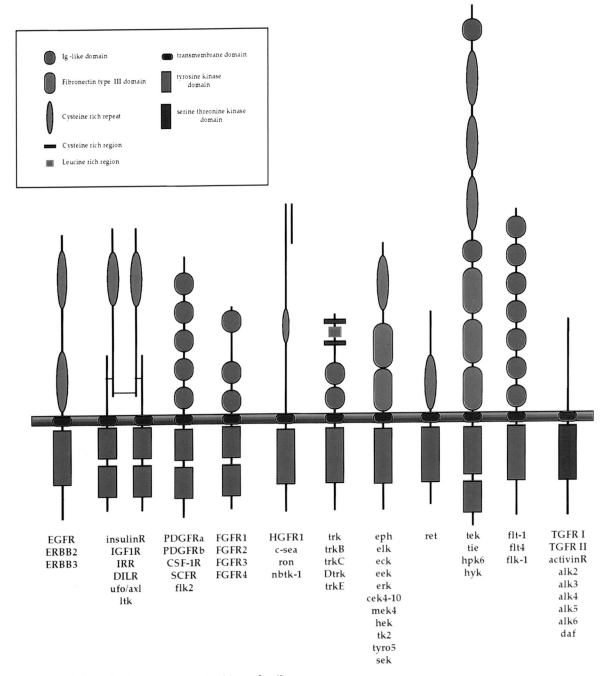

Plate 5. Schematic view of the receptor kinase family.
Abbreviations: EGF = epidermal growth factor, IGF = insulin-like growth factor, IRR = insulin-related receptor, PDGF = platelet-derived growth factor, CSF-1 = colony-stimulating factor -1, SCF = stem cell factor, FGF = fibroblast growth factor, HGF = hepatocyte growth factor, TGF = transforming growth factor, R = receptor.
(Kirilee Anne Wilson is thanked for help in creating this figure.)

An Introduction to the Cytokines

For the purposes of the present volume, cytokines are defined as secreted regulatory proteins that control the survival, growth, differentiation and effector function of tissue cells. Cytokines encompass those families of regulators variously known as growth factors, colony-stimulating factors, interleukins, lymphokines, monokines, and interferons. I have excluded most of the classical endocrine hormones because their general properties are different from those of the cytokines (see Table 1) and because they have been amply described in other texts. Nevertheless, there are no sharp boundaries that distinguish cytokines from other regulatory proteins and the selections in this book reflect my own bias.

The confusing nomenclature of cytokines has arisen because several different streams of investigation led to the discovery of different cytokines. The classical growth factors such as nerve growth factor (NGF) and epidermal growth factor (EGF) were first characterized in the sixties and this was made possible by their peculiarly high abundance in the male salivary gland (Cohen 1960, 1962). In the mid- to late sixties, a variety of *in vitro* bioassays were developed that measured the proliferation, differentiation and function of lymphoid and haemopoietic cells (Ichikawa *et al.* 1966; Bradley and Metcalf 1966; David 1966; Bloom and Bennett 1966; Williams and Granger 1968; Ruddle and Waksman 1968). The colony-stimulating factors were so named because they stimulated the formation of colonies of granulocytes and macrophages from precursor cells in the bone marrow (Robinson *et al.* 1967). The term 'lymphokine' was coined in 1969 (Dumonde *et al.* 1969) to describe the soluble mediators generated in immunological reactions between lymphocytes and antigen but the recognition that monocytes also produced lymphocyte-activating factors soon led to the complementary term 'monokine'. Interferon was first described as a protein conferring cellular resistance to viral infection (Isaacs and Lindenmann 1957) but it was not until nearly ten years later that the actions of interferons as immune mediators were recognized (Wheelock 1965). The term 'interleukin' and a numbering system of nomenclature were introduced in 1979 to reflect the fact that these regulators served as communication signals between leucocytes rather than just lymphocytes (Aarden *et al.* 1979). Similar reasons had led to the earlier term 'cytokine' (Cohen *et al.* 1974) and we have used that term in this book because it is the most general and is not restricted to the immunohaemopoietic system.

This group of molecules (cytokines) shares some features in common. They are all molecules found in the extracellular medium that interact with specific target cells to communicate information regarding the status of the animal and they result in an appropriate biological response in the target tissue. The producer cells therefore represent the biological sensors of the animals' condition, the cytokines, the method of signal transduction, the receptors, the means of signal reception and the responsive cells represent the means of effecting an appropriate biological response. The sensors include oxygen sensors for red blood cell production; sensors of various microbiological products for white blood cell production, function and chemotaxis; sensors of on-going immune reactions also for white blood cell function; and sensors of tissue damage products for epithelial and neural cells. In other cases, there may be a constitutive level of production of cytokines to maintain steady state levels of renewing tissues or the constitutive production of cytokine may be required for continued cell survival and selection (e.g. in the nervous system).

Cytokines, as can be seen from the entries in this book, represent a bewildering array of different molecules with sometimes puzzling sets of biological activities which often overlap with each other. Indeed, the two key words describing cytokine biological action are pleiotropy and redundancy. The extreme examples of pleiotropic cytokines are leukaemia inhibitory factor and transforming growth factor β which affect nearly all organs of the body; but it has become increasingly apparent that cytokines originally thought to be relatively specific in their action are now thought of as pleiotropic as additional biological tests are performed. On the other hand there appear to be very few biological responses that are mediated by only one cytokine and many responses that can be achieved by several different cytokines. Gene deletion experiments are revealing that few individual cytokines are absolutely essential for life or even for individual cellular functions (Tavernier 1993). Rather than suggesting that these cytokines are therefore unimportant, these experiments suggest that important cellular functions are usually backed up in a 'fail-safe' mechanism where one cytokine can compensate for the loss of another.

Table 1. Differences between cytokines and endocrine hormones

Property	Endocrine hormones	Cytokines
Sites of production	Few	Many
Cellular targets	Many	Few
Biological role	Homeostatis	Fighting infection tissue repair
Biological redundancy	Low	High
Biological pleiotropy	Low	High
In the circulation?	Yes	Rarely
Sphere of ionfluence	Widespread	Autocrine, paracrine (Local)
Inducers	Physiologic variation	External insults

The above might suggest that the induction of secondary cytokines by a given cytokine is important in the regulation of biological responses. Indeed, many examples have been documented of cytokines inducing the synthesis and release from their target cells of both positive and negative regulatory cytokines which may act in an autocrine, paracrine or intercrine manner (Arai *et al*. 1990). Nevertheless, it has been extremely difficult to document the details of cytokine production and secondary cytokine induction *in vivo* especially under steady-state conditions. This has led some to propose that most of the cytokines described in this book have little to do with the day-to-day functioning of the body but represent instead an emergency response to tissue damage, infection, or other insults. While the latter is undoubtedly true, there may be other reasons why many cytokines are hard to find in the serum under normal conditions as detailed below.

The key feature of cytokines compared to classical hormones is that they are rarely found in the circulation but rather are produced by cells that are widespread in the body. Given the pleiotropy of cytokines, the local production and action of cytokines may be a parsimonious way that the body can use the same cytokine/receptor system to effect multiple biological actions. For this system to work, the sphere of influence of the released cytokine must be extremely limited and systemic spread to other sites must be avoided. Many mechanisms appear to have evolved in the cytokine system to achieve this result. First, cytokine producer cells are often physically located immediately adjacent to the responder cells (Metcalf 1991). Second, cytokine producer cells generally secrete very small quantities of cytokine and, in some cases, this is directional towards the responder cells (Poo *et al*. 1988). In addition, the responder cells destroy the cytokine they respond to in the process of receptor mediated endocytosis (Nicola *et al*. 1988). Third, many cytokines bind to elements of the extracellular matrix around responder cells, further restricting their spread beyond the immediate site and increasing their bioavailability to the responder cells (Gordon 1991). An extreme example of this bound localization are the cell surface cytokines which presumably require cell–cell contact for their action (Gordon 1991). Finally, there are several examples of quite high levels of circulating soluble cytokine receptors, binding proteins and even receptor antagonists which may serve to inhibit the biological action of a cytokine that does find its way into the serum (Arend 1993; Fernandez-Bottran 1991).

Given the role of many cytokines in vigorous tissue reactions to a variety of insults, it is not surprising that the cytokines have very stable structures. Most cytokines are relatively small proteins containing protective carbohydrate and intramolecular disulphide bonds. These elements enhance the solubility, stability, and protease resistance of the cytokines.

Despite the lack of sequence similarity in most members of the cytokine family, it is possible to attempt to subclassify them on several different bases. These include similarities in biological responses, induction mechanisms, overall three-dimensional structure and similarities in the types of receptor subunits they utilize.

■ Structural classification of cytokines

Since the overall three-dimensional structure of a protein can often reflect its evolutionary origins to a greater degree than its amino acid sequence, it is useful to classify the cytokines according to the structural motifs they adopt (Figs 1–4) (Bazan 1990, 1991; Parry *et al*. 1991; Young 1992; Sprang and Bazan 1993; Boulay and Paul 1993).

■ Group 1 cytokines: 4-α-helical bundles

The largest family of cytokines adopt an antiparallel 4-α-helical bundle structure first observed for growth hormone (Abdel-Meguid *et al*. 1987). In these structures, two long over-hand loops (those between the A and B or the C and D helices) allow for helix direction reversal and for the antiparallel A, D helix pair to pack against the antiparallel B, C helix pair with a characteristic skew angle. These are tightly packed, amphipathic structures with the A–D helix pair commonly presenting the surface residues that interact with the receptors. This group of cytokines has been subclassified into short-chain (sc) or long-chain (lc) helical bundles. Sc cytokines (IL-2, IL-3, IL-4, IL-5, IL-7, IL-9, IL-13, GM-CSF, M-CSF, SCF, and IFNγ) are characterized by a shorter overall chain length, shorter (~15 aa) helices, a large skew angle (~35°) and two short twisted antiparallel β-strands in the A–B and C–D loops. The lc cytokines (IL-6, IL-12, Epo, G-CSF, LIF, OSM, CNTF, IL-11, GH, PRL, IFNα/β, IL-10) have a longer overall chain length (160–200 aa), longer (~25 aa) helices, a smaller skew angle (~18°) and usually contain additional short helical segments in the A–B and C–D loops. Sc cytokines have an overall oblate ellipsoid shape, while lc cytokines appear more as elongated cylinders. Some of these cytokines (IL-5, M-CSF, and IFNγ) function as covalent or non-covalent dimers. In the case of M-CSF, the monomers form a symmetric head to head orientation, while for IL-5 and IFNγ each subunit interdigitates such that it provides the C–D loop and D-helix for the opposite subunit 4-α helical bundle.

■ Group 2 cytokines: long chain β-sheet structures

The tumour necrosis factor (TNF) family of cytokines (TNFα, TNFβ, CD40, CD27, and Fas ligands) (Mallett and Barclay 1991; Farrah and Smith 1992) are often cell surface-associated, form symmetric homotrimers and the subunits take up the conformation of β-jelly rolls described previously for some viral coat proteins (Plate 1). The trimer is cone-shaped with each monomer contributing to the hydrophobic core and adjacent subunits each contributing residues that interact with the receptor. These composite binding sites at the subunit interfaces explain both the inactivity of isolated monomers and the symmetry of three receptor subunits interacting with the TNF trimer (Banner *et al*. 1993).

The interleukin-1 family of cytokines and the fibroblast growth factor family adopt a β-trefoil fold which consists of

12 strands of β-sheet forming six hairpins. Three sets of 4-β strands each of which takes up a Y or trefoil shape, make up the structure with three of the hairpins forming a β-barrel structure and the other three a triangular array (Murzin et al. 1992). Unlike most other cytokines, they are synthesised with no traditional leader sequence and their mechanism of secretion is still unclear.

The platelet-derived growth factors, transforming growth factor β and nerve growth factors display β-sheet-rich structures known as cystine knots (McDonald and Hendrickson 1993). These structures are characterized by an interlocking set of three disulphide bridges and at least four segments of twisted antiparallel β-sheet that give rise to an elongated asymmetric monomer. Structural variations in this group of growth factors occur in the loops connecting the four β-strands. Each of these growth factors forms a (usually) homodimeric complex although the orientation of the monomers in each type of dimer is different.

Group 3 cytokines: short chain α/β

The epidermal growth factor family of cytokines are produced as large transmembrane precursor molecules which each contain at least one EGF domain in the extracellular region. The EGF domain contains at least two antiparallel β-strands connected to the intervening loops by three disulphide bonds and is presumably released from the precursor molecules by proteolysis.

The chemokines fall into two major subgroups based on the amino acid sequence around the conserved cysteine residues (C–C or C–X–C); but all are thought to take up the conformation of an open-face β-sandwich with a C-terminal α-helical segment.

The insulin-related cytokines contain a conserved set of three disulphide bonds that link three short α-helices.

Group 4 cytokines: mosaic structures

In addition, other growth factors exhibit mosaic structures. The heregulins or neuregulins are cell surface proteins that contain both an EGF and an immunoglobulin-like domain on the extracellular side while glial growth factor contains both an EGF domain and a Kringle domain. Hepatocyte growth factor is a heterodimer with the α-subunit containing four Kringle domains (typical of plasminogen) and the β-subunit having homology to serine proteases, although it is probably devoid of enzymic activity. Interleukin-12 is also a heterodimer with the α-subunit having the features typical of long-chain 4-α-helical bundle cytokines and the β-subunit having the features of a complementary haemopoietin domain receptor.

Classification of cytokines based on receptor usage

There is a relatively strong relationship between the structural class of a cytokine and the type of receptor to which it binds. Almost all of the 4-α-helical cytokines bind to receptors which contain a 200-amino-acid ligand-binding domain known as the haemopoietin domain and which are devoid of intrinsic tyrosine kinase activity. The exceptions to this generalization are the interferons and IL-10 that bind to interferon receptor-like proteins and the homodimeric proteins M-CSF, SCF, and FLK-2-ligand which bind to class III tyrosine kinase receptors that contain immunoglobulin domains in the extracellular region. However, the haemopoietin domain is structurally related to the interferon receptor extracellular domain, and both are more distantly related to fibronectin III type repeats and immunoglobulin domains. This leads to a further subclassification of cytokines as shown in Fig. 1.

Some of the 4-α-helical cytokines appear to use homodimers of a single receptor chain (M-CSF, SCF, Epo, G-CSF, GH, PRL), but most require heterodimers of at least two distinct receptor subunits either for the generation of high-affinity binding or for biological responses or both. A major breakthrough in our understanding of cytokine action has occurred with the accumulation of data showing that many cytokine receptors consist of a unique ligand-binding α-chain and a shared or common β-chain. IL-2, IL-4, IL-7, IL-9, and IL-13 share a common chain called γ; IL-3, GM-CSF and IL-5 share a common chain call $β_c$; and IL-6, LIF, oncostatin-M, CNTF, and IL-11 share a common chain called gp130. These common β-chains provide a clear rationale for the biological redundancy of many cytokines. There is also evidence that interferon receptors require additional, probably related, receptor chains for biological activity, but it is unclear yet whether or not these may be shared by different interferon and IL-10 receptors (see Introduction to Cytokine Receptors, p. 8).

The group 2 cytokines bind to a more diverse group of receptor classes. The β-jelly-roll trimers bind to three receptor chains, all of which are related in sequence and probably structure to the TNF receptors which contain a 40-amino-acid cysteine repeat in their ligand binding domains. The signalling mechanisms of these receptors are unknown but can in some cases lead to cell death. The neurotrophic cystine knot cytokines can bind to two types of receptors—one a TNF receptor-like protein and the other a tyrosine kinase receptor of the TRK family. The PDGF-like cystine knot cytokines bind to receptors of the class III tyrosine kinase family that contain duplicated immunoglobulin domains in the extracellular region. On the other hand, TGFβ-related cytokines bind to a unique class of receptors with intrinsic serine/threonine kinase activity, as well as proteoglycans related to endoglin. The β-trefoil cytokines also bind to receptors with immunoglobulin-like repeats in the extracellular domain, but IL-1 receptors have no intrinsic tyrosine kinase activity while FGF receptors have intrinsic tyrosine kinase activity (type IV) and require heparin-bound FGF in order to generate high-affinity binding.

Group 3 and 4 cytokines cannot be further subclassified according to receptor class. All chemokines bind to serpentine receptors (seven transmembrane segments and coupled to G proteins), all EGF-related cytokines bind to class I tyrosine kinase receptors, and all insulin-related cytokines bind to class II covalent tyrosine kinase receptors. Insulin-like growth factors also bind to the mannose-6-phosphate receptor, but it is unclear whether or not this receptor transmits biological responses. Neuregulins bind to EGF-receptor-like proteins and hepatocyte growth factor binds to the met protooncogene (a class VI non-covalent heterodimeric tyrosine kinase receptor).

■ Classification of cytokines based on biological action

With the above subclassifications, common biological actions of groups of cytokines can be seen. Short chain 4-α helical cytokines are all involved in immunohaemopoietic regulation. Those that use the common γ-chain receptor are involved in acquired immunity along the T cell/B cell/ macrophage axis, while those using the common β-chain are involved in innate immunity involving the bacteriocidal cells—neutrophils, eosinophils, macrophages, and mast cells. IFNγ is also involved in immune regulation, while M-CSF and SCF are involved in generating immunohaemopoietic cells from the bone marrow. Long-chain 4-α cytokines are somewhat paradoxical with Epo and G-CSF being amongst the most specific of cytokines in generating erythroid or neutrophil cells, respectively, and the cytokines which use gp130 being amongst the most pleiotropic of cytokines. Most of these cytokines also have a role in the immunohaemopoietic system, except for growth hormone and prolactin which have a more classical endocrine role in whole-body metabolism.

Group 2 cytokines are involved primarily in the growth and differentiation of a variety of epithelial, endothelial, and neural tissues and many appear to play essential roles during tissue modelling in development (especially the

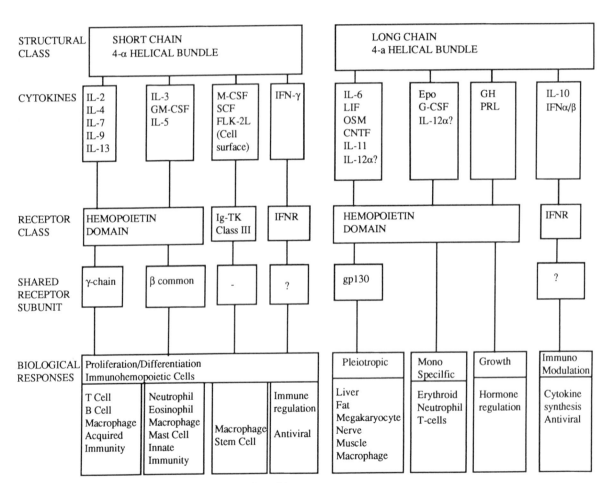

Figure 1. Subclassification of group 14-α-helical cytokines.
Abbreviations: IL, interleukin; CSF, colony-stimulating factor; G, granulocyte; M, macrophage; SCF, stem cell factor; FLK-2L, fetal liver kinase-2 ligand; LIF, leukaemia inhibitory factor; OSM, oncostatin M; Epo, erythropoietin; CNTF, ciliary neurotrophic factor; GH, growth hormone; PRL, prolactin; IFN, interferon; TK, tyrosine kinase; Ig, immunoglobulin-like; IFNR, interferon receptor.

TGFβ, NGF, and FGF families). The exceptions appear to be the TNF and interleukin-1 families that are involved in immune regulatory events and especially the acute phase response to injury and infection. They mediate many of the toxic reactions observed in acute septicaemia and inflammation.

The group 3 cytokines mediate quite separate biological responses according to the structural class. The chemokines are involved in innate immunity mediated by neutrophils, eosinophils, and macrophages and serve to attract and activate these cells at sites of inflammation. The EGF class of cytokines is involved in the proliferation of epithelial cells and in wound healing while the insulin-like growth factors are involved in the proliferation and differentiation of mesenchymal cells and in metabolic responses in tissue cells.

The group 4 mosaic cytokines have unique biological activities typical of at least one component of the mosaic structure. Neuregulins (or heregulins) have EGF-like activities and IL-12 has immunoregulatory activities typical of the 4-α-helical cytokines. Hepatocyte growth factor has a unique structure and is involved as a mitogen for hepatocytes, endothelial and epithelial cells.

■ Conclusions and outlook

The determination of cytokine and receptor structures has helped to reveal an order and evolutionary relationship amongst cytokines that was previously impossible to discern from the biological properties and even amino acid sequences of the myriad cytokines thus far described. Those of us embarking relatively early in the field of cytokines and growth factors could not have suspected that upwards of 100 cytokines would be described in molecular detail in such a short space of time. Given the extremely large number of cytokines that have already been described to affect the immunohaemopoietic system and the clear indication that many more remain to be discovered in this system, it is certain that many more cytokines will be found that affect tissues that are more difficult to study, such as the brain, kidney, lung, colon, and others including the development of fetal tissues.

With the burgeoning of new cytokines, it will be important to remember some of the lessons learnt in the past. The specificity of a cytokine cannot be inferred until extensive biological tests are performed since cytokines often display completely unexpected combinations of biological

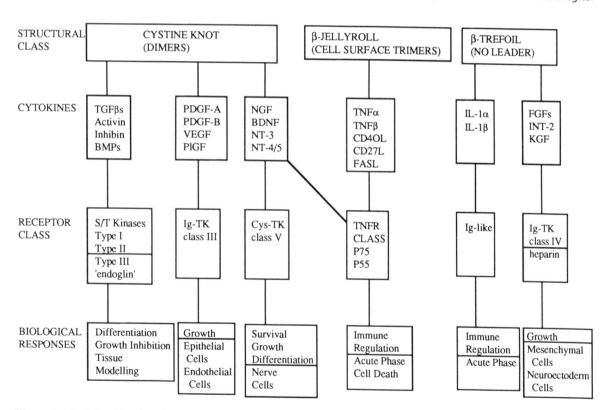

Figure 2. Subclassification of group 2 β-sheet rich cytokines.
Abbreviations: TGF, transforming growth factor; BMP, bone morphogenetic protein; PDGF, platelet-derived growth factor; VEGF, vascular endothelial cell growth factor; PlGF, placenta growth factor; NGF nerve growth factor; BDNF, brain-derived neurotrophic factor; NT neurotrophin; TNF, tumour necrosis factor; CD40L, CD27L, FASL, ligands for the cell surface antigens CD40, CD27 and Fas; IL, interleukin; FGF, fibroblast growth factor; KGF, keratinocyte growth factor; S/T, Serine/threonine; TK, tyrosine kinase; Ig, immunoglobulin-like; CYS, cysteine rich; TNFR, TNF receptor.

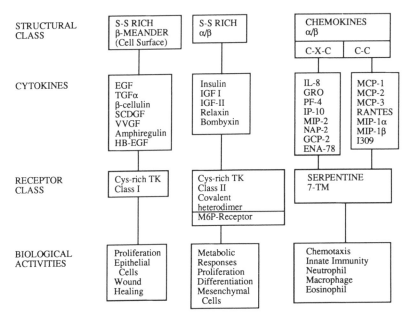

Figure 3. Subclassification of group 3 small α/β cytokines.
Abbreviations: EGF, epidermal growth factor; TGF, transforming growth factor; SCDGF, Schwann cell-derived growth factor; VVGF, vaccinia virus-derived growth factor; HB-EGF, heparin-binding EGF; IGF, insulin-like growth factor; IL, interleukin; PF, platelet factor; IP, interferon-inducible protein; MIP, macrophage inflammatory protein; GCP, granulocyte chemotactic peptide; MCP, monocyte chemotactic peptide; TK, tyrosine kinase; TM, transmembrane domain; GRO, growth related cytokine; NAP, neutrophil-activating peptide.

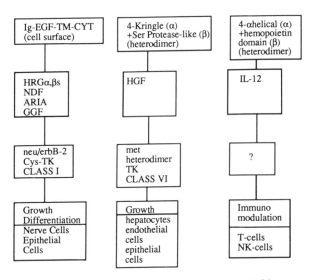

Figure 4. Subclassification of group 4 mosaic cytokines.
Abbreviations: HRG, heregulin; NDF, Neu differentiation factor; ARIA, acetylcholine receptor inducing activity; GGF, glial growth factor; HGF, hepatocyte growth factor; IL, interleukin; Cys, cysteine rich domain; TK, tyrosine kinase; Ig, immunoglobulin-like; EGF, epidermal growth factor-like; TM, transmembrane domain; CYT, cytoplasmic domain.

activities. *In vivo* testing and the use of transgenic animals that over-express or do not express particular cytokines will be important but not conclusive in helping to determine the physiological role for a cytokine. In the former case, the cytokine may have access to sites in the body from which it is normally restricted and, in the latter case, other cytokines may be able to compensate for the loss of any particular cytokine. It will therefore be important to also determine the usual sites of production of a cytokine within the body, the physiological inducers of its production and the sites to which it has access. Data of this type are lacking for most of the known cytokines.

The present analysis has been an attempt to arrive at a classification system for cytokines that may simplify the apparent complexity of the overall system. In this analysis, the most useful parameters were the predicted three-dimensional structure of the cytokine, the class of receptor to which it bound (particularly the identification of shared receptor subunits), and finally the types of biological responses elicited. Further analysis of these parameters for the known and newly discovered cytokines will be essential for unifying concepts to emerge. Two newer areas will also be very useful for this purpose—a detailed analysis of intracellular signalling pathways induced by each cytokine in which considerable progress is currently being made; and the evolutionary history of cytokine and receptor structures along with the evolution of biological function.

■ References

Aarden, L.A., Brunner, T.K., Cerottini, J.C., Dayer, J.-M., de Weck, A.L., Dinarello, C.A., Di Sabato, G., Farrar, J.J., Gery, I., Gillis, S., and 24 others. (1979). Revised nomenclature for antigen-non-specific T cell proliferation and helper factors. *J. Immunol.*, **123**, 2928–9.

Abdel-Meguid, S.S., Shieh, H.S., Smith, W.W., Dayringer, H.E., Violand, B.N., and Bentle, L.A. (1987). Three-dimensional structure of a genetically engineered variant of porcine growth hormone. *Proc. Natl. Acad. Sci. (USA)*, **84**, 6434–7.

Arai, K.I., Lee, F., Miyajima, A., Miyatake, S., Arai, N., and Yokota, T. (1990). Cytokines: coordinators of immune and inflammatory responses. *Ann. Rev. Biochem*, **59**, 783–836.

Arend, W.P. (1993). Interleukin-1 receptor antagonist. *Adv. Immunol.*, **54**, 167–227.

Banner, D.W., D'Arcy, A., Janes, W., Gentz, R., Schoenfeld, H.J., Broger, C., Loetscher, H., and Lesslauer, W. (1993). Crystal structure of the soluble human 55 kd TNF receptor-human TNF beta complex: implications for TNF receptor activation. *Cell*, **73**, 431–45.

Bazan, J.F. (1990). Haemopoietic receptors and helical cytokines. *Immunol. Today*, **11**, 350–4.

Bazan, J.F. (1991). Neuropoietic cytokines in the hematopoetic fold. *Neuron*, **7**, 197–208.

Bloom, B.R., and Bennett, B. (1966). Mechanism of reaction *in vitro* associated with delayed-type hypersensitivity. *Science*, **153**, 80–2.

Boulay, J., and Paul, W.E. (1993). Hemapoietin sub-family classification based on size, gene organization and sequence homology. *Current Biology*, **3**, 573–81.

Bradley, T.R., and Metcalf, D. (1966). The growth of mouse bone marrow cells in vitro. *Aust. J. Exp. Biol. Med. Sci.*, **44**, 287–300.

Cohen, S. (1960). Purification of a nerve-growth promoting protein for the mouse salivary gland and its neuro-cytotoxic antiserum. *Proc. Natl. Acad. Sci. (USA)*, **46**, 302–11.

Cohen, S. (1962). Isolation of a mouse submaxillary gland protein accelerating incisor eruption and eyelid opening in the new-born animal. *J. Biol. Chem.*, **237**, 1555–62.

Cohen, S., Bigazzi, P.E., and Yoshida, T. (1974). Similarities of T cell function in cell mediated immunity and antibody production. *Cell Immunol.*, **12**, 150–9.

David, J.R. (1966) Delayed hypersensitivity in vitro: Its mediation by cell-free substances formed by lymphoid cell-antigen interaction. *Proc. Natl. Acad. Sci. (USA)*, **56**, 73–7.

Dumonde, D.C., Wolstencroft, R.A., Panayi, G.S., Mathew, M., Marley, J., and Howson, W.T. (1969). 'Lymphokines': Non-antibody mediators of cellular immunity generated by lymphocyte activation. *Nature*, **224**, 38–42.

Farrah, T., and Smith, C.A. (1992). Emerging cytokine family. *Nature*, **358**, 26.

Fernandez-Botran, R. (1991). Soluble cytokine receptors: their role in immunoregulation. *Faseb J.* **5**, 2567–74.

Gordon, M.Y. (1991). Hemopoietic growth factors and receptors: bound and free. *Cancer Cells* **3**, 127–33.

Ichikawa, T., Pluznik, D.M., and Sachs, L. (1966). *In vitro* control of the development of macrophage and granulocyte colonies. *Proc. Natl. Acad. Sci. (USA)*, **56**, 488–95.

Isaacs, A., and Lindenmann, J. (1957). Virus interference. I. The interferon. *Proc. R. Soc. (London)*, Ser. B, **147**, 258–67.

Kraulis, P.J. (1991). Molscript: a program to produce both detailed and schematic plots of protein structure. *J. Appl. Cryst.*, **24**, 946–50.

Mallett, S., and Barclay, A.N. (1991). A new superfamily of cell surface proteins related to the nerve growth factor receptor. *Immunol. Today*, **12**, 220–3.

McDonald, N.Q., and Hendrickson, W.A. (1993). A structural superfamily of growth factors containing a cystine knot motif. *Cell*, **73**, 421–4.

Metcalf, D. (1991). The leukaemia inhibitory factor (LIF). *Int. J. Cell Cloning*, **9**, 95–108.

Murzin, A.G., Lesk, A.M., and Chothia, C. (1992). Beta-Trefoil fold. Patterns of structure and sequence in the Kunitz inhibitors, interleukins-1 beta and 1 alpha and fibroblast growth factors. *J. Mol. Biol.* **223**, 531–43.

Nicola, N.A., Peterson, L., Hilton, D.J., and Metcalf, D. (1988). Cellular processing of murine colony-stimulating factor (Multi-CSF, GM-CSF, G-CSF) receptors by normal haemopoietic cells and cell lines. *Growth Factors*, **1**, 41–9.

Parry, A.D., Minasian, E., and Leach, S.J. (1991). Cytokine conformations: predictive studies. *J. Molec. Recog.*, **4**, 63–75.

Poo, W.J., Conrad, L., and Janeway, C.A. Jr. (1988). Receptor-directed focusing of lymphokine release by helper T cells. *Nature*, **332**, 378–80.

Robinson, W.A., Metcalf, D., and Bradley, T.R. (1967). Stimulation by normal and leukaemic mouse sera of colony formation *in vitro* by mouse bone marrow cells. *J. Cell Comp. Physiol.*, **69**, 83.

Ruddle, N.H., and Waksman, B.H. (1968). Cytotoxicity mediated by soluble antigen and lymphocytes in delayed hypersensitivity III. Analysis of Mechanism. *J. Exp. Med.*, **128**, 1267–80.

Sprang, S.R., and Bazan, J.F. (1993). Cytokine structural taxonomy and mechanisms of receptor engagement. *Curr. Opin. Structural Biol.*, **3**, 815–27.

Tavernier, J. (1993). Transgenic mice in the study of cytokine function. *Int. J. Exp. Path.*, **74**, 525–46.

Wheelock, E.F. (1965). Interferon-like virus-inhibitor induced in human leukocytes by phytohemagglutin. *Science*, **149**, 310–11.

Williams, T.W., and Granger, T.W. (1968). Lymphocyte *in vitro* cytotoxicity: lymphotoxins of several mammalian species. *Nature*, **219**, 1076–7.

Young, P.R. (1992). Protein hormones and their receptors. *Curr. Opin. Biotech.*, **3**, 408–21.

Nicos A. Nicola:

The Walter and Eliza Hall Institute of Medical Research and, The Cooperative Research Centre for Cellular Growth Factors,
PO Royal Melbourne Hospital,
Victoria 3050, Australia

An Introduction to Cytokine Receptors

It is well-established that polypeptide hormones elicit their biological effects by binding to receptors expressed on the surface of responsive cells. The characterization of cell surface receptors has proceeded in two distinct phases. Initially, the majority of receptors were described on the basis of their ability to bind radiolabelled derivatives of a particular hormone. These studies allowed the specificity of receptors to be determined and the equilibrium and kinetic characteristics of binding to be investigated. In some cases, through the additional use of chemical cross-linking reagents, an estimate of the apparent molecular weight of receptors was also possible. In general, however, the low abundance of receptors and their integral membrane location hindered their detailed biochemical characterization. Only with the advent and application of appropriate molecular genetic techniques has the predicted primary structure of many receptors been determined. The aim of the following review is to summarize the important structural features that unite receptors of a given family and to highlight themes common to the function of receptors from different families.

■ Receptor families

At least four families of hormone receptors can be defined on the basis of similarity in primary sequence, predicted secondary and tertiary structure and biochemical function. These are the haemopoietin/interferon receptor family, the receptor kinase family, the tumour necrosis factor (TNF)/nerve growth factor (NGF) receptor family and the family of G-protein coupled, seven membrane-spanning receptors. The important properties of receptors in each of these groups are described below.

■ Haemopoietin/interferon receptors

Haemopoietin and interferon receptors are type I transmembrane glycoproteins that are defined on the basis of sequence similarity in their extracellular domains (Gearing et al. 1989; Bazan 1990a; Thoreau et al. 1991). Ligands that interact with haemopoietin/interferon receptors include interferons (IFNs) -α, -β, and -γ, interleukins (IL) -2, -3, -4, -5, -6, -7, -9, -10, -11, -13, leukaemia inhibitory factor (LIF), oncostatin-M (OSM), erythropoietin (epo), ciliary neurotrophic factor (CNTF), growth hormone and prolactin (Gearing and Ziegler 1993; Kishimoto et al. 1994). Each of these cytokines are active as secreted glycoproteins and are themselves thought to share a common structure comprising a four α-helical bundle (Bazan 1990b, 1991). The p40 subunit of IL-12 is also a member of the cytokine receptor family (Gearing and Cosman 1991) and there is a member, termed c-mpl, for which a ligand has yet to be described.

This 'orphan receptor' was isolated as the cellular homologue of the transforming gene of a derivative of the Friend virus that caused a myeloproliferative syndrome in mice (Souyri et al. 1990; Vigon et al. 1992).

While many haemopoietin and interferon receptors contain extracellular fibronectin type III domains and/or immunoglobulin-like domains (Plate 2), the unifying feature of this family is the presence of one or two domains of 200 amino acids, termed D200. D200 domains may, in turn, be divided into subdomains of 100 amino acids (SD100), which are themselves homologous (Plate 2) (Bazan 1990a; Thoreau et al. 1991). The major conserved features include a pair of cysteine residues in the first SD100 domain of each D200 domain, proline residues that precede each SD100 domain, and a series of conserved hydrophobic and hydrophilic residues (Bazan 1990a; Thoreau et al. 1991). Two subfamilies of receptors may also be defined. These are the interferon receptor family and the haemopoietin receptor family. Members of the latter contain other conserved residues including an additional pair of cysteine residues, a series of aromatic residues and the five-amino-acid motif Trp-Ser-Xaa-Trp-Ser (Bazan 1990a; Thoreau et al. 1991).

Although the level of primary sequence similarity between members of the haemopoietin/interferon receptor family is quite low it is thought to underlie a common tertiary structure (Bazan 1990a; Thoreau et al. 1991). The structure of the growth hormone receptor has been solved (de Vos et al. 1992) and has become a paradigm for the family in general. Each SD100 domain of the growth hormone receptor contains 7 β-strands arranged to form a barrel similar to those in the bacterial chaperone PapD, the cell surface molecule CD4 and the fibronectin type III domain (de Vos et al. 1992).

Surprisingly, the stoichiometry of the growth hormone receptor/ligand complex was 2:1, with each receptor utilizing similar residues to interact with the hormone (de Vos et al. 1992). Subsequent experiments demonstrated that formation of the ternary complex proceeds in a strict order. Growth hormone interacts first, using epitopes on the A helix, the D helix and the loop between helices A and B, and only after forming a 1:1 complex does a second receptor bind to epitopes on the A helix and C helix (de Vos et al. 1992; Cunningham et al. 1991).

Where examined in detail, the receptors for other cytokines have also been found to be multimeric. Unlike the growth hormone receptor many receptors form heterodimers (Kishimoto et al. 1994; Nicola and Metcalf 1991) yet the sequential nature of binding may be similar. GM-CSF, for example, binds initially to its receptor α-chain using residues in the D helix and then to the receptor β-chain using residues in the A helix (Kastelein and Shanafelt 1993). The situation for other receptor systems may be further complicated by the presence of three or more receptor polypeptides (Kishimoto et al. 1994).

The biological action of many cytokines appears to be pleiotropic, yet also redundant (Kishimoto *et al.* 1994; Metcalf 1992). Biological redundancy may be explained partly by the observation that one or more receptor subunits may be utilized by different cytokines (Plate 3) (Gearing and Ziegler 1993; Kishimoto *et al.* 1993; Nicola and Metcalf 1991) and also by regions of sequence homology in the cytoplasmic domains of some receptors, termed box 1 and box 2, which may underlie common signalling pathways (Murakami *et al.* 1991). Some examples of cytokines which share receptor subunits are described below.

IL-3, IL-5, and GM-CSF bind to specific α-subunits with low affinity. The generation of high-affinity receptor complexes capable of signal transduction requires interaction with a common β-subunit (Kishimoto *et al.* 1994; Nicola and Metcalf 1991; Miyajima *et al.* 1992). In the case of the mouse IL-3 receptor, there exists a specific β-subunit (AIC2A), in addition to that which is shared with IL-5 and GM-CSF (AIC2B) (Schreurs *et al.* 1991).

The receptors for IL-2, IL-4, IL-7, IL-9, and IL-13 contain a common component, isolated initially as the γ-chain of the IL-2 receptor (Kishimoto *et al.* 1994; Takeshita *et al.* 1992; Kondo *et al.* 1993; Noguchi *et al.* 1993). In addition, each cytokine interacts with a specific receptor component. In the case of IL-4, IL-7, and IL-13, the specific receptor subunits may exist as homodimers, but for IL-2, in contrast, there is a heterodimer containing an α- and β-subunit (Nikaido *et al.* 1984; Leonard *et al.* 1984; Hatakeyama *et al.* 1989). The α-subunit (tac antigen) is unusual among interleukin receptors in that it is not itself a member of the haemopoietin/interferon receptor family (Nikaido *et al.* 1984; Leonard *et al.* 1984).

The situation for IL-6, IL-11, CNTF, LIF, and OSM is more complex. The receptor for each cytokine contains gp130 which, although it does not bind to IL-6, was first identified because of its ability to interact with the complex between IL-6 and its specific, low-affinity receptor α-subunit to yield a high affinity receptor capable of signal transduction (Hibi *et al.* 1990). The stoichiometry of the IL-6 receptor appears to be two molecules of gp130 to a single α-chain (Murakami *et al.* 1993). CNTF, LIF, and OSM receptors differ in that they contain one molecule of gp130 and one molecule of a LIF receptor polypeptide (Davis and Yancopoulos 1993; Davis *et al.* 1993; Gearing *et al.* 1991, 1992; Gearing and Bruce 1992). CNTF also binds to a specific α-subunit that is most similar in primary sequence to the α-subunit of the IL-6 receptor, but which unlike other members of the haemopoietin/interferon receptor family is tethered to the plasma membrane by a glycophosphatidyl inositol (GPI) anchor rather than a classical transmembrane domain (Davis *et al.* 1991). Whether similar specific receptor α-subunits also exist for IL-11, LIF, and OSM remains unknown. The inclusion of the IL-11 receptor in this class is circumstantial. Although the structure of the receptor is not known, gp130 is implicated since neutralizing antibodies to this molecule inhibit the biological activity of IL-11 (Yin *et al.* 1993).

Homodimerization appears to be an important feature of the receptors for other cytokines, including those for growth hormone, prolactin, epo and G-CSF (de Vos *et al.* 1992; Fukunaga *et al.* 1991; Watowich *et al.* 1992).

■ NGF/TNF receptors

Members of the TNF/NGF receptor family have been identified in three distinct manners. The type I and type II TNF receptors, as well as the p75 subunit of the NGF receptors were isolated on the basis of their capacity to bind their cognate ligand (Johnson *et al.* 1986; Loetscher *et al.* 1990; Smith *et al.* 1990). CD-30, CD40, FAS, OX-40, and 4–1BB were cloned as cell surface antigens with no reference to their physiological ligand (Kwon and Weissman 1989; Stamenkovic *et al.* 1989; Mallett *et al.* 1990; Itoh *et al.* 1991; Durkop *et al.* 1992; Camerini *et al.* 1991), although recently the ligands for CD-30, CD-40, and FAS have been cloned (Suda *et al.* 1993; Armitage *et al.* 1992; Goodwin *et al.* 1993; Smith *et al.* 1993). Other members of this family, such as SalF19R, are of viral origin (Howard *et al.* 1991).

The defining features of members of the TNF-NGF receptor family are located in the extracellular domain and centre on four copies of a domain that contains 6 cysteine residues (Bazan 1993) (Plate 4). The crystal structure of the type I TNF receptor revealed that these domains exist as separate structural entities with little overlap (Banner *et al.* 1993). In addition to extracellular sequence similarity, the p55 TNF receptor, FAS and CD40 molecules share cytoplasmic sequence similarity, which may contribute to their related cytotoxic effects (Itoh *et al.* 1991; Tartaglia *et al.* 1993). The recognition that viruses produce proteins that are similar in sequence to receptors for regulators of cell death is interesting since it raises the possibility that such viral proteins are used as a means of evading the body's defence against virally infected cells (Howard *et al.* 1991).

Unlike the hormones which interact with the cytokine receptor family, the ligands for many of the TNF/NGF family members exist as type II transmembrane proteins, as well as secreted regulators. TNFα and TNFβ exist as homotrimers in solution and in crystals (Banner *et al.* 1993; Loetscher *et al.* 1991; Schoenfeld *et al.* 1991) and TNFβ has also been shown to exist on the cell surface as a heterotrimer in association with LTβ (Browning *et al.* 1993). It is therefore possible that heterotrimers of other family members may also exist. In solution three receptor molecules interact with a single ligand trimer (Loetscher *et al.* 1991) and this stoichiometry has recently been confirmed with the solution of the structure of the TNFβ/p55 TNF receptor complex (Banner *et al.* 1993). The TNF/TNF receptor complex exhibited perfect symmetry, with each receptor interacting with an interface of two of the three TNF molecules in an identical manner. This arrangement contrasts with the growth hormone receptor/growth hormone complex in which the two receptors bound to completely different faces of the ligand (de Vos *et al.* 1992).

■ Receptor kinases

While the regions of amino acid sequence similarity that unite members of the haemopoietin and TNF/NGF receptor families lie in their extracellular ligand-binding domains, receptor kinase consensus motifs are cytoplasmic, catalytic in nature and are shared with a large group of proteins that are not receptors (Hanks et al. 1988). Receptor kinases and kinases in general may be classified according to their substrate specificity, with one class phosphorylating tyrosine residues and others serine and threonine residues. Recent results suggest that in some cases this distinction may be blurred (Ben-David et al. 1991).

More than 50 receptor tyrosine kinases have been cloned and while at least ten distinct classes have been recognized on the basis of the structure of their extracellular ligand binding domains, the cognate ligands for members of four of these groups are yet to be defined. The extracellular domains of the receptor tyrosine kinases are composed of a wide range of modular structures including immunoglobulin domains, fibronectin type III domains and conserved cysteine containing repeats (Plate 5). In some cases, additional receptor diversity is generated by alternative splicing, the biological significance of which remains relatively unexplored (Yee et al. 1989; Petch et al. 1990; Attisano et al. 1992; Rodrigues et al. 1991).

While heteromerization and trimerization are central to the function of haemopoietin/interferon receptors and TNF/NGF receptors, respectively, homodimerization appears to be central to receptor kinase action (Ullrich and Schlessinger 1990). The majority of receptor tyrosine kinases exists as single polypeptide chains in the unoccupied state and dimerize upon interaction with their cognate ligand. The exceptions to this are the members of the insulin/IGF-1 receptor subfamily, which are tetramers containing two α-subunits and two β-subunits, and the receptors for NGF, brain-derived neurotrophic factor and neurotrophin-3 (the trk kinases), that associate with the low affinity NGF receptor (p75), which is a member of the TNF/NGF receptor family (Johnson et al. 1986; Park 1991; Ross 1991; Klein et al. 1991; Lamballe et al. 1991; Loeb et al. 1991; Ohmichi et al. 1991; Squinto et al. 1991; Soppet et al. 1991). The interaction between the trks and p75 has been claimed to be important for the generation of a high affinity neurotrophin receptor, however, this point remains controversial.

The receptors for the activin/inhibin/TGF-β class of cytokines are serine/threonine kinases rather than tyrosine kinases (Attisano et al. 1992; Matthews and Vale 1991; Lin et al. 1992; Massague 1992; ten-Dijke et al. 1993; Matsuzaki et al. 1993; Ebner et al. 1993; Shinozaki et al. 1992). The kinase domain of serine/threonine kinase receptors share many of the structural features found in receptor tyrosine kinases. Other receptors, such as the nematode daf-1 protein and four human activin-like receptor kinases have been described but the identity of their cognate ligands has not been determined (ten-Dijke et al. 1993; Georgi et al. 1990; Massague et al. 1992). The theme of multi-chain receptors appears to hold for receptor serine and threonine kinases. For example, compelling evidence exists for the interaction of type I and type II TGFβ-receptors (Ebner et al. 1993; Wrana et al. 1992). In addition, it has been suggested that the TGFβ type III receptor (Cheifetz et al. 1988; Wang et al. 1991), which is not a member of the serine/threonine family of receptors but rather a transmembrane proteogly-

can, also interacts with the type II receptor (Wang et al. 1991; Lopez-Casillas et al. 1993).

■ G-protein coupled receptors

Receptors that are coupled to G-proteins are fundamentally different from the classes of receptors considered above. While receptors in the haemopoietin, TNF/NGF and kinase families contain a single transmembrane domain, G-protein coupled receptors traverse the membrane seven times. The range of extracellular stimuli that utilize G-protein coupled receptors is large and includes light, odorants, neurotransmitters and an array of peptide and protein hormones and cytokines (Bourne et al. 1991; Iismaa and Shine 1992). Examples of the latter are IL-8 and endothelin, described in this handbook, as well as calcitonin, luteinizing hormone, parathyroid hormone, somatostatin, thyroid stimulating hormone and vasoactive intestinal polypeptide (Iismaa and Shine 1992). The structure of the ligand binding domain of G-protein coupled receptors is very much dependent on the nature of the ligand. While the transmembrane regions of adrenergic and muscarinic receptors are important for binding their neurotransmitters (and similar regions of rhodopsin are important for the interaction with cis-retinal), polypeptide hormones such as luteinizing hormone interact with extensive receptor extracellular domains. The cytoplasmic loops of the G-protein coupled receptors play a central role in coupling to the heterotrimeric G-proteins (Iismaa and Shine 1992).

■ Signal transduction

A great deal is known concerning the molecular consequences of the activation of G-protein coupled receptors (Bourne et al. 1991; Spiegel 1992), so only the briefest of overviews will be provided here. G-proteins are heterotrimeric (αβγ) and although they are not integral membrane proteins, they associate with the inner face of the plasma membrane (Spiegel 1992). Upon occupation of the receptor with ligand, the G-protein associates with the receptor cytoplasmic domain. As a consequence, although the β- and γ-subunits are primarily responsible for interaction with the receptor, it is the α-subunit that is activated. Activation of the α-subunit is intimately associated with the exchange of GTP for GDP (Bourne et al. 1991; Spiegel 1992). α-subunits are inactivated by hydrolysis of GTP and return to the basal GDP-bound state. The signal that emanates from activated G-proteins depends upon the identity of the α-subunit. Various $G_{\alpha s}$ subunits activate adenyl cyclase and therefore increase cAMP levels. Conversely, $G_{\alpha i}$ subunits inhibit adenyl cyclase and reduce cAMP levels. The biochemical events effected by the level of cAMP are numerous and include regulation of kinases and alteration in the net metabolic state of cells (Spiegel 1992). Other G-protein α-subunits serve to couple receptors to the regulation of

TYROSINE KINASE RECEPTORS

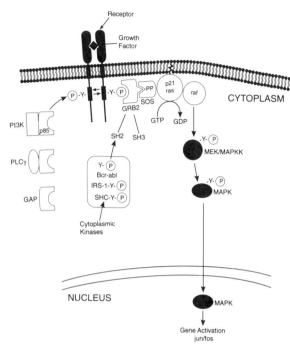

Figure 1. Signal transduction from tyrosine kinase receptors (see text for details).

homology 2 (SH2) domains (Heldin 1991; Koch *et al.* 1991). SH2 containing proteins that interact with phosphorylated receptor molecules include the p85 regulatory subunit of phosphatidylinositol-3'-kinase (PI3 kinase), members of the src family of kinases, phospholipase-C-γ, nck, GRB2 and shc (Escobedo *et al.* 1991; Kanner *et al.* 1991; Pelicci *et al.* 1992; Lowenstein *et al.* 1992).

In addition to its interaction with phosphorylated receptor through its SH2 domains, GRB2 appears to be constitutively bound to a second protein, son of sevenless (sos), through its src homology 3 (SH3) domain (Lowenstein *et al.* 1992; Skolnik *et al.* 1993; Buday and Downward 1993; Gale *et al.* 1993; Chardin *et al.* 1993; Egan *et al.* 1993; Olivier *et al.* 1993; McCormick 1993; Li *et al.* 1993; Rozakis-Adcock *et al.* 1993). Sos, in turn, interacts with membrane-bound ras and promotes the exchange of GTP for GDP (Egan *et al.* 1993; Baltensperger *et al.* 1993). The activated ras molecule then initiates a kinase cascade involving raf-1 and MAP kinase kinase that eventually results in activation of MAP kinase and thus allows transcriptional regulation through the phosphorylation of transcription factors (Nakajima *et al.* 1993; Wood *et al.* 1992).

Surprisingly, the consequence of bringing the cytoplasmic domains of haemopoietin/interferon receptors together is similar to the events that result from activation of receptor tyrosine kinases (Fig. 2). Despite their lack of an intrinsic tyrosine kinase activity, it has been known for

ion channels, cGMP phosphodiesterase and various forms of phospholipase (Spiegel 1992). Recent evidence suggests that the β- and γ-subunits of G-proteins may also contribute to signal transduction (Spiegel 1992).

As described above, cytokine-driven interaction between receptor subunits appears to be the initial event in signal transduction for haemopoietin receptors, receptor kinases, and TNF/NGF receptors. Multimerization allows information to pass from the extracellular domain to the cytoplasmic environment without the necessity of a conformational change being transmitted directly through the plasma membrane. Although relatively little is known at the molecular level of events that are important in signalling through TNF/NGF receptors, it has recently become apparent that receptors of the haemopoietin family and receptor tyrosine kinases share a number of signal transduction characteristics.

The initial events in the receptor kinase signalling pathway are beginning to become clear (Fig. 1). In the case of receptors such as those for EGF and PDGF, dimerization leads to the juxtaposition of two cytoplasmic domains and activation of their intrinsic kinase activity (Ullrich and Schlessinger 1990). As a result, receptors are phosphorylated on tyrosine residues, not in an autocatalytic reaction but by the other receptor in the dimer. Phosphorylated tyrosine residues in receptor cytoplasmic domains are the site of interaction for cytoplasmic proteins containing src

CYTOKINE RECEPTORS

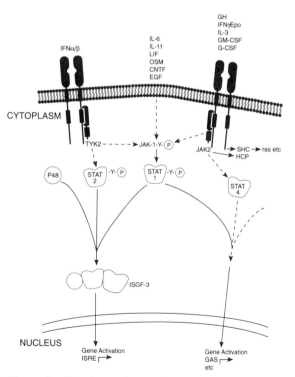

Figure 2. Signal transduction from cytokine receptors (see text for details).

many years that tyrosine phosphorylation occurs rapidly in response to cytokines that act through haemopoietin receptors (Garland 1988; Quelle and Wojchowski 1991; Kanakura *et al*. 1991) (Fig. 2). It has therefore been speculated that activation of a cytoplasmic tyrosine kinase is a pivotal early event in signal transduction initiated by these receptors. This notion is reinforced by the observation that cytokines such as IL-3 and GM-CSF, like hormones that activate receptors with intrinsic tyrosine kinase activity, activate ras, stimulate the phosphorylation of raf-1 and activate MAP kinase (Carroll *et al*. 1990; Satoh *et al*. 1991, 1992; Sakamaki *et al*. 1992). These cytokines also enhance the phosphorylation of shc which is consistent with a role for this SH2 containing protein in the coupling of cytokine receptor activation to ras activation (Burns *et al*. 1993).

The kinases that transduce signals from haemopoietin receptors have been difficult to define. Members of the src family of kinases are attractive candidates in this process (Hatakeyama *et al*. 1991; Torigoe *et al*. 1992; Horak *et al*. 1991). Recently, however, as a result of a series of elegant genetic studies into interferon signalling, interest has focused on the Jak family of kinases (Fig. 2). Three members of the Jak kinase family have been described: Jak1, Jak2, and Tyk2. Each plays a critical role in interferon signal transduction. Three mutant cell lines have been generated which fail to respond to interferons; U1 is IFNα unresponsive, γ-1 is IFNγ unresponsive, while U4 fails to respond to both classes of IFN (Stark and Kerr 1992). The defects in these cell lines may be complemented by expression of Tyk2, Jak2, and Jak1, respectively (Watling *et al*. 1993; Velazquez *et al*. 1992; Muller *et al*. 1993a). These kinases, however, could not be placed in a linear arrangement in the signalling cascade. Rather, although phosphorylation of Tyk2 and Jak1 occurs in response to IFNα, neither are phosphorylated in the cell lines U1 or U4. Similarly, Jak2 and Jak1 are phosphorylated in response to IFNγ, but neither are phosphorylated in γ-1 or U4. Light was shed on this problem when it was shown that activation and phosphorylation of the Jak family of kinases was not restricted to interferons. A wide range of other cytokines which bind to haemopoietin receptors also utilize Jak kinases (Silvennoinen *et al*. 1993a; Witthuhn *et al*. 1993; Argetsinger *et al*. 1993). In the case of epo and IL-3, for example, it can be shown that Jak-2 interacts with the epo receptor and the common IL-3, IL-5, and GM-CSF β-chain (Silvennoinen *et al*. 1993a; Witthuhn *et al*. 1993). Likewise, gp130 and the LIF receptor (components of the LIF, IL-6, IL-11, OSM, and CNTF receptors) associate with all three members of the Jak family (Stahl *et al*. 1994; Lutticken *et al*. 1994).

In each case, upon the interaction of the cytokine with its receptor, Jak kinases are phosphorylated and almost certainly activated. Given the homodimeric nature of the epo receptor, the presence of two gp130 molecules in the IL-6 receptor and gp130 and the LIF receptor in the LIF, OSM, and CNTF receptors, it is attractive to speculate that there is cross-phosphorylation of Jak kinases, in a similar way as there is cross-phosphorylation of receptors with intrinsic tyrosine kinase activity.

What is the consequence of activating members of the Jak kinase family? Again, somatic cell genetics has helped to provide some of the clues (Stark and Kerr 1992). U2, which is IFNα unresponsive, and U3, which is unresponsive to both IFNα and IFNγ, are two other mutant cell lines that have defects in the transcription factors p48 and p91/p84 (p91 and p84 are produced from the same gene by alternative splicing) (Schindler *et al*. 1992a; Muller *et al*. 1993b). Upon interaction of IFNα with its receptor, p91, in addition to p84 and p113, becomes phosphorylated, moves from the cytoplasm to the nucleus and assembles with p48 to form a complex known as interferon stimulated gene factor-3 (ISGF3) (Fu *et al*. 1990; Kessler *et al*. 1990; Fu 1992; David and Larner 1992; Schindler *et al*. 1992b). ISGF3 binds to the interferon response element present in the promoters of IFNα responsive genes and contributes to the transcriptional activation of these genes (Kessler *et al*. 1990). A similar but simpler scenario exists for IFNγ response where p91 has been shown to be the gamma-activated factor (GAF). p91 is phosphorylated in response to IFNγ and again migrates from the cytoplasm to the nucleus where it binds to the gamma-activated sequence (GAS) of IFNγ responsive promoters (Igarashi *et al*. 1993; Shuai *et al*. 1992).

Recent results show that activation of p91 and related molecules is not specific to the interferons. A similar set of events occurs in response to IL-10 and perhaps more surprisingly in response to EGF and PDGF (FU and Zhang 1993; Larner *et al*. 1993; Silvennoinen *et al*. 1993b; Ruff-Jamison *et al*. 1993; Sadowski *et al*. 1993). Stimulation of cells with EGF results in the direct interaction of the tyrosine phosphorylated receptor with p91 which contains an SH2 domain (Fu and Zhang 1993). As a result of this interaction, p91 is rapidly phosphorylated on tyrosine residues and is translocated to the nucleus, where it gains enhanced ability to bind to DNA (specifically the c-sis inducible element or SIE). As a consequence, the transcription of genes such as fos is activated. Early signalling events for IL-3, IL-5, and GM-CSF, as well as IL-6, IL-11, CNTF, LIF, and OSM, may follow a similar pattern (Lutticken *et al*. 1994; Larner *et al*. 1993). Like IFNγ, the cytokines IL-3, IL-5, and GM-CSF, which utilize a common receptor β-chain, trigger the phosphorylation of proteins capable of binding to the IFNγ responsive element of the FCγ receptor gene (Larner *et al*. 1993). In contrast to IFNγ and IL-10, the biological outcome of IL-3, IL-5, and GM-CSF stimulation is not transcriptional activation, but rather repression of the interferon inducible pathway. Moreover, unlike IFNγ and IL-10, the IL-3, IL-5, and GM-CSF-activated DNA-binding protein does not contain p91, nor does it contain the related p113, but rather an uncharacterized 80 kDa phosphoprotein is present (Larner *et al*. 1993).

■ Summary

In the past five years, amazing progress has been made in cloning the genes encoding cytokine receptors. As a result, it has become possible to gain an understanding of the relationships between receptor structure and function and to produce receptors in relatively large amounts. This has allowed the co-crystallization of receptor and ligand and, in the case of growth hormone and TNF, has excitingly led to the solution of the three dimensional structure of the complex.

Recent work has also provided a unifying framework with which to view receptor signal transduction, with common themes emerging in the intracellular pathways used by receptor tyrosine kinases and receptors of the haemopoietin/interferon family. It should be remembered, however, that the cytokines described in this handbook are bewildering in the variety of biological responses they elicit. The majority of signalling studies have focused on cells that proliferate or increase synthesis of specific proteins in response to cytokine stimulation. It is certain that in the ensuing months and years equally exciting progress will be made into the mechanisms that regulate other cellular functions including self renewal, cell differentiation, cell survival and apoptosis.

■ References

Argetsinger, L.S., Campbell, G.S., Yang, X., Witthuhn, B.A., Silvennoinen, O., Ihle, J.N., and Carter, S.C. (1993). Identification of JAK2 as a growth hormone receptor-associated tyrosine kinase. *Cell*, **74**, 237–44.

Armitage, R.J., Fanslow, W.C., Strockbine, L., Sato, T.A., Clifford, K.N., Macduff, B.M., Anderson, D.M., Gimpel, S.D., Davis-Smith, T., Maliszewski, C.R., Clark, E.A., Smith, C.A., Grabstein, K.H., Cosman, D., and Spriggs, M.K. (1992). Molecular and biological characterization of a murine ligand for CD40. *Nature*, **357**, 80–2.

Attisano, L., Wrana, J.L., Cheifetz, S., and Massague, J. (1992). Novel activin receptors: distinct genes and alternative splicing generate a repertoire of serine/threonine kinase receptors. *Cell*, **68**, 97–108.

Baltensperger, K., Kozma, L.M., Cherniack, A.D., Klarlund, J.K., Chawla, A., Banerjee, U., and Czech, M.P. (1993). Binding of the Ras activator son of sevenless to insulin receptor substrate-1 signaling complexes. *Science*, **260**, 1950–2.

Banner, D.W., D'Arcy, A., Janes, W., Gentz, R., Schoenfeld, H.J., Broger, C., Loetscher, H., and Lesslauer, W. (1993). Crystal structure of the soluble human 55 kd TNF receptor-human TNF β complex: implications for TNF receptor activation. *Cell*, **73**, 431–45.

Bazan, J.F. (1990a). Structural design and molecular evolution of a cytokine receptor superfamily. *Proc. Natl. Acad. Sci. (USA)*, **87**, 6934–8.

Bazan, J.F. (1990b). Hemopoietic receptors and helical cytokines. *Immunol. Today*, **11**, 350–4.

Bazan, J.F. (1991). Neuropoietic cytokines in the hematopoietic fold. *Neuron*, **7**, 197–208.

Bazan, J.F. (1993). Emerging families of cytokines and receptors. *Curr. Biol.* **3**, 603–6.

Ben-David, Y., Letwin, K., Tannock, L., Bernstein, A., and Pawson, T. (1991). A mammalian protein kinase with potential for serine/threonine and tyrosine phosphorylation is related to cell cycle regulators. *EMBO J.*, **10**, 317–25.

Bourne, H.R., Sanders, D.A., and McCormick, F. (1991). The GTPase superfamily: conserved structure and molecular mechanism. *Nature*, **349**, 117–27.

Browning, J.L., Ngam-ek, A., Lawton, P., DeMarinis, J., Tizard, R., Chow, E.P., Hession, C., O'Brine-Greco, B., Foley, S.F., and Ware, C.F. (1993). Lymphotoxin β, a novel member of the TNF family that forms a heterotrimeric complex with lymphotoxin at the cell surface. *Cell*, **72**, 847–56.

Buday, L., and Downward, J. (1993). Epidermal growth factor regulates p21 ras through the formation of a complex of receptor, Grb2 adapter protein, and Sos nucleotide exchange factor. *Cell*, **73**, 611–20.

Burns, L.A., Karnitz, L.M., Sutor, S.L., and Abraham, R.T. (1993). Interleukin-2-induced tyrosine phosphorylation of p52sch in T lymphocytes. *J. Biol. Chem.*, **268**, 17659–61.

Camerini, D., Waltz, G., Loenen, W.A.M., Borst, J., and Seed, B. (1991). The T cell activation antigen CD27 is a member of the nerve growth factor/tumour necrosis factor gene family. *J. Immunol.*, **147**, 3165–9.

Carroll, M.P., Clark-Lewis, I., Rapp, U.R., and May, W.S. (1990). Interleukin-3 and granulocyte-macrophage colony-stimulating factor mediate rapid phosphorylation and activation of cytosolic c-raf. *J. Biol. Chem.* **265**, 19812–17.

Chardin, P., Camonis, J.H., Gale, N.W., van-Aelst, L., Schlessinger, J., Wigler, M.H., and Bar-Sagi, D. (1993). Human Sos1: a guanine nucleotide exchange factor for Ras that binds to GRB2. *Science*, **260**, 1338–43.

Cheifetz, S., Andres, J.L., and Massague, J. (1988). The transforming growth factor-β receptor type III is a membrane proteoglycan. Domain structure of the receptor. *J. Biol. Chem.*, **263**, 16984–91.

Cunningham, B.C., Ultsch, M., De, V.A., Mulkerrin, M.G., Clauser, K.R., and Wells, J.A. (1991). Dimerization of the extracellular domain of the human growth hormone receptor by a single hormone molecule. *Science*, **254**, 821–25.

David, M., and Larner, A.C. (1992). Activation of transcription factor by interferon-α in a cell-free system. *Science*, **257**, 813–15.

Davis, S., and Yancopoulos, G.D. (1993). The molecular biology of the CNTF receptor. *Curr. Opin. Cell Biol.*, **5**, 281–5.

Davis, S., Aldrich, T.H., Valenzuela, D.M., Wong, V.V., Furth, M.E., Squinto, S.P., and Yancopoulos, G.D. (1991). The receptor for ciliary neurotrophic factor. *Science*, **253**, 59–63.

Davis, S., Aldrich, T.H., Stahl, N., Pan, L., Taga, T., Kishimoto, T., Ip, N.Y., and Yancopoulos, G.D. (1993). LIFRβ-and gp130 as heterodimerizing signal transducers of the tripartite CNTP receptor. *Science*, **260**, 1805–8.

de Vos, A.M., Ultsch, M., and Kossiakoff, A.A. (1992). Human growth hormone and extracellular domain of its receptor: crystal structure of the complex. *Science*, **255**, 306–12.

Durkop, H., Latza, U., Hummel, M., Eitelbach, F., Seed, B., and Stein, H. (1992). Molecular cloning and expression of a new member of the nerve growth factor receptor family that is characteristic for Hodgkin's disease. *Cell*, **68**, 421–7.

Ebner, R., Chen, R.H., Shum, L., Lawler, S., Zioncheck, T.F., Lee, A., Lopez, A.R., and Derynck, R. (1993). Cloning of a type I TGF-β receptor and its effect on TGF-β binding to the type II receptor. *Science*, **260**, 1344–8.

Egan, S.E., Giddings, B.W., Brooks, M.W., Buday, L., Sizeland, A.M., and Weinberg, R.A. (1993). Association of Sos Ras exchange protein with Grb2 is implicated in tyrosine kinase signal transduction and transformation. *Nature*, **363**, 45–51.

Escobedo, J.A., Navankasattusas, S., Kavanaugh, W.M., Milfay, D., Fried, V.A., and Williams, L.T. (1991). cDNA cloning of a novel 85 kd protein that has SH2 domains and regulates binding of PI3-kinase to the PDGF β-receptor. *Cell*, **65**, 75–82.

Fu, X.Y. (1992). A transcription factor with SH2 and SH3 domains is directly activated by an interferon α-induced cytoplasmic protein tyrosine kinase(s). *Cell*, **70**, 323–35.

Fu, X.-Y., and Zhang, J.-J. (1993). Transcription factor p91 interacts with the epidermal growth factor receptor and mediates activation of the c-fos gene promoter. *Cell*, **74**, 1135–45.

Fu, X.-Y., Kessler, D.S., Veals, S.A., Levy, D.E., and Darnell, J.E. (1990). ISGF3, the transcriptional activator induced by interferon-α, consists of multiple interacting polypeptides. *Proc. Natl. Acad. Sci. (USA)*, **87**, 8555–9.

Fukunaga, R., Ishizaka, I.E., Pan, C.X., Seto, Y., and Nagata, S. (1991). Functional domains of the granulocyte colony-stimulating factor receptor. *EMBO J.*, **10**, 2855–65.

Gale, N.W., Kaplan, S., Lowenstein, E.J., Schlessinger, J., and Bar-Sagi, D. (1993). Grb2 mediates the EGF-dependent activation of guanine nucleotide exchange on Ras. *Nature*, **363**, 88–92.

Garland, J.M. (1988). Rapid phosphorylation of a specific 33-kDa protein (p33) associated with growth stimulated by murine and rat IL3 in different IL3-dependent cell lines, and its constitutive expression in a malignant independent clone. *Leukaemia*, **2**, 94–102.

Gearing, D.P., and Bruce, A.G. (1992). Oncostatin M binds the high-affinity leukaemia inhibitory factor receptor. *New Biol.*, **4**, 61–5.

Gearing, D.P., and Cosman, D. (1991). Homology of the p40 subunit of natural killer cell stimulatory factor (NKSF) with the extracellular domain of the interleukin-6 receptor. *Cell*, **66**, 9–10.

Gearing, D.P., and Ziegler, S.F. (1993). The haematopoietic growth factor receptor family. *Curr. Opin. Hemat.*, **1993**, 138–48.

Gearing, D.P., King, J.A., Gough, N.M., and Nicola, N.A. (1989). Expression cloning of a receptor for human granulocyte-macrophage colony-stimulating factor. *EMBO J.* **8**, 3667–76.

Gearing, D.P., Thut, C.J., VandeBos, T., Gimpel, S.D., Delaney, P.B., King, J., Price, V., Cosman, D., and Beckmann, M.P. (1991). Leukemia inhibitory factor receptor is structurally related to the IL-6 signal transducer, gp130. *EMBO J.*, **10**, 2839–48.

Gearing, D.P., Comeau, M.R., Friend, D.J., Gimpel, S.D., Thut, C.J., McGourty, J., Brasher, K.K., King, J.A., Gillis, S., Mosley, B., Ziegler, S.F., and Cosman, D. (1992). The IL-6 signal transducer, gp130: an oncostatin M receptor and affinity converter for the LIF receptor. *Science*, **255**, 1434–7.

Georgi, L.L., Albert, P.S., and Riddle, D.L. (1990). daf-1, a C. elegans gene controlling dauer larva development, encodes a novel receptor protein kinase. *Cell*, **61**, 635–45.

Goodwin, R.G., Alderson, M.R., Smith, C.A., Armitage, R.J., Vanden Bos, T., Jerzy R., Tough, T.W., Schoenborn, M.A., Davis-Smith, T., Hennen, K., Falk, B., Cosman, D., Baker, E., Sutherland, G.R., Grabstein, K.H., Farrah, T., Giri, J.G., and Beckmann, M.P. (1993). Molecular and biological characterization of a ligand for CD27 defines a new family of cytokines with homology to tumour necrosis factor. *Cell*, **73**, 447–56.

Hanks, S.K., Quinn, A.M., and Hunter, T. (1988). The protein kinase family: conserved features and deduced phylogeny of the catalytic domains. *Science*, **241**, 42–52.

Hatakeyama, M., Tsudo, M., Minamoto, S., Kono, T., Doi, T., Miyata, T., Miyasaka, M., and Taniguchi, T. (1989). Interleukin-2 receptor β chain gene: generation of three receptor forms by cloned human α and β chain cDNA's. *Science*, **244**, 551–6.

Hatakeyama, M., Kono, T., Kobayashi, N., Kawahara, A., Levin, S.D., Perlmutter, R.M., and Taniguchi, T. (1991). Interaction of the IL-2 receptor with the src-family kinase p561ck: identification of novel intermolecular association. *Science*, **252**, 1523–8.

Heldin, C.H. (1991). SH2 domains: elements that control protein interactions during signal transduction. *Trends Biochem. Sci.*, **16**, 450–2.

Hibi, M., Murakami, M., Saito, M., Hirano, T., Taga, T., and Kishimoto, T. (1990). Molecular cloning and expression of an IL-6 signal transducer, gp130. *Cell*, **63**, 1149–57.

Horak, I.D., Gress, R.E., Lucas, P.J., Horak, E.M., Waldmann, T.A., and Bolen, J.B. (1991). T-lymphocyte interleukin 2-dependent tyrosine protein kinase signal transduction involves the activation of p56lck. *Proc. Natl. Acad. Sci (USA)*, **88**, 1996–2000.

Howard, S.T., Chan, Y.S., and Smith, G.L. (1991). Vaccinia virus homologues of the Shope fibroma virus inverted terminal repeat proteins and a discontinuous ORF related to the tumour necrosis factor receptor family. *Virology*, **180**, 633–47.

Igarashi, K., David, M., Finbloom, D.S., and Larner, A.C. (1993). *In vitro* activation of the transcription factor gamma interferon activating factor by gamma interferon: evidence for a tyrosine phosphatase/kinase signalling cascade. *Mol. Cell. Biol.*, **13**, 1634–40.

Iismaa, T.P., and Shine, J. (1992). G protein-coupled receptors. *Curr. Opin. Cell Biol.*, **4**, 195–202.

Itoh, N., Yonehara, S., Ishii, A., Yonehara, M., Mizushima, S.-I., Sameshima, M., Hase, A., Seto, Y., and Nagata, S. (1991). The polypeptide encoded by the cDNA for the human cell surface antigen Fas can mediate apoptosis. *Cell*, **66**, 233–43.

Johnson, D., Lanahan, A., Buck, C.R., Sehgal, A., Morgan, C., Mercer, E., Bothwell, M., and Chao, M. (1986). Expression and structure of the human NGF receptor. *Cell*, **47**, 545–54.

Kanakura, Y., Druker, B., Wood, K.W., Mamon, H.J., Okuda, K., Roberts, T.M., and Griffin, J.D. (1991). Granulocyte-macrophage colony-stimulating factor and interleukin-3 induce rapid phosphorylation and activation of the proto-oncogene Raf-1 in a human factor-dependent myeloid cell line. *Blood*, **77**, 243–8.

Kanner, S.B., Reynolds, A.B., Wang, H.C., Vines, R.R., and Parsons, J.T. (1991). The SH2 and SH3 domains of pp60src direct stable association with tyrosine phosphorylated proteins p130 and p110. *EMBO J.*, **10**, 1689–98.

Kastelein, R.A., and Shanafelt, A.B. (1993). GM-CSF receptor: interactions and activation. *Oncogene*, **8**, 231–6.

Kessler, D.S., Veals, S.A., Fu, X.-Y., and Levy, D.E. (1990). IFN-α regulates nuclear transcription and DNA binding affinity of ISGF3, a multimeric transcriptional activator. *Gene Dev.*, **4**, 1753–65.

Kishimoto, T., Taga, T., and Akira, S. (1994). Cytokine signal transduction. *Cell*, **76**, 253–62.

Klein, R., Jing, S.Q., Nanduri, V., O'Rourke, E., and Barbacid, M. (1991). The trk proto-oncogene encodes a receptor for nerve growth factor. *Cell*, **65**, 189–97.

Koch, C.A., Anderson, D., Moran, M.F., Ellis, C., and Pawson, T. (1991). SH2 and SH3 domains: elements that control interactions of cytoplasmic signaling proteins. *Science*, **252**, 668–74.

Kondo, K., Takeshita, T., Ishii, N., Nakamura, M., Watanabe, S., Arai, K.-I., and Sugamura, K. (1993). Sharing of the interleukin-2 (IL-2) receptor γ-chain between receptors for IL-2 and IL-4. *Science*, **262**, 1874–7.

Kwon, B., and Weissman, S.M. (1989). cDNA sequences of two inducible T-cell clones. *Proc. Natl. Acad. Sci. (USA)*, **86**, 1963–7.

Lamballe, F., Klein, R., and Barbacid, M. (1991). trkC, a new member of the trk family of tyrosine protein kinases, is a receptor for neurotrophin-3. *Cell*, **66**, 967–79.

Larner, A.C., David, M., Feldman, G.M., Igarashi, K., Hackett, R.G., Webb, D.S., Sweitzer, S.M., Petricoin, E.F. 3rd, and Finbloom, D.S. (1993). Tyrosine phosphorylation of DNA binding proteins by multiple cytokines. *Science*, **261**, 1730–3.

Leonard, W.J., Depper, J.M., Crabtree, G.R., Rudikoff, S., Pumphrey, J., Robb, R.J., Kronke, M., Svetlik, P.B., Peffer, N.J., Waldmann, T.A., and Greene, W.C. (1984). Molecular cloning and expression of cDNAs for the human interleukin-2 receptor. *Nature*, **311**, 626–31.

Li, N., Batzer, A., Daly, R., Yajnik, V., Skolnik, E., Chardin, P., Bar-Sagi, D., Margolis, B., and Schlessinger, J. (1993). Guanine-nucleotide-releasing factor hSos 1 binds to Grb2 and links receptor tyrosine kinases to Ras signalling. *Nature*, **363**, 85–8.

Lin, H.Y., Wang, X.F., Ng, E.E., Weinberg, R.A., and Lodish, H.F. (1992). Expression cloning of the TGF-β type II receptor, a functional transmembrane serine/threonine kinase. *Cell*, **68**, 775–85.

Loeb, D.M., Maragos, J., Martin-Zanca, D., Chao, M.V., Parada, L.F., and Greene, L.A. (1991). The trk proto-oncogene rescues NGF responsiveness in mutant NGF-nonresponsive PC12 cell lines. *Cell*, **66**, 961–6.

Loetscher, H., Pan, Y.C., Lahm, H.W., Gentz, R., Brockhaus, M., Tabuchi, H., and Lesslauer, W. (1990). Molecular cloning and expression of the human 55 kd tumour necrosis factor receptor. *Cell*, **61**, 351–9.

Loetscher, H., Gentz, R., Zulauf, M., Lustig, A., Tabuchi, H., Schlaeger, E.J., Brockhaus, M., Gallati, H., Manneberg, M., and Lesslauer, W. (1991). Recombinant 55-kDa tumour necrosis factor (TNF) receptor. Stoichiometry of binding to TNF α and TNF β and inhibition of TNF activity. *J. Biol. Chem.*, **266**, 18324–9.

Lopez-Casillas, F., Wrana, J.L., and Massague, J. (1993). Betaglycan presents ligand to the TGF β signaling receptor. *Cell*, **73**, 1435–44.

Lowenstein, E.J., Daly, R.J., Batzer, A.G., Li, W., Margolis, B., Lammers, R., Ullrich, A., Skolnik, E.Y., Bar-Sagi D., and Schlessinger, J. (1992). The SH2 and SH3 domain-containing protein GRB2 links receptor tyrosine kinases to ras signaling. *Cell*, **70**, 431–42.

Lutticken, C., Wegenka, U.M., Yuan, J., Buschmann, J., Schindler, C., Ziemiecki, A., Harpur, A.G., Wilks, A.F., Yasukawa, K., Taga, T., Kishimoto, T., Barbieri, G., Pelligrini, S., Sendtner, M., Heinrich, P.C., and Horn, F. (1994). Association of transcription factor APRF and protein kinase JAK1 with the interleukin-6 signal transducer gp130. *Science*, **263**, 89–92.

Mallett, S., Fossum, S., and Barclay, A.N. (1990). Characterization of the MRC OX40 antigen of activated CD4 positive T-lymphocytes—a molecule related to the nerve growth factor receptor. *EMBO J.*, **9**, 1063–8.

Massague, J. (1992). Receptors for the TGF-β family. *Cell*, **69**, 1067–70.

Massague, J., Andres, J., Attisano, L., Cheifetz, S., Lopez-Casillas, F., Ohtsuki, M., and Wrana, J.L. (1992). TGF-β receptors. *Mol. Reprod. Dev.*, **32**, 99–104.

Matthews, L.S., and Vale, W.W. (1991). Expression cloning of an activin receptor, a predicted transmembrane serine kinase. *Cell*, **65**, 973–82.

Matsuzaki, K., Xu, J., Wang, F., McKeehan, W.L., Krummen, L., and Kan, M. (1993). A widely expressed transmembrane serine/threonine kinase that does not bind activin, inhibin, transforming growth factor β, or bone morphogenic factor. *J. Biol. Chem.*, **268**, 12719–23.

McCormick, F. (1993). Signal transduction. How receptors turn Ras on. *Nature*, **363**, 15–16.

Metcalf, D. (1992). Hemopoietic regulators. *Trends Biochem. Sci.*, **17**, 286–9.

Miyajima, A., Kitamura, T., Harada, N., Yokota, T., and Arai, K. (1992). Cytokine receptors and signal transduction. *Annu. Rev. Immunol.*, **10**, 295–331.

Muller, M., Briscoe, J., Laxton, C., Guschin, D., Ziemiecki, A., Silvennoinen, O., Harpur, A.G., Barbieri, G., Witthuhn, B.A., Schindler, C., Pelligrini, S., Wllks, A.F., Ihle, J.N., Stark, G.R., and Kerr, I.M. (1993a). The protein tyrosine kinase JAK1 complements defects in interferon-α/β and -γ signal transduction. *Nature*, **366**, 129–35.

Muller, M., Laxton, C., Briscoe, J., Schindler, C., Improta, T., Darnell, J.E., Stark, G.R., and Kerr, I.M. (1993b). Complementation of a mutant cell line: central role of the p91 kDa polypeotide of ISGF3 in the interferon-α and -γ signal transduction pathways. *EMBO J.*, **12**, 4221–8.

Murakami, M., Narazaki, M., Hibi, M., Yawata, H., Yasukawa, K., Hamaguchi, M., Taga, T., and Kishimoto, T. (1991). Critical cytoplasmic region of the interleukin 6 signal transducer gp130 is conserved in the cytokine receptor family. *Proc. Natl. Acad. Sci. (USA)*, **88**, 11349–53.

Murakami, M., Hibi, M., Nakagawa, N., Nakagawa, T., Yasukawa, K., Yamanishi, K., Taga, T., and Kishimoto T. (1993). IL-6-induced homodimerization of gp130 and associated activation of a tyrosine kinase. *Science*, **260**, 1808–10.

Nakajima, T., Kinoshita, S., Sasagawa, T., Sasaki, K., Naruto, M., Kishimoto, T., and Akira, S. (1993). Phosphorylation at threonine-235 by a ras-dependent mitogen-activated protein kinase cascade is essential for transcription factor NF-IL6. *Proc. Natl. Acad. Sci. (USA)*, **90**, 2207–11.

Nicola, N.A., and Metcalf, D. (1991). Subunit promiscuity among haemopoietic growth factor receptors. *Cell*, **67**, 1–4.

Nikaido, T., Shimizu, A., Ishida, N., Sabe, H., Teshigawara, K., Maeda, M., Uchiyama, T., Yodoi, J., and Honjo, T. (1984). Molecular cloning of cDNA encoding human interleukin-2 receptor. *Nature*, **311**, 631–5.

Noguchi, M., Nakamura, Y., Russel, S.M., Ziegler, S.F., Tsang, M., Cao, X., and Leonard, W.J. (1993). Interleukin-2 receptor γ-chain: a functional component of the interleukin-7 receptor. *Science*, **262**, 1877–80.

Ohmichi, M., Decker, S.J., Pang, L., and Saltiel, A.R. (1991). Nerve growth factor binds to the 140 kd trk proto-oncogene product and stimulates its association with the src homology domain of phospholipase C gamma 1. *Biochem. Biophys. Res. Commun.*, **179**, 217–23.

Olivier, J.P., Raabe, T., Henkemeyer, M., Dickson, B., Mbamalu, G., Margolis, B., Schlessinger, J., Hafen, E., and Pawson, T. (1993). A Drosophila SH2-SH3 adaptor protein implicated in coupling the sevenless tyrosine kinase to an activator of Ras guanine nucleotide exchange, Sos. *Cell*, **73**, 179–91.

Park, M. (1991). Lonesome receptors find their mates. The identification of the met and trk protein tyrosine kinases as the receptors for hepatocyte growth factor and neurotrophic factors, will allow rapid advances in the understanding of these factors. *Current Biology*, **1**, 248–50.

Pelicci, G., Lanfrancone, L., Grignani, F., McGlade, J., Cavallo, F., Forni, G., Nicoletti, I., Grignani, F., Pawson, T., and Pelicci, P.G. (1992). A novel transforming protein (SHC) with an SH2 domain is implicated in mitogenic signal transduction. *Cell*, **70**, 93–104.

Petch, L.A., Harris, J., Raymond, V.W., Blasband, A., Lee, D.C., and Earp, H.S. (1990). A truncated, secreted form of the epidermal growth factor receptor is encoded by an alternatively spliced transcript in normal rat tissue. *Mol. Cell Biol.*, **10**, 2973–82.

Quelle, F.W., and Wojchowski, D.M. (1991). Proliferative action of erythropoietin is associated with rapid protein tyrosine phosphorylation in responsive B6SUt.EP cells. *J. Biol. Chem.*, **266**, 609–14.

Rodrigues, G.A., Naujokas, M.A., and Park, M. (1991). Alternative splicing generates isoforms of the met receptor tyrosine kinase which undergo differential processing. *Mol. Cell Biol.*, **11**, 2962–70.

Ross, A.H. (1991). Identification of tyrosine kinase Trk as a nerve growth factor receptor. *Cell Regul.*, **2**, 685–90.

Rozakis-Adcock, M., Fernley, R., Pawson, T., and Bowtell, D. (1993). The SH2 and SH3 domains of mammalian Grb2 couple the EGF receptor to the Ras activator mSos1. *Nature*, **363**, 83–5.

Ruff-Jamison, S., Chen, K., and Cohen, S. (1993). Induction by EGF and interferon-gamma of tyrosine phosphorylated DNA binding proteins in mouse liver nuclei. *Science*, **261**, 1733–6.

Sadowski, H.B., Shuai, K., Darnell, J.E., and Gilman, M.Z. (1993). A common nuclear signal transduction pathway activated by growth factor and cytokine receptors. *Science*, **261**, 1739–44.

Sakamaki, K., Miyajima, I., Kitamura, T., and Miyajima A. (1992). Critical cytoplasmic domains of the common β subunit of the human GM-CSF, IL-3 and IL-5 receptors for growth signal transduction and tyrosine phosphorylation. *EMBO J.*, **11**, 3541–9.

Satoh, T., Nakafuku, M., Miyajima, A., and Kaziro, Y. (1991). Involvement of ras p21 protein in signal-transduction pathways from interleukin 2, interleukin 3, and granulocyte/macrophage colony-stimulating factor, but not from interleukin 4. *Proc. Natl. Acad. Sci. (USA)*, **88**, 3314–18.

Satoh, T., Uehara, Y., and Kaziro, Y. (1992). Inhibition of interleukin 3 and granulocyte-macrophage colony-stimulating factor stimulated increase of active ras.GTP by herbimycin A, a specific inhibitor of tyrosine kinases. *J. Biol. Chem.*, **267**, 2537–41.

Schindler, C., Fu, X.Y., Improta, T., Aebersold, R., and Darnell, J. Jr. (1992a). Proteins of transcription factor ISGF-3: one gene encodes the 91-and 84-kDa ISGF-3 proteins that are activated by interferon α. *Proc. Natl. Acad. Sci. (USA)*, **89**, 7836–9.

Schindler, C., SHuai, K., Prezioso, V.R., and Darnell, J.E. (1992b). Interferon-dependent tyrosine phosphorylation of a latent cytoplasmic transcription factor. *Science*, **257**, 809–12.

Schoenfeld, H.J., Poeschl, B., Frey, J.R., Loetscher, H.R., Hunziker, W., Lustig, A., and Zulauf, M. (1991). Efficient purification of recombinant human tumour necrosis factor β from E.coli yields biologically active protein with a trimeric structure that binds both tumour necrosis factor receptors. J. Biol. Chem., **266**, 3863–9.

Schreurs, J., Hung, P., May, W.S., Arai, K., and Miyajima, A. (1991). AIC2A is a component of the purified high affinity mouse IL-3 receptor: temperature-dependent modulation of AIC2A structure. Int. Immunol., **3**, 1231–42.

Shinozaki, H., Ito, I., Hasegawa, Y., Nakamura, K., Igarashi, S., Nakamura, M., Miyamoto, K., Eto, Y., Ibuki, Y., and Minegishi, T. (1992). Cloning and sequencing of a rat type II activin receptor. FEBS Lett., **312**, 53–6.

Shuai, K., Schindler, C., Prezioso, V.R., and Darnell, J. Jr. (1992). Activation of transcription by IFN-gamma: tyrosine phosphorylation of a 91-kD DNA binding protein. Science, **258**, 1808–12.

Silvennoinen, O., Witthuhn, B.A., Quelle, F.W., Cleveland, J.L., Yi, T., and Ihle, J.N. (1993a). Structure of the murine Jak2 protein-tyrosine kinase and its role in interleukin 3 signal transduction. Proc. Natl. Acad. Sci. (USA), **90**, 8429–33.

Silvennoinen, O., Schindler, C., Schlessinger, J., and Levy, D.E. (1993b). Ras-independent growth factor signaling by transcription factor tyrosine phosphorylation. Science, **261**, 1736–9.

Skolnik, E.Y., Batzer, A., Li, N., Lee, C.H., Lowenstein, E., Mohammadi, M., Margolis, B., and Schlessinger, J. (1993). The function of GRB2 in linking the insulin receptor to Ras signaling pathways. Science, **260**, 1953–5.

Smith, C.A., Davis, T., Anderson, D., Solam, L., Beckmann, M.P., Jerzy, R., Dower, S.K., Cosman, D., and Goodwin, R.G. (1990). A receptor for tumour necrosis factor defines an unusual family of cellular and viral proteins. Science, **248**, 1019–23.

Smith, C.A., Gruss, H.J., Davis, T., Anderson, D., Farrah, T., Baker, E., Sutherland, G.R., Brannan, C.I., Copeland, N.G., Jenkins, N.A., Grabstein, K.H., Gliniak, B., McAlister, I.B., Fanslow, W., Alderson, M., Falk, B., Gimpel, S., Gillis, S., Din, W.S., Goodwin, R.G., and Armitage, R.J. (1993). CD30 antigen, a marker for Hodgkin's lymphoma, is a receptor whose ligand defines an emerging family of cytokines with homology to TNF. Cell, **73**, 1349–60.

Soppet, D., Escandon, E., Maragos, J., Middlemas, D.S., Reid, S.W., Blair, J., Burton, L.E., Stanton, B.R., Kaplan, D.R., Hunter, T., Nikolics, K., and Parada, L.F. (1991). The neurotrophic factors brain-derived neurotrophic factor and neurotrophin-3 are ligands for the trkB tyrosine kinase receptor. Cell, **65**, 895–903.

Souyri, M., Vigon, I., Pencoilelli, J.F., Heard, J.M., Tambourin, P., and Wendling, F. (1990). A putative truncated cytokine receptor gene transduced by the myeloproliferative leukaemia virus immortalizes hematopoietic progenitors. Cell, **63**, 1137–47.

Spiegel, A.M. (1992). G proteins in cellular control. Curr. Opin. Cell Biol., **4**, 203–11.

Squinto, S.P., Stitt, T.N., Aldrich, T.H., Davis, S., Bianco, S.M., Radziejewski, C., Glass, D.J., Masiakowski, P., Furth, M.E., Valenzuela, D.M., DiStefano, P.S., and Yancopoulos, G.D. (1991). trkB encodes a functional receptor for brain-derived neurotrophic factor and neurotrophin-3 but not nerve growth factor. Cell, **65**, 885–93.

Stahl, N., Boulton, T.G., Farruggella, T., IP, N.Y., Davis, S., Witthuhn, B.A., Quelle, F.W., Silennionen, O., Barbieri, G., Pelligrini, S., Ihle, J.N., and Yancopoulos, G.D. (1994). Association and activation of Jak/Tyk kinases by CNTF-LIF-OSM-IL-6 β receptor components. Science, **263**, 92–4.

Stamenkovic, I., Clark, E.A., and Seed, B. (1989). A B-lymphocyte activation molecule related to the nerve growth factor receptor and induced by cytokines in carcinomas. EMBO J., **8**, 1403–10.

Stark, G.R., and Kerr, I.M. (1992). Interferon-dependent signaling pathways: DNA elements, transcription factors, mutations, and effects of viral proteins. J. Interferon Res., **12**, 147–51.

Suda, T., Takahashi, T., Golstein, P., and Nagata, S. (1993). Molecular cloning and expression of the Fas ligand, a novel member of the tumour necrosis factor family. Cell, **75**, 1169–78.

Takeshita, T., Asao, H., Ohtani, K., Ishii, N., Kumaki, S., Tanaka, N., Munakata, H., Nakamura, M., and Sugamura, K. (1992). Cloning of the gamma chain of the human IL-2 receptor. Science, **257**, 379–82.

Tartaglia, L.A., Ayres, T.M., Wong, G.H.W., and Goeddel, D.V. (1993). A novel domain within the 55 kd TNF receptor signals cell death. Cell, **74**, 845–53.

ten-Dijke, P., Ichijo, H., Franzen, P., Schulz, P., Saras, J., Toyoshima, H., Heldin, C.H., and Miyazono, K. (1993). Activin receptor-like kinases: a novel subclass of cell-surface receptors with predicted serine/threonine kinase activity. Oncogene, **8**, 2879–87.

Thoreau, E., Petridou, B., Kelly, P.A., Djiane, J., and Mornon, J.P. (1991). Structural symmetry of the extracellular domain of the cytokine/growth hormone/prolactin receptor family and interferon receptors revealed by hydrophobic cluster analysis. FEBS Lett., **282**, 26–31.

Torigoe, T., O'Connor, R., Santoli, D., and Reed, J.C. (1992). Interleukin-3 regulates the activity of the LYN protein-tyrosine kinase in myeloid-committed leukemic cell lines. Blood, **80**, 617–24.

Ullrich, A., and Schlessinger, J. (1990). Signal transduction by receptors with tyrosine kinase activity. Cell, **61**, 203–12.

Velazquez, L., Fellous, M., Stark, G.R., and Pellegrini, S. (1992). A protein tyrosine kinase in the interferon α/β signaling pathway. Cell, **70**, 313–22.

Vigon, I., Mornon, J.-P., Cocault, L., Mitjavila, M.-T., Tambourin, P., Gisselbrecht, S., and Souyri, M. (1992). Molecular cloning and characterization of mpl, the human homology of the v-mpl oncogene: Identification of a member of the haemopoietic growth factor receptor superfamily. Proc. Natl. Acad. Sci. (USA), **89**, 5640–4.

Wang, X.F., Lin, H.Y., Ng-Eaton, E., Downward, J., Lodish, H.F., and Weinberg, R.A. (1991). Expression cloning and characterization of the TGF-β type III receptor. Cell, **67**, 797–805.

Watling, D., Guschin, D., Muller, M., Silvennoinen, O., Witthuhn, B.A., Quelle, F.W., Rogers, N.C., Schindler, C., Stark, G.R. Ihle, J.N., and Kerr I.M. (1993). Complementation by the protein tyrosine kinase JAK2 of a mutant cell line defective in the interferon-γ signal transduction pathway. Nature, **366**, 166–70.

Watowich, S.S., Yoshimura, A., Longmore, G.D., Hilton, D.J., Yoshimura, Y., and Lodish, H.F. (1992). Homodimerization and constitutive activation of the erythropoietin receptor. Proc. Natl. Acad. Sci. (USA), **89**, 2140–4.

Witthuhn, B.A., Quelle, F.W., Silvennoinen, O., Yi, T., Tang, B., Miura, O., and Ihle, J.N. (1993). JAK2 associates with the erythropoietin receptor and is tyrosine phosphorylated and activated following stimulation with erythropoietin. Cell, **74**, 227–36.

Wood, K.W., Sarnecki, C., Roberts, T.M., and Blenis, J. (1992). ras mediates nerve growth factor receptor modulation of three signal-transducing protein kinases: MAP kinase, Raf-1, and RSK. Cell, **68**, 1041–50.

Wrana, J.L., Attisano, L., Carcamo, J., Zentella, A., Doody, J., Laiho, M., Wang, X.F., and Massague, J. (1992). TGF β signals through a heteromeric protein kinase receptor complex. Cell, **71**, 1003–14.

Yee, D., Lebovic, G.S., Marcus, R.R., and Rosen, J. (1989). Identification of an alternate type I insulin-like growth factor receptor β subunit mRNA transcript. J. Biol. Chem., **264**, 21439–41.

Yin, T., Taga, T., Tsang, M.L., Yasukawa, K., Kishimoto, T., and Yang, Y.C. (1993). Involvement of IL-6 signal transducer gp 130 in IL-11 mediated signal transduction. J. Immunol., **151**, 2555–61.

Douglas J. Hilton:
The Walter and Eliza Hall Institute of Medical Research, and The Cooperative Research Centre for Cellular Growth Factors,
PO Royal Melbourne Hospital,
Victoria 3050, Australia

Interleukin-1α, Interleukin-1β, and Interleukin-1 Receptor Antagonist (IL-1α, IL-1β, and IL-1ra)

The interleukin-1 triad of cytokines are 17–20 kDa polypeptides that have a wide range of biological activities centrally involved in the genesis and maintenance of inflammatory responses (Dower et al. 1992a; Dinarello 1991; Oppenheim et al. 1986). IL-1α and IL-1β, which have agonist activity, are initially synthesized as c. 30 kDa intracellular precursors, lacking a hydrophobic leader sequence, and released from cells by a mechanism as yet incompletely understood, but possibly involving apoptosis. IL-1β is processed during release by a specific protease, IL-1β converting enzyme (ICE). IL-1ra, a pure receptor antagonist, which binds to the type I IL-1R but does not trigger signals, is synthesized with a conventional leader sequence, and is secreted as a c. 20 kDa glycoprotein via the normal Golgi pathway. The major physiological source of IL-1s is the activated macrophage. Production of the IL-1s is elicited by a variety of pro-inflammatory stimuli, including bacterial cell wall products, zymosan, leukotrienes, activated complement components, tumour necrosis factor, immune complexes, GM-CSF, and IL-1 itself. Production can be downregulated by steroids (e.g. dexamethasone), TGFβ and retinoic acid. Since IL-1 is a central mediator of inflammation, clinical efforts are focused on inhibition of IL-1 action.

■ Alternative names

Catabolin, endogenous pyrogen (EP), osteoclast activating factor (OAF), lymphocyte activating factor (LAF), epidermal cell derived thymocyte activating factor (ETAF), serum amyloid A inducer or hepatocyte stimulating factor (HSF), leucocyte endogenous mediator (LEM), fibroblast activating factor (FAF), B cell activating factor (BAF), proteolysis inducing factor (PIF), haemopoietin-1 (H-1).

■ IL-1 sequences

IL-1s from several species have been cloned and sequenced (Fig. 1); a list follows: human pro-IL-1α (271 amino acid residues; mature protein starts at L119, GenBank accession number A23385) (March et al. 1985), human pro-IL-1β (269 amino acid residues, mature protein starts at A117; GenBank accession number A29019) (March et al. 1985; Auron et al. 1984), human IL-1ra (177 residues, mature protein starts at R25, GenBank accession number A40956) (March et al. 1985; Auron et al. 1984; Eisenberg et al. 1990; Hannum et al. 1990), human icIL-1ra (alternately spliced mRNA) (159 residues, GenBank accession number A39386) (Haskill et al. 1991), murine pro-IL-1α (270 amino acid residues; GenBank accession number A01846) (Lomedico et al. 1984); murine pro-IL-1β (269 amino acid residues; GenBank accession number A24719) (Gray et al. 1986); murine IL-1ra (179 residues, GenBank accession number M57525); rat pro-IL-1α (270 amino acid residues; GenBank accession number JX0064),

rat pro-IL-1β (269 amino acid residues; GenBank accession number M98820); rat IL-1ra (178 residues, GenBank accession number C40956); bovine pro-IL-1α (268 amino acid residues; GenBank accession number M36182), bovine pro-IL-1β (266 amino acid residues; GenBank accession number M35589); sheep pro-IL-1α (267 amino acid residues; GenBank accession number X60167), sheep pro-IL-1β (266 amino acid residues; GenBank accession number X54796); rabbit pro-IL-1α (268 amino acid residues; GenBank accession number B24073); rabbit pro-IL-1β (268 amino acid residues; GenBank accession number A27714); rabbit IL-1ra (177 residues, GenBank accession number P26980); porcine pro-IL-1α (266 amino acid residues; GenBank accession number S10532); porcine pro-IL-1β (267 amino acid residues; GenBank accession number P26889). (For species other than mouse or man, see Dower et al. 1992b.) The mature forms of IL-1α and IL-1β are generated from the pro forms by proteolytic cleavage at approximately position 115, and comprise the C-terminal 150 amino acids (Fig. 1(b)). Separate sequence comparisons show that the pro pieces are approximately as conserved as the mature C-terminal regions (Fig. 2 and Table 1). The IL-1ra precursors have hydrophobic signal peptides of c. 20 residues. For the mature portions of the human proteins, IL-1α is 26 per cent identical to IL-1β and 19 per cent identical to IL-1ra; IL-1β is 26 per cent identical to IL-1ra. The mature forms of all three human proteins show detectable affinity for both types of human IL-1 receptor (IL-1RI and IL-1RII), with IL-1ra binding with highest affinity to IL-1RI, and IL-1β binding with highest affinity to IL-1RII (Dower et al. 1992b). In rat and mouse, IL-1RII does not bind IL-1ra at all (Dripps et al. 1991). The pro form of IL-1β does not bind to IL-1RI, and hence has no detectable biological activity. The receptor binding and

(a)

```
                1                                                                                          100
Mu  IL1α    ..PYTYQSDLRYKLMKLVRQ KFVMNDSLNQTIYQDVDKHY LSTTWLN..DLQQEVKFDMY AYSSGGDDSKYPVTLKISDS QLFVS.AQGEDQPVLLKELP
Rat IL1α    SAPHSFQNNLRYKLIRIVKQ EFIMNDSLNQNIYVDMDRIH LKAASLN..DLQLEVKFDMY AYSSGGDDSKYPVTLKVSNT QLFVS.AQGEDKPVLLKEIP
Cow IL1α    SAHYSFQSNVKYNFMRVIHQ ECILNDALNQSIIRDMSGPY LTATTLN..NLEEAVKFDMV AY.VSEEDSQLPVTLRISKT QLFVS.AQNEDEPVLLKEMP
She IL1α    SAHYSFQSNVKYNFMRVIHQ ECILNDALNQSIIRDMSGPY MTAATLN..NLEEAVKFDMV AY.VSEEDSQLPVTLRISKT QLFVS.AQNEDEPVLHKEMP
Pig IL1α    .ATYSFQSNMKYNFMRVINH QCILNDARNQSIIRDPSGQY LMAAVLN..NLDEAVKFDMA AY.TSNDDSQLPVTLRISET RLFVS.AQNEDEPVLLKELP
Hu  IL1α    SSPFSFLSNVKYNFMRIIKY EFILNDALNQSIIRA.NDQY LTAAALH..NLDEAVKFDMG AYKSSKDDAKITVILRISKT QLYVT.AQDEDQPVLLKEMP
Rab IL1α    SVPYTFQRNMRYKYLRIIKQ EFTLNDALNQSLVRDTSDQY LRAAPLQ..NLGDAVKFDMG VYMTSKEDSILPVTLRISQT PLFVS.AQNEDEPVLLKEMP

Mu  IL1ra   ........RPSGKRPCKMQ AFRIWDTNQKTFYLRNN.Q. LIAGYLQGPNIKLEEKIDMV PIDLH......SVFLGIHGG KLCLSCAKSGDDIKLQLEEV
Rat IL1ra   ........HPAGKRPCKMQ AFRIWDTNQKTFYLRNN.Q. LIAGYLQGPNIKLEEKIDMV PIDFR......NVFLGIHGG KLCLSCVKSGDDTKLQLEEV
HuicIL1ra   ..MALETICRPSGRKSSKMQ AFRIWDVNQKTFYLRNN.Q. LVAGYLQGPNVNLEEKIDVV PIEPH......ALFLGIHGG KMCLSCVKSGDETRLQLEAV
Hu  IL1ra   ........RPSGRKSSKMQ AFRIWDVNQKTFYLRNN.Q. LVAGYLQGPNVNLEEKIDVV PIEPH......ALFLGIHGG KMCLSCVKSGDETRLQLEAV
Rab IL1ra   ........RPSGKRPCRMQ AFRIWDVNQKTFYLRNN.Q. LVAGYLQGPNAKLEERIDVV PLEPQ......LLFLGIQRG KLCLSCVKSGDKMKLHLEAV

Mu  IL1β    ............VPIRQL HYRLRDEQQKSLVLSDPYE. LKALHLNGQNINQQVIFSMS FVQGEPSNDKIPVALGLKGK NLYLSCVMKDGTPTLQLESV
Rat IL1β    ............VPIRQL HCRLRDEQQKSLVLSDPCE. LKALHLNGQNISQQVVFSMS FVQGETSNDKIPVALGLKGL NLYLSCVMKDGTPTLQLESV
Hu  IL1β    ............APVRSL NCTLRDSQQKSLVMSGPYE. LKALHLQGQDMEQQVVFSMS FVQGEESNDKIPVALGLKEK NLYLSCVLKDDKPTLQLESV
Rab IL1β    ............AVRSL HCRLQDAQQKSLVLSGTYE. LKALHLNAENLNQQVVFSMS FVQGEESNDKIPVALGLRGK NLYLSCVMKDDKPTLQLESV
Cow IL1β    ............APVQSI KCKLQDREQKSLVLASPCV. LKALHLLSQEMNREVVFCMS FVQGEERDNKIPVALGIRDK NLYLSCVKKGDTPTLQLEEV
She IL1β    ............AAVQSV KCKLQDREQKSLVLDSPCV. LKALHLLSQEMSREVVFCMS FVQGEERDNKIPVALGIRDK NLYLSCVKKGDTPTLQLEEV
Pig IL1β    ............ANVQSM ECKLQDKDHKSLVLAGPHM. LKALHLLTGDLKREVVFCMS FVQGDDSNNKIPVTLGIKGK NLYLSCVMKDNTPTLQLEDI
```

```
                101                                                 170
Mu  IL1α    ETPKLIT..GSETDLIFFWK SINSKNYFTSAAYPELFIAT KEQS.....RVHLARGLPSM TDFQIS....
Rat IL1α    ETPKLIT..GSETDLIFFWE KINSKNYFTSAAFPELLIAT KEQS.....QVHLARGLPSM IDFQIS....
Cow IL1α    ETPKIIK..D.ETNLLFFWE KHGSMDYFKSVAHPKLFIAT KQEK.....LVHMASGPPSI TDFQILEK..
She IL1α    ETPKIIK..D.ETNLLFFWE KHGSMDYFKSVAHPKLFIAT KQEK.....LVHMASGPPSI TDFLILEK..
Pig IL1α    ETPKTIK..D.ETSLLFFWE KHGNMDYFKSAAHPKLFIAT RQEK.....LVHMAAPGLPSV TDFQILENQS
Hu  IL1α    EIPKTIT..GSETNLLFFWE THGTKNYFTSVAHPNLFIAT KQDY.....WVCLAGGPPSI TDFQILENQA
Rab IL1α    ETPRIIT..DSESDILFFWE TQGNKNYFKSAANPQLFIAT KPEH.....LVHMARGLPSM TDFQIS....

Mu  IL1ra   NITDLSKNKEEDKRFTFIRS EKGPTTSFESAACPGWFLCT TLEADRPVSLTNTPEEPLIV TKFYFQEDQ.
Rat IL1ra   NITDLNKNKEEDKRFTFIRS ETGPTTSFESLACPGWFLCT TLEADHPVSLTNTPKEPCTV TKFYFQEDQ.
HuicIL1ra   NITDLSENRKQDKRFAFIRS DSGPTTSFESAACPGWFLCT AMEADQPVSLTNMPDEGVMV TKFYFQEDE.
Hu  IL1ra   NITDLSENRKQDKRFAFIRS DSGPTTSFESAACPGWFLCT AMEADQPVSLTNMPDEGVMV TKFYFQEDE.
Rab IL1ra   NITDLGKNKEQDKRFTFIRS NSGPTTTFESASCPGWFLCT ALEADQPVSLTNTPDDSIVV TKFYFQEDQ.

Mu  IL1β    DPKQYPK.KKMEKRFVFNKI EVKSKVEFESAEFPNWYIST SQAEHKPVFLGNNS.GQ.DI IDFTMESVSS
Rat IL1β    DPKQYPK.KKMEKRFVFNKI EVKTKVEFESAQFPNWYIST SQAEHRPVFLGNSN.GR.DI VDFTMEPVSS
Hu  IL1β    DPKNYPK.KKMEKRFVFNKI EINNKLEFESAQFPNWYIST SQAENMPVFLGGTKGGQ.DI TDFTMQFVSS
Rab IL1β    DPNRYPK.KKMEKRFVFNKI EIKDKLEFESAQFPNWYIST SQTEYMPVFLGNNSGGQ.DL IDFSMEFVSS
Cow IL1β    DPKVYPK.RNMEKRFVFYKT EIKNTVEFESVLYPNWYIST SQIEERPVFLGHFRGGQ.DI TDFRMETLSP
She IL1β    DPKVYPK.RNMEKRFVFYKT EIKNTVEFESVLYPNWYIST SQIEEKPVFLGRFRGGQ.DI TDFRMETLSP
Pig IL1β    DPKRYPK.RDMEKRFVFYKT EIKNRVEFESALYPNWYIST SQAEQKPVFLGNSKGRQ.DI TDFTMEVLSP
```

(b)

```
                1                                                                                          100
Mu  IL1α    MAKVPDLFEDLKNCYSENE. DYSSAIDHLSLNQKSFYDAS YGSLHETCTDQFVSLRTSET SKMSNFTFKESRVTVSATSS NGKILKKRRLSFSETFTEDD
Rat IL1α    MAKVPDLFEDLKNCYSENE. EYSSAIDHLSLNQKSFYDAS YGSLHENCTDKFVSLRTSET SKMSTFTFKESRVVVSATSN KGKILKKRRLSFNQPFTEDD
Cow IL1α    MAKVPDLFEDLKNCYSENE. DYSSEIDHLSLNQKSFYDAS YEPLREDHMNKFMSLDTSET SKTSKLSFKENVVMVAA... SGKILKKRRLSLNQFITDDD
She IL1α    MAKVPDLFEDLKNCYSENE. DYSSEIDHLSLNQKSFYDAS YEPLREDHMNKFMSLDTSET SKTSRLSFKENVVMVTA... NGKILKKRRLSLNQFITDDD
Pig IL1α    MAKVPDLFEDLKNCYSENE. EYSSDIDHLSLNQKSFYDAS YEPLPGDGMDKFMPLSTSKT SKTSRLNFKDSVVMAAA... NGKILKKRRLSLNQFITDDD
Hu  IL1α    MAKVPDMFEDLKNCYSENE. EDSSSIDHLSLNQKSFYHVS YGPLHEGCMDQSVSLSISET SKTSKLTFKESMVVVAT... NGKVLKKRRLSLSQSITDDD
Rab IL1α    MAKVPDLFEDLKNCFSENE. EYSSAIDHLSLNQKSFYDAS YEPLHEDCMNKVVSLSTSET SVSPNLTFQENVVAVTA... SGKILKKRRLSLNQPITDVD

Cow IL1β    MATVPEPINEMMAYYSD.EN ELLFEADDPKQMKSCIQHLD LGSMGDGNIQLQISHQFYNK S......FRQVVSVIVAMEK ....LRNSA.YAHVFHDDD
She IL1β    MATVPEPINEVMACYSD.EN ELLFEVDGPKQMKSCIQHLD LGSMGDGNIQLKIQSHKLYNK S......FRQAVSVIVAMEK ....LRSRA.YEHVFRDDD
Pig IL1β    MAIVPEPAKEVMANYGDNNN DLLFEADGPKEMKCCTQNLD LGSLRNGSIQLQISHQLWNK S......IRQMVSVIVAVEK ....PMKNP.SSQAFCDDD
Mu  IL1β    MATVPELNCEMPPFDS.DEN DLFFEVDGPQKMKGCFQTFD LGCP.DESIQLQISSQHINK S......FRQAVSLIVAVEK ....LWQLPVSFPWTFQDED
Rat IL1β    MATVPELNCEIAAFDS.EEN DLFFEADRPQKIKDCFQALD LGCP.DESIQLQISQQHLDK S......FRKAVSLIVAVEK ....LWQLPMSCPWSFQDED
Hu  IL1β    MAEVPELASEMMAYYSGNED DLFFEADGPKQMKCSFQDLD LCPL.DGGIQLRISDHHYSK G......FRQVLSVVVALEK ....LRQKAVPCPQAFQDDG
Rab IL1β    MATVPELTSEMMAYHSGNEN DLFFEADGPNYMKSCFQDLD LCCP.DEGIQLRISCQPYNK S......FRQVLSVVVALEK ....LRQKAVPCPQAFQDDG
```

```
                101                   129
Mu  IL1α    LQSITH...DLEET.IQPRS A........
Rat IL1α    LEAIAH...DLEET.IQPRS .........
Cow IL1α    LEAIAN...NTEEEIIKPRS .........
She IL1α    LEAIAN...DTEEEIIKPRS .........
Pig IL1α    LEAIAN...DTEEEIIKPRS .........
Hu  IL1α    LEAIAN...DSEEEIIKPR. .........
Rab IL1α    LETNVS...DPEEGIIKPRS .........

Cow IL1β    LRSILSFIFEEEPVIFETSS DE...FLCD
She IL1β    LRSILSFIFEEEPVIFETSS DE...LLCD
Pig IL1β    QKSIFSFIFEEEPIILETCN DD...FVCD
Mu  IL1β    MSTFFSFIFEEEPILCDSWD DDDNLLVCD
Rat IL1β    PSTFFSFIFEEEPVLCDSWD DDD.LLVCD
Hu  IL1β    LSTFFPFIFEEEPIFFDTWD NEA..YVHD
Rab IL1β    LRTFFSLIFEEEPVLCNTWD DYS..LECD
```

Figure 1. Sequence alignments for (a) mature IL-1 and (b) pro segment.

Table 1.1. Summary similarity scores for IL-1 family members

		All IL-1α	All IL-1ra	All IL-1β
All IL-1α	Complete CDS	1.17	0.14	0.28
	Mature sequence	1.01	0.20	0.27
	Pro piece	1.21		0.28
All IL-1ra	Complete CDS		0.51	0.33
	Mature sequence		1.12	0.39
All IL-1β	Complete CDS			1.13
	Mature sequence			1.03
	Pro piece			1.01

Scores were calculated with the UWGCG program DIS-TANCES and are scaled for the average length of the multiple sequence alignments shown in figure 1. The lower numbers for IL-1ra in the complete CDS comparison are a consequence of the shorter complete CDS for these members of the IL-1 family.

biological activities of the pro and mature forms of IL-1α appear indistinguishable (Mosley et al. 1987).

■ IL-1 proteins

IL-1α and IL-1β are released from cells concomitantly with proteolytic processing as c. 17 000 Da molecular weight proteins with no attached oligosaccharide. Human IL-1ra is secreted as three distinct glycosylated forms of M_r 18 000, 22 000 (α) and 22 000 (β) Da, the last two differing in pI, all with the same polypeptide core of c. 17 000 Da (Hannum et al. 1990). Mature human IL-1α is composed of two pI forms (5.2 and 5.4); the 5.2 form begins at L119; by analogy with murine IL-1α it has been suggested that the pI 5.4 form begins at S113, but the N-terminus of this form is blocked and this has not been confirmed by sequencing (Cameron et al. 1986). The pI of human IL-1β is 6.8–7.0 (Kronheim et al. 1985; Schmidt 1984); that of IL-1ra is variable depending on the extent of glycosylation (Hannum et al. 1990). None of the IL-1s contain any intramolecular disulphide bonds. The three-dimensional structures of human IL-1α and IL-1β have been solved and both are approximately tetrahedral structures composed of a tightly packed core of β-strands connected by less structured loops (Graves et al. 1990; Priestle et al. 1989; Tate et al. 1992). The three dimensional structure of IL-1ra has not been solved, but is presumed to be similar. Mutagenesis studies suggest that residues involved in receptor binding and activation are scattered widely over the surface of IL-1s (Auron et al. 1992; Simon et al. 1993). There is as yet no good structural explanation for the lack of biological activity of IL-1ra (Ju et al. 1991).

All three forms of IL-1 can be purified by conventional chromatography, although the high biological specific activity of IL-1α and IL-1β leads in general to little protein being isolated, even from sources high in biological activity.

■ IL-1 genes and control of expression

The human IL-1α gene (GenBank accession number X03833) spans approximately 11 kb at 2q13; there are seven exons, and transcription gives rise to a 2.1 kb mRNA. The human IL-1β gene (GenBank accession number X04500) spans approximately 7.5 kb in the region 2q13–2q21; there are seven exons and transcription gives rise to a 1.6 kb mRNA. The IL-1α and IL-1β genes appear to be related by duplication. The human IL-1ra gene maps to the same region (2q14–21), and spans approximately 15 kb. The IL-1ra gene contains four exons, and the mRNAs encoding the two forms are generated by alternate splicing of two different 5′ exons. The exon encoding the N-terminus of icIL-1ra is the most upstream and has not yet been linked to the rest of the gene. mRNAs for both forms are c. 1.8 kb. This general region of the long arm of chromosome 2 in man contains not only the genes for IL-1α, IL-1β, and IL-1ra, but also the genes for the type I and type II IL-1 receptors (see chapter on IL-1 receptors, p. 23).

IL-1 gene transcription in monocytes can be induced by a wide range of stimuli including adhesion to surfaces, LPS (E. coli, serotype 055:B5), Staphylococcus aureus, interleukin-1, leukotrienes, phorbol esters, zymosan, tumour necrosis factor, interferon-γ, calcium ionophores, complement components (C5a), indomethacin and GM-CSF. Neither gene contains any upstream binding sites for well-documented transcription factors, although there is evidence that IL-1β gene expression in monocytes may be regulated by an NF-IL-6-like transcription factor. It appears, based on patterns of mRNA expression, that IL-1ra expression is regulated quite differently from that of IL-1α and IL-1β.

■ Biological functions

IL-1α, IL-1β, and IL-1ra are centrally involved in the effector phase of immune and inflammatory responses (Dower et al. 1992a; Dinarello 1991; Oppenheim et al. 1986). This view is supported by the observation that inhibitors of IL-1 action, IL-1ra or soluble IL-1 receptors, when administered at high doses, are capable of suppressing such responses in vivo (Fanslow et al. 1991; Ohlsson et al. 1990; Wakabayashi et al. 1991). In addition, the acquisition by some viruses of soluble IL-1 receptors and inhibitors of the IL-1β converting enzyme (ICE) (Alcami and Smith 1992; Ray et al. 1992; Spriggs et al. 1992) suggests that the production of IL-1 inhibitors during infection confers a selective advantage on such pathogens by subverting host defenses. Consistent with a central role in host defence, IL-1 has a wide range of target cells including fibroblasts, where it causes proliferation (by induction of PDGFAA synthesis and secretion [Raines et al. 1989]) and induces collagenase, stromelysin, IL-6, and G-CSF secretion, induces cyclooxygenase synthesis and hence prostaglandin release, and suppresses expression of mRNA for matrix proteins (Qwarnstrom et al. 1993); smooth muscle cells, where it causes proliferation by induction of

A. Entire Protein Sequence

IL-1α block (cross-species):

	MouseIl1α	RatIl1α	BovineIl1α	SheepIl1α	PigIl1α	HumanIl1α	RabbitIl1α
Mouse Il1α	1.444						
Rat Il1α	1.277	1.444					
Bovine Il1α	0.986	1.012	1.422				
Sheep Il1α	0.981	1.012	1.387	1.422			
Pig Il1α	0.966	0.987	1.256	1.259	1.422		
Human Il1α	1.002	1.001	1.097	1.093	1.256	1.422	
Rabbit Il1α	1.006	1.026	1.094	1.093	1.093	1.079	1.433

IL-1ra columns (α rows and ra–ra block):

	MouseIl1ra	RatIl1ra	HuicIl1ra	HumanIl1ra	RabbitIl1ra
Mouse Il1α	0.158	0.172	0.154	0.166	0.129
Rat Il1α	0.144	0.157	0.133	0.146	0.125
Bovine Il1α	0.158	0.163	0.123	0.123	0.104
Sheep Il1α	0.168	0.153	0.153	0.132	0.116
Pig Il1α	0.154	0.152	0.115	0.115	0.085
Human Il1α	0.183	0.152	0.136	0.144	0.127
Rabbit Il1α	0.152	0.156	0.155	0.153	0.120
Mouse Il1ra	0.670				
Rat Il1ra	0.611	0.666			
HumIc Il1ra	0.419	0.398	0.387		
Human Il1ra	0.534	0.511	0.498	0.653	
Rabbit Il1ra	0.519	0.508	0.440	0.549	0.653

IL-1β columns (α rows, ra rows, and β–β block):

	BovineIl1β	SheepIl1β	PigIl1β	MouseIl1β	RatIl1β	HumanIl1β	RabbitIl1β
Mouse Il1α	0.298	0.281	0.284	0.274	0.222	0.315	0.305
Rat Il1α	0.289	0.271	0.281	0.285	0.232	0.309	0.318
Bovine Il1α	0.293	0.276	0.267	0.259	0.225	0.302	0.296
Sheep Il1α	0.291	0.274	0.267	0.278	0.225	0.296	0.297
Pig Il1α	0.290	0.283	0.278	0.271	0.246	0.314	0.310
Human Il1α	0.295	0.270	0.278	0.272	0.234	0.305	0.304
Rabbit Il1α	0.278	0.263	0.281	0.272	0.231	0.304	0.308
Mouse Il1ra	0.360	0.326	0.350	0.421	0.393	0.392	0.353
Rat Il1ra	0.342	0.344	0.337	0.384	0.371	0.383	0.346
HumIc Il1ra	0.340	0.339	0.329	0.373	0.343	0.369	0.347
Human Il1ra	0.357	0.354	0.334	0.397	0.369	0.386	0.363
Rabbit Il1ra	0.318	0.315	0.307	0.363	0.360	0.386	0.332
Bovine Il1β	1.404						
Sheep Il1β	1.327	1.404					
Pig Il1β	1.120	1.098	1.410				
Mouse Il1β	0.947	0.919	0.924	1.422			
Rat Il1β	0.912	0.897	0.903	1.289	1.416		
Human Il1β	1.008	0.975	0.989	1.084	1.045	1.422	
Rabbit Il1β	0.992	0.958	0.964	1.118	1.071	1.155	1.416

B. Mature Protein sequence

IL-1α block:

	MuIl1α	RatIl1α	CowIl1α	SheepIl1α	PigIl1α	HuIl1α	Rab Il1α
Mouse Il1α	1.262						
Rat Il1α	1.077	1.297					
Bovine Il1α	0.803	0.846	1.279				
Sheep Il1α	0.785	0.841	1.262	1.279			
Pig Il1α	0.783	0.799	1.078	1.073	1.279		
Human Il1α	0.822	0.838	0.946	0.942	0.899	1.315	
Rabbit Il1α	0.849	0.884	0.933	0.928	0.942	0.934	1.297

IL-1ra columns (α rows and ra–ra block):

	MuIl1ra	RatIl1ra	HuicIl1ra	HuIl1ra	RabIl1ra
Mouse Il1α	0.231	0.221	0.234	0.223	0.193
Rat Il1α	0.215	0.205	0.194	0.202	0.188
Bovine Il1α	0.211	0.203	0.174	0.183	0.164
Sheep Il1α	0.214	0.198	0.193	0.194	0.176
Pig Il1α	0.219	0.214	0.185	0.200	0.180
Human Il1α	0.223	0.223	0.200	0.209	0.194
Rabbit Il1α	0.228	0.202	0.204	0.211	0.179
Mouse Il1ra	1.200				
Rat Il1ra	1.200	1.200			
HumIc Il1ra	1.059	1.022	1.324		
Human Il1ra	1.039	1.013	1.240	1.200	
Rabbit Il1ra	1.033	1.015	1.065	1.041	1.200

IL-1β columns (α rows, ra rows, β–β block):

	MuIl1β	RatIl1β	HuIl1β	RabIl1β	CowIl1β	SheIl1β	PigIl1β
Mouse Il1α	0.258	0.250	0.291	0.275	0.282	0.280	0.277
Rat Il1α	0.278	0.250	0.289	0.290	0.284	0.278	0.289
Bovine Il1α	0.234	0.229	0.266	0.256	0.282	0.278	0.261
Sheep Il1α	0.238	0.254	0.269	0.258	0.281	0.292	0.264
Pig Il1α	0.259	0.240	0.300	0.291	0.298	0.272	0.291
Human Il1α	0.253	0.256	0.291	0.272	0.288	0.285	0.271
Rabbit Il1α	0.274	0.256	0.310	0.299	0.286	—	0.302
Mouse Il1ra	0.416	0.405	0.413	0.390	0.402	0.394	0.393
Rat Il1ra	0.403	0.389	0.401	0.376	0.401	0.394	0.378
HumIc Il1ra	0.434	0.408	0.425	0.413	0.418	0.414	0.398
Human Il1ra	0.404	0.379	0.395	0.389	0.392	0.387	0.372
Rabbit Il1ra	0.387	0.371	0.405	0.371	0.369	0.362	0.356
Mouse Il1β	1.209						
Rat Il1β	1.148	1.209					
Human Il1β	1.029	1.021	1.218				
Rabbit Il1β	1.033	1.004	1.200	1.200			
Bovine Il1β	0.883	0.875	1.029	1.046	1.218		
Sheep Il1β	0.866	0.865	0.918	0.879	1.181	1.218	
Pig Il1β	0.918	0.910	0.952	0.919	1.016	1.007	1.218

C. Pro Fragment Comparison for IL-1α and IL-1β

IL-1α block:

	MuIl1α	RatIl1α	CowIl1α	SheepIl1α	PigIl1α	HuIl1α	Rab Il1α
Mouse Il1α	1.448						
Rat Il1α	1.306	1.448					
Bovine Il1α	1.021	1.092	1.420				
Sheep Il1α	1.039	1.102	1.380	1.420			
Pig Il1α	1.009	1.080	1.281	1.291	1.420		
Human Il1α	1.061	1.072	1.129	1.135	1.132	1.420	
Rabbit Il1α	1.010	1.068	1.149	1.157	1.099	0.968	1.420

IL-1β columns (α rows and β–β block):

	MuIl1β	RatIl1β	HuIl1β	RabIl1β	CowIl1β	SheIl1β	PigIl1β
Mouse Il1α	0.276	0.290	0.231	0.215	0.213	0.343	0.305
Rat Il1α	0.308	0.326	0.252	0.212	0.222	0.345	0.313
Bovine Il1α	0.306	0.318	0.255	0.217	0.223	0.328	0.310
Sheep Il1α	0.312	0.323	0.258	0.221	0.226	0.329	0.311
Pig Il1α	0.290	0.308	0.220	0.194	0.194	0.311	0.290
Human Il1α	0.322	0.334	0.281	0.205	0.223	0.334	0.309
Rabbit Il1α	0.282	0.300	0.229	0.205	0.216	0.314	0.300
Mouse Il1β	1.341						
Rat Il1β	1.232	1.341					
Human Il1β	0.996	0.989	1.355				
Rabbit Il1β	0.793	0.772	0.700	1.359			
Bovine Il1β	0.763	0.722	0.695	1.184	1.358		
Sheep Il1β	0.877	0.853	0.830	0.918	0.866	1.371	
Pig Il1β	0.916	0.868	0.820	0.980	0.942	1.058	1.371

Figure 2. Sequence similarities in the IL-1 family.

PDGFAA; keratinocytes, where it causes proliferation and release of other cytokines, for example, IL-6; endothelial cells, where it induces TNF release and adhesion molecule expression/activation; osteoclasts, where it activates proton pumping, leading to bone resorption (Civitelli et al. 1989); chondrocytes, where it induces metalloprotease secretion leading to cartilage breakdown and proteoglycan release; hepatocytes, where it induces secretion of acute-phase proteins; haematopoetic precursors, where it synergizes with GM-CSF causing proliferation and an increase in CFU-GM; pre B-cells, where it induces differentiation and surface IgM expression; mature T cells, where it induces proliferation of Th2 cells in combination with stimulation through the T cell antigen receptor (TCR); mature B cells, where it induces proliferation and immunoglobulin secretion; monocytes and neutrophils, where it induces secretion of several cytokines including IL-8 and IL-1 itself and prolongs survival of neutrophils *in vitro*; and islet cells, where it induces insulin secretion and, with a slower time course, cell death (reviewed by Dinarello 1991).

Both IL-1α and IL-1β bind to the signal transducing type I IL-1 receptor with similar affinity (Dower et al. 1986). As a consequence of this, these two cytokines show an identical spectrum of biological activities *in vitro*. To date the only difference that may be of significance is the selective inhibition of IL-1β by soluble IL-1RII (Slack et al. 1992), suggesting that cells actively releasing this form of receptor might respond selectively to IL-1α.

Physiology and pathology

The actions described above underlie the broader *in vivo* responses that IL-1 can mediate. Thus, IL-1 will induce fever, cause bone resorption, muscle proteolysis, and cachexia (this may be indirect and a consequence of TNF induction), and induce increases in circulating levels of acute phase proteins such as C-reactive protein (CRP) and decreases in plasma iron and zinc. The actions of IL-1 on connective tissue cells suggest a central role in recovery of tissue after injury, promoting extracellular matrix turnover and allowing proliferation and migration of cells. In this context, IL-1 has angiogenic activity in the hamster cheek pouch assay. Also relevant is the finding that epidermis contains large deposits of preformed IL-1.

IL-1 has also been implicated *in vivo* in the pathogenesis of a variety of acute and chronic inflammatory diseases such as septic shock, rheumatoid arthritis, and atherosclerosis; much of our view of the physiological role of IL-1 is based on *in vitro* studies of the type outlined above and the effects of injection of IL-1 into animals (Dower et al. 1992a; Dinarello 1989).

The physiology of the IL-1 system is far from understood, since the potential molecular interactions between the components of the system are complex (Dower et al. 1992b). There are three forms of IL-1 with agonist activity (pro-IL-1α, IL-1α, and IL-1β), two forms of antagonist (IL-1ra and icIL-1ra) and an inactive precursor (pro-IL-1β) which can accumulate to high levels in some tissues; for example, epidermis. In addition, there are two IL-1 receptors which bind the various forms of IL-1 differentially and independently (Slack et al. 1992). Only one of these receptors (IL-1R type I) can signal (Sims et al. 1993), the other (IL-1R type II) functions as a decoy receptor and hence a second type of IL-1 antagonist (Colotta et al. 1993). Indeed it is unclear whether IL-1ra actually acts as an antagonist *in vivo*, since at sites of ongoing inflammation, for example synovium of patients with rheumatoid arthritis, there are high levels of IL-1ra present. By contrast, the agonist forms of IL-1 and the type I receptor which mediates cellular responses to them are usually expressed at very low levels, making it difficult to correlate cytokine expression with disease pathology. Finally, it remains to be established what the physiological role of the IL-1 system is in the normal development and function of an organism. It is presumed to be crucial, however, since the divergence of IL-1α, IL-1β and IL-1ra from one another in a species is far greater than that between species (Fig. 1) and IL-1-like molecules can be found in primitive organisms, suggesting that IL-1s are ancient molecules and that the duplicated genes and their products have been conserved through evolution.

References

Alcami, A., and Smith, G.L. (1992). A soluble receptor for interleukin-1 beta encoded by vaccinia virus: a novel mechanism of virus modulation of the host response to infection. *Cell*, **71**, 153–67.

Auron, P.E., Webb, A.C., Rosenwasser, L.J., Mucci, S.F., Rich, A., Wolff, S.M., and Dinarello C.A. (1984). Nucleotide sequence of human monocyte interleukin 1 precursor cDNA. *Proc. Natl. Acad. Sci. (USA)*, **81**, 7907–11.

Auron, P.E., Quigley, G.J., Rosenwasser, L.J., and Gehrke, L. (1992). Multiple amino acid substitutions suggest a structural basis for the separation of biological activity and receptor binding in a mutant interleukin-1 beta protein. *Biochemistry*, **31**, 6632–8.

Cameron, P.M., Limjuco, G.A., Chin, J., Silberstein, L., and Schmidt, J.A. (1986). Purification to homogeneity and amino acid sequence analysis of two anionic species of human interleukin 1. *J. Exp. Med.*, **164**, 237–50.

Civitelli, R., Teitelbaum, S.L., Hruska, K.A., and Lacey, D.L. (1989). IL-1 activates the Na$^+$/H$^+$ antiport in a murine T cell. *J. Immunol.*, **143**, 4000–8.

Colotta, F., Re, F., Muzio, M., Bertini, R., Polentarutti, N., Sironi, M., Giri, J.G., Dower, S.K., Sims, J.E., and Mantovani, A. (1993). Interleukin-1 type II receptor: a decoy target for IL-1 that is regulated by IL-4. *Science*, **261**, 472–5.

Dinarello, C.A. (1989). Interleukin-1 and its biologically related cytokines. *Adv. Immunol.*, **44**, 153–205.

Dinarello, C.A. (1991). Interleukin-1 and interleukin-1 antagonism. *Blood*, **77**, 1627–52.

Dower, S.K., Kronheim, S.R., Hopp, T.P., Cantrell, M., Deeley, M., Gillis, S., Henney, C.S., and Urdal, D.L. (1986). The cell surface receptors for interleukin-1 alpha and interleukin-1 beta are identical. *Nature*, **324**, 266–8.

Dower, S.K., Bird, T.A., and Sims, J.E. (1992a). *Adv. Neuroimmunol.*, **2**, 1–16.

Dower, S.K., Sims, J.E., Cerretti, D.P., and Bird, T.A. (1992b). In *Interleukins: Molecular biology and immunology* (ed. K. Ishizaka, P.J. Lachman, R. Lerner, and B.H. Waksman, pp. 33–64. S. Karger, Basel.

Dripps, D.J., Brandhuber, B.J., Thompson, R.C., and Eisenberg, S.P. (1991). Interleukin-1 (IL-1) receptor antagonist binds to the 80-kDa IL-1 receptor but does not initiate IL-1 signal transduction. *J. Biol. Chem.*, **266**, 10331–6.

Eisenberg, S.P., Evans, R.J., Arend, W.P., Verderber, E., Brewer, M.T., Hannum, C.H., and Thompson, R.C. (1990). Primary structure and functional expression from complementary DNA of a human interleukin-1 receptor antagonist. *Nature*, **343**, 341–6.

Fanslow, W.C., Clifford, K.N., Park, L.S., Rubin, A.S., Voice, R.F., Beckmann, M.P., and Widmer, M.B. (1991). Regulation of alloreactivity *in vivo* by IL-4 and the soluble IL-4 receptor. *J. Immunol.*, **147**, 535–40.

Graves, B.J., Hatada, M.H., Hendrickson, W.A., Miller, J.K., Madison, V.S., and Satow, Y. (1990). Structure of interleukin 1 alpha at 2.7 Å resolution. *Biochemistry*, **29**, 2679–84.

Gray, P.W., Glaister, D., Chen, E., Goeddel, D.V., and Pennica, D. (1986). Two interleukin 1 genes in the mouse: cloning and expression of the cDNA for murine interleukin 1 beta. *J. Immunol.*, **137**, 3644–8.

Hannum, C.H., Wilcox, C.J., Arend, W.P., Joslin, F.G., Dripps, D.J., Heimdal, P.L., Armes, L.G., Sommer, A., Eisenberg, S.P., and Thompson, R.C. (1990). Interleukin-1 receptor antagonist activity of a human interleukin-1 inhibitor. *Nature*, **343**, 336–40.

Haskill, S., Martin, G., Van Le, L., Morris, J., Peace, A., Bigler, C.F., Jaffe, G.J., Hammerberg, C., Sporn, S.A., Fong, S., Arend, W.P., and Ralph, P. (1991). cDNA cloning of an intracellular form of the human interleukin 1 receptor antagonist associated with epithelium. *Proc. Natl. Acad. Sci. (USA)*, **88**, 3681–5.

Ju, G., Labriola-Tompkins, E., Campen, C.A., Benjamin, W.R., Karas, J., Plocinski, J., Biondi, D., Kaffka, K.L., Kilian, P.L., Eisenberg, S.P., and Evans, R.J. (1991). Conversion of the interleukin 1 receptor antagonist into an agonist by site-specific mutagenesis. *Proc. Natl. Acad. Sci. (USA)*, **88**, 2658–62.

Kronheim, S.R., March, C.J., Erb, S.K., Conlon, P.J., Mochizuki, D.Y., and Hopp, T.P. (1985). Human interleukin 1. Purification to homogeneity. *J. Exp. Med.*, **161**, 490–502.

Lomedico, P.T., Gubler, U., Hellman, C.P., Dukovich, M., Giri, J.G., Pan, Y.-C.E., Collier, K., Semionow, R., Chua, A.O., and Mizel, S.B. (1984). Cloning and expression of murine interleukin-1 cDNA in *Escherichia coli. Nature*, **312**, 458–62.

March, C.J., Mosley, B., Larsen, A., Cerretti, D.P., Braedt, G., Price, V., Gillis, S., Henney, C.S., Kronheim, S., Grabstein, K., Conlon, P.J., Hopp, T.P., and Cosman, D.C. (1985). Cloning, sequence and expression of two distinct human interleukin-1 complementary DNAs. *Nature*, **315**, 641–7.

Mosley, B., Urdal, D.L., Prickett, K.S., Larsen, A., Cosman, D., Conlon, P.J., Gillis, S., and Dower, S.K. (1987). The interleukin-1 receptor binds the human interleukin-1 alpha precursor but not the interleukin-1 beta precursor. *J. Biol. Chem.*, **262**, 2941–4.

Ohlsson, K., Bjork, P., Bergenfeldt, M., Hageman, R., and Thompson, R.C. (1990). Interleukin-1 receptor antagonist reduces mortality from endotoxin shock. *Nature*, **348**, 550–2.

Oppenheim, J.J., Kovacs, E.J., Matsushima, K., and Durum, S.K. (1986). There is more than one interleukin 1. *Immunol. Today*, **7**, 45–55.

Priestle, J.P., Schar, H.-P., and Grutter, M.G. (1989). Crystallographic refinement of interleukin 1 beta at 2.0 Å resolution. *Proc. Natl. Acad. Sci. (USA)*, **86**, 9667–71.

Qwarnstrom, E.E., Jarvelainen, H.T., Kinsella, M.G., Ostberg, C.O., Sandell, L.J., Page, R.C., and Wight, T.N. (1993). Interleukin-1 beta regulation of fibroblast proteoglycan synthesis involves a decrease in versican steady-state mRNA levels. *Biochem. J.*, **294**, 613–20.

Raines, E.W., Dower, S.K., and Ross, R. (1989). Interleukin-1 mitogenic activity for fibroblasts and smooth muscle cells is due to PDGF-AA. *Science*, **243**, 393–6.

Ray, C.A., Black, R.A., Kronheim, S.R., Greenstreet, T.A., Sleath, P.R., Salvesen, G.S., and Pickup, D.J. (1992). Viral inhibition of inflammation: cowpox virus encodes an inhibitor of the interleukin-1 beta converting enzyme. *Cell*, **69**, 597–604.

Schmidt, J.A. (1984). Purification and partial biochemical characterization of normal human interleukin 1. *J. Exp. Med.*, **160**, 772–87.

Simon, P.L., Kumar, V., Lillquist, J.S., Bhatnagar, P., Einstein, R., Lee, J., Porter, T., Green, D., Sathe, G., and Young, P.R. (1993). Mapping of neutralizing epitopes and the receptor binding site of human interleukin 1 beta. *J. Biol. Chem.*, **268**, 9771–9.

Sims, J.E., Gayle, M.A., Slack, J.L., Alderson, M.R., Bird, T.A., Giri, J.G., Colotta, F., Re, F., Mantovani, A., Shanebeck, K., Grabstein, K.H., and Dower, S.K. (1993). Interleukin 1 signaling occurs exclusively via the type I receptor. *Proc. Natl. Acad. Sci. (USA)*, **90**, 6155–9.

Slack, J., McMahan, C.J., Waugh, S., Schooley, K., Spriggs, M.K., Sims, J.E., and Dower, S.K. (1992). Independent binding of interleukin-1 alpha and interleukin-1 beta to type I and type II interleukin-1 receptors. *J. Biol. Chem.*, **268**, 2513–24.

Spriggs, M.K., Hruby, D.E., Maliszewski, C.R., Pickup, D.J., Sims, J.E., Buller, R.M.L., and VanSlyke, J. (1992). Vaccinia and cowpox viruses encode a novel secreted interleukin-1-binding protein. *Cell*, **71**, 145–52.

Tate, S., Kikumoto, Y., Ichikawa, S., Kaneko, M., Masui, Y., Kamogashira, T., Ouchi, M., Takahashi, S., and Inagaki, F. (1992). Stable isotope aided nuclear magnetic resonance study to investigate the receptor-binding site of human interleukin 1 beta. *Biochemistry*, **31**, 2435–42.

Wakabayashi, G., Gelfand, J.A., Burke, J.F., Thompson, R.C., and Dinarello, C.A. (1991). A specific receptor antagonist for interleukin 1 prevents *Escherichia coli*-induced shock in rabbits. *FASEB J.*, **5**, 338–43.

S.K. Dower and J.E. Sims:
Immunex R&D Corporation, Seattle, WA 98101 USA

There are two cell-surface binding proteins for interleukin 1 (IL-1). The type I IL-1 receptor is an 80 kDa molecule which mediates the known biological responses to IL-1. The type II IL-1 receptor appears not to deliver biological signals, and therefore serves to trap IL-1 and inhibit its action by preventing its interaction with the type I receptor. The type II receptor can perform this role either on the cell surface or as a soluble receptor, subsequent to shedding.

■ Type I receptor

The human type I IL-1 receptor is predicted from the cDNA sequence to be a protein of 569 amino acids (GenBank accession number M27492; mouse, M20658; rat, M95578; chicken, M81846) (Sims *et al*. 1989). It contains a 20-amino-acid signal peptide, an extracellular region of 317 amino acids, a single 22-amino-acid transmembrane region, and a cytoplasmic portion of 210 amino acids (see Figs 1 and 2). The extracellular region is comprised entirely of three immunoglobulin-like domains. All of the six potential sites for N-linked glycosylation are used, giving an apparent molecular weight on SDS gels of approximately 80 000. The protein is encoded by an mRNA of approximately 5 kb which is present at a low level in a wide variety of cells. While type I receptor expression is capable of being regulated (being induced upon T-cell activation [Sims *et al*. 1989], for example, and downregulated in T cells treated with IL-1 [Ye *et al*. 1992]), this phenomenon has not been widely studied, and at the moment it is unclear to what extent regulation of type I receptor expression is used as a means of controlling IL-1 responses. By *in situ* hybridization, a few murine tissues naturally express high levels of type I receptor mRNA (Deyerle *et al*. 1992). These include the dentate gyrus of the hippocampus, beta cells in the pancreatic islets of Langerhans, developing oocytes, and follicular granulosa cells subsequent to ovulation. This natural high-level expression suggests that IL-1 may play a role in the normal physiology of these tissues.

The type I receptor appears to mediate all of the responses to IL-1 (Curtis *et al*. 1989; Heguy *et al*. 1991; Stylianou *et al*. 1992; Sims *et al*. 1993*a*). This conclusion is based both on antibody-blocking and on transfection experiments. The cytoplasmic domain of the receptor is necessary for its signalling function. A specific region of about 10 amino acids (Phe 513 to Pro 522) has been identified by point mutation as being critical for induction of transcription from the IL-2 promoter in human T-cells, whereas induction of transcription from the IL-8 promoter in the same cells can be prevented by any of a number of small internal deletions spanning a large portion of the cytoplasmic domain (Heguy *et al*. 1992; Kuno *et al*. 1993). The mechanism of signalling is not known (Sims *et al*. 1993*b*). There are at least two signalling pathways, one leading to induction of NFκB DNA binding activity and the other resulting in increased MAP kinase activity (Bird *et al*. 1992). Interestingly, the type I IL-1 receptor cytoplasmic region shows significant sequence similarity to the cytoplasmic portion of the *Drosophila* gene *Toll* (GenBank accession number M19969). *Toll* is a transmembrane receptor protein whose activation leads to induction of DNA binding activity by *dorsal*, the Drosophila equivalent of mammalian NFκB (Gay and Keith 1991; Schneider *et al*. 1991).

Recombinant soluble type I IL-1 receptor is an effective antagonist of IL-1 action, both *in vitro* (Maliszewski *et al*. 1990) and *in vivo* (Fanslow *et al*. 1990; Jacobs *et al*. 1991; Mullarkey *et al*. 1994). In murine models, it can delay rejection of cardiac allografts, eliminate hypertrophy of draining lymph nodes near the site of injection of allogeneic cells, and alleviate the symptoms of experimental allergic encephalitis. It is also effective at eliminating the late phase reaction in a human model of cutaneous allergy. Further tests of the clinical utility of this molecule in human volunteers are currently in progress.

■ Type II receptor

The human type II IL-1 receptor protein is predicted from the cDNA sequence (GenBank accession number X59770; mouse, X59769) to contain 398 amino acids (McMahan *et al*. 1991). These are organised into a 13-amino-acid signal peptide, an external portion of 334 amino acids, a single transmembrane region of 22 amino acids, and a short cytoplasmic tail of 29 amino acids. The extracellular, ligand-binding portion is generally similar to that of the type I receptor, being comprised of three immunoglobulin-like domains. The type I and II receptors share 28 per cent amino acid identity in this region, and are clearly more related to one another than to other Ig family members. The transmembrane and cytoplasmic portions of the type I and type II receptors share no similarity (see Figs 1 and 2).

Type II IL-1 receptor distribution appears to be more restricted than that of the type I receptor. It is the predominant receptor expressed by myeloid and B lymphoid cells (Bomsztyk *et al*. 1989; Chizzonite *et al*. 1989), although it can be strongly induced in some T cells by T-cell receptor cross-linking (McMahan *et al*. 1991). It is also present at high levels in basal epithelium (of skin, vagina, and urethra), and at low levels in a limited number of other cell types, at least in culture (Deyerle *et al*. 1992; McMahan *et al*. 1991). It seems to be much more actively regulated than is the type I receptor. For example, glucocorticoids lead to strong

signal peptide

```
I    MKVLLRLICFIALLISSLEA                                      20
II   MLRLYVLVMGVSA                                             13
```

extracellular

```
I            DKCKER-EEKIILVSSANEIDVRPCPLNP------NEH            51
II   FTLQPAAHTGAARSCRFRGRHYKREFRLEGEPVALRCPQVPYWLWASVSP        63

I    KGTITWYKDDSKTPVSTEQASRIHQHKEKLWFVPAKVEDSGHYYCVVRNS       101
II   RINLTWHKNDSARTVPGEEETRMWAQDGALWLLPALQEDSGTYVCTTRNA       113

I    SYCLRIKISAKFVENEPNLCYNAQAIFKQKLPVAGDGGLVCPYMEFFKNE       151
II   SYCDKMSIELRVFENTDA--FLPFISYPQILTLSTSGVLVCPDLSEFTRD       161

I    NNELPKLQWYKDCKPLLLD--NIHFSGVKD--RLIVMNVAEKHRGNYTCH       197
II   KTDV-KIQWYKDS--LLLDKDNEKFLSVRGTTHLLVHDVALEDAGYYRCV       208

I    ASYTYLGKQYPITRVIEFITLEENKPTRPVIVSPANETMEVDLGSQIQLI       247
II   LTFAHEGQQYNITRSIELRIKKKKEETIPVIISPLK-TISASLGSRLTIP       257

I    CNVT----GQLSDIAYWKWNGSVIDEDDPVLGEDYYSVENPANKRRSTLI       293
II   CKVFLGTGTPLTTMLWWTANDTHIESAYPGGRVTEGPRQEYSENNENYIE       307

I    TVLNISEIESRFYKHPFTCFAKNTHGIDAAYIQLIYPVTNFQKH            337
II   VPLIFDPVTREDLHMDFKCVVHNTLSFQTLRTTVKEASST               347
```

transmembrane

```
I    MIGICVTLTVIIVCSVFIYKIF                                   359
II   FSWGIVLAPLSLAFLVLGGIWM                                   369
```

cytoplasmic

```
I    KIDIVLWYRDSCYDFLPIKASDGKTYDAYILYPKTVGEGSTSDCDIFVFK       409
II   HRRCKHRTGKADGLTVLWPHHQDFQSYPK*                           398

I    VLPEVLEKQCGYKLFIYGRDDYVGEDIVEVINENVKKSRRLIIILVRETS       459

I    GFSWLGGSSEEQIAMYNALVQDGIKVVLLELEKIQDYEKMPESIKFIKQK       509

I    HGAIRWSGDFTQGPQSAKTRFWKNVRYHMPVQRRSPSSKHQLLSPATKEK       559

I    LQREAHVPLG*                                              569
```

Figure 1. Predicted amino acid sequences of the human type I and type II IL-1 receptors.

upregulation on various leucocytes, as does IL-4 (Akahoshi *et al.* 1988; Colotta *et al.* 1993).

The type II receptor does not appear to be a signalling receptor (Stylianou *et al.* 1992; Sims *et al.* 1993a). This conclusion is based primarily on looking at responses to IL-1 in cells which express type II receptors as the vast majority of their IL-1 binding sites (cells which express type II receptors in the absence of any type I receptor have never been found, with the possible exception of the human B cell line

Raji, and Raji does not respond to IL-1). When antibodies which specifically block binding of IL-1 to either receptor are used, it can be shown that all responses examined are mediated via the type I receptor, even when type I receptor expression is as low as 10–20 receptors per cell and greatly exceeded by type II receptor expression. No example of an IL-1 response which is mediated by the type II receptor has been reported in the literature.

The current hypothesis is that the type II IL-1 receptor, by

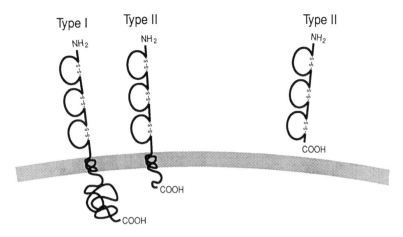

Figure 2. The IL-1 receptor gene family.

binding IL-1 but failing to signal, acts negatively to regulate the actions of IL-1, particularly IL-1β. It can do so either as a surface receptor or as a soluble molecule subsequent to shedding (most likely by proteolytic cleavage). The shed receptor may be more effective, in part because it can accumulate to higher levels than surface receptor. Naturally existing soluble type II IL-1R has been demonstrated (Giri *et al.* 1990; Symons and Duff 1990), and shedding can be induced from neutrophils by agents such as IL-4 and glucocorticoids (Colotta *et al.* 1993). Induction of type II receptor expression on human neutrophils by IL-4 is capable of blocking the response of these cells to IL-1 (Colotta *et al.* 1993).

A soluble type II IL-1R gene is encoded in the genome of certain pox viruses (Spriggs *et al.* 1992; Alcamí and Smith 1992). Deletion of the IL-1R from the viral genome leads to significant differences in the course of infection compared to wild-type virus, including more severe symptoms of illness. The lethality of the virus is also altered, although in which direction depends on the route of injection. Nevertheless, blockade of IL-1 by the virally-encoded receptor clearly affects viral pathogenicity.

■ Binding characteristics

The human type I IL-1 receptor binds human IL-1α with a K_D of about 5×10^{-10} M, IL-1β with a K_D of about 5×10^{-9} M, and the IL-1ra with a K_D of about 10^{-10} M (McMahan *et al.* 1991; Slack *et al.* 1993). The binding characteristics are essentially the same regardless of whether one examines natural type I receptor *in situ*, recombinant type I receptor transfected into monkey kidney cells, or purified soluble receptor (recombinant extracellular region) in a microtitre plate, suggesting that no other protein contributes to the binding energy. The human type II IL-1 receptor binds human IL-1β with a somewhat higher affinity (K_D about 10^{-9} M) than it binds IL-1α (K_D about 10^{-8} M) or IL-1ra (K_D about 10^{-8} M), which is consistent with the hypothesis that its primary role is to regulate the action of systemic IL-1.

Both receptors are capable of generating either linear or complex binding isotherms, depending on the particular experiment. The reasons for this are not understood, but the behaviour can be mimicked by transfected, overexpressed recombinant receptor and therefore may be inherent to the protein rather than due to interaction with other components.

■ Genetic mapping

The human type I and type II IL-1 receptors both map to chromosome 2q12–13 (McMahan *et al.* 1991; Copeland *et al.* 1991). Their mouse counterparts map to the centromere-proximal region of chromosome 1. Interestingly, a cDNA which bears considerable resemblance to the IL-1 receptors has been cloned from cells recently stimulated to divide (Tominaga 1989; Klemenz *et al.* 1989). This cDNA, alternatively termed ST2 or T1, also maps to the same location of mouse chromosome 1 (Tominaga *et al.* 1991), and presumably is derived from the same ancestor as the type I and type II IL-1 receptors. Whether the ST2/T1 protein binds any of the IL-1s is unknown at the present time.

■ References

Akahoshi, T., Oppenheim, J.J., and Matsushima, K. (1988). Induction of high-affinity interleukin 1 receptor on human peripheral blood lymphocytes by glucocorticoid hormones. *J. Exp. Med.*, **167**, 924–36.

Alcamí, A., and Smith, G.L. (1992). A soluble receptor for interleukin-1β encoded by vaccinia virus: a novel mechanism of virus modulation of the host response to infection. *Cell*, **71**, 153–67.

Bird, T.A., Schule, H.D., Delaney, P.B., Sims, J.E., Thoma, B., and Dower, S.K. (1992). Evidence that MAP (mitogen-activated protein) kinase activation may be a necessary but not sufficient signal for a restricted subset of responses in IL-1-treated epidermoid cells. *Cytokine*, **4**, 429–40.

Bomsztyk, K., Sims, J.E., Stanton, T.H., Slack, J., McMahan, C.J., Valentine, M.A., and Dower, S.K. (1989). Evidence for different interleukin 1 receptors in murine B- and T-cell lines. *Proc. Natl. Acad. Sci. (USA)*, **86**, 8034–8.

Chizzonite, R., Truitt, T., Kilian, P.L., Stern, A.S., Nunes, P., Parker, K.P., Kaffka, K.L., Chua, A.O., Lugg, D.K., and Gubler, U. (1989). Two high-affinity interleukin 1 receptors represent separate gene products. *Proc. Natl. Acad. Sci. (USA)*, **86**, 8029–33.

Colotta, F., Re, F., Muzio, M., Bertini, R., Polentarutti, N., Sironi, M., Giri, J.G., Dower, S.K., Sims, J.E., and Mantovani, A. (1993). Interleukin-1 type II receptor: a decoy target for IL-1 that is regulated by IL-4. *Science*, **261**, 472–5.

Copeland, N.G., Silan, C.M., Kingsley, D.M., Jenkins, N.A., Cannizzaro, L.A., Croce, C.M., Huebner, K., and Sims, J.E. (1991). Chromosomal location of murine and human IL-1 receptor genes. *Genomics*, **9**, 44–50.

Curtis, B.M., Gallis, B., Overell, R.W., McMahan, C.J., DeRoos, P., Ireland, R., Eisenman, J., Dower, S.K., and Sims, J.E. (1989). T-cell interleukin 1 receptor cDNA expressed in Chinese hamster ovary cells induces functional responses to interleukin 1. *Proc. Natl. Acad. Sci. (USA)*, **86**, 3045–9.

Deyerle, K.L., Sims, J.E., Dower, S.K., and Bothwell, M.A. (1992). Pattern of IL-1 receptor gene expression suggests role in noninflammatory processes. *J. Immunol.*, **149**, 1657–65.

Fanslow, W.C., Sims, J.E., Sassenfeld, H., Morrissey, P.J., Gillis, S., Dower, S.K., and Widmer, M.B. (1990). Regulation of alloreactivity *in vivo* by a soluble form of the interleukin-1 receptor. *Science*, **248**, 739–42.

Gay, N.J., and Keith, F.J. (1991). *Drosophila* Toll and IL-1 receptor. *Nature*, **351**, 355–6.

Giri, J.G., Newton, R.C., and Horuk, R. (1990). Identification of soluble interleukin-1 binding protein in cell-free supernatants. Evidence for soluble interleukin-1 receptor. *J. Biol. Chem.*, **265**, 17416–19.

Heguy, A., Baldari, C., Bush, K., Nagele, R., Newton, R.C., Robb, R.J., Horuk, R., Telford, J.L., and Melli, M. (1991). Internalization and nuclear localization of interleukin 1 are not sufficient for function. *Cell Growth Diff.*, **2**, 311–15.

Heguy, A., Baldari, C.T., Macchia, G., Telford, J.L., and Melli, M. (1992). Amino acids conserved in interleukin-1 receptors (IL-1Rs) and the *Drosophila* toll protein are essential for IL-1R signal transduction. *J. Biol. Chem.*, **267**, 2605–9.

Jacobs, C.A., Baker, P.E., Roux, E.R., Picha, K.S., Toivola, B., Waugh, S., and Kennedy, M.K. (1991). Experimental autoimmune encephalomyelitis is exacerbated by IL-1α and suppressed by soluble IL-1 receptor. *J. Immunol.*, **146**, 2983–9.

Klemenz, R., Hoffmann, S., and Werenskiold, A.-K. (1989). Serum- and oncoprotein-mediated induction of a gene with sequence similarity to the gene encoding carcinoembryonic antigen. *Proc. Natl. Acad. Sci. (USA)*, **86**, 5708–12.

Kuno, K., Okamoto, S., Hirose, K., Murakami, S., and Matsushima, K. (1993). Structure and function of the intracellular portion of the mouse interleukin 1 receptor (type I). Determining the essential region for transducing signals to activate the interleukin 8 gene. *J. Biol. Chem.*, **268**, 13510–18.

Maliszewski, C.R., Sato, T.A., Vanden Bos, T., Waugh, S., Dower, S.K., Slack, J., Beckmann, M.P., and Grabstein, K.H. (1990). Cytokine receptors and B cell functions. I. Recombinant soluble receptors specifically inhibit IL-1- and IL-4-induced B cell activities *in vitro*. *J. Immunol.*, **144**, 3028–33.

McMahan, C.J., Slack, J.L., Mosley, B., Cosman, D., Lupton, S.D., Brunton, L.L., Grubin, C.E., Wignall, J.M., Jenkins, N.A., Brannan, C.I., Copeland, N.G., Huebner, K., Croce, C.M., Cannizzaro, L.A., Benjamin, D., Dower, S.K., Spriggs, M.K., and Sims, J.E. (1991). A novel IL-1 receptor, cloned from B cells by mammalian expression, is expressed in many cell types. *EMBO J.*, **10**, 2821–32.

Mullarkey, M.F., Leiferman, K.M., Peters, M.S., Cara, I., Roux, E.R., Hanna, R.K., Rubin, A.S., and Jacobs, C.A. (1994). Human cutaneous allergic late-phase response is inhibited by soluble IL-1 receptor. *J. Immunol.*, **152**, 2033–41.

Schneider, D.S., Hudson, K.L., Lin, T.-Y., and Anderson, K.V. (1991). Dominant and recessive mutations define functional domains of Toll, a transmembrane protein required for dorsal-ventral polarity in the Drosophila embryo. *Genes Dev.*, **5**, 797–807.

Sims, J.E., Acres, R.B., Grubin, C.E., McMahan, C.J., Wignall, J.M., March, C.J., and Dower, S.K. (1989). Cloning the interleukin 1 receptor from human T cells. *Proc. Natl. Acad. Sci. (USA)*, **86**, 8946–50.

Sims, J.E., Gayle, M.A., Slack, J.L., Alderson, M.R., Bird, T.A., Giri, J.G., Colotta, F., Re, F., Mantovani, A., Shanebeck, K., Grabstein, K.H., and Dower, S.K. (1993a). Interleukin 1 signaling occurs exclusively via the type I receptor. *Proc. Natl. Acad. Sci. (USA)*, **90**, 6155–9.

Sims, J.E., Bird, T.A., Giri, J.G., and Dower, S.K. (1993b). *Signal transduction through growth factor receptors* (eds Y. Kitagawa and R. Sasaki). (In press.)

Slack, J., McMahan, C.J., Waugh, S., Schooley, K., Spriggs, M.K., Sims, J.E., and Dower, S.K. (1993). Independent binding of interleukin-1α and interleukin-1β to type I and type II interleukin-1 receptors. *J. Biol. Chem.*, **268**, 2513–24.

Spriggs, M.K., Hruby, D.E., Maliszewski, C.R., Pickup, D.J., Sims, J.E., Buller, R.M., and VanSlyke, J. (1992). Vaccinia and cowpox viruses encode a novel secreted interleukin-1-binding protein. *Cell*, **71**, 145–52.

Stylianou, E., ONeill, L.A., Rawlinson, L., Edbrooke, M.R., Woo, P., and Saklatvala, J. (1992). Interleukin 1 induces NF-κB through its type I but not its type II receptor in lymphocytes. *J. Biol. Chem.*, **267**, 15836–41.

Symons, J.A., and Duff, G.W. (1990). A soluble form of the interleukin-1 receptor produced by a human B cell line. *FEBS Lett.*, **272**, 133–6.

Tominaga, S. (1989). A putative protein of a growth specific cDNA from BALB/c-3T3 cells is highly similar to the extracellular portion of mouse interleukin 1 receptor. *FEBS Lett.*, **258**, 301–4.

Tominaga, S., Jenkins, N.A., Gilbert, D.J., Copeland, N.G., and Tetsuka, T. (1991). Molecular cloning of the murine ST2 gene. Characterization and chromosomal mapping. *Biochim. Biophys. Acta*, **1090**, 1–8.

Ye, K., Koch, K.C., Clark, B.D., and Dinarello, C.A. (1992). Interleukin-1 down-regulates gene and surface expression of interleukin-1 receptor type I by destabilizing its mRNA whereas interleukin-2 increases its expression. *Immunology*, **75**, 427–34.

John E. Sims* and Steven K. Dower†:
*Department of Molecular Genetics,
†Research Investigator,
Immunex Research and Development Corporation,
51 University Street,
Seattle, Washington 98101, USA

Interleukin-2 (IL-2)

Interleukin-2 was first recognized as a growth factor for T lymphocytes, but it is now known to deliver signals both for differentiation and for proliferation of a variety of haemopoietic cell types. The molecular basis of ligand binding to its cognate multimeric receptor has emerged through a combination of molecular and structural studies. Some immune deficiency states may benefit from reconstitution of the IL-2/IL-2 receptor system, and certain clinical syndromes involving over-expression of IL-2 receptors may benefit from targeting this receptor system with therapeutic antagonists.

■ Alternative name

T cell growth factor (TCGF).

■ IL-2 protein

IL-2 is a 15.5 kDa glycoprotein secreted primarily by activated T lymphocytes. The mature human IL-2 protein contains 133 amino acid residues, while the homologous proteins in other species contain 133 to 149 amino acids. The IL-2 primary translation product undergoes several post-translational processing steps that include removal of an N-terminal signal peptide, glycosylation at Thr 3, and disulphide linkage of Cys 58 and Cys 105. Glycosylation appears to be dispensable for IL-2 function, but formation of the disulphide bond is critical for bioactivity of the resultant molecule (Robb *et al*. 1983). IL-2 appears to be a very stable protein, exhibiting resistance to both heat and acidic pH. An X-ray crystallographic study of human IL-2 demonstrated that it is a largely helical molecule, with α-helices representing approximately 67 per cent of the entire protein (Brandhuber *et al*. 1987). A proposed structure derived from these studies contains six distinct helical domains involving 89 amino acids (Fig. l(a)). However, an alternative three-dimensional structure was subsequently described which appeared to accommodate the carbon backbone data provided by the above analysis and simultaneously to incorporate the results of mutational studies defining the site of ligand–receptor binding (Bazan 1992). This alternate model contains four core α-helices and two connecting loops containing β-strands (Fig. l(b)). Because of the improved accommodation of existing mutagenesis data, the latter model may more accurately reflect the true structure of IL-2.

■ IL-2 sequence

IL-2 cDNAs have been cloned from seven species, and the nucleotide sequences of each have been reported to the GenBank. The deduced primary sequences of human (Taniguchi *et al*. 1983) (accession number V00564), mouse (Yokota *et al*. 1985; Kashima *et al*. 1985) (accession numbers K02292 and X01772), rat (McKnight *et al*. 1989) (accession

number M22899), cow (Cerretti *et al*. 1986; Reeves *et al*. 1986) (accession number M12791), sheep (Seow *et al*. 1990) (accession number X55641), pig (Goodall *et al*. 1991) (accession number X56750), and gibbon ape (Chen *et al*. 1985) (accession number K03174) IL-2 are highly similar, including complete conservation of all three cysteine residues (Fig. 2). Of some potential interest is the unique insertion within the murine IL-2 molecule of a twelve-residue polyglutamine segment near the N-terminus.

■ Receptor binding

Antibody blocking studies and mutational analyses have suggested that the N-terminal A helix of IL-2 is responsible for binding to the β subunit (p70/75) of the IL-2 receptor (IL-2R) complex (Ju *et al*. 1987; Kuo and Robb 1986; Collins *et al*. 1988; Zurawski and Zurawski 1989). Similar molecular studies have suggested that the interloop connecting the A and B helices of IL-2 is responsible for contacting the α (p55, Tac) subunit of the IL-2 receptor (Weigel *et al*. 1989; Sauve *et al*. 1991; Grant *et al*. 1992), and that the C-terminal D α-helix may be responsible for interactions with the γ chain (Zurawski *et al*. 1990). Confirmation of these preliminary conclusions awaits more detailed molecular and structural analysis. By analogy to the recently crystallized growth hormone receptor (deVos *et al*. 1992), it is plausible that activation of the IL-2 receptor complex is determined by the precise stereochemical assembly of the receptor–ligand complex as determined by specific IL-2/IL-2 receptor contacts.

■ IL-2 gene and expression

The human IL-2 gene has been localized to chromosome 4q bands 26–28 (Siegel *et al*. 1987). *In vivo* expression of this gene is inducible, and it is regulated primarily at the transcription level *via* a 5′ enhancer element (for reviews, see Crabtree 1989; Muegge and Durum 1989; Ullman et al. 1990). Cis-acting regulatory sequences that have been identified within the IL-2 enhancer include binding sites for NFAT-1, NF-κB, AP-1, and octamer proteins. IL-2 gene expression is also regulated post-transcriptionally through a mechanism that involves stabilization of the IL-2 mRNA,

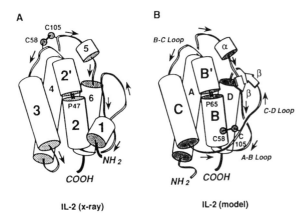

A

IL-2 (x-ray)

B

IL-2 (model)

Figure 1. Two models of the three-dimensional structure of IL-2. (a) Original model deduced from X-ray crystallography contains six *a*-helices. (b) An alternative structure, based upon both the X-ray scaffold data as well as published molecular analyses of ligand binding to its receptor, contains four core *a*-helices. The second model permits the uncrowded contact of receptor to *a*-helices A and D and to the interloop connecting helices A and B. (Reprinted with permission from Bazan 1992.)

which is apparently mediated through the AU-rich sequence motifs present in the 3'-untranslated region of the IL-2 message (see above reviews).

■ Biological functions

The *de novo* synthesis and secretion of IL-2 is triggered primarily by antigen- or mitogen-induced activation of mature T lymphocytes. The subsequent binding of IL-2 to its cognate receptor, an event which occurs either through paracrine or autocrine mechanisms, promotes the clonal expansion of antigen-specific effector T cells (for a review,

see Smith 1980). This growth-promoting property of IL-2 is the best characterized and perhaps most important function of this lymphokine. In addition to its role in expanding mature T cell populations, some evidence has emerged suggesting that IL-2 may play a role in thymic development. Expression of both IL-2 and IL-2 receptor subunits within the thymus has been detected (Raulet 1985; Ceredig *et al.* 1985; Habu *et al.* 1985; Hardt *et al.* 1985; von Boehmer *et al.* 1985; Carding *et al.* 1989; Zuniga-Pflucker *et al.* 1990), and proliferative effects of IL-2 on thymocytes have been reported (Zuniga-Pflucker *et al.* 1990). Similarly, antibodies directed against the IL-2 receptor have, in some systems, demonstrated inhibition of thymic development (Jenkinson *et al.* 1987; Tentori *et al.* 1988). Surprisingly, however, mice lacking a functional IL-2 gene demonstrate grossly normal thymocyte and peripheral T cell development, thereby calling into question the importance of IL-2 in normal thymic development (Schorle *et al.* 1991; Kündig *et al.* 1993).

IL-2 also appears to be a T-cell differentiation factor, as it is able to induce production of other lymphokines such as interferon-γ and IL-4 (Farrar *et al.* 1982; Howard *et al.* 1983). Indeed, some cells of a non-T lymphocyte lineage respond to IL-2. For example, IL-2 can promote the growth of B cells, the induction of immunoglobulin secretion by B cells, and the induction of J chain synthesis leading to assembly and secretion of IgM (Jelinek and Lipsky 1987; Nakanishi *et al.* 1992; Tigges *et al.* 1989). Natural killer (NK) cells also express receptors for IL-2, and the growth, production of interferon-γ, and cytolytic activity of these cells is induced by IL-2 (Trinchieri *et al.* 1984; Ortaldo *et al.* 1984). Similarly, myeloid cell populations such as macrophage precursors and primary peripheral blood monocytes display cell surface receptors for IL-2 (Baccarini *et al.* 1989; Ohashi *et al.* 1989). In such cells IL-2 promotes proliferation, differentiation, and enhancement of cytolytic activity and macrophage antibody-dependent tumouricidal activity (Baccarini *et al.* 1989; Malkovsky *et al.* 1987; Ralph *et al.* 1988). Finally, oligodendrocytes also display IL-2 receptors and the growth of these cells is modulated by IL-2 (Saneto *et al.* 1986; Benveniste and Merrill 1986).

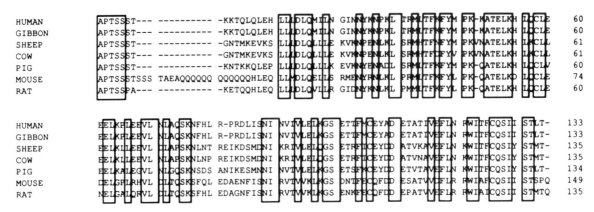

Figure 2. Alignment of deduced amino acid sequences of the IL-2 molecule from seven species (see text). Boxed residues are completely conserved, although many other similarities are also evident.

The most conventional bioassay of IL-2 activity depends upon the induction of proliferation of IL-2-dependent T cell lines (Bottomly *et al.* 1991). For example, CTLL-2 cells grow in an IL-2-dependent fashion, and a quantitative assay of biological function uses proliferation of these cells as an indicator of bioactivity.

Pathology

The IL-2/IL-2 receptor system may play a pathogenetic role in several human clinical syndromes and animal models of disease. For example, the Tax product derived from the human retrovirus HTLV-I appears to cause dysregulation of cellular gene expression. Upon infection of human T lymphocytes by HTLV-I, altered expression of cellular products may underlie transformation of the cells and induction of the fatal adult T-cell leukaemia/lymphoma syndrome. Among the genes that are dysregulated upon infection with HTLV-I is that which encodes the α subunit of the IL-2 receptor, suggesting a pathogenetic role of the IL-2/IL-2 receptor system in this disease (for a review, see Yodoi and Uchiyama 1992). Perhaps a more compelling demonstration of the role of IL-2 and its receptor in human immune function is found in recent studies of X-linked severe combined immunodeficiency (X-SCID), in which the genetic defect has been mapped to the IL-2Rγ gene locus present on chromosome Xq1 (Noguchi *et al.* 1993). X-SCID patients characterized thus far contain mutations that result in premature termination of translation of the IL-2Rγ chain, resulting in expression of truncated receptor subunits. Thus, although reconstitution experiments are still needed to prove this hypothesis, the severe impairment of cell-mediated and humoral immunity in patients with X-SCID appears to result from alterations in this receptor subunit. It is unknown whether the immune deficits are attributable entirely to alteration in responsiveness to IL-2, since it is possible that the IL-2Rγ chain participates in receptor complexes for other lymphokines as well. Finally, selective agonists or antagonists of the IL-2 receptor system have been envisioned to offer therapeutic value in some immunodeficiency or autoimmune disorders, investigations of which are currently in progress.

Acknowledgements

The authors acknowledge the excellent assistance of Ms Leticia Chand and Ms Kathleen Raneses in the preparation of this manuscript. M.A.G. and W.C.G. are supported by the J. David Gladstone Institutes. M.A.G. is also supported in part by the National Institutes of Health through the University of California, San Francisco AIDS Program.

References

Baccarini, M., Schwinzer, R., and Lohmann-Matthes, M.L. (1989). Effect of human recombinant IL-2 on murine macrophage precursors. Involvement of a receptor distinct from the p55 (Tac) protein. *J. Immunol.*, **142**, 118–25.

Bazan, J.F. (1992). Unraveling the structure of IL-2 [letter]. *Science*, **257**, 410–3.

Benveniste, E.N., and Merrill, J.E. (1986). Stimulation of oligodendroglial proliferation and maturation by interleukin-2. *Nature*, **321**, 610–3.

Bottomly, K., Davis, L.S., and Lipsky, P.E. (1991). Measurement of human and murine interleukin-2 and interleukin-4. In *Current protocols in immunology* (eds J.E. Coligan, A.M. Kruisbeek, D.H. Margulies, E.M. Shevach, and W. Strober), pp. 6.3.1–6.3.12. Greene Pub. Assoc. and Wiley-Interscience, New York.

Brandhuber, B.J., Boone, T., Kenney, W.C., and McKay, D.B. (1987). Three-dimensional structure of interleukin-2. *Science*, **238**, 1707–9.

Carding, S.R., Jenkinson, E.J., Kingston, R., Hayday, A.C., Bottomly, K., and Owen, J.J. (1989). Developmental control of lymphokine gene expression in fetal thymocytes during T-cell ontogeny. *Proc. Natl. Acad. Sci. (USA)*, **86**, 3342–5.

Ceredig, R., Lowenthal, J.W., Nabholz, M., and MacDonald, H.R. (1985). Expression of interleukin-2 receptors as a differentiation marker on intrathymic stem cells. *Nature*, **314**, 98–100.

Cerretti, D.P., McKereghan, K., Larsen, A., Cantrell, M.A., Anderson, D., Gillis, S., Cosman, D., and Baker, P.E. (1986). Cloning, sequence, and expression of bovine interleukin 2. *Proc. Natl. Acad. Sci. (USA)*, **83**, 3223–7.

Chen, S.J., Holbrook, N.J., Mitchell, K.F., Vallone, C.A., Greengard, J.S., Crabtree, G.R., and Lin, Y. (1985). A viral long terminal repeat in the interleukin 2 gene of a cell line that constitutively produces interleukin 2. *Proc. Natl. Acad. Sci. (USA)*, **82**, 7284–8.

Collins, L., Tsien, W.H., Seals, C., Hakimi, J., Weber, D., Bailon, P., Hoskings, J., Greene, W.C., Toome, V., and Ju, G. (1988). Identification of specific residues of human interleukin 2 that affect binding to the 70-kDa subunit (p70) of the interleukin 2 receptor. *Proc. Natl. Acad. Sci. (USA)*, **85**, 7709–13.

Crabtree, G.R. (1989). Contingent genetic regulatory events in T lymphocyte activation. *Science*, **243**, 355–61.

deVos, A., Ultsch, M., and Kossiakoff, A.A. (1992). Human growth hormone and extracellular domain of its receptor: crystal structure of the complex. *Science*, **255**, 306–12.

Farrar, J.J., Benjamin, W.R, Hilfiker, M.L., Howard, M., Farrar, W.L., and Fuller-Farrar, J. (1982). The biochemistry, biology, and role of interleukin 2 in the induction of cytotoxic T cell and antibody-forming B cell responses. *Immunol. Rev.*, **63**, 129–66.

Goodall, J.C., Emery, D.C., Bailey, M., English, L.S., and Hall, L. (1991). cDNA cloning of porcine interleukin 2 by polymerase chain reaction. *Biochim. Biophys. Acta*, **1089**, 257–8.

Grant, A.J., Roessler, E., Ju, G., Tsudo, M., Sugamura, K., and Waldmann, T.A. (1992). The interleukin 2 receptor (IL-2R): the Il-2R alpha subunit alters the function of the Il-2R beta subunit to enhance IL-2 binding and signaling by mechanisms that do not require binding of IL-2 to IL-2R alpha subunit. *Proc. Natl. Acad. Sci. (USA)*, **89**, 2165–9.

Habu, S., Okumura, K., Diamantstein, T., and Shevach, E.M. (1985). Expression of interleukin 2 receptor on murine fetal thymocytes. *Eur. J. Immunol.*, **15**, 456–60.

Hardt, C., Diamantstein, T., and Wagner, H. (1985). Developmentally controlled expression of IL 2 receptors and of sensitivity to IL 2 in a subset of embryonic thymocytes. *J. Immunol.*, **134**, 3891–4.

Howard, M., Matis, L., Malek, T.R., Shevach, E., Kell, W., Cohen, D., Nakanishi, K., and Paul, W.E. (1983). Interleukin 2 induces antigen-reactive T cell lines to secrete BCGF-I. *J. Exp. Med.*, **158**, 2024–39.

Jelinek, D.F., and Lipsky, P.E. (1987). Regulation of human B lymphocyte activation, proliferation, and differentiation. *Adv. Immunol.*, **40**, 1–59.

Jenkinson, E.J., Kingston, R., and Owen, J.J. (1987). Importance of IL-2 receptors in intra-thymic generation of cells expressing T-cell receptors. *Nature*, **329**, 160–2.

Ju, G., Collins, L., Kaffka, K.L., Tsien, W.H., Chizzonite, R., Crowl, R., Bhatt, R., and Kilian, P.L. (1987). Structure-function analysis of human interleukin-2. Identification of amino acid residues required for biological activity. *J. Biol. Chem.*, **262**, 5723–31.

Kashima, N., Nishi-Takaoka, C., Fujita, T., Taki, S., Yamada, G., Hamuro, J., and Taniguchi, T. (1985). Unique structure of murine interleukin-2 as deduced from cloned cDNAs. *Nature*, **313**, 402–4.

Kündig, T.M., Schorle, H., Bachmann, M.F., Hengartner, H., Zinkernagel, R.M., and Horak, I. (1993). Immune responses in interleukin-2-deficient mice. *Science*, **262**, 1059–61.

Kuo, L.M., and Robb, R.J. (1986). Structure–function relationships for the IL 2-receptor system. I. Localization of a receptor binding site on IL 2. *J. Immunol.*, **137**, 1538–43.

Malkovsky, M., Loveland, B., North, M., Asherson, G.L., Gao, L., Ward, P., and Fiers, W. (1987). Recombinant interleukin-2 directly augments the cytotoxicity of human monocytes. *Nature*, **325**, 262–5.

McKnight, A.J., Mason, D.W., and Barclay, A.N. (1989). Sequence of rat interleukin 2 and anomalous binding of a mouse interleukin 2 cDNA probe to rat MHC class II-associated invariant chain mRNA. *Immunogenetics*, **30**, 145–7.

Muegge, K., and Durum, S.K. (1989). From cell code to gene code: cytokines and transcription factors. *New Biol.*, **1**, 239–46.

Nakanishi, K., Hirose, S., Yoshimoto, T., Ishizashi, H., Hiroishi, K., Tanaka, T., Kono, T., Miyasaka, M., Taniguchi, T., and Higashino, K. (1992). Role and regulation of interleukin (IL)-2 receptor alpha and beta chains in IL-2-driven B-cell growth. *Proc. Natl. Acad. Sci. (USA)*, **89**, 3551–5.

Noguchi, M., Yi, H., Rosenblatt, H.M., Filipovich, A.H., Adelstein, S., Modi, W.S., McBride, O.W., and Leonard, W.J. (1993). Interleukin-2 receptor gamma chain mutation results in X-linked severe combined immunodeficiency in humans. *Cell*, **73**, 147–57.

Ohashi, Y., Takeshita, T., Nagata, K., Mori, S., and Sugamura, K. (1989). Differential expression of the IL-2 receptor subunits, p55 and p75 on various populations of primary peripheral blood mononuclear cells. *J. Immunol.*, **143**, 3548–55.

Ortaldo, J.R., Mason, A.T., Gerard, J.P., Henderson, L.E., Farrar, W., Hopkins, R.F. 3rd, Herberman, R.B., and Rabin, H. (1984). Effects of natural and recombinant IL 2 on regulation of IFN gamma production and natural killer activity: lack of involvement of the Tac antigen for these immunoregulatory effects. *J. Immunol.*, **133**, 779–83.

Ralph, P., Nakoinz, I., and Rennick, D. (1988). Role of interleukin 2, interleukin 4, and alpha, beta, and gamma interferon in stimulating macrophage antibody-dependent tumouricidal activity. *J. Exp. Med.*, **167**, 712–17.

Raulet, D.H. (1985). Expression and function of interleukin-2 receptors on immature thymocytes. *Nature*, **314**, 101–3.

Reeves, R., Spies, A.G., Nissen, M.S., Buck, C.D., Weinberg, A.D., Barr, P.J., Magnuson, N.S., and Magnuson, J.A. (1986). Molecular cloning of a functional bovine interleukin 2 cDNA. *Proc. Natl. Acad. Sci. (USA)*, **83**, 3228–32.

Robb, R.J., Kutny, R.M., and Chowdhry, V. (1983). Purification and partial sequence analysis of human T-cell growth factor. *Proc. Natl. Acad. Sci. (USA)*, **80**, 5990–4.

Saneto, R.P., Altman, A., Knobler, R.L., Johnson, H.M., and de Vellis, J. (1986). Interleukin 2 mediates the inhibition of oligodendrocyte progenitor cell proliferation *in vitro*. *Proc. Natl. Acad. Sci. (USA)*, **83**, 9221–5.

Sauve, K., Nachman, M., Spence, C., Bailon, P., Campbell, E., Tsien, W.H., Kondas, J.A., Hakimi, J., and Ju, G. (1991). Localization in human interleukin 2 of the binding site to the alpha chain (p55) of the interleukin 2 receptor. *Proc. Natl. Acad. Sci. (USA)*, **88**, 4636–40.

Schorle, H., Holtschke, T., Hunig, T., Schimpl, A., and Horak, I. (1991). Development and function of T cells in mice rendered interleukin-2 deficient by gene targeting. *Nature*, **352**, 621–4.

Seow, H.F., Rothel, J.S., Radford, A.J., and Wood, P.R. (1990). The molecular cloning of ovine interleukin 2 gene by the polymerase chain reaction. *Nucleic Acids Res.*, **18**, 7175.

Siegel, J.P., Sharon, M., Smith, P.L., and Leonard, W.J. (1987). The IL-2 receptor beta chain (p70): role in mediating signals for LAK, NK, and proliferative activities. *Science*, **238**, 75–8.

Smith, K.A. (1980). T-cell growth factor. *Immunol. Rev.*, **51**, 337–57.

Taniguchi, T., Matsui, H., Fujita, T., Takaoka, C., Kashima, N., Yoshimoto, R., and Hamuro, J. (1983). Structure and expression of a cloned cDNA for human interleukin-2. *Nature*, **302**, 305–10.

Tentori, L., Longo, D.L., Zuniga-Pflucker, J.C., Wing, C., and Kruisbeek, A.M. (1988). Essential role of the interleukin 2-interleukin 2 receptor pathway in thymocyte maturation *in vivo*. *J. Exp. Med.*, **168**, 1741–7.

Tigges, M.A., Casey, L.S., and Koshland, M.E. (1989). Mechanism of interleukin-2 signaling: mediation of different outcomes by a single receptor and transduction pathway. *Science*, **243**, 781–6.

Trinchieri, G., Matsumoto-Kobayashi, M., Clark, S.C., Seehra, J., London, L., and Perussia, B. (1984). Response of resting human peripheral blood natural killer cells to interleukin 2. *J. Exp. Med.*, **160**, 1147–69.

Ullman, K.S., Northrop, J.P., Verweij, C.L., and Crabtree, G.R. (1990). Transmission of signals from the T lymphocyte antigen receptor to the genes responsible for cell proliferation and immune function: the missing link. *Annu. Rev. Immunol.*, **8**, 421–52.

von Boehmer, H., Crisanti, A., Kisielow, P., and Haas, W. (1985). Absence of growth by most receptor-expressing fetal thymocytes in the presence of interleukin-2. *Nature*, **314**, 539–40.

Weigel, U., Meyer, M., and Sebald, W. (1989). Mutant proteins of human interleukin 2. Renaturation yield, proliferative activity and receptor binding. *Eur. J. Biochem.*, **295**, 295–300.

Yodoi, J., and Uchiyama, T. (1992). Diseases associated with HTLV-I: virus, IL-2 receptor dysregulation and redox regulation. *Immunol. Today*, **13**, 405–11.

Yokota, T., Arai, N., Lee, F., Rennick, D., Mosmann, T., and Arai, K. (1985). Use of a cDNA expression vector for isolation of mouse interleukin 2 cDNA clones: expression of T-cell growth-factor activity after transfection of monkey cells. *Proc. Natl. Acad. Sci. (USA)*, **82**, 68–72.

Zuniga-Pflucker, J.C., Smith, K.A., Tentori, L., Pardoll, D.M., Longo, D.L., and Kruisbeek, A.M. (1990). Are the IL-2 receptors expressed in the murine fetal thymus functional? *Dev. Immunol.*, **1**, 59–66.

Zurawski, S.M., and Zurawski, G. (1989). Mouse interleukin-2 structure-function studies: substitutions in the first alpha-helix can specifically inactivate p70 receptor binding and mutations in the fifth alpha-helix can specifically inactivate p55 receptor binding. *EMBO J.*, **8**, 2583–90.

Zurawski, S.M., Imler, J.L., and Zurawski, G. (1990). Partial agonist/antagonist mouse interleukin-2 proteins indicate that a third component of the receptor complex functions in signal transduction. *EMBO J.*, **9**, 3899–905.

Mark A. Goldsmith and Warner C. Greene:
Gladstone Institute of Virology and Immunology,
San Francisco General Hospital,
University of California, San Francisco, CA, USA

Receptors for Interleukin-2 (IL-2)

IL-2 receptors consist of three ligand-binding subunits (a, β, γ) which combine in various configurations having a variety of affinities. The β and γ chains belong to the haemopoietin receptor family characterized by a highly conserved extracellular domain. The cytoplasmic segments of both chains play an essential role in coupling ligand binding to signalling pathways involving tyrosine kinases and other intracellular elements. The a chain is a distinct structure and its short cytoplasmic domain does not appear to play a functional role. The expression of a chain is tightly controlled by mitogenic and antigenic stimulation, however, and it combines with the β and γ chains to form a high-affinity receptor complex. The IL-2 receptor is expressed on lymphocytes, natural killer (NK) cells, monocytes, and neutrophils as well as some non-haemopoietic cells. Its functions include both promotion of proliferation and certain forms of cellular maturation.

■ Protein structure of the receptor subunits

■ IL-2 receptor a subunit protein

The a chain (p55, Tac) of the human IL-2 receptor, encoded by a single gene on chromosome 10, p14–15 (Leonard et al. 1985a), is produced as a 272 amino acid (aa) precursor with a 21 aa signal sequence (GenBank accession number K63122) (Leonard et al. 1985b). The mature protein contains a 219 aa extracellular domain, a 19 aa membrane-spanning region, and an intracellular, carboxy-terminal domain of 13 residues (Fig. 1) (Leonard et al. 1984, 1985b). The difference between the observed molecular weight of 55 000 Da (SDS-PAGE) and the predicted molecular weight of 28 437 Da is due to N- (Asn 49 and 68) and O-linked glycosylation and sulphation (Leonard et al. 1985c). The a chain is characterized by an internal repeat of $\sim$ 65 residues corresponding to exons 2 and 4 of the gene (Leonard et al. 1985b). These segments have 25 per cent homology with each other and limited homology to the Ba fragment of complement factor B. Exon 4 is deleted in a fraction of a chain mRNA (Leonard et al. 1984). The corresponding protein is synthesized, but does not bind IL-2 or traffic efficiently to the cell surface. There are five internal disulphide bonds critical to the functional integrity of the a receptor, as well as two free Cys residues, one (position 192 of the mature protein) located close to, and the other (position 225) located within the membrane-spanning segment (Neeper et al. 1987; Rusk et al. 1988; Miedel et al. 1988). Soluble a chain capable of binding IL-2 is released from the cell surface by an apparent proteolytic cleavage carboxy-terminal to Cys 192 (Rubin et al. 1985; Robb and Kutny 1987; Loughnan et al. 1988). a chain can form non-covalent homodimers on the cell surface and in its soluble form (Jacques et al. 1990). Amino acid residues particularly important to maintaining an active IL-2 binding site map by mutational analysis to two small segments (residues 1–6 and 35–43) in the portion of the a chain encoded by exon 2 (Robb et al. 1988).

The murine IL-2 receptor a chain, encoded by a single gene on chromosome 2 (GenBank accession numbers M30856, K02891) shows 61 per cent homology with the human protein (Miller et al. 1985; Webb et al. 1990), with a similar short (11 aa) cytoplasmic domain. In addition to regions with high degrees of local homology, thirteen of fourteen cysteine residues are conserved between the two species.

■ IL-2 receptor β subunit protein

The human β subunit (p70), which is encoded by a single gene on chromosome 22, q11.2–12 (Gnarra et al. 1990; Shibuya et al. 1990), is synthesized as a 551 residue precursor with a 26 aa signal sequence (GenBank accession number M26062). The mature protein, which has a calculated molecular weight of 58 358 Da, migrates as a 70–75 000 Da protein on SDS-PAGE due to N-linked glycosylation. It contains a 214 aa extracellular domain, a 25 aa transmembrane region, and a 286 aa cytoplasmic domain (Fig. 1) (Hatakeyama et al. 1989a). The extracellular segment shows distinct homology to members of the haemopoietin receptor family (Bazan 1990; Taga and Kishimoto 1992) with a characteristic spacing of four conserved cysteine residues and an element consisting of Trp-Ser-Pro-Trp-Ser (the WSXWS motif) (Hatakeyama et al. 1989a). Based on X-ray crystallography of the human growth hormone receptor, the extracellular region of such receptors forms two fibronectin type-III-like modules, each a β-barrel of seven antiparallel β strands (de Vos et al. 1992). The junction of the two barrels is predicted to define the ligand-binding site.

The cytoplasmic portion of the β subunit contains no apparent catalytic motifs characteristic of kinase consensus sequences. Two segments within the cytoplasmic region, however, have been defined on the basis of structural and functional characteristics (Fig. 1) (Minami et al. 1993; Satoh et al. 1992; Hatakeyama et al. 1989b), a 'serine-rich' segment which shows homology with a similar region of the erythropoietin receptor known to be required for signal

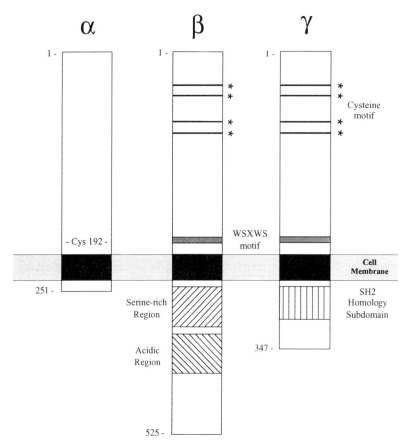

Figure 1. Schematic representation of the IL-2 receptor α, β, and γ subunits. Functional significance of the 'serine-rich' and 'acidic' regions, as well as the cysteine repeat and WXSWS motifs, are described more fully in the text. Each of the indicated regions/motifs are found in both human and murine IL-2 receptor chains.

transduction, and an 'acidic- amino-acid rich' region implicated as the interaction site with various src-family tyrosine kinases.

The murine β chain (GenBank accession number M28052), encoded by a gene on chromosome 15 (Malek et al. 1993), has 58 per cent homology to the human protein. The mature receptor chain consists of 513 amino acids and contains an extracellular WSXWS motif and conservation of seven of the nine extracellular cysteine residues found in the human protein (Kono et al. 1990). In addition, there is a high degree of homology within the 'serine-rich' segment of the cytoplasmic domain, a region similarly conserved between murine IL-2R β and the murine erythropoietin receptor.

■ IL-2 receptor γ subunit protein

The γ chain (p64) of the human IL-2 receptor is encoded by a single gene located on the X chromosome, Xq13 (Noguchi et al. 1993a). The γ subunit is produced as a 369 aa protein consisting of a 22 aa signal sequence, an extracellular

domain of 232 aa, a transmembrane domain of 29 aa, and an 86 aa cytoplasmic domain (GenBank accession number D11086) (Fig. 1) (Takeshita et al. 1992). The mature protein migrates as a 64 000 Da protein on SDS-PAGE due to N-linked glycosylation of the predicted 39 918 Da molecule. The extracellular domain contains the four conserved cysteine residues and the WSXWS motif characteristic of the haemopoietin receptor family. The small cytoplasmic domain contains a segment homologous to the src homology domain (SH2), but no apparent catalytic motifs. The human IL-2R γ chain has also recently been shown to be a component of the human IL-4 and IL-7 receptors, leading to its proposed designation as the common γ chain (γ_c) (Noguchi et al. 1993b; Russell et al. 1993).

The murine γ chain is encoded by a single gene on the mouse X chromosome (Cao et al. 1993) between the *Rsvp* and *Plp* loci. The murine γ chain cDNA sequence (GenBank accession numbers D13565, L20048) predicts a mature protein with 70 per cent homology to the human sequence, including the conserved cysteine residues, WSXWS motif and intracytoplasmic SH2 domain, although the SH2 domain is less conserved in the murine sequence (Cao et al.

1993; Kumaki *et al.* 1993). Surprisingly, murine $\beta\gamma$ complexes reconstituted by cDNA transfection, in contrast to their human counterparts, fail to bind IL-2 (Kumaki *et al.* 1993). Although this finding must be confirmed in other systems, it suggests a possible requirement for the α subunit or other proteins to form a functional murine receptor complex. The murine IL-2R γ chain is also utilized as a common receptor subunit and has been identified as a component of the murine IL-4R complex (Kondo *et al.* 1993).

■ Receptor binding of IL-2

The α, β, and γ subunits of the IL-2 receptor associate in various combinations to yield receptors with a variety of ligand binding characteristics (Table 1) (Ringheim *et al.* 1991; Wang and Smith 1987; Lowenthal and Greene 1987). By itself, the α chain binds IL-2 with a low affinity characterized by rapid association and dissociation. Binding of IL-2 to the α chain does not lead to signal transduction or ligand internalization. The β chain binds IL-2 with an even lower affinity characterized by relatively slow association. Receptors consisting of α and β chains, on the other hand, bind with high affinity and kinetics characterized by a fast association rate and an intermediate dissociation rate. The γ chain by itself does not appear to bind IL-2 with any appreciable affinity. On cells containing both β and γ chains, the binding of IL-2 to the β chain appears to induce a conformational (in the chain or IL-2, or both) or chemical (such as phosphorylation) modification which leads to association of the γ chain with the complex. Once this trimolecular complex is formed, the β chain can be artificially removed, revealing a stable association of the γ chain with IL-2 (Voss *et al.* 1993). Receptors consisting of β and γ chains bind IL-2 with an intermediate affinity characterized by the slow on-rate of the β chain, but an even slower off-rate, which appears to be based on the induced association of IL-2 with the γ chain. Both $\beta\gamma$ and $\alpha\beta\gamma$ receptors internalize IL-2 with a $t_{1/2}$ of $\sim$ 10 min. The latter receptor complexes

bind IL-2 with high affinity and possess fast association kinetics characteristic of the α subunit and very slow dissociation kinetics similar to $\beta\gamma$ receptors. Such high-affinity receptors have been reconstituted on cells transfected with recombinant versions of the three chains (Takeshita *et al.* 1992). In addition to interactions of IL-2 with each of the subunits, direct interactions between the three subunits appear to play an important role in determining the overall affinity for IL-2 (Grant *et al.* 1992; Arima *et al.* 1991). Ligand binding and signalling by the receptor complex may be further influenced by non-covalent associations with such other cell surface proteins as HLA class I antigen or ICAM-1 (Waldmann 1991).

Putative binding sites on the IL-2 molecule have been assigned for each of the three receptor subunits based on antibody and mutational mapping studies (Kuo and Robb 1986; Robb *et al.* 1987; Ju *et al.* 1987; Collins *et al.* 1988;

Figure 2. Model of IL-2 interactions with α, β, and γ receptor subunits. Predicted sites of interaction between IL-2 and the respective receptor subunits are based on mutational and epitope mapping studies (Grant *et al.* 1992; Arima *et al.* 1991; Waldmann 1991; Kuo and Robb 1986; Robb *et al.* 1987; Ju *et al.* 1987; Collins *et al.* 1988; Sauvé *et al.* 1991; Imler and Zurawski 1992; Bazan and McKay 1992) and the demonstration of a direct interaction between IL-2 and the γ chain (Voss *et al.* 1993).

Table 1. Receptor binding of IL-2*

Receptor	K_D	$t_{1/2}$ association	$t_{1/2}$ disassociation
α	3.2 nM	5 s	35 s
β	70 nM	20–30 m	1.6 m
$\alpha\beta$	5.0 pM	5 s	18.5 m
$\beta\gamma$	1.2 nM	30–40 m	255 m
$\alpha\beta\gamma$	2.5 pM	20–40 s	255 m

*Values (seconds, s; minutes, m) were determined at 4 °C using human lymphoblastoid cell lines (HUT 102, MT-1, and YT), or CHO cells transfected with recombinant human receptor chains. K_D and $t_{1/2}$ values are based on previously published data (Ringheim *et al.* 1991; Wang and Smith 1987; Lowenthal and Greene 1987), as well as unpublished observations.

Sauvé et al. 1991; Imler and Zurawski 1992). IL-2, like other ligands of the haemopoietin receptor family, consists of four left-handed α-helices connnected by loops (Bazan and McKay 1992). Interactions between IL-2 and the β receptor subunit were localized to the amino terminal helix A of IL-2 while interactions with the α subunit were localized to a loop structure between helices A and B/B′ (Fig. 2). The interaction site for the γ subunit was tentatively assigned to C-terminal helix D.

■ Cellular distribution and function of IL-2 receptors

IL-2 receptor expression has been demonstrated for most haemopoietic cells (Smith 1989; Waldmann 1989), including B, T, and NK lymphocytes, monocytes and neutrophils, and certain non-haemopoietic cells (Plaisance et al. 1992; Weidmann et al. 1992). Activated lymphocytes transiently secrete IL-2 and proliferate in response to IL-2 through high-affinity receptors. Resting T lymphocytes, however, neither produce IL-2 nor express high-affinity receptors. The absence of IL-2 production and high-affinity IL-2 receptor expression on the latter cells is due to tight control of IL-2 and IL-2R α expression by related cis-acting DNA regulatory sequences located upstream of both genes. NF-κB is the most important element controlling α chain expression (Greene et al. 1989). The sustained high level of α expression characteristic of HTLV-1 induced adult T-cell leukaemias is caused by induction of κB-specific transcription factors by the HTLV-1-encoded Tax transactivator protein.

In contrast to the regulated expression of the IL-2R α subunit, both β and γ subunits are expressed constitutively on resting lymphocytes, due in part to the fact that the promoter elements of the β and γ chain genes lack the cis-acting regulatory motifs which control the inducibility of the IL-2 and IL-2R α genes (Gnarra et al. 1990; Shibuya et al. 1990; Noguchi et al. 1993a). The upstream regulatory sequences of both β and γ chain genes are also notable for their lack of TATA- or CAAT-box motifs upstream of the sites for transcription initiation (Gnarra et al. 1990; Shibuya et al. 1990; Noguchi et al. 1993a). Within the T lymphocyte population, β chain is expressed primarily on the CD8-positive subset. Higher expression levels occur on NK cells and monocytes (Ohashi et al. 1989). Following lymphocyte activation, moderate increases in expression of both β and γ chain genes occur through as yet uncharacterized regulatory interactions (Hatakeyama et al. 1989a; Takeshita et al. 1992).

Resting NK cells appear to be the only lymphocyte subset that relies primarily on interactions with intermediate-affinity IL-2R for proliferative and cytolytic responses, although a very small percentage of resting NK cells express high-affinity IL-2R (Robertson and Ritz 1990). A functional distinction between these two populations of NK cells, however, remains controversial. In vivo expansion of NK cells by continuous infusion of IL-2, while clearly activating NK cells and LAK activity in vivo, does not cause

upregulation of high-affinity IL-2R and these cells respond through intermediate-affinity receptors (Voss et al. 1990).

Monocytes and neutrophils have also been found to express primarily intermediate-affinity IL-2 receptors, and functional responses to IL-2 have been documented following exposure of isolated populations of these cells to IL-2 (Espinoza-Delgado et al. 1990; Djeu et al. 1992). B cells, in contrast, appear to respond to IL-2 through high-affinity receptors with the regulation of this response being controlled in part by IL-4 and cell-surface IgM interactions (Nakanishi et al. 1992).

The importance of IL-2 in a variety of biological functions can be inferred indirectly from the expression of IL-2 receptors on cells of numerous haemopoietic lineages. None the less, the full impact of the importance of the IL-2/IL-2R pathway was not appreciated until the identification of the IL-2R γ chain as the site of molecular defects in X-linked severe combined immunodeficiency (X-linked SCID) (Noguchi et al. 1993c; Puck et al. 1993). The finding that 'IL-2 knockout mice', in which the IL-2 gene had been experimentally disrupted, had normal thymus development and normal thymocyte and peripheral T-cell subsets led to the conclusion that IL-2 did not play a critical role in immunologic development (Schorle et al. 1991; Kündig et al. 1993). In fact, in certain animals the elimination of IL-2 resulted not in immune deficiency, but rather the development of autoimmune ulcerative colitis (Sadlack et al. 1993). The apparent discrepancy between these separate observations led to speculation that the IL-2R γ chain may play a broader role beyond its function in the IL-2R complex (Voss et al. 1994; Taniguchi and Minami 1993). Consistent with these speculations, the IL-2R γ chain has now been identified as a component of IL-4, IL-7, and possibly IL-13 receptor complexes (Kondo et al. 1993; Russell et al. 1993; Noguchi et al. 1993b; Zurawski et al. 1993). A murine equivalent to X-linked SCID has not yet been described. The X-linked immunodeficient mouse strain, xid, corresponds to the human disorder, X-linked agammaglobulinaemia, due to a defect in the BTK kinase gene rather than the IL-2R γ gene.

■ IL-2 receptor signalling

Early events associated with IL-2 signal transduction include protein phosphorylation, activation of a Na^+/H^+ antiport and an increase in PI_3-kinase activity (Minami et al. 1993). The cytoplasmic domains of the IL-2R α, β, and γ subunits do not contain a recognizable catalytic domain for tyrosine kinase activity, yet IL-2 induces tyrosine phosporylation of a variety of cellular proteins including the β and γ subunits themselves. Src-family kinases, associated both physically and functionally with the IL-2R (Minami et al. 1993), have been implicated together with other putative tyrosine kinase molecules (Farrar et al. 1990). In addition, distinct serine/threonine kinases such as the p70 S6 kinase, are critical in regulating the cellular response to IL-2 (Kuo et al. 1992).

The available data suggest that IL-2 receptor-induced signalling occurs through at least two non-overlapping pathways (Minami *et al.* 1993; Taniguchi and Minami 1993). Src-family kinase activation, mediated through an association with the acidic domain of the β chain and in addition requiring the adjacent 'serine-rich' segment, is associated with p21^{ras} activation and leads to an induction of the c-jun and c-fos transcription factors. Induction of c-myc expression, which is associated with cell-cycle progression, follows a separate pathway requiring the 'serine-rich' cytoplasmic region and as yet undetermined signalling elements. The distinction between the two pathways is emphasized by the observation that deletion of the acidic region did not interfere with the ability of the IL-2 receptor to stimulate cellular proliferation (Hatakeyama *et al.* 1989*b*).

The importance of the γ chain in the activation of these two signalling pathways is demonstrated by the fact that IL-2 βγ receptors containing γ chain mutants deficient in the C-terminal 30 amino acids (but containing the SH2 homology domain) were able to induce tyrosine kinase activity and c-myc activation but not expression of c-fos and c-jun (Asao *et al.* 1993). Deletion of a larger cytoplasmic segment from the γ chain (the C-terminal 68 aa) further eliminated the ability of receptor to induce tyrosine phosphorylation and c-myc expression, yet IL-2 binding and internalization were preserved.

It is not clear whether the domains in the β and γ subunits important for mediating IL-2 receptor-induced signalling directly interact with separate kinase molecules, whether they interact together with the same kinase molecules, or whether the β and γ subunits interact first with each other and then with associated signalling molecules. Receptor hetero- and homodimerization have been documented for other members of the haematopoietin receptor family as the initial event leading to intracellular signalling (Stahl and Yancopoulos 1993). These related receptors function through cytokine-specific α subunits and common β subunits (β$_c$) and it is likely that the IL-2R γ chain (γ$_c$) corresponds to the β$_c$ subunits of these other related receptor complexes. Whether the γ$_c$ subunit will similarly be involved in anchoring receptor homo- and heterodimerization remains to be determined. None the less, the data predict that ligand-facilitated subunit association within the IL-2R complex will be important in signal initiation and that both β and γ chains will contribute directly in this process.

Acknowledgements

Dr Paul M. Sondel is thanked for his continued encouragement and support of the work performed in his laboratory by S. V. Nancy Farner is thanked for helpful discussions and for careful reviewing of the manuscript. S. V. is supported by a Life and Health Insurance Medical Research Fund/Lutheran Brotherhood MD., PhD. training fellowship, the University of Wisconsin MD., PhD. Integrated Degree Program, American Cancer Society Grant CH237, and NIH grants CA-32685, CA-53441, and RR-03186.

References

Arima, N., Kamio, M., Okuma, M., Ju, G., and Uchiyama, T. (1991). The IL-2 receptor alpha-chain alters the binding of IL-2 to the beta-chain. *J. Immunol.*, **147**, 3396–401.

Asao, H., Takeshita, T., Ishii, N., Kumaki, S., Nakamura, M., and Sugamura, K. (1993). Reconstitution of functional interleukin 2 receptor complexes on fibroblastoid cells: involvement of the cytoplasmic domain of the gamma chain in two distinct signaling pathways. *Proc. Natl. Acad. Sci. (USA)*, **90**, 4127–31.

Bazan, J.F. (1990). Structural design and molecular evolution of a cytokine receptor superfamily. *Proc. Natl. Acad. Sci. (USA)*, **87**, 6934–8.

Bazan, J.F., and McKay, D.B. (1992). Unraveling the structure of IL-2. *Science*, **257**, 410–13.

Cao, X., Kozak, C.A., Liu, Y.-J., Noguchi, M., O'Connel, E., and Leonard, W.J. (1993). Characterization of cDNAs encoding the murine interleukin 2 receptor (IL-2R) γ chain: Chromosomal mapping and tissue specificity of IL-2R γ chain expression. *Proc. Natl. Acad. Sci. (USA)*, **90**, 8464–8.

Collins, L., Tsien, W.H., Seals, C., Hakimi, J., Weber, D., Bailon, P., Hoskings, J., Greene, W.C., Toome, V., and Ju, G. (1988). Identification of specific residues of human interleukin 2 that affect binding to the 70-kDa subunit (p70) of the interleukin 2 receptor. *Proc. Natl. Acad. Sci. (USA)*, **85**, 7709–13.

de Vos, A.M., Ultsch, M., and Kossiakoff, A.A. (1992). Human growth hormone and extracellular domain of its receptor: crystal structure of the complex. *Science*, **255**, 306–12.

Djeu, J.Y., Liu, J.H., Wei, S., Rui, H., Pearson, C.A., Leonard, W.J., and Blanchard, D.K. (1992). Function associated with IL-2 receptor-beta on human neutrophils. Mechanism of activation of antifungal activity against *Candida albicans* by IL-2. *J. Immunol.*, **150**, 960–70.

Espinoza-Delgado, I., Ortaldo, J.R., Winkler-Pickett, R., Sugamura, K., Varesio, L., and Longo, D.L. (1990). Expression and role of p75 interleukin 2 receptor on human monocytes. *J. Exp. Med.*, **171**, 1821–6.

Farrar, W.L., Garcia, G., Evans, G., Michiel, D., and Linnekin, D. (1990). Cytokine regulation of protein phosphorylation. *Cytokine*, **2**, 77–91.

Gnarra, J.R., Otani, H., Wang, M.G., McBride, O.W., Sharon, M., and Leonard, W.J. (1990). Human interleukin 2 receptor beta-chain gene: chromosomal localization and identification of 5' regulatory sequences. *Proc. Natl. Acad. Sci. (USA)*, **87**, 3440–4.

Grant, A.J., Roessler, E., Ju, G., Tsudo, M., Sugamura, K., and Waldmann, T.A. (1992). The interleukin 2 receptor (IL-2R): the IL-2R alpha subunit alters the function of the IL-2R beta subunit to enhance IL-2 binding and signaling by mechanisms that do not require binding of IL-2 to IL-2R alpha subunit. *Proc. Natl. Acad. Sci. (USA)*, **89**, 2165–9.

Greene, W.C., Bohnlein, E., and Ballard, D.W. (1989). HIV-1, HTLV-1 and normal T-cell growth: transcriptional strategies and surprises. *Immunol. Today*, **10**, 272–8.

Hatakeyama, M., Tsudo, M., Minamoto, S., Kono, T., Doi, T., Miyata, T., Miyasaka, M., and Taniguchi, T. (1989*a*). Interleukin-2 receptor beta chain gene: generation of three receptor forms by cloned human alpha and beta chain cDNAs. *Science*, **244**, 551–6.

Hatakeyama, M., Mori, H., Doi, T., and Taniguchi, T. (1989*b*). A restricted cytoplasmic region of IL-2 receptor beta chain is essential for growth signal transduction but not for ligand binding and internalization. *Cell*, **59**, 837–45.

Imler, J.L., and Zurawski, G. (1992). Receptor binding and internalization of mouse interleukin-2 derivatives that are partial agonists. *J. Biol. Chem.*, **267**, 13185–90.

Jacques, Y., Le Mauff, B., Godard, A., Naulet, J., Concino, M., Marsh,

H., Ip, S., and Soulillou, J.-P. (1990). Biochemical study of a recombinant soluble interleukin-2 receptor. Evidence for a homodimeric structure. *J. Biol. Chem.*, **265**, 20252–8.

Ju, G., Collins, L., Kaffka, K.L., Tsien, W.H., Chizzonite, R., Crow, R., Bhatt, R., and Kilian, P.L. (1987). Structure–function analysis of human interleukin-2. Identification of amino acid residues required for biological activity. *J. Biol. Chem.*, **262**, 5723–31.

Kondo, M., Takeshita, T., Ishii, N., Nakamura, M., Watanabe, S., Arai, K., and Sugamura, K. (1993). Sharing of the interleukin-2 (IL-2) receptor γ chain between receptors for IL-2 and IL-4. *Science*, **262**, 1874–7.

Kono, T., Doi, T., Yamada, G., Hatakeyama, M., Minamoto, S., Tsudo, M., Miyasaka, M., Miyata, T., and Taniguchi, T. (1990). Murine interleukin 2 receptor beta chain: dysregulated gene expression in lymphoma line EL-4 caused by a promoter insertion. *Proc. Natl. Acad. Sci. (USA)*, **87**, 1806–10.

Kumaki, S., Kondo, M., Takeshita, T., Asao, H., Nakamura, M., and Sugamura, K. (1993). Cloning of the mouse interleukin 2 receptor gamma chain: demonstration of functional differences between the mouse and human receptors. *Biochem. Biophys. Res. Comm.*, **193**, 356–63.

Kündig, T.M., Schorle, H., Bachmann, M.F., Hengartner, H., Zinkernagel, R.M., and Horak, I. (1993). Immune responses in interleukin-2-deficient mice. *Science*, **262**, 1059–61.

Kuo, L.-M., and Robb, R.J. (1986). Structure-function relationships for the IL 2-receptor system. I. Localization of a receptor binding site on IL 2. *J. Immunol.*, **137**, 1538–43.

Kuo, C.J., Chung, J., Fiorentino, D.F., Flanagan, W.M., Blenis, J., and Crabtree, G.R. (1992). Rapamycin selectively inhibits interleukin-2 activation of p70 S6 kinase. *Nature*, **358**, 70–3.

Leonard, W.J., Depper, J.M., Crabtree, G.R., Rudikoff, S., Pumphrey, J., Robb, R.J., Krönke, M., Svetlik, P.B., Peffer, N.J., Waldmann, T.A., and Greene, W.C. (1984). Molecular cloning and expression of cDNAs for the human interleukin-2 receptor. *Nature*, **311**, 626–31.

Leonard, W.J., Donlon, T.A., Lebo, R.V., and Greene, W.C. (1985a). Localization of the gene encoding the human interleukin-2 receptor on chromosome 10. *Science*, **228**, 1547–9.

Leonard, W.J., Depper, J.M., Kanehisa, M., Krönke, M., Peffer, N.J., Svetlik, P.B., Sullivan, M., and Greene, W.C. (1985b). Structure of the human interleukin-2 receptor gene. *Science*, **230**, 633–9.

Leonard, W.J., Depper, J.M., Krönke, M., Robb, R.J., Waldmann, T.A., and Greene, W.C. (1985c). The human receptor for T-cell growth factor. Evidence for variable post-translational processing, phosphorylation, sulfation, and the ability of precursor forms of the receptor to bind T-cell growth factor. *J. Biol. Chem.*, **260**, 1872–80.

Loughnan, M.S., Sanderson, C.J., and Nossal, G.J.V. (1988). Soluble interleukin 2 receptors are released from the cell surface of normal murine B lymphocytes stimulated with interleukin 5. *Proc. Natl. Acad. Sci. (USA)*, **85**, 3115–9.

Lowenthal, J.W., and Greene, W.C. (1987). Contrasting interleukin 2 binding properties of the alpha (p55) and beta (p70) protein subunits of the human high-affinity interleukin 2 receptor. *J. Exp. Med.*, **166**, 1156–61.

Malek, T.R., Vincek, V., Gatalica, B., and Bucan, M. (1993). The IL-2 receptor beta chain gene (IL-2rb) is closely linked to the Pdgfb locus on mouse chromosome 15. *Immunogenet.*, **38**, 154–6.

Miedel, M.C., Hulmes, J.D., Weber, D.V., Bailon, P., and Pan, Y.-C. (1988). Structural analysis of recombinant soluble human interleukin-2 receptor. Primary structure, assignment of disulphide bonds and core IL-2 binding structure. *Biochem. Biophys. Res. Commun.*, **154**, 372–9.

Miller, J., Malek, T.R., Leonard, W.J., Greene, W.C., Shevach, E.M., and Germain, R.N. (1985). Nucleotide sequence and expression of a mouse interleukin 2 receptor cDNA. *J. Immunol.*, **134**, 4212–7.

Minami, Y., Kono, T., Miyazaki, T., and Taniguchi, T. (1993). The IL-2 receptor complex: its structure, function, and target genes. *Annu. Rev. Immunol.*, **11**, 245–68.

Nakanishi, K., Hirose, S., Yoshimoto, T., Ishizashi, H., Hiroishi, K., Tanaka, T., Kono, T., Miyasaka, M., Taniguchi, T., and Higashino, K. (1992). Role and regulation of interleukin (IL)-2 receptor alpha and beta chains in IL-2-driven B-cell growth. *Proc. Natl. Acad. Sci. (USA)*, **89**, 3551–5.

Neeper, M.P., Kuo, L.-M., Kiefer, M.C., and Robb, R.J. (1987). Structure function relationships for the IL 2 receptor system. III. Tac protein missing amino acids 102 to 173 (exon 4) is unable to bind IL 2: detection of spliced protein after L cell transfection. *J. Immunol.*, **138**, 3532–8.

Noguchi, M., Adelstein, S., Cao, X., and Leonard, W.J. (1993a) Characterization of the human interleukin-2 receptor gamma chain gene. *J. Biol. Chem.*, **268**, 13601–8.

Noguchi, M., Nakamura, Y., Russell, S.M., Ziegler, S.F., Tsang, M., Cao, X., and Leonard, W.J. (1993b). Interleukin-2 receptor γ chain: A functional component of the interleukin-7 receptor. *Science*, **262**, 1877–80.

Noguchi, M., Yi, H., Rosenblatt, H.M., Filipovich, A.H., Adelstein, S., Modi, W.S., McBride, O.W., and Leonard, W.J. (1993c). Interleukin-2 receptor gamma chain mutation results in X-linked severe combined immunodeficiency in humans. *Cell*, **73**, 147–57.

Ohashi, Y., Takeshita, T., Nagata, K., Mori, S., and Sugamura, K. (1989). Differential expression of the IL-2 receptor subunits, p55 and p75 on various populations of primary peripheral blood mononuclear cells. *J. Immunol.*, **143**, 3548–55.

Plaisance, S., Rubinstein, E., Alileche, A., Sahraoui, Y., Krief, P., Augery-Bourget, Y., Jasmin, C., Suarez, H., and Azzarone, B. (1992). Expression of the interleukin-2 receptor on human fibroblasts and its biological significance. *Int. Immunol.*, **4**, 739–46.

Puck, J.M., Deschênes, S.M., Porter, J.C., Dutra, A.S., Brown, C.J., Willard, H.F., and Henthorn, P.S. (1993). The interleukin-2 receptor γ chain maps to Xq13.1 and is mutated in X-linked severe combined immunodeficiency, SCIDX1. *Hum. Molec. Genet.*, **2**, 1099–104.

Ringheim, G.E., Freimark, B.D., and Robb, R.J. (1991). Quantitative characterization of the intrinsic ligand-binding affinity of the interleukin 2 receptor beta chain and its modulation by the alpha chain and a second affinity-modulating element. *Lymphokine Cytokine Res.*, **10**, 219–24.

Robb, R.J., and Kutny, R.M. (1987). Structure-function relationships for the IL 2-receptor system. IV. Analysis of the sequence and ligand-binding properties of soluble Tac protein. *J. Immunol.*, **139**, 855–62.

Robb, R.J., Rusk, C.M., Yodoi, J., and Greene, W.C. (1987). Interleukin 2 binding molecule distinct from the Tac protein: analysis of its role in formation of high-affinity receptors. *Proc. Natl. Acad. Sci. (USA)*, **84**, 2002–6.

Robb, R.J., Rusk, C.M., and Neeper, M.P. (1988). Structure-function relationships for the interleukin 2 receptor: location of ligand and antibody binding sites on the Tac receptor chain by mutational analysis [published erratum appears in *Proc. Natl. Acad. Sci. (USA)*, (1988) **85**, 8226]. *Proc. Natl. Acad. Sci. (USA)*, **85**, 5654–8.

Robertson, M.J., and Ritz, J. (1990). Biology and clinical relevance of human natural killer cells. *Blood*, **76**, 2421–38.

Rubin, L.A., Kurman, C.C., Fritz, M.E., Biddson, W.E., Boutin, B., Yarchoan, R., and Nelson, D.L. (1985). Soluble interleukin 2 receptors are released from activated human lymphoid cells in vitro. *J. Immunol.*, **135**, 3172–7.

Rusk, C.M., Neeper, M.P., Kuo, L.-M., Kutny, R.M., and Robb, R.J. (1988). Structure–function relationships for the IL-2 receptor system. V. Structure–activity analysis of modified and truncated forms of the Tac receptor protein: site-specific mutagenesis of cysteine residues. *J. Immunol.*, **140**, 2249–59.

Russell, S.M., Keegan, A.D., Harada, N., Nakamura, Y., Noguchi, M., Leland, P., Friedmann, M.C., Miyajima, A., Puri, R.K., Paul, W.E., and Leonard, W.J. (1993). Interleukin-2 receptor γ chain: A functional component of the interleukin-4 receptor. *Science*, **262**, 1880–3.

Sadlack, B., Merz, H., Schorle, H., Schimpl, A., Feller, A.C., and Horak, I. (1993). Ulcerative colitis-like disease in mice with a disrupted interleukin-2 gene. *Cell*, **75**, 253–61.

Satoh, T., Minami, Y., Kono, T., Yamada, K., Kawahara, A., Taniguchi, T., and Kaziro, Y. (1992). Interleukin 2-induced activation of Ras requires two domains of interleukin 2 receptor beta subunit, the essential region for growth stimulation and Lck-binding domain. *J. Biol. Chem.*, **267**, 25423–7.

Sauvé, K., Nachman, M., Spence, C., Bailon, P., Campbell, E., Tsien, W.H., Kondas, J.A., Hakimi, J., and Ju, G. (1991). Localization in human interleukin 2 of the binding site to the alpha chain (p55) of the interleukin 2 receptor. *Proc. Natl. Acad. Sci. (USA)*, **88**, 4636–40.

Schorle, H., Holtschke, T., Hunig, T., Schimpl, A., and Horak, I. (1991). Development and function of T cells in mice rendered interleukin-2 deficient by gene targeting. *Nature*, **352**, 621–4.

Shibuya, H., Yoneyama, M., Nakamura, Y., Harada, H., Hatakeyama, M., Minamota, S., Kono, T., Doi, T., White, R., and Taniguchi, T. (1990). The human interleukin-2 receptor beta-chain gene: genomic organization, promoter analysis and chromosomal assignment. *Nucl. Acids Res.*, **18**, 3697–703.

Smith, K.A. (1989). The interleukin 2 receptor. *Annu. Rev. Cell Biol.*, **5**, 397–425.

Stahl, N., and Yancopoulos, G.D. (1993). The alphas, betas, and kinases of cytokine receptor complexes. *Cell*, **74**, 587–90.

Taga, T., and Kishimoto, T. (1992). Cytokine receptors and signal transduction. *FASEB J.*, **6**, 3387–96.

Takeshita, T., Asao, H., Ohtani, K., Ishii, N., Kumaki, S., Tanaka, N., Munakata, H., Nakamura, M., and Sugamura, K. (1992). Cloning of the gamma chain of the human IL-2 receptor. *Science*, **257**, 379–82.

Taniguchi, T., and Minami, Y. (1993). The IL-2/IL-2 receptor system: a current overview. *Cell*, **73**, 5–8.

Voss, S.D., Robb, R.J., Weil-Hillman, G., Hank, J.A., Sugamura, K., Tsudo, M., and Sondel, P.M. (1990). Increased expression of the interleukin 2 (IL-2) receptor beta chain (p70) on CD56+ natural killer cells after *in vivo* IL-2 therapy: p70 expression does not alone predict the level of intermediate affinity IL-2 binding. *J. Exp. Med.*, **172**, 1101–14.

Voss, S.D., Leary, T.P., Sondel, P.M., and Robb, R.J. (1993). Identification of a direct interaction between interleukin 2 and the p64 interleukin 2 receptor gamma chain. *Proc. Natl. Acad. Sci. (USA)*, **90**, 2428–32.

Voss, S.D., Hong, R., and Sondel, P.M. (1994). Severe combined immunodeficiency, interleukin-2 (IL-2), and the IL-2 receptor: Experiments of nature continue to point the way. *Blood*, **83**, 626–35.

Waldmann, T.A. (1989). The multi-subunit interleukin-2 receptor. *Annu. Rev. Biochem.*, **58**, 875–911.

Waldmann, T.A. (1991). The interleukin-2 receptor. *J. Biol. Chem.*, **266**, 2681–4.

Wang, H.-M., and Smith, K.A. (1987). The interleukin 2 receptor. Functional consequences of its bimolecular structure. *J. Exp. Med.*, **166**, 1055–69.

Webb, G.C., Campbell, H.D., Lee, J.S., and Young, I.G. (1990). Mapping the gene for murine T-cell growth factor, Il-2, to bands B-C on chromosome 3 and for the alpha chain of the IL2-receptor, Il-2ra, to bands A2–A3 on chromosome 2. *Cytogenet. Cell Genet.*, **54**, 164–8.

Weidmann, E., Sacchi, M., Plaisance, S., Heo, D.S., Yasumura, S., Lin, W.-C., Johnson, J.T., Herberman, R.B., Azzarone, B., and Whiteside, T.L. (1992). Receptors for interleukin 2 on human squamous cell carcinoma cell lines and tumour *in situ*. *Cancer Res.*, **52**, 5963–70.

Zurawski, S.M., Vega, F., Huyghe, B., and Zurawski, G. (1993). Receptors for interleukin-13 and interleukin-4 are complex and share a novel component that functions in signal transduction. *EMBO J.*, **12**, 2663–70.

Stephan D. Voss:
Department of Human Oncology,
University of Wisconsin Clinical Sciences Center,
600 Highland Avenue,
Madison, WI 53792 USA

Richard J. Robb:
Onco Therapeutics, Inc.,
1002 Eastpark Blvd.,
Cranbury, NJ 08512 USA

Interleukin-3 (IL-3)

IL-3 is a 20–32 kDa glycoprotein that stimulates the proliferation, differentiation, and survival of pluripotential haemopoietic stem cells and haemopoietic progenitor cells and their mature progeny of multiple cell lineages. IL-3 is produced and secreted primarily in response to immunological stimuli by T lymphocytes activated by specific antigens, or mast cells activated by cross-linking of cytophilic antibodies bound to their surface by specific antigens. The potential clinical utility of IL-3 lies in its ability to enhance the recovery of haemopoiesis following cytotoxic cancer therapy or bone-marrow transplantation. IL-3 has an important role in diseases characterized by increases in mast cells and basophils, and IL-3 antagonists may provide a new approach to the treatment of allergy and asthma.

■ Alternative names

Multi-lineage colony-stimulating factor, persisting (P)-cell-stimulating factor, mast-cell growth factor, burst-promoting activity, Thy-1-stimulating activity, Histamine-producing cell stimulating activity, haemopoietic cell growth factor, multilineage haemopoietic growth factor, CFUs-stimulating activity, stem cell-activating factor, haemopoietin-2, synergistic activity, pan specific haemopoietin, and CSF-2α and -2β.

■ IL-3 sequence

The nucleotide sequences of IL-3 cDNAs have GenBank accession numbers M14743 for human-IL-3, and K03233 for murine IL-3. The predicted amino acid sequences are only 29 per cent identical. (Yang *et al.* 1986). There is no cross-reactivity between murine and human IL-3 in bioactivity or receptor binding. The IL-3 gene has a similar overall structure to the GM-CSF gene and the two are located close together in the genome (Yang *et al.* 1988; Barlow *et al.* 1987), reflecting their probable origin from duplication of an ancestral gene.

■ IL-3 protein

IL-3 is a secreted glycoprotein, the peptide core being comprised of 140 amino acids in the mouse and a predicted 133 amino acids in the human. The molecular weight of the polypeptide core is about 14000. Like GM-CSF, IL-2 and IL-4, IL-3 is a monomer with two intramolecular disulphide bonds. The β subunit of the IL-3 receptor is also shared by GM-CSF and IL-5 receptors, suggesting an overall similarity in the three-dimensional structure of these three cytokines. The three-dimensional structure of IL-3 is predicted to resemble that determined by X-ray crystallography for GM-CSF and to be based on an antiparallel four α-helical bundle

with an up-up-down-down arrangement of the helices. IL-3 is a highly stable protein resistant to heat and low pH. Studies with antibodies and mutated analogues of IL-3 suggest the importance for biological activity of residues in the N-terminal, A α-helix, and in helices C and D.

Natural IL-3 is extensively modified by glycosylation and is a heterogeneous mixture of different glycoforms. In the mouse there are three main molecular species with apparent M_r 22 kDa, 28 kDa, and 36 kDa (Ziltener *et al.* 1988). After removal of N-linked carbohydrates, murine IL-3 has an M_r of 18 kDa, the difference in mobility from the polypeptide core (M_r 14 kDa) reflecting O-linked glycosylation. Glycosylation has no detectable effect on the specific biological activity either *in vitro* (Ziltener 1993) or *in vivo* (Ziltener, unpublished data), and the function of the extensive modifications is obscure .

■ IL-3 gene and transcription

The human IL-3 gene is located on chromosome 5q21–q32, very close to the GM-CSF gene (Yang *et al.* 1988). In the mouse the two genes are on chromosome 11 (Barlow *et al.* 1987). The size of the IL-3 gene is about 2.2 kb, and there are five exons. The mRNA is about 1 kb and like that encoding many other cytokines, has an AU-rich sequence in the 3' untranslated region that is responsible for the relative instability of the mRNA.

The IL-3 gene is activated in T lymphocytes following engagement of the T cell antigen receptor (Schrader 1981; Clark and Kamen 1987) and in mast cells following cross-linking of Fc receptors by the binding of multivalent antigens to cytophilic antibodies (Wodnar-Filipowicz *et al.* 1989). Calcium ionophores and phorbol esters can mimic these events, as can other agents, e.g. antibodies or lectins that cross-link the relevant receptor structures.

Signals from activation of the T cell antigen-receptor appear to act on an enhancer shared with the GM-CSF gene that is 3 kb upstream of the GM-CSF gene (Cockerill *et al.* 1993). Cyclosporin A inhibits activation of the IL-3 gene in T cells (Cockerill *et al.* 1993) and mast cells (Wodnar-Filipowicz *et al.* 1989). Cyclosporin A blocks the Ca^{2+}/calcineurin mediated translocation to the nucleus of a cytoplasmic component, termed NFATc, of the transcription factor NFAT. The IL-3/GM-CSF enhancer contains NFAT binding sites (Cockerill et al. 1993).

◼ Biological functions

IL-3 is a potent growth factor and, depending on the nature of the bioassay, has 50 per cent maximal activity at concentrations of 10^{-12}–10^{-13} M (Schrader 1992). IL-3 has the broadest spectrum of activity of all the haemopoietic growth factors, with activity on pluripotential stem cells, the progenitors of all lineages with the exception of lymphocytes, and mature cells of many lineages (Table 1). Like all haemopoietic growth factors, IL-3 not only stimulates the growth of most of its cellular targets, but also enhances survival by blocking apoptosis and regulates many important functions of mature effector cells.

◼ Pathology

IL-3 can be detected in the serum in graft-versus-host disease, where there is massive activation of T lymphocytes (Crapper and Schrader 1986), and in mice that have been rechallenged with a parasite to which they have been sensitized (Abbud-Filho et al. 1983). In the latter case, the IL-3 is probably released by mast cells that have been activated by cross-linking of antibodies bound to the mast cell surface by parasite antigens. IL-3 may play an important role in allergic disease or asthma, promoting the production, increased survival, and enhanced function of the two cell types that mediate much of the pathology: mast cells and eosinophils.

In mice, IL-3 has been shown to have a causal role in a proportion of myeloid leukaemias in which mutations have resulted in constitutive activation of an IL-3 gene (Schrader and Crapper 1983; Leslie and Schrader 1989). The consequent production of IL-3 leads to autonomy by maintaining survival and continuous growth through an autostimulatory mechanism. Dysregulated IL-3 production, however, does not seem to be an important factor in human acute myeloid leukaemia.

◼ References

Abbud-Filho, M., Dy, M., Lebel, B., Luffau, G., and Hamburger, J. (1983). In vitro and in vivo histamine-producing cell-stimulating factor (or IL3) production during Nippostrongylus brasiliensis infection: coincidence with self-cure phenomenon. Eur. J. Immunol., 13, 841–5.

Barlow, D.P., Bucan, M., Lehrach, H., Hogan, B.L.M., and Gough, N.M. (1987). Close genetic and physical linkage between the murine haemopoietic growth factor genes GM-CSF and Multi-CSF (IL3). EMBO J., 6, 617.

Clark, S.C., and Kamen, R. (1987). The human hematopoietic colony-stimulating factors. Science, 236, 1229–37.

Cockerill, P.N., Shannon, M.F., Bert, A.G., Ryan, G.R., and Vadas, M.A. (1993). The granulocyte-macrophage colony-stimulating factor/interleukin 3 locus is regulated by an inducible cyclosporin A-sensitive enhancer. Proc. Soc. Natl. Acad. Sci. (USA), 90, 2466–70.

Crapper, R.M., and Schrader, J.W. (1986). Evidence for the in vivo production and release into the serum of a T-cell lymphokine, persisting-cell stimulating factor (PSF), during graft-versus-host reactions. Immunology, 57, 553–8.

Leslie, K.B., and Schrader, J.W. (1989). Growth factor gene activation and clonal heterogeneity in an autostimulatory myeloid leukaemia. Mol. Cell Biol., 6, 2414–23.

Schrader, J.W. (1981). In in vitro production and cloning of the P cell, a bone marrow-derived null cell that expresses H-2 and Ia-antigens, has mast cell-like granules, and is regulated by a factor released by activated T cells. J. Immunol., 126, 452–8.

Table 1. Biological actions of IL-3

General properties
Enhances survival
Regulates function
Stimulates proliferation (most targets)

Actions on specific targets

Haemopoietic stem cells with long-term reconstituting activity and those that form splenic colonies	Stimulates survival and proliferation, and differentiation in some cases
Progenitors of erythrocytes, megakaryocytes, mast cells, basophils, eosinophils, neutrophils	Stimulates survival and proliferation
Mature mast cells, particularly mast cells associated with mucosal surfaces	Enhances survival and proliferation; regulates expression of certain mast cell proteases
Megakaryocytes	Enhances survival
Macrophages	Stimulates proliferation, increased phagocytic activity, increased level of class II major histocompatibility antigens and LFA-1, increased levels of IL-1, IL-6, TNFα mRNA
Basophils	Enhances release of histamine
Eosinophils	Enhances survival, antibody-dependent cell-mediated cytotoxicity, superoxide production in response to f-met-leu-phe

Schrader, J.W. (1992). Biological effects of myeloid growth factors. *Baillières Clin. Haematol.*, **5**, 509–31.

Schrader, J.W., and Crapper, R.M. (1983). Autogenous production of a haemopoietic growth factor, persisting-cell-stimulating factor, as a mechanism for transformation of bone marrow-derived cells. *Proc. Soc. Natl. Acad. Sci. (USA)*, **80**, 6892–6.

Wodnar-Filipowicz, A., Heusser, C.H., and Moroni, C. (1989). Production of the haemopoietic growth factors GM-CSF and interleukin-3 by mast cells in response to IgE receptor-mediated activation. *Nature*, **339**, 150–2.

Yang, Y.C., Ciarletta, A.B., Temple, P.A., Chung, M.P., Kovacic, S., Witek-Giannotti, J.S., Leary, A.C., Kriz, R., Donahue, R.E., Wong, G.G., and Clark, S.C. (1986). Human IL-3 (multi-CSF): identification by expression cloning of a novel hematopoietic growth factor related to murine IL-3. *Cell*, **47**, 3–10.

Yang, Y.C., Kovacic, S., Kriz, R., Wolf, S., Clark, S.C., Wellems, T.E., Nienhuis, A., and Epstein N. (1988). The human genes for GM-CSF and IL 3 are closely linked in tandem on chromosome 5. *Blood*, **71**, 958–61.

Ziltener, H. (1993). Glycosylation does not affect *in vitro* activity of inter-leukin-3. *Cytokine*, **5**, 291–7.

Ziltener, H.J., Fazekas de St. Groth, B., Leslie, K.B., and Schrader, J.W. (1988). Close genetic and physical linkage between the murine haemopoietic growth factor genes GM-CSF and Multi-CSF (IL3). *J. Biol. Chem.*, **263**, 14511–17.

John W. Schrader:
Biomedical Research Centre,
Vancouver, B.C., Canada

Receptors for Interleukin-3 (IL-3)

The IL-3 receptor (IL-3R) consists of α and β subunits. The α subunit binds IL-3 with low affinity by itself. Two homologous β subunits (β_c and β_{IL-3}) are known in the mouse. Whereas the β_c subunit (originally termed AIC2B) does not bind mouse IL-3 (mIL-3), the β_{IL-3} subunit (originally termed AIC2A) binds mIL-3 with low affinity. The mIL-3R α subunit forms a high affinity receptor with either β_c or β_{IL-3}. The β_c subunit is also involved in the formation of the high affinity GM-CSF and IL-5 receptors. The α and β subunits are 60–70 kDa and 120–140 kDa, respectively. Both are members of the class I cytokine receptor family and the β subunits have two repeats of the common motif of the cytokine receptor family. In contrast to the mouse receptors, only one β subunit (β_c, originally termed KH97) is present. The human β_c does not bind any cytokine by itself but participates in the formation of the high-affinity receptors for IL-3, IL-5, and GM-CSF. Neither α nor β subunit has any intrinsic enzymatic activity such as kinase or phosphatase, but both are required for signalling.

■ The α subunit protein

The murine and human IL-3R α subunit cDNAs (GenBank accession numbers X64534 and M74782) encode a protein of 396 and 378 amino acids including a putative signal peptide of 16 and 18 amino acids, an extracellular domain of 315 and 288 amino acids, a single transmembrane spanning segment of 24 and 20 amino acids, and a cytoplasmic domain of 41 and 52 amino acids, respectively (Fig. 1). The sequence identity between these two α subunits is 47 per cent at the DNA level and 30 per cent at the amino-acid level. The deduced molecular weights of the murine and human IL-3R α subunits are 41 215 Da and 41 254 Da, respectively. The observed molecular weights on SDS-PAGE are 60–70 kDa due to glycosylation at several of the predicted five N-glycosylation sites. The extracellular domains contain common features of the class I cytokine receptor family that include four conserved cysteine residues and the WSXWS motif near the transmembrane domain. There is a short strech of amino acid sequence that is conserved among the α subunits of the IL-3, IL-5, and GM-CSF receptors (Kitamura *et al.* 1991; Hara and Miyajima 1992).

■ The β subunits

Two distinct β subunits are present in the mouse. The β_{IL-3} cDNA (also known as AIC2A; GenBank accession number M29855) encodes a protein of 878 amino acids composed of a signal sequence, an extracellular domain, a transmembrane domain, and a cytoplasmic domain of 22, 417, 26, and 413 amino acid residues, respectively (Itoh *et al.* 1990) (Fig. 2). The β_c subunit (also known as AIC2B; GenBank accession number M34397) consists of 896 amino acids, including a signal sequence, and is 91 per cent identical to β_{IL-3} at the amino acid level (Gorman *et al.* 1990). Mature β subunits are glycoproteins of 120–140 kDa. Both β subunits have two repeats of the common motif of the class I cytokine receptors in their extracellular domain. Both β subunits form high affinity IL-3 receptors with the same IL-3R α subunit (Hara and Miyajima 1992). β_{IL-3} binds IL-3 with low affinity by itself, but β_c alone does not bind any cytokine. In contrast, whereas β_c also forms high affinity receptors for GM-CSF and IL-5 with the respective α subunit, β_{IL-3} does not (Takaki *et al.* 1991; Park *et al.* 1992). Thus, β_{IL-3} is specific to the IL-3 receptor (Fig. 3). In contrast to the mouse receptors, only the β_c subunit is present in the human high-affinity IL-3 receptor (Hayashida *et al.* 1990). The human β_c (originally termed KH97; GenBank accession number M38275) consisting of 897 amino acids, including a signal sequence, is 55 per cent identical to the mouse β_c subunit and is shared with the GM-CSF and IL-5 receptors (Hayashida *et al.* 1990; Tavernier *et al.* 1991) (Fig. 3).

■ Binding properties

The mIL-3R α subunit binds IL-3 with low affinity ($K_D = 40$ nM) by itself (Hara and Miyajima 1992) and the murine β_{IL-3} also binds IL-3 with low affinity ($K_D = 10$ nM) (Itoh *et al.* 1990). In contrast, the murine β_c alone does not bind IL-3 at a detectable level (Gorman *et al.* 1990). However, there is no significant difference in the binding affinity between the two high affinity receptors formed with either β_{IL-3} or β_c ($K_D = 300$ pM) (Hara and Miyajima 1992). The human IL-3R α subunit binds IL-3 with very low affinity ($K_D \sim 100$ nM)

(Kitamura *et al.* 1991) and the human β_c does not bind IL-3 (Hayashida *et al.* 1990). A combination of these two subunits confers high-affinity IL-3 binding ($K_D = 140$ pM). While the murine and human β_c subunits do not bind IL-3, they are cross-linked with IL-3 when they form a high-affinity receptor.

■ Genes and expression

The human IL-3R α subunit gene is linked to the GM-CSF receptor α subunit gene on the pseudoautosomal region of the X and Y chromosomes (Milatovich *et al.* 1993). In contrast, the murine IL-3R α and GM-CSF receptor α subunit genes are on chromosomes 14 and 19, respectively (Miyajima *et al.* 1994; Disteche *et al.* 1992). The two murine β subunit genes are tightly linked on chromosome 15 (Gorman *et al.* 1992) and the human β_c gene is on chromosome 22q12–13 (Shen *et al.* 1992). Exon–intron structures of the

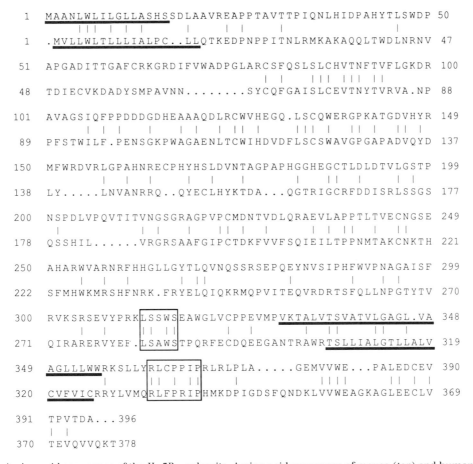

Figure 1. Amino acid sequences of the IL-3R α subunits. Amino acid sequences of mouse (*top*) and human (*bottom*) IL-3R α subunits are shown. Putative signal sequences and the transmembrane domains are indicated by underlines. The WS motif and the conserved sequence in the cytoplasmic domains of the α subunits are shown by boxes.

Signal Peptide

```
hβc      1  MVLAQGLLSMALLALCWERSLAGAEETIPLQTLRCYNDYTSHITCRWADTQDAQRLVNVTLIRRVNEDLLEPVSCDLSDDMPWSACPHP.RCVPRRCVIPCQSFVV
mβIL3    1  MDQQ*A*TW**CY***V***GHEVTEE****K**E****NR*I*S***E**E***G*I*M**LYHQLDKI QS***E**EKLM**E**SSH********YTR*SN
mβc      1  MDQQ*A*TW**CY***V***GHGVTE*****V**K**Q*******N**I*S****E***G*I*M* YHQL*KK Q***E**EKLM**E**SSH********YTR*SI

hβc    106  TDVDYFSFQPDRPLGTRLTVTLTQHVQPEPRDLQISTDQDHFLLTWSVALGSPQSHWLSPGDLEFEVVYKRLQDSWEDAAILLSNTSQATLGPEHLMPSSTYVARVRTRLAPGSRLSGR
mβIL3  110  G*N**Y****D**IQ*M*P*A*********P*K*IH**PSG****E***s**DS*VS**SK*I***A*********SS*HTSNF*VN*E*KLFL*N*I*A*****N**SA*S****
mβc    109  *NE**Y**R**SD**IQ*M*P*A*N****I*KNVS***SSE*R***E***S**DA*VS**SK*I****A*********YS*HTSKF*VNFE*KLFL*N*I*AP*****Y***S****

hβc    226  PSKWSPEVCWDSQPGDEAQPQNLECFFDGAAVLSCSWEVRKEVASSVSFGLFYKPSPDAGEEECSPVLREGLG SLHTRHHCQIPVPDPATHGQYIVSVQPRRAEKHIKSSVNIQMAPPS
mβIL3  230  *R***********H*********Q****IQS*H****WTQTTG*****R**A*P**K****VK*PQ A*VY*YR*SL***E*SA*S**T***KHLEQG*F*M*YYH**E**I
mβc    229  **AH********K*********H****IQS*H****WTQTTG*****R**V*P**K****VK*PP*A*VY**Y**SL**S*E*SA*S**T***KHLEQG*F*M*YNH**E**T
```

Trans-Membrane

```
hβc    345  LNVTKDGDSYSLRWETMKMRYEHIDHTFEIQYRKDTATWKDSKTETLQNAHSMALPALEPSTRYWARVRVRTSRTGYNGIWSEWSEARSWDTESVLPMWVLALIVLFLTTAVLLALRFCG
mβIL3  349  *Q**NR*****H**Q*IP KY****Q**K*KSES*****N*GRVN**D*Q***D*S*C****KPI SD*D****NEYT*T*DW*M*TLWIV**LV*IFTL**I**H*GF
mβc    349  **L**NR****H**Q**SF*E***Q**K*KSDS*E****N*DR****D*SQ***D*S*C****KPI SN*D**K**EYT*K*DW*M*TLWIV**LV*ILTL**I***GC

hβc    465  IYGYRLRRKWEEKIPNPSKSHLFQNGSAELWPPGSMSAFTSGSPPHQGPWGSRFPELEGVFPVGFGDSEVSPLTIEDPKHVCDPPSGPDTTPAASDLPTEQPPSPQPGPPAASHTPEKQA
mβIL3  467  V****TY****K******L******D*GKG******L**D*****A*ATKN*AL**QSRLLA*QQ*SYEHLE*NN*******R******NIIR*****SES**L*NV*VEG*IP*SR*R**L
mβc    468  VSV**TY****K******L******D*GKG******L**D*****A*ATKN*AL**QSRLLA*QQ*ESYAHLE*NN*********NIIRV*****SES**L*NV*VEG*TPN R*R**L

hβc    585  SSFDFNGPYLGPHSRSLPDILGQPEPPQEGGSQKSPPPGSLEYLCLPAGGVQLVPLAQAMGPGQAVEVERRPSQGAAGSPSLESGGGPAPPALGPRVGGQDKDSPVAIPMSSGDTED
mβIL3  587  P*********Q*H*****LP**LGS**V***L*PAL***********M**P*******S*V**Q***MD*QCGS*LETT****V*PKEN *PVELS*EK*EAR*N*MTL*I***GP*G
mβc    587  P*********Q*H*****LPD*LGS**V***L*PAL***********M**AP*****S*V**Q***MD*QCGS*LETS****V*PKEN *PVELSMEE*EAR*N**TL*I***GP*G

hβc    705  PGVASGYVSSADLVFTPNSGASSVSLVPSLGLPSDQTPSLCPGLASGPPGAPGPVKSGFEGYVELPPIEGRSPRSPRNNPVPPEAKSPVLNPGERPADVSPTSPQPEGLLVLQQVGDYCF
mβIL3  704  SMM**D**TPG*P*L*LPT*PL*T**G********A*S****LK*PRV*S*S*ALGPP***D*****SVSQAAT**PGH*A*V*S**TVI***PREE*G*A**H********R*******
mβc    704  SMM**D**TPG*P*L*LPT*PL*T**G********A*S****LK*PRV*S*S*ALGPP***D*****SVSQAAK**PGH*A*V*S**TVI***PREE*G*A**H********R*******

hβc    825  LPGLGPGPLSLRSKPSSPGPGPEIKNLDQAFQVKKPPGQAVPQVPVIQLFKALKQODYLSLPPWEVNKPGEVC  897
mβIL3  824  *******S*PH***P**SLCS*TED***DLS***F*Y*PL**A*A**F***S**Y  878
mβc    824  *******S*PH***P**SLCS*TED*V*DLS***F*Y*PM**A*A**F***S**H*********DNSQS**K**  896
```

Figure 2. Amino acid sequences of the β subunits. Amino-acid sequences of the human β_c, mouse β_c, mouse β_{IL-3}, and mouse β_c subunits are shown. The identical amino acids are shown by asterisks and the WS motifs are shown by boxes.

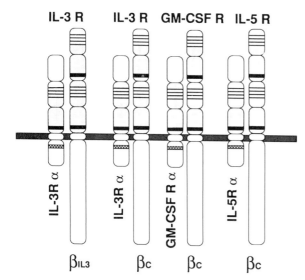

IL-3 R	IL-3 R	GM-CSF R	IL-5 R
IL-3R α	IL-3R α	GM-CSF R α	IL-5R α
β_{IL3}	β_c	β_c	β_c

Figure 3. Subunit structures of the high-affinity receptors. The high-affinity receptors for IL-3, GM-CSF, and IL-5 are shown. The domains with the cytokine receptor motif are shaded. The conserved cysteine residues and the WS motifs are indicated by the lines and bars. The conserved sequence in the cytoplasmic domains of the a subunits is shown by a hatched bar.

two murine β subunit genes are almost identical and the 5' flanking sequences are more than 90 per cent identical, suggesting that the two β genes arose by duplication probably after the divergence between human and mouse (Gorman *et al.* 1992).

Monoclonal antibodies against the receptor subunits as well as cDNA probes were used to examine the expression pattern of the receptor subunits. The two murine β subunits are coexpressed in various haemopoietic cells (Gorman *et al.* 1992). Both IL-3 R a and β_c are expressed on various myeloid cell lineages which express the CD13[+], CD14[+], CD15[+], or CD33[+] cell surface antigens, as well as early progenitors which express CD34[+]. Neither a nor β_c subunit is expressed in CD3[+] T cells, but a proportion of CD19[+] B cells express both subunits (Sato *et al.* 1993a). The expression of β_c is upregulated by IL-1, TNF-a, or IFN-γ in CD34+ cells (Sato *et al.* 1993a; Watanabe *et al.* 1992; Hallek *et al.* 1992).

■ Signal transduction

Heterodimerization of the a and β subunits is an essential step for receptor activation (Kitamura and Miyajima 1992). Cytoplasmic domains of both subunits are required for signal transduction. Although neither subunit has an intrinsic kinase activity, receptor activation is associated with rapid tyrosine phosphorylation of several proteins including the β subunit itself. As a consequence of the shared β

subunit, IL-3 and GM-CSF induce similar signals and their tyrosine phosphorylation patterns are almost identical (Isfort and Ihle 1990; Kanakura *et al.* 1990). Deletion analysis of the β_c subunit defined two cytoplasmic regions that are responsible for different signalling. The membrane proximal region of about 60 amino acid residues is responsible for induction of c-*myc* and *pim*-1, whereas the distal region is responsible for activation of ras, raf-1, MAP kinase, and S6 kinase, as well as induction of c-*fos* and c-*jun*. The distal region is also responsible for the major tyrosine phosphorylation and PI3 kinase activation (Sakamaki *et al.* 1992; Sato *et al.* 1993b). It is reported that several non-receptor tyrosine kinases such as lyn, fyn and jak2 are activated by IL-3 (Torigoe *et al.* 1992; Kobayashi *et al.* 1993; Silvennoinen *et al.* 1993).

■ References

Disteche, C.M., Brannan, C.I., Larsen, A., Adler, D.A., Schorderet, D.F., Gearing, D., Copeland, N.G., Jenkins, N.A., and Park, L.S. (1992). The human pseudoautosomal GM-CSF receptor alpha subunit gene is autosomal in mouse. *Nature Genetics*, **1**, 333–6.

Gorman, D.M., Itoh, N., Kitamura, T., Schreurs, J., Yonehara, S., Yahara, I., Arai, K., and Miyajima, A. (1990). Cloning and expression of a gene encoding an interleukin 3 receptor-like protein: identification of another member of the cytokine receptor gene family. *Proc. Natl. Acad. Sci. (USA)*, **87**, 5459–63.

Gorman, D.M., Itoh, N., Jenkins, N.A., Gilbert, D.J., Copeland, N.G., and Miyajima, A. (1992). Chromosomal localization and organization of the murine genes encoding the beta subunits (AIC2A and AIC2B) of the interleukin 3, granulocyte/macrophage colony-stimulating factor, and interleukin 5 receptors. *J. Biol. Chem.*, **267**, 15842–8.

Hallek, M., Lepisto, E.M., Slattery, K.E., Griffin, J.D., and Ernst, T.J. (1992). Interferon-gamma increases the expression of the gene encoding the beta subunit of the granulocyte-macrophage colony-stimulating factor receptor. *Blood*, **80**, 1736–42.

Hara, T., and Miyajima, A. (1992). Two distinct functional high affinity receptors for mouse interleukin-3 (IL-3). *EMBO J.*, **11**, 1875–84.

Hayashida, K., Kitamura, T., Gorman, D.M., Arai, K., Yokota, T., and Miyajima, A. (1990). Molecular cloning of a second subunit of the receptor for human granulocyte-macrophage colony-stimulating factor (GM-CSF): reconstitution of a high-affinity GM-CSF receptor. *Proc. Natl. Acad. Sci. (USA)*, **87**, 9655–9.

Isfort, R.J., and Ihle, J.N. (1990). Multiple hematopoietic growth factors signal through tyrosine phosphorylation. *Growth Factors*, **2**, 213–20.

Itoh, N., Yonehara, S., Schreurs, J., Gorman, D.M., Maruyama, K., Ishii, A., Yahara, I., Arai, K., and Miyajima, A. (1990). Cloning of an interleukin-3 receptor gene: a member of a distinct receptor gene family. *Science*, **247**, 324–7.

Kanakura, Y., Druker, B., Cannistra, S.A., Furukawa, Y., Torimoto, Y., and Griffin, J.D. (1990). Signal transduction of the human granulocyte-macrophage colony-stimulating factor and interleukin-3 receptors involves tyrosine phosphorylation of a common set of cytoplasmic proteins. *Blood*, **76**, 706–15.

Kitamura, T., and Miyajima, A. (1992). Functional reconstitution of the human interleukin-3 receptor. *Blood*, **80**, 84–90.

Kitamura, T., Sato, N., Arai, K., and Miyajima, A. (1991). Expression cloning of the human IL-3 receptor cDNA reveals a shared beta subunit for the human IL-3 and GM-CSF receptors. *Cell*, **66**, 1165–74.

Kobayashi, N., Kono, T., Hatakeyama, M., Minami, Y., Miyazaki, T.,

Perlmutter, R.M., and Taniguchi, T. (1993). Functional coupling of the src-family protein tyrosine kinases p59fyn and p53/56lyn with the interleukin 2 receptor: implications for redundancy and pleiotropism in cytokine signal transduction. *Proc. Natl. Acad. Sci. (USA)*, **90**, 4201–05.

Milatovich, T., Kitamura, T., Miyajima, A., and Francke, U. (1993). Gene for the α-subunit of the human interleukin-3 receptor (IL3RA) localized to the X-Y pseudoautosomal region. *Amer. J. Human Genet.*, **53**, 1146–53.

Miyajima, I., Levitt, L., Hara, T., Bedell, M.A., Copeland, N.G., Jenkins, N., and Miyajima, A. (1994). The murine interleukin-3 receptor α subunit gene: Chromosomal localization, genomic structure, and promoter function. (Submitted.)

Park, L. S., Martin, U., Sorensen, R., Luhr, S., Morrissey, P. J., Cosman, D., and Larsen, A. (1992). Cloning of the low-affinity murine granulocyte-macrophage colony-stimulating factor receptor and reconstitution of a high-affinity receptor complex. *Proc. Natl. Acad. Sci. (USA)*, **89**, 4295–9.

Sakamaki, K., Miyajima, I., Kitamura, T., and Miyajima, A. (1992). Critical cytoplasmic domains of the common beta subunit of the human GM-CSF, IL-3 and IL-5 receptors for growth signal transduction and tyrosine phosphorylation. *EMBO J.*, **11**, 3541–9.

Sato, N., Caux, C., Kitamura, T., Watanabe, Y., Arai, K., Banchereau, J., and Miyajima, A. (1993a). Expression and factor-dependent modulation of the interleukin-3 receptor subunits on human hematopoietic cells. *Blood*, **82**, 752–61.

Sato, N., Sakamaki, K., Terada, N., Arai, K., and Miyajima, A. (1993b). Signal transduction by the high-affinity GM-CSF receptor: two distinct cytoplasmic regions of the common β subunit responsible for different signaling. *EMBO J.*, **12**, 4181–9.

Shen, Y., Baker, E., Callen, D.F., Sutherland, G.R., Willson, T.A., Rakar, S., and Gough, N.M. (1992). Localization of the human GM-CSF receptor beta chain gene (CSF2RB) to chromosome 22q12.2–:q13.1. *Cytogenet. Cell Genet.*, **61**, 175–7.

Silvennoinen, O., Witthuhn, B., Quelle, F. W., Cleveland, J. L., Yi, T., and Ihle, J. N. (1993). Structure of the murine jak2 protein-tyrosine kinase and its role in interleukin 3 signal transduction. *Proc. Natl. Acad. Sci. (USA)*, **90**, 8429–33.

Takaki, S., Mita, S., Kitamura, T., Yonehara, S., Yamaguchi, N., Tominaga, A., Miyajima, A., and Takatsu, K. (1991). Identification of the second subunit of the murine interleukin-5 receptor: interleukin-3 receptor-like protein, AIC2B is a component of the high affinity interleukin-5 receptor. *EMBO J.*, **10**, 2833–8.

Tavernier, J., Devos, R., Cornelis, S., Tuypens, T., Van der Heyden, J., Fiers, W., and Plaetinck, G. (1991). A human high affinity interleukin-5 receptor (IL5R) is composed of an IL5-specific alpha chain and a beta chain shared with the receptor for GM-CSF. *Cell*, **66**, 1175–84.

Torigoe, T., O'Connor, R., Santoli, D., and Reed ,J.C. (1992). Interleukin-3 regulates the activity of the LYN protein-tyrosine kinase in myeloid-committed leukemic cell lines. *Blood*, **80**, 617–24.

Watanabe, Y., Kitamura, T., Hayashida, K., and Miyajima, A. (1992). Monoclonal antibody against the common beta subunit (beta c) of the human interleukin-3 (IL-3), IL-5, and granulocyte-macrophage colony-stimulating factor receptors shows upregulation of beta c by IL-1 and tumour necrosis factor-alpha. *Blood*, **80**, 2215–20.

Toshiya Ogorochi and Atsushi Miyajima:
DNAX Research Institute of Molecular and Cellular Biology,
901 California Avenue,
Palo Alto, CA, USA
*(Deceased)

Interleukin-4 (IL-4)

IL-4 is an 18–20 kDa glycoprotein produced primarily by activated T lymphocytes. Its many immuno-regulatory properties include induction of immunoglobulin isotype switching to IgE expression, and induction of the subpopulation of helper T cells (designated Th2 cells) which regulate humoral immunity, eosinophilia, and inflammatory macrophage deactivation. IL-4 has also exhibited striking anti-tumour activity in a variety of animal models and is currently in clinical trials as an anti-cancer therapeutic.

■ Alternative names

B-cell growth factor I (BCGFI); B cell stimulatory factor I (BSF-1); T cell growth factor II (TCGF II); mast cell growth factor II (MCGF II).

■ IL-4 sequence

The nucleotide sequences of IL-4 cDNAs cloned from mouse and man (GenBank accession numbers M13238 and M13982) predict a protein of 153 (human) or 140 (mouse) amino acids, each containing a 20-amino-acid leader sequence (Lee *et al.* 1986; Noma *et al.* 1986; Yokota *et al.* 1986). These proteins are approximately 50 per cent homologous at the amino acid level, and there is no cross-reaction either biologically or at the receptor level between the two species of cytokine.

■ IL-4 protein

Recombinant IL-4 contains either three (mouse) or two (human) sites for potential N-glycosylation. A murine protein of identical N-terminal sequence to that of recombinant IL-4 has been purified from natural cellular sources by two independent groups (Grabstein *et al.* 1986; Paul and

Ohara 1987). Natural IL-4 is produced as a glycoprotein of apparent molecular weight 18–20 000. The glycosylation of IL-4 is not required for biological activity, since both endoglycosidase F-treated natural IL-4 and *E. coli*-expressed recombinant IL-4 retain all known biological activities.

■ IL-4 gene and transcription

The genes for mIL-4 and hIL-4 are located on chromosomes 11 and 5, respectively (reviewed by Arai *et al.* 1990). Thus, IL-4 belongs to the cluster of cytokine genes which map to the long arm of human chromosome 5 or mouse chromosome 11, and is located in close proximity to the gene for IL-5 (Arai *et al.* 1990). The genes for mIL-4 and hIL-4 comprise 4 exons arranged over 6 kb and 10 kb of DNA, respectively (Arai *et al.* 1990). While IL-2 and IL-4 are frequently expressed in different subsets of helper T cells, there are several regions of sequence homology in the 200 bp upstream of the IL-2 and IL-4 genes. Of particular interest is the region between -100 and -50 in the IL-4 gene, which has a high degree of homology to the site A sequence of the IL-2 gene but contains an insertion of 11 bp (CGAAAATTTCC, termed the P-element) in the middle of the site A-like sequence (Abe *et al.* 1992). The P-element does not share any homology with the IL-2 gene, but does have weak homology with the IL-5 gene. Insertion of the P-element in the site A-like sequence of the IL-4 gene abolishes the binding of the transcription factor NF-IL-2-A with the IL-4 enchancer. NF-IL-2-A recognizes site A in the IL-2 enhancer and is indistinguishable from the ubiquitous octamer binding protein Oct-1, suggesting that this sequence may contribute to the differential expression of IL-2 and IL-4 in different T cell subsets. The P-element is essential for phorbol ester/calcium ionophore stimulated reponses in T cells, and a protein designated NF(P), which binds to this sequence, has been found. However, NF(P) is present in different T cell subsets so the mechanism of the differential expression of IL-2 and IL-4 in different T cell subsets remains unknown.

■ IL-4 sources

To date, five natural sources of IL-4 have been identified: (1) some subpopulations of CD4+ peripheral T cells (Mosmann and Coffman 1989); (2) a minor subpopulation of T lymphocytes in spleen and thymus with the phenotype of CD4− CD8− $\alpha\beta$+ T cell receptor (Zlotnik *et al.* 1992); (3) a subpopulation of CD3+CD4+CD8− thymocytes that briefly continue to secrete IL-4 after they emigrate to the spleen (Bendelac and Schwartz 1992; Fischer *et al.* 1991); (4) cells of the basophil/mast cell lineage (Plaut *et al.* 1989); and (5) some subpopulations of CD8+ T cell clones (Seder *et al.* 1992; Yamamura *et al.* 1992). Induction of IL-4 secretion in each of these cases requires cross linkage of either the T cell receptor in the case of T lineage cells, or IgE Fc receptors in the case of basophil/mast cells.

■ Biological functions

Like many of the cytokines, IL-4 exhibits a multitude of *in vitro* properties (summarized in Table 1). Of these many *in vitro* properties, at least two have been amply confirmed by *in vivo* animal model experiments. The first of these relates to the class-switching properties of IL-4, specifically its ability to cause switching to IgE and augmentation of IgG_1 production in B-lymphocytes. The first *in vivo* support for these properties came from experiments showing that neutralizing anti-IL-4 antibodies blocked the development of IgE in a variety of immune responses (reviewed by Finkelman *et al.* 1990). Ultimate *in vivo* confirmation has now been provided by the generation of mutant mice in which the IL-4 gene has been disrupted by homologous recombination (Kuhn *et al.* 1991). These animals are completely unable to produce IgE, and have impaired IgG_1 development .

A second *in vitro* property of IL-4 that has been confirmed *in vivo* in animal models is the ability of IL-4 to drive precursor T cells into a subpopulation of T helper effectors known as Th2 cells. These helper T cells are important regulators of humoral immunity, eosinophilia, and mastocytosis (Mosmann and Coffman 1989). In some infectious disease models, administration of anti-IL-4 at the time of infection will divert the ensuing response away from Th2 cells toward the Thl subpopulation of T helper effectors (Sadick *et al.* 1990). Similarly, administration of IL-4 at the time of infection will have the reverse effect (Chatelain *et al.* 1992).

An important property of IL-4 that was initially discovered in *in vivo* animal model experiments is its potent anti-tumour activity (Tepper *et al.* 1989; Bosco *et al.* 1990; Golumbek *et al.* 1991; Tepper *et al.* 1992). In these experiments, local subcutaneous administration of small amounts of IL-4 around tumour-draining lymph nodes (Bosco *et al.* 1990), or transfection of the IL-4 gene into tumour explants which are then reintroduced into the animal (Tepper *et al.* 1989; Golumbek *et al.* 1991; Tepper *et al.* 1992), induced striking anti-tumour effects in a range of

Table 1. *In vitro* properties of IL-4

1. Growth co-stimulator for B cells, T cells, mast cells, myeloid progenitors, and erythroid progenitors.
2. Upregulates numerous activation antigens on B cells and T cells.
3. Induces activated B cells to switch to IgE synthesis, and specifically enhances secretion of human IgG4 and mouse IgG antibodies.
4. Augments cytolytic T cells differentiation.
5. Regulates induction of Th2 cells, a subpopulation of helper T cells that regulates humoral immunity, eosinophilia, mast cell development, and deactivation of inflammatory macrophages.
6. Inhibits release of inflammatory mediators (e.g. TNF, IL-1, IL-8, PGE_2) from activated monocytes.
7. Exerts direct anti-tumour activity on fresh tumour explants in *in vitro* clonogenic assays.

poorly or apparently non-immunogenic tumours such as renal carcinomas and melanomas. The IL-4 induced anti-tumour activity appears to be due to rapid tumour infiltration by cytotoxic eosinophils (Tepper *et al.* 1992), followed by a long-lived systemic immunity involving tumour-specific cytotoxic T cells (Golumbek *et al.* 1991). Recently, the potential anti-tumour activity of IL-4 has been further extended by a series of *in vitro* experiments demonstrating that IL-4 has direct growth suppressive activity on a variety of human malignancies (e.g. Karray *et al.* 1988).

■ Clinical applications

Based on the anti-tumour activities described above, IL-4 entered phase I/II clinical trials in 1991 as an anti-tumour therapeutic. While these studies are still at an early stage of evaluation, information to date appears promising. While clinical trials utilizing IL-4 in other clinical settings have not yet been initiated, one can readily speculate on potential applications. The monokine-suppressing property of IL-4 suggests possible utilization as an anti-inflammatory agent in diseases such as sepsis, rheumatoid arthritis, and inflammatory bowel disease. Similarly, the ability of IL-4 to downregulate Th1 mediated immunity in *in vitro* and animal model experiments suggests possible utilization in management of T cell mediated autoimmune diseases such as Type 1 diabetes, as well as in acute graft versus host disease.

■ References

Abe, E., de Waal Malefyt, R., Matsuda, I., Arai, K., and Arai, N. (1992). An 11-base-pair DNA sequence motif apparently unique to the human interleukin 4 gene confers responsiveness to T-cell activation signals. *Proc. Natl. Acad. Sci. (USA)*, **89**, 2864–8.

Arai, K.I., Lee, F., Miyajima, A., Miyatake, S., Arai, N., and Yokota, T. (1990). Cytokines: coordinators of immune and inflammatory responses. *Annu. Rev. Biochem.*, **59**, 783–836.

Bendelac, A., and Schwartz, R.H. (1992). CD4+ and CD8+ T cells acquire specific lymphokine secretion potentials during thymic maturation. *Nature*, **353**, 68–71.

Bosco, M., Giovarelli, M., Forni, M., Modesti, A., Scarpa, S., Masuelli, L., and Forni, G. (1990). Low doses of IL-4 injected perilymphatically in tumour-bearing mice inhibit the growth of poorly and apparently nonimmunogenic tumours and induce a tumour-specific immune memory. *J. Immunol.*, **145**, 3136–43.

Chatelain, R., Varkila, K., and Coffman, R.L. (1992). IL-4 induces a Th2 response in Leishmania major-infected mice. *J. Immunol.*, **148**, 1182–7.

Finkelman, F.D., Holmes, J., Katona, I.M., Urban, J.F. Jr., Beckmann, M.P., Park, L.S., Schooley, K.A., Coffman, R.L., Mosmann, T.R., and Paul, W.E. (1990). Lymphokine control of in vivo immunoglobulin isotype selection. *Annu. Rev. Immunol.*, **8**, 303–33.

Fischer, M., MacNeil, I., Suda, T., Cupp, J.E., Shortman, K., and Zlotnik, A. (1991). Cytokine production by mature and immature thymocytes. *J. Immunol.*, **146**, 3452–6.

Golumbek, P.T., Lazenby, A.J., Levitsky, H.I., Jaffee, L.M., Karasuyama, H., Baker, M., and Pardoll, D.M. (1991). Treatment of established renal cancer by tumour cells engineered to secrete interleukin-4. *Science*, **254**, 713–6.

Grabstein, K., Eisenman, J., Mochizuki, D., Shanebeck, K., Conlon, P., Hopp, T., March, C., and Gillis, S. (1986). Purification to homogeneity of B cell stimulating factor. A molecule that stimulates proliferation of multiple lymphokine-dependent cell lines. *J. Exp. Med.*, **163**, 1405–14.

Karray, S., DeFrance, T., Merle-Beral, H., Banchereau, J., Debre, P., and Galanaud, P. (1988). Interleukin 4 counteracts the interleukin 2-induced proliferation of monoclonal B cells. *J. Exp. Med.*, **168**, 85–94.

Kuhn, R., Rajewsky, K., and Muller, W. (1991). Generation and analysis of interleukin-4 deficient mice. *Science*, **254**, 707–10.

Lee, F., Yokota, T., Otsuka, T., Meyerson, P., Villaret, D., Coffman, R., Mosmann, T., Rennick, D., Roehm, N., Smith, C., Zlotnik, A., and Arai, K. (1986). Isolation and characterization of a mouse interleukin cDNA clone that expresses B-cell stimulatory factor 1 activities and T-cell- and mast-cell-stimulating activities. *Proc. Natl. Acad. Sci. (USA)*, **83**, 2061–5.

Mosmann, T.R., and Coffman, R.L. (1989). TH1 and TH2 cells: different patterns of lymphokine secretion lead to different functional properties. *Ann. Rev. Immunol.*, **7**, 145–73.

Noma, Y., Sideras, P., Naito, T., Bergstedt-Lindquist, S., Azuma, C., Severinson, E., Tanabe, T., Kinashi, T., Matsuda, F., Yaoita, Y., and Honjo, T. (1986). Cloning of cDNA encoding the murine IgG1 induction factor by a novel strategy using SP6 promoter. *Nature*, **319**, 640–6.

Paul, W.E., and Ohara, J. (1987). B-cell stimulatory factor-1/interleukin 4. *Annu. Rev. Immunol.*, **5**, 429–59.

Plaut, M., Pierce, J.H., Watson, C.J., Hanley-Hyde, J., Nordan, R.P., and Paul, W.E. (1989). Mast cell lines produce lymphokines in response to cross-linkage of Fc epsilon RI or to calcium ionophores. *Nature*, **339**, 64–7.

Sadick, M.D., Heinzel, F.P., Holaday, B.J., Pu, R.T., Dawkins, R.S., and Locksley, R.M. (1990). Cure of murine leishmaniasis with anti-interleukin 4 monoclonal antibody. Evidence for a T cell-dependent, interferon gamma-independent mechanism. *J. Exp. Med.*, **171**, 115–27.

Seder, R.A., Boulay, J.L., Finkelman, F., Barbier, S., Ben-Sasson, S.Z., Le Gros, G., and Paul, W.E. (1992) CD8+ T cells can be primed in vitro to produce IL-4. *J. Immunol.*, **148**, 1652–6.

Tepper, R.I., Pattengale, P.K., and Leder, P. (1989). Murine interleukin-4 displays potent anti-tumour activity in vivo. *Cell*, **57**, 503–12.

Tepper, R.I., Coffman, R.L., and Leder, P. (1992). An eosinophil-dependent mechanism for the antitumour effect of interleukin-4. *Science*, **257**, 548–51.

Yamamura, M., Wang, X.-H., Ohmen, J.D., Uyemura, K., Rea, T.H., Bloom, B.R., and Modlin, R.L. (1992). Cytokine patterns of immunologically mediated tissue damage. *J. Immunol.*, **149**, 1470–5.

Yokota, T., Otsuka, T., Mosmann, T., Banchereau, J., DeFrance, T., Blanchard, D., de Vries, J.E., Lee, F., and Arai, K. (1986). Isolation and characterization of a human interleukin cDNA clone, homologous to mouse B-cell stimulatory factor 1, that expresses B-cell- and T-cell-stimulating activities. *Proc. Natl. Acad. Sci. (USA)*, **83**, 5894–8.

Zlotnik, A., Godfrey, D.I., Fischer, M., and Suda, T. (1992). Cytokine production by mature and immature CD4-CD8- T cells. Alpha beta-T cell receptor+ CD4-CD8- T cells produce IL-4. *J. Immunol.*, **149**, 1211–15.

Maureen Howard and Nobuyuki Harada:
DNAX Research Institute of Molecular and Cellular Biology Inc.,
901 California Avenue,
Palo Alto, CA 94304–1104 USA

Interleukin-4 Receptors

The Interleukin-4 (IL-4) receptor is a type 1 membrane glycoprotein which is expressed on a wide variety of cell types of both human and murine origin. The IL-4 receptor belongs to the haemopoietin superfamily of cytokine receptors whose members share homology within their extracellular domains and generally appear to exist as homodimers or heterodimers in their biologically active high affinity forms. Binding of IL-4 to cells reveals a single high affinity binding site, which can be recreated by introduction of the cloned IL-4 receptor subunit alone into cells which do not bind IL-4, suggesting that the high affinity form may be a homodimer. As with other members of the haemopoietin receptor superfamily, this receptor is also characterized by the lack of a tyrosine kinase domain in its cytoplasmic portion, suggesting that additional receptor associated proteins are likely to be necessary to elicit a biological response.

■ The IL-4 receptor protein

Both the murine (Mosley et al. 1989) and human (Idzerda et al. 1990) IL-4 receptors have been molecularly cloned. The murine IL-4 receptor, which was cloned first (Mosley et al. 1989), is produced as an 810-amino-acid precursor with a predicted 25-amino-acid signal peptide (GenBank accession number M27959). It contains a single transmembrane domain of 24 amino acids, an extracellular domain of 208 amino acids and a cytoplasmic domain of 553 amino acids. The extracellular domain contains seven cysteine residues and five potential N-linked glycosylation sites. A cDNA clone for the human IL-4 receptor was isolated utilizing cross-species hybridization and found to encode an 825-amino-acid protein (Idzerda et al. 1990). The predicted protein sequence exhibited 53 per cent sequence identity with the murine IL-4 receptor and contained a 25-amino-acid signal peptide, 207-amino-acid extracellular domain, 24-amino-acid transmembrane domain, and a 569-amino-acid cytoplasmic domain. Five of the seven cysteine residues, and three of the six potential N-linked glycosylation sites in the extracellular domain have been conserved between mouse and human receptors.

The IL-4 receptor shows significant homology to a number of other cytokine receptors which together form a superfamily of receptors known as the haemopoietin receptor superfamily (Idzerda et al. 1990; Cosman et al. 1990). This family has two highly conserved features in the extracellular domain which are a pair of conserved cysteine residues and a classic Trp-Ser-X-Trp-Ser motif proximal to the transmembrane domain. The structure of a complex of the growth hormone receptor, which belongs to this family, and its ligand has recently been solved by X-ray crystallography (de Vos et al. 1992). The structure revealed that the receptor extracellular binding domain formed two β-barrels, each consisting of seven antiparallel β-sheets. In addition, the receptor/ligand complex was found to consist of a single molecule of growth hormone bound to two growth hormone receptor molecules. It is likely that IL-4 and IL-4 receptor are also capable of forming a similar complex.

An additional feature of the haemopoietin receptor superfamily is that many of its members are found to exist in both membrane bound and naturally occurring soluble forms. In all cases, it appears that these two forms are generated by an alternative splicing mechanism. The murine IL-4 receptor was the first member of the superfamily for which a cDNA clone encoding a soluble form was isolated (Mosley et al. 1989). In two different murine T cell libraries, receptor clones were detected which contained a 114 bp insertion in the 3′ end of the nucleic acid sequence coding for the extracellular region of the molecule. This insertion led to a shift in the reading frame such that an additional six amino acids were encoded at the C-terminus, and termination of translation occurred prior to the transmembrane domain due to an introduced stop codon. When these clones were expressed in COS-7 cells, secretion of a soluble IL-4 receptor into the medium could be detected (Mosley et al. 1989). Efforts to find a cDNA encoding a soluble human IL-4 receptor, either by screening cDNA libraries or utilizing the polymerase chain reaction (PCR) and various sources of RNA, have to date been unsuccessful (Beckmann et al. 1992). Whether a soluble form of the human IL-4 receptor does not exist, or whether it has not been detected for other reasons, such as a restricted expression pattern, is not known.

■ IL-4 receptor gene

The murine IL-4 receptor gene maps to the distal region of chromosome 7 and the human gene was localized to 16p11.2–16p12.1 (Pritchard et al. 1991). These results suggest that the IL-4 receptor locus is not linked to other members of the haemopoietin receptor superfamily. The location of the human IL-4 receptor gene raises the possibility that it may be a candidate for alteration in myxoid liposarcomas which are often associated with 12;16 translocations (Pritchard et al. 1991). The structure of the murine IL-4 receptor gene has also been determined (Wrighton et al. 1992). The gene spans approximately 25 kb and is composed of 12 exons. The overall intron–exon organization of the IL-4 receptor gene bears marked similarity to that of the murine erythropoietin receptor, human IL-2 receptor and the human growth hormone receptor, all members of the haemopoietin receptor superfamily.

■ Expression and binding characteristics of the IL-4 receptor

IL-4 has numerous biological activities due to the variety of cellular targets which display its specific receptor. In general, a single class of high affinity receptors has been detected on nearly all murine and human cells tested (Ohara and Paul 1987; Park et al. 1987a,b; Lowenthal et al. 1988). Receptors are expressed on T and B lymphocytes, mast cells, cells of myeloid and monocytic origin, fibroblasts and endothelial cells. Almost without exception, both primary cells and cell lines express low numbers of receptors ranging from less than 100 receptors per cell up to a few thousand at most. Although most investigators have observed only high affinity IL-4 sites, there have been reports that low affinity IL-4 binding sites can also be detected (Foxwell et al. 1989). As discussed above, evidence gained from binding studies with the cloned IL-4 receptor subunit suggests that this protein alone is sufficient to generate a high affinity binding site for IL-4, probably in the form of a homodimer. It is possible that binding of IL-4 to an IL-4 receptor monomer might be detected in some cases as low affinity binding. No substantial evidence exists that any additional proteins are required to generate a high affinity binding site. Cross-linking studies utilizing radiolabelled IL-4 (Park et al. 1987b) and immunoprecipitations using antibodies generated against the murine IL-4 receptor (Beckmann et al. 1990) suggest that the cell surface form of the IL-4 receptor has a molecular weight of 130–140 kDa. This estimate is supported by studies of the recombinant receptor expressed in COS-7 cells as well as by the predicted molecular weights determined from the sequences of the cDNA clones.

■ Signalling mechanisms

The mechanism by which IL-4 transmits a signal is still unknown. Analysis of the intracellular domain of the IL-4 receptor has failed to reveal any structural features common to protein kinases or any other catalytic functions. Thus, it is likely that signal transduction through the IL-4 receptor involves associated proteins, either membrane associated or cytoplasmic, as has recently been shown for other members of the haemopoietin receptor superfamily (Cosman 1993). A number of studies have been done with both murine and human cells looking at the possible involvement in IL-4 signalling of calcium mobilization, phosphoinositol metabolism, phosphorylation of cytoplasmic proteins, and other classic pathways associated with signal transduction. These studies, which have been reviewed recently (Beckmann et al. 1992), have not yet resulted in any consensus view of what metabolic events may be critical to signal transduction in response to IL-4. Additional information is being generated by the approach of mutagenesis analysis of the cytoplasmic domain. It has recently been reported (Harada et al. 1992) that a critical region for signal transduction is located between amino acid residues 433 and 473 numbering from the C-terminus. This region is highly conserved between the murine and human IL-4 receptors, but lacks homology with other members of the haemopoietin receptor superfamily.

■ References

Beckmann, M.P., Schooley, K.A., Gallis, B., VandenBos, T., Friend, D., Alpert, A.R., Raunio, R., Prickett, K.S., Baker, P.E., and Park, L.S. (1990). Monoclonal antibodies block murine IL-4 receptor function. J. Immunol., **144**, 4212–17.

Beckmann, M.P., Cosman, D., Fanslow, W., Maliszewski, C.R., and Lyman, S.D. (1992). The interleukin-4 receptor: structure, function, and signal transduction. Chem. Immunol., **51**, 107–34.

Cosman, D. (1993). The hematopoietin receptor superfamily. Cytokine, **5**, 95–106.

Cosman, D., Lyman, S.D., Idzerda, R.L., Beckmann, M.P., Park, L.S., Goodwin, R.G., and March, C.J. (1990). A new cytokine receptor superfamily. Trends Biochem. Sci., **15**, 265–70.

de Vos, A.M., Ultsch, M., and Kossiakoff, A.A. (1992). Human growth hormone and extracellular domain of its receptor: crystal structure of the complex. Science, **255**, 306–12.

Foxwell, B.M., Woerly, G., and Ryffel, B. (1989). Identification of interleukin 4 receptor-associated proteins and expression of both high- and low-affinity binding on human lymphoid cells. Eur. J. Immunol., **19**, 1637–41.

Harada, N., Yang, G., Miyajima, A., and Howard, M. (1992). Identification of an essential region for growth signal transduction in the cytoplasmic domain of the human interleukin-4 receptor. J. Biol. Chem., **267**, 22752–8.

Idzerda, R.L., March, C.J., Mosley, B., Lyman, S.D., VandenBos, T., Gimpel, S.D., Din, W.S., Grabstein, K.H., Widmer, M.B., Park, L.S., Cosman, D., and Beckmann, M.P. (1990). Human interleukin 4 receptor confers biological responsiveness and defines a novel receptor superfamily. J. Exp. Med., **171**, 861–73.

Lowenthal, J.W., Castle, B.E., Christiansen, J., Shreurs, J., Rennick, D., Arai, N., Hoy, P., Takebe, Y., and Howard, M. (1988). Expression of high affinity receptors for murine interleukin 4 (BSF-1) on haemopoietic and nonhaemopoietic cells. J. Immunol., **140**, 456–64.

Mosley, B., Beckmann, M.P., March, C.J., Idzerda, R.L., Gimpel, S.D., VandenBos, T., Friend, D., Alpert, A., Anderson, D., Jackson, J., Wignall, J.M. Smith, C., Gallis, B., Sims, J.E., Urdal, D., Widmer, M.B., Cosman, D., and Park, L.S. (1989). The murine interleukin-4 receptor: molecular cloning and characterization of secreted and membrane bound forms. Cell, **59**, 335–48.

Ohara, J., and Paul, W.E. (1987). Receptors for B-cell stimulatory factor-1 expressed on cells of haemopoietic lineage. Nature, **325**, 537–40.

Park, L.S., Friend, D., Grabstein, K., and Urdal, D. (1987a). Characterization of the high-affinity cell-surface receptor for murine B-cell-stimulating factor 1. Proc. Natl. Acad. Sci. (USA), **84**, 1669–73.

Park, L.S., Friend, D., Sassenfeld, H.M., and Urdal, D.L. (1987b). Characterization of the human B cell stimulatory factor 1 receptor. J. Exp. Med., **166**, 476–88.

Pritchard, M.A., Baker, E., Whitmore, S.A., Sutherland, G.R., Idzerda, R.L., Park, L.S., Cosman, D., Jenkins, N.A., Gilbert, D.J., Copeland, N.G., and Beckmann, M.P. (1991). The interleukin-4 receptor gene (IL4R) maps to 16p11.2–16p12.1 in human and to the distal region of mouse chromosome 7. Genomics, **10**, 801–6.

Wrighton, N., Campbell, L.A., Harada, N., Miyajima, A., and Lee, F. (1992). The murine interleukin-4 receptor: molecular cloning and characterization of secreted and membrane bound forms. *Growth Factors*, **6**, 103–18.

Linda S. Park:
Immunex Research and Development Corporation, Seattle, WA, USA

Interleukin-5 (IL-5)

IL-5 is produced by T lymphocytes as a glycoprotein with an M_r of 40 000–45 000 Da and is unusual among the T-cell-produced cytokines in being a disulphide-linked homodimer. It is the most highly conserved member of a group of evolutionarily related cytokines, including also IL-3, IL-4 and granulocyte/macrophage colony stimulating factor (GM-CSF), which are closely linked on human chromosome 5. IL-5 appears to be the main cytokine involved in the control of eosinophilia. In the mouse it also has activity on B cells, but in man the analogous activity is undetectable. The critical role of eosinophils in allergy indicates that IL-5 is a key cytokine in the development of allergic diseases such as asthma.

Alternative names

Eosinophil differentiation factor (EDF); eosinophil colony stimulating factor (Eo-CSF). In the mouse: B-cell growth factor II (BCGFII), T-cell replacing factor (TRF).

Sequence

The cDNA sequence for human and mouse (GenBank accession numbers X04688 and X06270) predict proteins of 134 and 133 amino acids, respectively. Sequence information is available for both the human and mouse genomic genes (GenBank accession numbers J02971 and XO6271). The coding sequence of the IL-5 gene forms four exons. The introns show areas of similarity between the mouse and human sequences, although the mouse has a considerable amount of sequence (including repeat sequences) which are not present in the human gene. The mouse gene includes a 738 bp segment in the 3' untranslated region which is not present in the human gene, giving a transcript of 1.6 kb compared to 0.9 kb in man. The gross gene structure are also shared by IL-3, IL-4, and GM-CSF (Sanderson 1994).

Protein

IL-5 is a glycoprotein which is highly homologous between species (Fig. 1), and shows cross-reactivity of the protein across a variety of mammalian species. The human mature protein consists of 115 amino acids (M_r of 12 000 and 24 000 for the dimer), while the mouse is truncated by two residues at the N-terminus.

The crystal structure of human IL-5 (Milburn *et al.* 1993) shows it to be similar to the structures of other cytokines and most closely resembles IL-4 and GM-CSF which consist of a bundle of four α-helices (A,B,C,D from the N-terminus) with two overconnecting loops. The dimer structure of IL-5 forms an elongated ellipsoidal disc, made up of two domains about a twofold axis (Fig. 2). Each domain is made up of three helices from one monomer (A,B,C) and one helix (D') from the other. The two monomers are held together by two disulphide bridges connecting Cys44 of one molecule with Cys86 of the other. In addition, residues 32 to 35 form an antiparallel β-sheet with residues 89 to 92 of the other monomer.

The monomer has no biological activity and has no inhibitory activity, suggesting that they do not form high affinity interactions with the IL-5 receptor. Studies with human–mouse hybrid molecules suggest that residues in the C-terminal third of the molecule are responsible for species specificity and hence interaction with the receptor (McKenzie *et al.* 1991). These residues correspond to part of helix C and helix D.

The secreted material has a M_r of 40 000 to 45 000, thus nearly half the native material consists of carbohydrate. Human IL-5 has one N-linked carbohydrate chain at position Asn28, and one O-linked carbohydrate at position Thr3 (Minamitake *et al.* 1990). Mouse IL-5 has an additional N-linked carbohydrate at Asn55, this site does not exist in hIL-5. The potential N-linked site at Asn71 is apparently not glycosylated in either species. Deglycosylated IL-5 has been reported to have full activity *in vitro*.

Activity on eosinophils

IL-5 induces the production of eosinophils from both mouse and human bone marrow (Sanderson 1993). Transgenic mice which constitutively express IL-5 have high level,

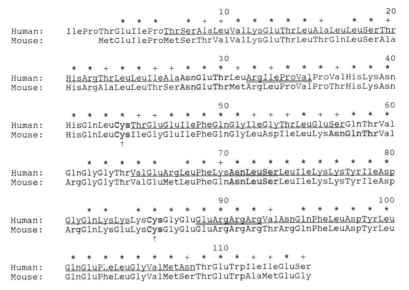

```
                                   10                        20
                        *  *  *        *  +  +  *  *  *  *  *  +  *        *  *  +
Human:   IleProThrGluIleProThrSerAlaLeuValLysGluThrLeuAlaLeuLeuSerThr
Mouse:      MetGluIleProMetSerThrValValLysGluThrLeuThrGlnLeuSerAla

                                   30                        40
           *  *  +  *  *  +  +  *  *  *  +  *  +  *  *  *  +  *  *  *
Human:   HisArgThrLeuLeuIleAlaAsnGluThrLeuArgIleProValProValHisLysAsn
Mouse:   HisArgAlaLeuLeuThrSerAsnGluThrMetArgLeuProValProThrHisLysAsn

                                   50                        60
           *  *  *  *  *  +  +  *  *  *  *  +  +  *  +  +  *  *  *  *
Human:   HisGlnLeuCysThrGluGluIlePheGlnGlyIleGlyThrLeuGluSerGlnThrVal
Mouse:   HisGlnLeuCysIleGlyGluIlePheGlnGlyLeuAspIleLeuLysAsnGlnThrVal
                    ↑

                                   70                        80
           *  *  *  *  *     *  *  +  *  *  *  *  *  *  *  *  *  *
Human:   GlnGlyGlyThrValGluArgLeuPheLysAsnLeuSerLeuIleLysLysTyrIleAsp
Mouse:   ArgGlyGlyThrValGluMetLeuPheGlnAsnLeuSerLeuIleLysLysTyrIleAsp

                                   90                        100
              *     *     *  *  *  *  *  *  *  *  +     *  *  *  *  *
Human:   GlyGlnLysLysLysCysGlyGluGluArgArgArgValAsnGlnPheLeuAspTyrLeu
Mouse:   ArgGlnLysGluLysCysGlyGluGluArgArgArgThrArgGlnPheLeuAspTyrLeu
                          ↑

                                   110
           *  *  *  *  *  *  *  *  +  *  *  *  *  +  +  *  +
Human:   GlnGluPheLeuGlyValMetAsnThrGluTrpIleIleGluSer
Mouse:   GlnGluPheLeuGlyValMetSerThrGluTrpAlaMetGluGly
```

Figure 1. Alignment of human and mouse mature IL-5 amino-acid sequences. The numbering is based on the human sequence. An asterisk indicates identity a plus sign indicates a conservative change; cysteines are arrowed. Predicted N-linked glycosylation sites are tinted. Residues forming α-helices are underlined. Residues involved in antiparallel β-sheets are double underlined.

lifelong eosinophilia (Dent *et al.* 1990). The administration of anti-IL-5 neutralizing antibody to mice infected with *Trichinella spiralis* totally blocks the production of eosinophils (Coffman *et al.* 1989). These experiments illustrate the unique role of IL-5 in the control of eosinophilia in this parasite infection.

The ability of eosinophils to perform in functional assays can be increased markedly by incubation with a number of different agents, including IL-5 (Sanderson 1992*a*), and this is called 'activation'. IL-5 is a potent inducer of IgA-induced eosinophil degranulation (Fujisawa *et al.* 1990).

Eosinophils appear to migrate into tissues where IL-5 is expressed (Sanderson 1992*a*). While IL-5 is reported to be chemotactic for eosinophils, the activity is relatively weak, and it is not clear what role this could play *in vivo*. On the other hand, IL-5 has been shown to upregulate adhesion molecule integrin CD11b on human eosinophils and this increased expression was accompanied by an increased adhesion to endothelial cells (Walsh *et al.* 1990), providing a possible mechanism for eosinophil localization in the tissues.

Activity on basophils

IL-5 induces the production of eosinophils with little or no effect on other lineages, but it primes basophils for increased histamine production and leucotriene generation (Bischoff *et al.* 1990), and basophils in the blood clearly express the IL-5 receptor (Lopez *et al.* 1990).

Activity on mouse B cells

IL-5 is a late-acting factor in the differentiation of primary B cells, requiring a priming stimulus to make resting B cells responsive (O'Garra *et al.* 1986). It has been suggested that IL-5 plays a key role in the production of IgA, but more recent studies have indicated that its effect is minor compared to the activity of other cytokines, and that it may only augment these activities (Beagley *et al.* 1989). A possible mechanism for the effect of IL-2 on B cells is suggested by the observation that IL-5 increases the expression of the IL-2 receptor (Loughnan and Nossal 1989).

Sources of IL-5

T cells are an important source of the cytokine, and the development of eosinophilia is T cell dependent. IL-5, as well as other cytokine mRNAs, are produced by mouse mast cell lines. Eosinophils themselves have been demonstrated to produce IL-5. It is not clear what biological role the non-T-cell derived IL-5 plays.

Pathology

The correlation between the development of eosinophilia and detectable levels of serum IL-5 or IL-5 mRNA in the tissues in a wide range of disease states is remarkable, and include parasite infections, idiopathic eosinophilia, eosinophilic myalgia, Hodgkin's lymphoma and asthma

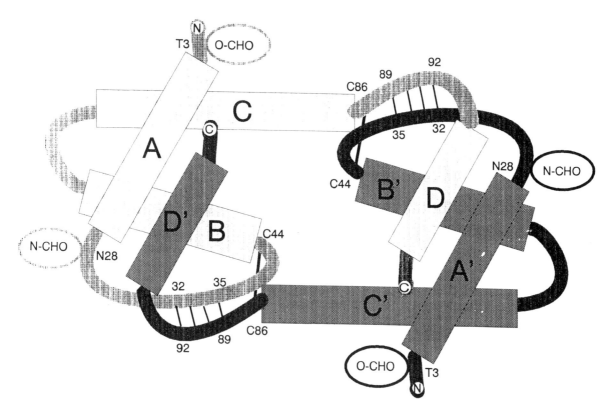

Figure 2. Diagram based on the crystal structure (Milburn *et al.* 1993) of hIL-5 showing the main structural features. One monomer is shown in light grey and the other in dark grey. Helices are indicated A–D, and A′–D′, respectively, starting at the N-terminus (N). The disulphide bridges are shown as thick lines. The interactions between antiparallel β-sheets are shown by fine lines.

(Sanderson 1992*a*). These studies are part of an expanding body of evidence linking IL-5 expression with eosinophilia in a wide variety of diseases.

Antigen challenge of asthmatic patients results in accumulation of eosinophils in the lung, and the number of eosinophils correlates with the severity of the late asthmatic reaction. Evidence that eosinophils are producing tissue damage in the asthmatic lung is based on the demonstration of eosinophil major basic protein and eosinophil-derived neurotoxin indicating degranulation at sites of injury. These types of data have caused a reappraisal of the aetiology of the late-phase asthmatic reaction, from a disease mediated by mast cells and IgE, to a disease which is primarily a T cell response (delayed type hypersensitivity reaction) accompanied by eosinophilia (Corrigan and Kay 1992). The possibility that antagonists to IL-5 would block the production of eosinophils provides a new therapeutic approach to the treatment of allergic diseases such as asthma (Sanderson 1992*a,b*).

Eosinophilic inflammation after renal transplantation is an adverse prognostic factor for graft survival. Eosinophils in liver grafts were reported to be a sensitive and specific indicator for acute rejection. Experimental lung allograft in the rat causes eosinophil infiltration. It is not clear from

these observations whether the eosinophils are involved in graft rejection, or whether they simply represent a marker for an active T cell response (Sanderson 1992*a*).

The presence of eosinophils in association with tumours has been noted for many years, and a local infiltration of eosinophils is associated with a positive prognosis in primary lung cancers, gastric carcinoma, colonic carcinoma, and Hodgkin's disease (Sanderson 1992*a*).

■ References

Beagley, K.W., Eldridge, J.H., Lee, F., Kiyono, H., Everson, M.P., Koopman, W.J., Hirano, T., Kishimoto, T., and McGhee, J.R. (1989). Interleukins and IgA synthesis. Human and murine interleukin 6 induce high rate IgA secretion in IgA-committed B cells. *J. Exp. Med.*, **169**, 2133–48.

Bischoff, S.C., Brunner, T., De Weck, A.L., and Dahinden, C.A. (1990). Interleukin 5 modifies histamine release and leukotriene generation by human basophils in response to diverse agonists. *J. Exp. Med.*, **172**, 1577–82.

Coffman, R.L., Seymour, B.W., Hudak, S., Jackson, J., and Rennick, D. (1989). Antibody to interleukin-5 inhibits helminth-induced eosinophilia in mice. *Science*, **245**, 308–10.

Corrigan, C.J., and Kay, A.B. (1992). T cells and eosinophils in the pathogenesis of asthma. *Immunol. Today*, **13**, 501–7.

Dent, L.A., Strath, M., Mellor, A.L., and Sanderson, C.J. (1990). Eosinophilia in transgenic mice expressing interleukin 5. *J. Exp. Med.*, **172**, 1425–31.

Fujisawa, T., Abu-Ghazaleh, R., Kita, H., Sanderson, C.J., and Gleich, G.J. (1990). Regulatory effect of cytokines on eosinophil degranulation. *J. Immunol.*, **144**, 642–6.

Lopez, A.F., Eglinton, J.M., Lyons, A.B., Tapley, P.M., To, L.B., Park, L.S., Clark, S.C., and Vadas, M.A. (1990). Human interleukin-3 inhibits the binding of granulocyte-macrophage colony-stimulating factor and interleukin-5 to basophils and strongly enhances their functional activity. *J. Cell Physiol.*, **145**, 69–77.

Loughnan, M.S., and Nossal, G.J. (1989). Interleukins 4 and 5 control expression of IL-2 receptor on murine B cells through independent induction of its two chains. *Nature*, **340**, 76–9.

McKenzie, A.N., Barry, S.C., Strath, M., and Sanderson, C.J. (1991). Structure–function analysis of interleukin-5 utilizing mouse/human chimeric molecules. *EMBO J.*, **10**, 1193–9.

Milburn, M.V., Hassell, A.M., Lambert, M.H., Jordan, S.R., Proudfoot, A.E.I., Grabar, P., and Wells, T.N.C. (1993). A novel dimer configuration revealed by the crystal structure at 2.4. A resolution of human interleukin-5. *Nature*, **363**, 172–6.

Minamitake, Y., Kodama, S., Katayama, T., Adachi, H., Tanaka, S., and Tsujimoto, M. (1990). Structure of recombinant human interleukin 5 produced by Chinese hamster ovary cells. *J. Biochem.*, **107**, 292–7.

O'Garra, A., Warren, D.J., Holman, M., Popham, A.M., Sanderson, C.J., and Klaus, G.G. (1986). Interleukin 4 (B-cell growth factor II eosinophil differentiation factor) is a mitogen and differentiation factor for preactivated murine B lymphocytes. *Proc. Natl. Acad. Sci. (USA)*, **83**, 5228–32.

Sanderson, C.J. (1992a). Interleukin-5, eosinophils, and disease. *Blood*, **79**, 3101–9.

Sanderson, C.J. (1992b). Pharmacological implications of interleukin-5 in the control of eosinophilia. *Adv. Pharmacol.*, **23**, 163–77.

Sanderson, C.J. (1993). In *Immunopharmacology of eosinophils* (eds H. Smith and R. Cook), pp. 11–24. Academic Press, London.

Sanderson, C.J. (1994). In *Immunology and molecular biology of cytokines* (ed. A.W. Thomson). Academic Press, London. (In press.)

Walsh, G.M., Hartnell, A., Wardlaw, A.J., Kurihara, K., Sanderson, C.J., and Kay, A.B. (1990). IL-5 enhances the *in vitro* adhesion of human eosinophils, but not neutrophils, in a leucocyte integrin (CD11/18)-dependent manner. *Immunology*, **71**, 258–65.

Colin J. Sanderson:
Western Australian Research Institute for Child Health,
Princess Margaret Hospital,
Perth,
Western Australia

The Interleukin-5 Receptor

The interleukin-5 receptor is a heterodimer consisting of two subunits, both belonging to the cytokine/haemopoietin receptor superfamily. The IL-5Rα chain is ligand-specific and its expression is restricted to eosinophils and basophils in man (and in addition activated B cells in the mouse). It binds IL-5 with intermediate affinity. Association with the IL-5Rβ chain (β_c) confers a 2–3-fold increase in human IL-5 binding affinity to the IL-5R complex (~100-fold in the mouse). This β-subunit is required for signal transduction and its use is shared with IL-3 and GM-CSF, in the latter cases also through association with ligand-specific receptor α-subunits. In addition to the membrane-anchored receptor, eosinophils express two additional IL-5Rα mRNA isoforms, both encoding secreted receptor variants. One such variant is abundant, and has antagonistic properties through association with one IL-5 dimer in solution. Both receptor subunits are required for signalling. Deletion of the cytoplasmic domain of either subunit abolishes a proliferative signal. Activation of the IL-5R complex leads to tyrosine phosphorylation of several cytoplasmic proteins through putative associated tyrosine kinase(s). As the β-subunit is shared with the IL-3 and GM-CSF receptors, a high overlap of the signalling cascade is likely. However, ligand-specific signalling through the specific α-subunits cannot be ruled out at present.

■ The IL-5 receptor α-subunit protein

The nucleotide sequence of the human IL-5Rα cDNA (GenBank accession numbers M75914, M96651, M96652, X61176, X61177, X61178, and X62156) predicts a polypeptide of 420 amino acids (Fig. 1) (Tavernier *et al.* 1991, 1992; Murata *et al.* 1992). It is characterized by a 20 residue N-terminal signal peptide, followed by a 322-amino-acid-long extracellular domain, a membrane anchor spanning 20 residues, and a 58-amino-acid-long cytoplasmic tail. The predicted molecular mass is 45.5 kDa, which is slightly below the 60 kDa value observed on SDS-PAGE gels indicating N-linked glycosylation of one or more of the six potential N-glycosylation sites (and perhaps O-linked glycosylation). The IL-5Rα subunit belongs to the cytokine/haemopoietin receptor superfamily which is characterized by a modular structure, each module having a seven β-sheet scaffold (Bazan 1990; deVos *et al.* 1992). The IL-5Rα contains three such domains: a juxtamembrane module containing a canonical Trp-Ser-Xxx-Trp-Ser motif (WS-WS box); a central module containing four conserved cysteines, involved in two disulphide bridges; and a third N-terminal module, which is related to the WS-WS containing module (Figs 1, 2). This subunit structure is reflected in the gene organization (Tuypens *et al.* 1992) (Fig. 3). The integrity of all three domains is required for ligand binding.

The murine IL5-Rα polypeptide (Takaki *et al.* 1990) has 71 per cent sequence identity with its human counterpart (Fig. 1) (GenBank accession number D90205).

■ The associated receptor β-subunit protein (β_c)

Cross-linking studies reveal the existence of a 120–130 kDa accessory receptor protein, both in man and mouse. Co-expression of this receptor β-subunit (also known under the acronym KH97; GenBank accession numbers M59941, M38275) (Hayashida *et al.* 1990) with the receptor α-chain on Cos1 cells causes a 2–3-fold increase in binding affinity of IL-5 (see below). On itself, it does not bind IL-5 with detec-

table affinity. This β-subunit also belongs to the cytokine/haemopoietin receptor superfamily, having a tandem array of both the WS-WS and C-C modules (Fig. 2). It totals 897 amino acids, including a predicted signal sequence of 16 residues, a 27-amino-acid-long membrane-spanning domain, and a cytoplasmic domain of 350 residues. It is shared with the receptor complexes of GM-CSF (Tavernier *et al.* 1991) and IL-3 (Kitamura *et al.* 1991) which explains the overlap of biological activities and the cross-competition observed for these cytokines on haemopoietic cells (Lopez *et al.* 1992). The β-subunit is required for signal transduction.

The homologous murine polypeptide has 55 per cent amino acid sequence identity, and is referred to as AIC2B (GenBank accession number M34397) (Gorman *et al.* 1990; Devos *et al.* 1991). In the mouse, a second highly related β-chain (AIC2A; 91 per cent amino acid sequence identity to AIC2B; GenBank accession number M29855) has been identified (Itoh *et al.* 1990). This latter β-chain can bind IL-3 with low affinity and associates with the IL-3 receptor complex only. A scheme of the different receptor complexes is shown in Fig. 2.

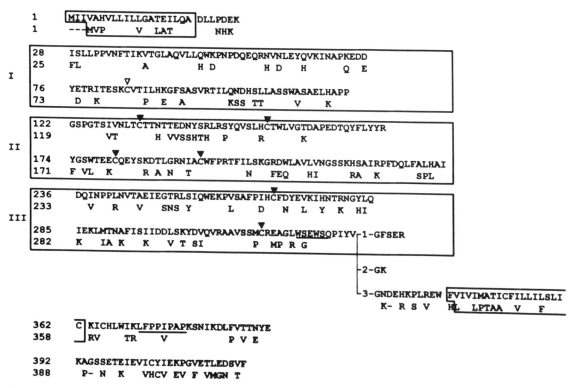

Figure 1. Alignment of human and mouse IL-5rα amino acid sequences. The full sequence is shown for the human protein (*on top*); for the mouse sequence, only diverging residues are printed below; gaps are indicated by dashes. Each sequence line corresponds to one exon (for the human gene, see Fig. 3). Boxed sequences respectively represent the signal peptide, the three extracellular cytokine receptor submodules, and the transmembrane region. For the human protein, the sequences of two secreted isoforms are also shown. Conserved motifs (WS–WS box, and a proline rich 'box 1' in the cytoplasmic tail) are underlined. Triangles indicate conserved cysteines; full triangles indicate the involvement in disulphide bridges.

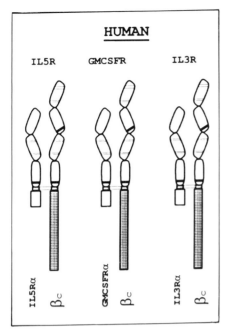

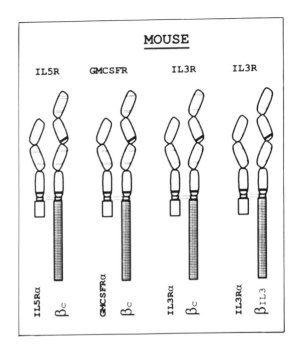

Figure 2. Model of IL-5R complex and related receptors. The receptor a- and β-subunits contain three and four conserved cytokine receptor submodules, respectively. Conserved cysteine residues are shown by fine lines, WS–WS motifs by thick lines. Open and hatched boxes represent the cytoplasmic tails of the receptor a- and β-subunits, respectively.

■ The IL-5 receptor a-subunit gene and its expression

The human IL5-Ra gene is located on chromosome 3 in the region 3p26, which is syntenic with the murine chromosome 6 location (Tuypens *et al.* 1992). The human and mouse IL-5Rβ locus is at chromosome 22q12.3–13.1 and at

chromosome 15, respectively (Shen *et al.* 1992; Gorman *et al.* 1992). AIC2A and AIC2B are closely linked. The hIL-5Ra (and -β [Gorman *et al.* 1992]) gene organization reflects its relationship to the cytokine/haemopoietin receptor superfamily (Tuypens *et al.* 1992) (Fig. 3). Human eosinophils express through differential splicing, three different transcripts from the same IL-5Ra locus (Fig. 3). As a result, in addition to the membrane-anchored receptor, two soluble

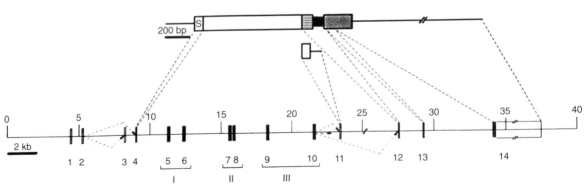

Figure 3. Structure and transcription pattern of the human IL-5Ra gene. Diagrams representing two different isoforms are shown on top (above: membrane-anchored variant; below: a major soluble variant). The genomic organization of the human IL-5Ra gene is shown at the bottom. Boxes represent exons and are numbered underneath. Exons marked pairwise by roman numbering represent the three cytokine receptor modules building up the extracellular domain. Dashed lines correlate other functional domains on the protein structures with the corresponding exons. 5′ and 3′ untranslated regions are shown by open boxes. Alternative splicing patterns are also indicated. (Adapted from Tuypens *et al.* 1992.)

isoforms can be produced, one of which is abundantly expressed at the mRNA level (> 90 per cent). In the latter case, splicing to a soluble-specific exon is required (Tuypens et al. 1992). The purified soluble receptor variant, produced in heterologous systems, has antagonistic properties (Tavernier et al. 1991). It associates with IL-5 dimer in solution. In the mouse, soluble isoforms have also been reported (Takaki et al. 1990), but are associated with an alternative splicing pattern which causes skipping of the membrane-anchor encoding exon (Takaki et al. 1990; Tavernier et al. 1992).

The expression of the IL-5Rα subunit is restricted to eosinophils and basophils in man, and in addition peritoneal Ly^{1+} and splenic Ly^{1-} B cells in the mouse. In contrast, the β-subunit is expressed on various haemopoietic cell types. AIC2A and AIC2B are invariably co-expressed, albeit at different levels. Upregulation of the human β-subunit has been reported upon treatment with IL-1 or TNF (Watanabe et al. 1992).

■ Binding characteristics of the IL-5R complex

The hIL-5Rα binds hIL-5 with intermediate affinity (k_D ~500pM–1nM). Association with the β-subunit results in a moderate 2–3-fold affinity increase. The β-subunit does not bind IL-5 in the absence of the α-subunit. In the mouse, a more pronounced affinity conversion is observed, with low and high binding affinities of 20–50 nM and 150 pM, respectively. IL-5 binds to its receptor with unidirectional species-specificity: mIL-5 binds equally well to both murine and human α-subunits but hIL-5 has a 100-fold lower affinity for the murine α-chain than for its human counterpart. The use of the β-subunit is not species-specific. The dissociation rate of mIL-5 from the $\alpha\beta$-receptor complex is much slower ($t_{1/2}$ > 60 min) than from the low affinity site ($t_{1/2}$ < 2 min) (Devos et al. 1991). When the IL-5Rα and β chains are co-expressed with the α subunits of IL-3 or GM-CSF, all three ligands will cross-compete for high affinity binding. Such cross-competition follows a hierarchical pattern: IL-3 = GM-CSF>IL-5. However, in the mouse, mIL-5 and mIL-3 do not cross-compete per se due to the presence of AIC2A, a unique β-subunit for mIL-3.

■ Signalling mechanisms

Neither the α- nor β-subunit alone can mediate a signal by IL-5. Furthermore, signalling requires the cytoplasmic tails of both subunits. Neither subunit has a consensus sequence for cytoplasmic kinases, phosphatases, or nucleotide-binding signalling proteins. On the other hand, both cytoplasmic domains contain a motif conserved throughout the cytokine receptor family: box 1 or proline cluster (Murakami et al. 1991) (residues 371–378 of the α-subunit, within residues 456–487 of the β-subunit). Mutations in this proline cluster of either subunit completely abrogate a proliferative signalling response (Sakamaki et al. 1992; our own unpublished results). In addition, the β-subunit contains a second conserved motif (box 2, within residues 518–544)

(Murakami et al. 1991), which is also involved in signalling (Sakamaki et al. 1992). Receptor triggering causes the rapid tyrosine phosphorylation of several substrates, including the β-subunit and many of which are shared with the IL-3 or GM-CSF signalling cascade (Takatsu 1992; Miyajima et al. 1993). A non-receptor tyrosine kinase which becomes physically associated upon ligand binding (shown for GM-CSF and IL-3, but likely also the case for IL-5), is the c-fps/fes proto-oncogene (Hanazono et al. 1993).

Likewise, the Jak2 tyrosine kinase was also shown to be activated by IL-5, IL-3, or GM-CSF. The association with the β-subunit requires integrity of the box 1 motif (Silvennoinen et al. 1993; our own unpublished results). In general a high overlap of the signalling cascade within the cytokine system is apparent. How IL-5-dependent, cell-type specific signals are generated is still unclear at present.

■ References

Bazan, J.F. (1990). Structural design and molecular evolution of a cytokine receptor superfamily. Proc. Natl. Acad. Sci. (USA), 87, 6934–8.

Devos, R., Plaetinck, G., Van der Heyden, J., Cornelis, S., Vandekerckhove, J., Fiers, W., and Tavernier, J. (1991). Molecular basis of a high affinity murine interleukin-5 receptor. EMBO J., 10, 2133–7.

deVos, A.M., Ultsch, M., and Kossiakoff, A.A. (1992). Human growth hormone and extracellular domain of its receptor: crystal structure of the complex. Science, 255, 306–12.

Gorman, D.M., Itoh, N., Jenkins, N.A., Gilbert, D.J., Copeland, N.G., and Miyajima, A. (1992). Chromosomal localization and organization of the murine genes encoding the beta subunits (AIC2A and AIC2B) of the interleukin 3, granulocyte/macrophage colony-stimulating factor, and interleukin 5 receptors. J. Biol. Chem., 267, 15842–8.

Gorman, D.M., Itoh, N., Kitamura, T., Schreurs, J., Yonehara, S., Yahara, I., Arai, K., and Miyajima, A. (1990). Cloning and expression of a gene encoding an interleukin 3 receptor-like protein: identification of another member of the cytokine receptor gene family. Proc. Natl. Acad. Sci. (USA), 87, 5459–63.

Hanazono, Y., Chiba, S., Sasaki, K., Mano, H., Miyajima, A., Arai, K., Yazaki, Y., and Hirai, H. (1993). c-fps/fes protein-tyrosine kinase is implicated in a signaling pathway triggered by granulocyte-macrophage colony-stimulating factor and interleukin-3. EMBO J., 12, 1641–6.

Hayashida, K., Kitamura, T., Gorman, D.M., Arai, K., Yokota, T., and Miyajima, A. (1990). Molecular cloning of a second subunit of the receptor for human granulocyte-macrophage colony-stimulating factor (GM-CSF): reconstitution of a high-affinity GM-CSF receptor. Proc. Natl. Acad. Sci. (USA), 87, 9655–9.

Itoh, N., Yonehara, S., Schreurs, J., Gorman, D.M., Maruyama, K., Ishii, A., Yahara, I., Arai, K., and Miyajima, A. (1990). Cloning of an interleukin-3 receptor gene: a member of a distinct receptor gene family. Science, 247, 324–7.

Kitamura, T., Sato, N., Arai, K., and Miyajima, A. (1991). Expression cloning of the human IL-3 receptor cDNA reveals a shared beta subunit for the human IL-3 and GM-CSF receptors. Cell, 66, 1165–74.

Lopez, A.F., Elliott, M.J., Woodcock, J., and Vadas, M.A. (1992). GM-CSF, IL-3 and IL-5: cross-competition on human haemopoietic cells. Immunol. Today, 13, 495–500.

Murakami, M., Narazaki, M., Hibi, M., Yawata, H., Yasukawa, K., Hamaguchi, M., Taga, T., and Kishimoto, T. (1991). Critical cytoplasmic region of the interleukin 6 signal transducer gp130 is conserved in the cytokine receptor family. *Proc. Natl. Acad. Sci. (USA)*, **88**, 11349–53.

Murata, Y., Takaki, S., Migita, M., Kikuchi, Y., Tominaga, A., and Takatsu, K. (1992). Molecular cloning and expression of the human interleukin 5 receptor. *J. Exp. Med.*, **175**, 341–51.

Miyajima, A., Mui, A., Ogorochi, T., and Sakamaki, K. (1993). Receptors for granulocyte-macrophage colony-stimulating factor, interleukin-3, and interleukin-5. *Blood*, **82**, 1960–74.

Sakamaki, K., Miyajima, I., Kitamura, T., and Miyajima, A. (1992). Critical cytoplasmic domains of the common beta subunit of the human GM-CSF, IL-3 and IL-5 receptors for growth signal transduction and tyrosine phosphorylation. *EMBO J.*, **11**, 3541–9.

Shen, Y., Baker, E., Callen, D.F., Sutherland, G.R., Willson, T.A., Rakar, S., and Gough, N.M. (1992). Localization of the human GM-CSF receptor beta chain gene (CSF2RB) to chromosome 22q12.2–:q13.1. *Cytogenet. Cell. Genet.*, **61**, 175–7.

Silvennoinen, O., Witthuhn, B., Quelle, F., Cleveland, J., Yi, T., and Ihle, J. (1993). Structure of the murine Jak2 protein-tyrosine kinase and its role in interleukin 3 signal transduction. *Proc. Natl. Acad. Sci. (USA)*, **90**, 8429–33.

Takaki, S., Tominaga, A., Hitoshi, Y., Mita, S., Sonoda, E., Yamaguchi, N., and Takatsu, K. (1990). Molecular cloning and expression of the murine interleukin-5 receptor. *EMBO J.*, **9**, 4367–74.

Takatsu, K. (1992). Interleukin-5. *Curr. Opin. Immunol.*, **4**, 299–306.

Tavernier, J., Devos, R., Cornelis, S., Tuypens, T., Van der Heyden, J., Fiers, W., and Plaetinck, G. (1991). A human high affinity interleukin-5 receptor (IL5R) is composed of an IL5-specific alpha chain and a beta chain shared with the receptor for GM-CSF. *Cell*, **66**, 1175–84.

Tavernier, J., Tuypens, T., Plaetinck, G., Verhee, A., Fiers, W., and Devos, R. (1992). Molecular basis of the membrane-anchored and two soluble isoforms of the human interleukin 5 receptor alpha subunit. *Proc. Natl. Acad. Sci. (USA)*, **89**, 7041–5.

Tuypens, T., Plaetinck, G., Baker, E., Sutherland, G., Brusselle, G., Fiers, W., Devos, R., and Tavernier, J. (1992). Organization and chromosomal localization of the human interleukin 5 receptor alpha-chain gene. *Eur. Cytokine Network*, **3**, 451–9.

Watanabe, Y., Kitamura, T., Hayashida, K., and Miyajima, A. (1992). Monoclonal antibody against the common beta subunit (beta c) of the human interleukin-3 (IL-3), IL-5, and granulocyte-macrophage colony-stimulating factor receptors shows upregulation of beta c by IL-1 and tumour necrosis factor-alpha. *Blood*, **80**, 2215–20.

Jan Tavernier:
Roche Research,
Gent, Belgium

Interleukin-6 (IL-6)

Interleukin-6 (IL-6) was originally identified as a T-cell-derived lymphokine that induces the final maturation step of B cells into antibody-producing cells. Subsequent studies revealed that IL-6 possesses pleiotropic activities that play a central role in host defence. These activities include (1) immunoglobulin secretion in B cells, (2) production of various acute-phase proteins in liver cells, (3) growth promotion on various B cells such as myeloma, plasmacytoma and hybridoma cells, (4) maturation of megakaryocytes, (5) neuroneeal differentiation, and (6) osteoclast activation. Reflecting its wide variety of activities, IL-6 has been implicated in the pathology of many diseases including Castleman's disease, multiple myeloma, rheumatoid arthritis, and postmenopausal osteoporosis. Selective inhibition of IL-6 action may have therapeutic benefit against the IL-6-associated diseases. On the other hand, IL-6 has clinical utility for the treatment of radiation or chemotherapy induced myelosuppression.

■ Alternative names

B cell stimulatory factor 2 (BSF-2), B cell differentiation factor (BCDF), interferon-β2 (IFNβ2), 26-kDa protein, hybridoma growth factor (HGF), hepatocyte-stimulating factor (HSF).

■ Sequence

Human IL-6 consists of 212 amino acids including a 28 amino acid signal peptide (Hirano *et al.* 1986) (GenBank accession number X04602), whereas mouse and rat IL-6 consist of 211 amino acids with a 24-amino-acid signal peptide (Van Snick *et al.* 1988; Northemann *et al.* 1989) (GenBank accession number X06203 for mouse IL-6, M26744 for rat IL-6). Comparison of the amino acid sequence of human IL-6 with that of mouse IL-6 shows a homology of only 42 per cent, although the mouse and rat protein sequence are 93 per cent identical. Human IL-6 acts on murine cells with the same biological activity as mouse IL-6, but not vice versa. The central portion of IL-6 is more conserved (57 per cent for the region spanning residues 42–102) and contains the four cysteine residues of the protein in perfect alignment. This motif is also found in human and mouse G-CSF and in chicken myelomonocytic growth factor (MGF). .

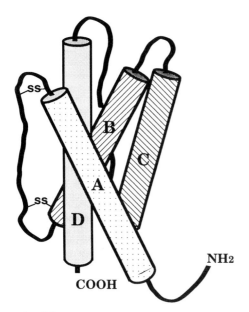

Figure 1. Schematic structure of IL-6.

Protein

IL-6 is a secreted glycoprotein containing 184 (human) or 187 (mouse) amino acids in the mature protein. The molecular weight of the core protein is about 20 000 Da. The two disulphide bridges have been located between Cys44–Cys50 and Cys73–Cys83 in human IL-6 and between Cys46–Cys52 and Cys75–Cys85 in mouse IL-6 (Simpson *et al.* 1988). The molecular weight of natural IL-6 is 21–26 000 Da depending on the cellular source. Its heterogeneity results from post-translational modifications such as N- and O-linked glycosylation and phosphorylation (position 45 and 144 in human IL-6 are N-glycosylated). IL-6 is predicted to have a tertiary fold (Fig. 1) which is similar to the four-α-helix bundle structure found in growth hormone, despite little similarity in amino acid sequence (Bazan 1990*a*,*b*). The four α-helices (labelled A to D in Fig. 1) and loops (two long A–B and C–D loops, and a short B–C loop) predicted in the IL-6 protein are adopted in other cytokines including G-CSF, MGF, prolactin (PRL) and erythropoietin (EPO). Struc-

ture/function studies have suggested that the surface of the predicted helix D is the primary receptor-binding structure.

Gene and transcription

The human and mouse IL-6 genes are approximately 5 and 7 kb long, respectively, and both consist of five exons and four introns (Yasukawa *et al.* 1987; Tanabe *et al.* 1988). The gene organization of IL-6 is very similar to that of G-CSF. The genes for human and mouse IL-6 are located on the short arm of chromosome 7 (7p21) and the proximal region of chromosome 5, respectively.

The IL-6 gene is constitutively active in a number of tumour cells such as cardiac myxoma, cervical carcinoma, renal carcinoma, and bladder carcinoma cells. Normal cells do not usually produce IL-6 unless appropriately stimulated. Lipopolysaccharide (LPS) enhances IL-6 production in monocytes and fibroblasts. Various viruses, adenyl cyclase activators (forskolin, cholera toxin, etc.), calcium ionophore, and cycloheximide induce IL-6 production in fibroblasts. Cytokines such as IL-1, TNF, IFNβ, and PDGF enhance IL-6 production in fibroblasts (Akira *et al.* 1993).

Comparison of the human and mouse IL-6 genes showed that the 300 bp sequences of the 5' flanking region from the translational initiation site are highly conserved. Within this region several potential transcriptional control elements are observed (Isshiki *et al.* 1990) (Fig. 2), such as an NF-κB binding site, a retinoblastoma control element, a c-fos serum-responsive element, a cAMP-responsive element, and an AP-1 binding site. Three DNA sequences in the IL-6 promoter have been identified as functional elements for IL-6 gene expression; a 23 bp multiresponse element that might be involved in IL-6 induction by IL-1, TNF, forskolin, and TPA; a 14 bp palindromic sequence that is recognized by NF-IL6 which is a member of the C/EBP family of transcription factors (Akira *et al.* 1990); and an NF-κB binding site that is responsive for IL-6 induction by IL-1 and TNF.

Biological functions

IL-6 was first recognized as a T-cell-derived factor acting on B cells to induce immunoglobulin secretion. Thus, it was

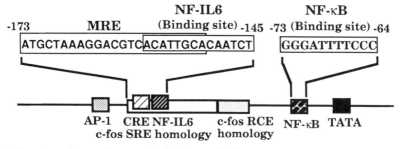

Figure 2. Transcriptional regulatory elements in the human IL-6 promoter.

purified and molecularly cloned based on the assay for B cell differentiation factor (BCDF) activity, in which 1 unit/ml was defined as the activity that induces 50 per cent of the maximum response of IgM production in a B cell line SKW6-CL4 (Hirano *et al.* 1986). IL-6 is a pleiotropic cytokine and acts on a wide variety of tissues (Kishimoto 1989; Kishimoto *et al.* 1992). For example, IL-6 promotes the growth of myeloma/plasmacytoma/hybridoma cells, Kaposi's sarcoma, T cells, keratinocytes, and renal mesangial cells. On the other hand, IL-6 inhibits the growth of myeloid leukaemic cell lines. In addition to its influence on cell growth, IL-6 also induces differentiation of myeloid leukaemic cell lines into macrophages, cytotoxic T cell differentiation, megakaryocyte maturation, neural differentiation of PC12 cells, immunoglobulin production in B cells, and acute-phase protein synthesis in hepatocytes.

■ Pathology

In accordance with its pleiotropic functions, IL-6 has been suggested to be involved in the pathogenesis of various diseases (Akira *et al.* 1993; Kishimoto 1989)—polyclonal B cell activation with autoimmune symptoms in cardiac myxoma, uterine cervical carcinoma, rheumatoid arthritis, Castleman's disease, systemic lupus erythematosus (SLE), and AIDS; lymphoid malignancies, such as multiple myeloma and Lennert's T cell lymphoma. IL-6 is also involved in the generation of mesangial proliferative glomerulonephritis and in the development of AIDS Kaposi's sarcoma. In mice, ovariectomy (which causes oestrogen loss) results in an IL-6-mediated stimulation of osteoclastogenesis, leading to osteoporosis (Jilka *et al.* 1992). In transgenic mice carrying the IL-6 gene, high levels of serum IL-6 and IgG1 were detected, and a massive plasmacytosis developed in thymus, lymph node, and spleen. They also showed a mesangial proliferative glomerulonephritis and an increase of megakaryocytes in the bone marrow (Suematsu *et al.* 1989).

■ References

Akira, S., Isshiki, H., Sugita, T., Tanabe, O., Kinoshita, S., Nishio, Y., Nakajima, T., Hirano, T., and Kishimoto, T. (1990). A nuclear factor for IL-6 expression (NF-IL6) is a member of a C/EBP family. *EMBO J.*, **9**, 1897–906.

Akira, S., Taga, T., and Kishimoto, T. (1993). Interleukin-6 in biology and medicine. *Adv. Immunol.*, **54**, 1–78.

Bazan, J.F. (1990*a*). Haemopoietic receptors and helical cytokines. *Immunol. Today*, **11**, 350–4.

Bazan, J.F. (1990*b*). Structural design and molecular evolution of a cytokine receptor superfamily. *Proc. Natl. Acad. Sci. (USA)*, **87**, 6934–8.

Hirano, T., Yasukawa, K., Harada, H., Taga, T., Watanabe, Y., Matsuda, T., Kashiwamura, S., Nakajima, K., Koyama, K., Iwamatsu, A., Tsunasawa, S., Sakiyama, F., Matsui, H., Takahara, Y., Taniguchi, T., and Kishimoto, T. (1986). Complementary DNA for a novel human interleukin (BSF-2) that induces B lymphocytes to produce immunoglobulin. *Nature*, **324**, 73–6.

Isshiki, H., Akira, S., Tanabe, O., Nakajima, T., Shimamoto, T., Hirano, T., and Kishimoto, T. (1990). Constitutive and interleukin-1 (IL-1)-inducible factors interact with the IL-1-responsive element in the IL-6 gene. *Mol. Cell. Biol.* **10**, 2757–64.

Jilka, R.L., Hangoc, G., Girasole, G., Passeri, G., Williams, D.C., Abrams, J.S., Boyce, B., Broxmeyer, H., and Manolagas, S.C. (1992). Increased osteoclast development after estrogen loss: mediation by interleukin-6. *Science*, **257**, 88–91.

Kishimoto, T. (1989). The biology of interleukin-6. *Blood*, **74**, 1–10.

Kishimoto, T., Akira, S., and Taga, T. (1992). Interleukin-6 and its receptor: a paradigm for cytokines. *Science*, **258**, 593–7.

Northemann, W., Braciak, T.A., Hattori, M., Lee, F., and Fey, G.H. (1989). Structure of the rat interleukin 6 gene and its expression in macrophage-derived cells. *J. Biol. Chem.*, **264**, 16072–82.

Simpson, R.J., Moritz, R.L., Van Roost, E., and Van Snick, J. (1988). Characterization of a recombinant murine interleukin-6: assignment of disulphide bonds. *Biochem. Biophys. Res. Commun.*, **157**, 364–72.

Suematsu, S., Matsuda, T., Aozasa, K., Akira, S., Nakano, N., Ohno, S., Miyazaki, J., Yamamura, K., Hirano, T., and Kishimoto, T. (1989). IgG1 plasmacytosis in interleukin 6 transgenic mice. *Proc. Natl. Acad. Sci. (USA)*, **86**, 7547–51.

Tanabe, O., Akira, S., Kamiya, T., Wong, G.G., Hirano, T., and Kishimoto, T. (1988). Genomic structure of the murine IL-6 gene. High degree conservation of potential regulatory sequences between mouse and human. *J. Immunol.*, **141**, 3875–81.

Van Snick, J., Cayphas, S., Szikora, J.-P., Renauld, J.-C., Van Roost, E., Boon, T., and Simpson, R.J. (1988). cDNA cloning of murine interleukin-HP1: homology with human interleukin 6. *Eur. J. Immunol.*, **18**, 193–7.

Yasukawa, K., Hirano, T., Watanabe, Y., Muratani, K., Matsuda, T., and Kishimoto, T. (1987). Structure and expression of human B cell stimulatory factor-2 (BSF-2/IL-6) gene. *EMBO J.*, **6**, 2939–45.

Masashi Narazaki:
Institute for Molecular and Cellular Biology,
Osaka University 1–3,
Yamada-oka, Suita, Osaka 565, Japan

Tadamitsu Kishimoto:
Department of Medicine III,
Osaka University Medical School 2–2,
Yamada-oka, Suita, Osaka 565, Japan

Receptors for Interleukin-6 (IL-6)

The IL-6 receptor system consists of two membrane proteins, a ligand-binding receptor (IL-6R) and a non-binding signal transducer (gp130), both of which belong to the haemopoietic cytokine receptor family. Binding of IL-6 to IL-6R triggers the association of IL-6R and gp130, forming a high affinity IL-6-binding site, and gp130, in turn, transduces the signal. gp130 is also essential for signal transduction by leukaemia inhibitory factor (LIF), oncostatin M (OSM), ciliary neurotrophic factor (CNTF) and IL-11, in addition to IL-6. Functional redundancy of these five cytokines could be explained by a model in which gp130 interacts with different receptor chains of each cytokine and serves as a common signal transducer.

■ The IL-6 receptor

The human IL-6 receptor consists of 468 amino acids, including a signal peptide of 19 amino acids, an extracellular region of 339 amino acids, a membrane-spanning region of 28 amino acids, and a cytoplasmic region of 82 amino acids (Yamasaki *et al.* 1988) (GenBank accession numbers M20566, X12830). The predicted molecular weight of IL-6R is 50 000 Da, although the observed molecular weight is 80 000 Da due to N-glycosylation. In the amino terminus of the extracellular region, IL-6R has a domain of about 90 amino acids which fulfils the criteria for the constant 2 (C2) set of the immunoglobulin supergene family. The remaining extracellular region of IL-6R has the features characteristic of the haemopoietic cytokine receptor family; four conserved cysteine residues distributed in its amino-terminal half and a Trp-Ser-X-Trp-Ser motif at the carboxy-terminal end (Bazan 1990*a,b*) (Fig. 1). This homologous module comprises two tandemly placed fibronectin type III domains

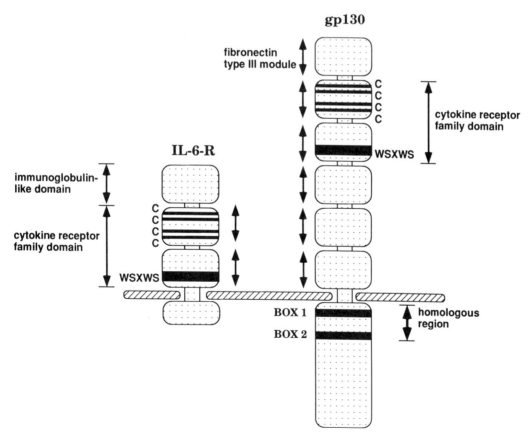

Figure 1. Structure of IL-6R and gp130.

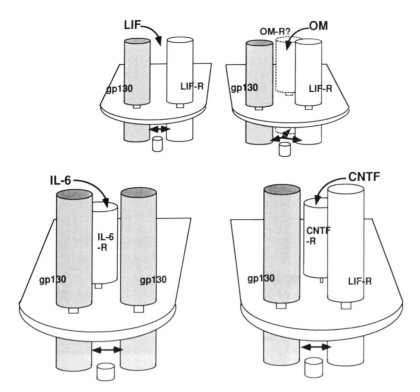

Figure 2. Receptor complexes for IL-6, CNTF, LIF, and OSM.

each of which is composed of seven antiparallel β-sheets and forms two β-barrels. The cytoplasmic region of IL-6R is relatively short, and deletion of this region did not affect the cellular responsiveness to IL-6.

The mature murine IL-6R is 441 amino acids long (Sugita et al. 1990) (GenBank accession number X51975), and has 69 per cent identity at the amino acid level with human IL-6R.

■ The IL-6 signal transducer, gp130

Upon binding of IL-6, IL-6R becomes associated with a signal transducing receptor component, gp130 (Taga et al. 1989). gp130 has no intrinsic IL-6 binding capability, but is involved in the formation of high-affinity IL-6 binding sites. gp130 consists of 918 amino acids, including a leader sequence of 22 amino acids, an extracellular region of 597 amino acids, a membrane-spanning region of 22 amino acids, and a cytoplasmic region of 277 amino acids (Hibi et al. 1990) (GenBank accession number M57230). The extracellular region of human as well as mouse gp130 comprises six repeats of the fibronectin type III domain. The second and third type III domains compose the haemopoietic cytokine receptor family module (Fig. 1). The predicted molecular weight of the mature gp130 of 896 amino acids is 101 000 Da, but its native molecular weight is 130 000 Da because of glycosylation. The gp130 protein serves as a signal transducer not only for IL-6 but also for leukaemia

inhibitory factor (LIF), oncostatin M (OSM), ciliary neurotrophic factor (CNTF), and IL-11 (Taga and Kishimoto 1992; Yin et al. 1993). Stimulation by these cytokines induces oligomerization of the receptor components (Davis et al. 1993; Murakami et al. 1993) (Fig. 2).

The mature murine gp130 is composed of 895 amino acids (Saito et al. 1992) (GenBank accession number M83336), and has the overall 77 per cent identity at the amino acid level with human gp130. Especially a stretch of 116 amino acid residues on both ends of transmembrane region is identical with that of human gp130.

■ Binding characteristics

IL-6R has an intrinsic capacity for IL-6 but the association of IL-6R and gp130 is necessary for signal transduction. In the presence of IL-6 at 37 °C, the association of IL-6R and gp130 took place within five minutes and was stable for at least 40 minutes but was not observed at 0 °C. Although IL-6R has a low IL-6 binding affinity (K_D = 5nM), after association with gp130 the affinity becomes higher (K_D = 40–70pM) (Hibi et al. 1990).

■ IL-6R and gp130 expression

The IL-6R is expressed in human and mouse peripheral blood mononuclear cells, monocytes, and T cells (both CD4+

and CD8+ subpopulations). B cells do not usually express IL-6R, but they are induced to express IL-6R when stimulated *in vitro* with mitogen. Among cell lines, IL-6R is expressed in the EB-virus transformed B cell line (CESS), the T cell line (KT-3), the myeloma cell line (U266), hepatoma cell lines (HepG2, Hep3B) and the histiocytoma cell line (U937) (Taga *et al.* 1987). gp130 is expressed in nearly all human and mouse cell lines, exceptionally absent in a mouse pro-B cell line BAF.BO3 (Hibi *et al.* 1990; Saito *et al.* 1992). In the mouse, such tissues as brain, thymus, heart, lung, spleen, liver, and kidney have been shown to express gp130. Naturally produced soluble forms of IL-6R and gp130 exist in human serum (Narazaki *et al.* 1993).

■ Signalling mechanism

After binding of IL-6 to IL-6R, this complex interacts with gp130 and induces disulphide-linked homodimerization of gp130 (Fig. 2). Although gp130 has no obvious enzymatic motifs in its structure, homodimerized gp130 associates with a tyrosine kinase activity and tyrosine phosphorylation of gp130 occurs (Murakami *et al.* 1993). Recently, it has been shown that Janus kinase (JAK) family, including JAK1, JAK2, and Tyk2, are activated after stimulation of gp130 with cytokines such as IL-6, LIF, and CNTF, whose receptors share the signal transducer gp130. Both of JAK1 and JAK2 can associate with gp130 prior to stimulation with ligand (Stahl *et al.* 1994). In addition, APRF (acute-phase response factor), a transcription factor that binds to the promotor region of various acute phase protein genes, is tyrosin-phosphorylated and rapidly migrates to the nucleus after stimulation of gp130 (Akira *et al.* 1994). Mutational analysis of gp130 showed that a 61-amino-acid region proximal to the transmembrane domain of gp130 was sufficient for growth signalling (Murakami *et al.* 1991). This region contains two highly conserved segments among haemopoietic cytokine receptor family members. One segment (box 1; Ile651 to Pro658) is conserved in almost all members of the family, and the other (box 2; Val691 to Lys702) is found in G-CSF-R, IL-2Rβ, EPO-R, GM-CSF-R, and IL-3R. It is suggested that a similar mechanism might be involved in the signalling processes of various cytokines.

■ References

Akira, S., Nishio, Y., Inoue, M., Wang, X.-J., Wei, S., Matsusaka, T., Yoshida, K., Sudo, T., Naruto, M., and Kishimoto, T. (1994). Molecular cloning of APRF, a novel IFN-stimulated gene factor 3 p91-related transcription factor involved in the gp130-mediated signalling pathway. *Cell*, **77**, 63–71.

Bazan, J.F. (1990a). Haemopoietic receptors and helical cytokines. *Immunol. Today*, **11**, 350–4.

Bazan, J.F. (1990b). Structural design and molecular evolution of a cytokine receptor superfamily. *Proc. Natl. Acad. Sci. (USA)*, **87**, 6934–8.

Davis, S., Aldrich, T.H., Stahl, N., Pan, L., Taga, T., Kishimoto, T., Ip, N.Y., and Yancopoulos, G.D. (1993). LIFR beta and gp130 as heterodimerizing signal transducers of the tripartite CNTF receptor. *Science*, **260**, 1805–8.

Hibi, M., Muakami, M., Saito, M., Hirano, T., Taga, T., and Kishimoto, T. (1990). Molecular cloning and expression of an IL-6 signal transducer, gp130. *Cell*, **63**, 1149–57.

Murakami, M., Narazaki, M., Hibi, M., Yawata, H., Yasukawa, K., Hamaguchi, M., Taga, T., and Kishimoto, T. (1991). Critical cytoplasmic region of the interleukin 6 signal transducer gp130 is conserved in the cytokine receptor family. *Proc. Natl. Acac. Sci. (USA)*, **88**, 11349–53.

Murakami, M., Hibi, M., Nakagawa, N., Nakagawa, T., Yasukawa, K., Yamanishi, K., Taga, T., and Kishimoto, T. (1993). IL-6-induced homodimerization of gp130 and associated activation of a tyrosine kinase. *Science*, **260**, 1808–10.

Narazaki, M., Yasukawa, K., Saito, T., Ohsugi, Y., Fukui, H., Koishihara, Y., Yancopoulos, G.D., Taga, T., and Kishimoto, T. (1993). Soluble forms of the interleukin-6 signal-transducing receptor component gp130 in human serum possessing a potential to inhibit signals through membrane-anchored gp130. *Blood*, **82**, 1120–6.

Saito, M., Yoshida, K., Hibi, M., Taga, T., and Kishimoto, T. (1992). Molecular cloning of a murine IL-6 receptor-associated signal transducer, gp130, and its regulated expression in vivo. *J. Immunol.*, **148**, 4066–71.

Stahl, N., Boulton, T.G., Farruggella, T., Ip, N.Y., Davis, S., Witthuhn, B.A., Quelle, F.W., Silvannoinen, O., Barbieri, G., Pellegrini, S., Ihle, J.N., and Yancopoulos, G.D. (1994). Association and activation of Jak-Tyk kinase by CNTF-LIF-OSM-IL-6 β receptor components. *Science*, **263**, 92–5.

Sugita, T., Totsuka, T., Saito, M., Yamasaki, K., Taga, T., Hirano, T., and Kishimoto, T. (1990). Functional murine interleukin 6 receptor with the intracisternal A particle gene product at its cytoplasmic domain. Its possible role in plasmacytomagenesis. *J. Exp. Med.*, **171**, 2001–9.

Taga, T., and Kishimoto, T. (1992). Cytokine receptors and signal transduction. *FASEB J.*, **6**, 3387–96.

Taga, T., Kawanishi, Y., Hardy, R.R., Hirano, T., and Kishimoto, T. (1987). Receptors for B cell stimulatory factor 2. Quantitation, specificity, distribution, and regulation of their expression. *J. Exp. Med.*, **166**, 967–81.

Taga, T., Hibi, M., Hirata, Y., Yamasaki, K., Yasukawa, K., Matsuda, T., Hirano, T., and Kishimoto, T. (1989). Interleukin-6 triggers the association of its receptor with a possible signal transducer, gp130. *Cell*, **58**, 573–81.

Yamasaki, K., Taga, T., Hirata, Y., Yawata, H., Kawanishi, Y., Seed, B., Taniguchi, T., Hirano, T., and Kishimoto, T. (1988). Cloning and expression of the human interleukin-6 (BSF-2/IFN beta 2) receptor. *Science*, **241**, 825–8.

Yin, T., Taga, T., Tsang, M.L.-S., Yasukawa, K., Kishimoto, T., and Yang, Y.-C. (1993). Involvement of IL-6 signal transducer gp130 in IL-11-mediated signal transduction. *J. Immunol.*, **151**, 2555–61.

Masashi Narazaki:
Institute for Molecular and Cellular Biology,
Osaka University 1–3,
Yamada-oka, Suita, Osaka 565, Japan

Tadamitsu Kishimoto:
Department of Medicine III,
Osaka University Medical School 2–2
Yamada-oka, Suita, Osaka 565, Japan

Interleukin-7 (IL-7)

Interleukin-7 is an approximately 25 kDa glycoprotein whose primary biological activity is concerned with development of lymphocytes. It has been well established that IL-7 is a potent proliferative stimulus for both B cell and T cell progenitors. In addition, IL-7 has been shown to act as a growth factor for mature functional T cells and to promote the generation of lytically active antigen-specific cytotoxic T cells and lymphokine activated killer (LAK) cells.

■ Alternative names

Lymphopoeitin-1 (LP-1) and pre-B-cell growth factor (PBGF).

■ IL-7 sequence

The nucleotide sequence of murine IL-7 was derived from a cDNA isolated from a bone marrow stromal cell line and predicted a protein composed of 129 amino acids (Namen *et al.* 1988) (GenBank accession number X07962). The human cDNA was isolated by cross-hybridization from a human hepatoma cell line (Goodwin *et al.* 1989). The cDNA sequence predicted a protein containing 152 amino acids (GenBank accession number J04156). Comparison of the murine and human cDNAs shows a high homology in the coding region (81 per cent) and a lesser homology in the 5′ noncoding (73 per cent) and 3′ non-coding region (63 per cent).

■ IL-7 protein

Native IL-7 is a 25 kDa glycoprotein which is secreted as a monomer from some stromal cell lines. Both murine and human IL-7 contain three N-linked glycosylation sites, which are not essential for bioactivity. Each of the proteins also contains six cysteine residues which are highly conserved and absolutely essential for bioactivity. Unlike a number of other growth factors and cytokines, IL-7 does not exhibit a species specificity as both murine and human IL-7 are cross-reactive with comparable specific activities (Goodwin and Namen 1992). Human IL-7 exhibits two major bands at 22 and 25 kDa by SDS-PAGE, and exhibits an isoelectric point of approximately 9.0. The purified protein is quite stable and resistant to denaturation (in 1 per cent SDS) as long as the disulphide bridges remain undisturbed. Since murine and human IL-7 are cross-reactive, the major focus has been the production and purification of human IL-7 (Armitage *et al.* 1990).

■ IL-7 gene

The murine IL-7 gene consists of greater than 56 kbp and transcription results in the production of RNA transcripts of 1.5, 1.7, 2.6, and 2.9 kb. The human IL-7 gene consists of greater than 33 kbp and is located at 8q12–13. The gene contains six exons and five introns and yields RNA transcripts of 1.8 and 2.4 kb. Human IL-7 contains an 18 amino acid segment which has not been found in any preparations of murine IL-7 (Lupton *et al.* 1990). This additional segment is encoded by exon 5 of the human gene. Removal of this 54 bp human-specific exon had no effect upon the biological activity. The physiological role of this additional exon is unknown.

Adherent stromal cells from a number of tissues (including bone marrow, spleen, thymus, and kidney) constitutively produce low levels of IL-7. IL-7 production is highly controlled and refractory to stimulation with mitogens and other growth factors. This tight regulation may be due to the unusual 5′ non-coding region, which contains multiple initiation sites (8) distributed over a 200 bp regulatory region. The canonical TATA sequences and other common regulatory sequences usually found in eukaryotic promoters are absent from the 5′ flanking region. There is, however, conservation of the 'helix–loop–helix' recognition sequences for DNA-binding proteins in the 5′ portion of both the murine and human genes.

■ Biological functions

IL-7 was originally identified as a pre-B cell growth factor isolated from bone marrow stroma. More recent studies have shown that IL-7 is a potent growth stimulus for thymocytes and, in particular, the earlier stages of developing thymocytes (Conlon *et al.* 1989; Henney 1989; Suda *et al.* 1990). It is also now well established that IL-7 acts as a growth factor for mature peripheral T cells and T cell clones; however, a similar effect on mature IgM+B cells has not been observed (Grabstein *et al.* 1990; Londei *et al.* 1990). This T cell stimulation includes both the CD4+ and the CD8+ subsets of T cells. Additionally, IL-7 enhances the generation of cytotoxic T cells (Alderson *et al.* 1990) and lymphokine-activated killer cells (Lynch and Miller 1990) in an IL-2 independent mechanism. While most of the biological effects of IL-7 involve lymphocytes, it was recently demonstrated that IL-7 has potent effects on human peripheral blood monocytes. The treatment of purified monocytes with IL-7 induced the secretion of high levels of IL-1α, IL-1β, IL-6, and TNFα. IL-7 treatment also strongly stimulated the monocyte-mediated lysis of the human melanoma cell line A375 (Alderson *et al.* 1991). Thus, IL-7 may have a broader range of activities than previously believed and may play an important role in the inflammatory immune response as well as lymphocyte development and activation. The current known biological activities of IL-7 are summarized in Table 1.

Table 1. Biological activities of IL-7

1. Stimulates the proliferation of progenitor B cells.
2. Stimulates the proliferation and differentiation of T cell progenitors.
3. Is a potent proliferative stimulus of mature peripheral T cells (both CD+ and CD8+) with or without a co-mito-genic signal.
4. Induces the generation of cytotoxic T cells.
5. Induces the generation of lymphokine activated killer cells.
6. Stimulates the lytic activity of peripheral blood monocytes.
7. Induces peripheral blood monocytes to secrete high levels of cytokines, including IL-1α, IL-1β, IL-6, and TNFα.
8. Induces the proliferation of some B and T cell leukae-mias (Touw *et al.* 1990).

■ Pathology

Transgenic mice have been established which have targeted expression of IL-7 in the lymphoid compartment (Samaridis *et al.* 1991). The transgene is expressed in bone marrow, spleen, and thymus, but not in liver, kidney, brain, or heart. These transgenic mice exhibit elevated levels of B cell precursors and mature B cells in the bone marrow and spleen. Additionally, the numbers of thymocytes and pe-ripheral T cells is also elevated, but a normal phenotype and distribution of T cell subsets is observed. Both of the lym-phoid compartments are functionally competent in the transgenic mice. Interestingly, the myeloid compartment (granulocytes and monocytes) is indistinguishable from lit-termate controls. Another *in vivo* study utilized an anti-IL-7 neutralizing monoclonal antibody. Anti-IL-7 treatment completely inhibits the production of B cell precursors from the pro-β cell stage onwards (Grabstein *et al.* 1993). Also, anti-IL-7 treatment severely depleted the thymic cellularity, although the remaining thymocytes exhibited a normal subset distribution. These studies indicate that IL-7 plays a pivotal role in development of both B cells and T cells.

■ Therapy

IL-7 is only just entering experimentation exploring its po-tential as a therapeutic in tumour immunity and gene trans-fer. In a number of different tumour models (Jicha *et al.* 1991; Lynch *et al.* 1991; Jicha *et al.* 1992), it has been demon-strated that IL-7 can mediate the generation of anti-tu-mour cytotoxic lymphocytes which display an enhanced

immunotherapeutic efficacy upon cellular adoptive immu-notherapy. Other studies (Hock *et al.* 1991; McBride *et al.* 1992; Aoki *et al.* 1992) have indicated that the introduction of the IL-7 gene into a number of different tumour cell lines resulted in a marked induction of tumour immunity when injected into susceptible animals, with a resultant rejection of the tumour cells. While the preliminary results look promising, further study will be required to clarify the usefulness of IL-7 as a therapeutic in tumour-associated pathologies.

■ References

Alderson, M.R., Sassenfeld, H.M., and Widmer, M.B. (1990). Interleu-kin 7 enhances cytolytic T lymphocyte generation and induces lymphokine-activated killer cells from human peripheral blood. *J. Exp. Med.*, **172**, 577–87.

Alderson, M.R., Tough, T.W., Ziegler, S.F., and Grabstein, K.H. (1991). Interleukin 7 induces cytokine secretion and tumouricidal activity by human peripheral blood monocytes. *J. Exp. Med.*, **173**, 923–30.

Aoki, T., Tashiro, K., Miyatake, S., Kinashi, T., Nakano, T., Oda, Y., Kikuchi, H., and Honjo, T. (1992). Expression of murine interleukin 7 in a murine glioma cell line results in reduced tumourigenicity *in vivo*. *Proc. Natl. Acad. Sci. (USA)*, **89**, 3850–4.

Armitage, R.J., Namen, A.E., Sassenfeld, H.M., and Grabstein, K.H. (1990). Regulation of human T cell proliferation by IL-7. *J. Immu-nol.*, **144**, 938–41.

Conlon, P.J., Morrissey, P.J., Nordan, R.P., Grabstein, K.H., Prickett, K.S., Reed, S.G., Goodwin, R., Cosman, D., and Namen, A.E. (1989). Murine thymocytes proliferate in direct response to interleukin-7. *Blood*, **74**, 1368–73.

Goodwin, R.G., and Namen, A.E. (1992). In *Human cytokines: hand-book for basic and clinical research* (eds B.B. Aggarwal and J.V. Gutterman), pp. 168–80. Blackwell Scientific, Boston.

Goodwin, R.G., Lupton, S., Schmierer, A., Hjerrild, K.J., Jerzy, R., Clevenger, W., Gillis, S., Cosman, D., and Namen, A.E. (1989). Human interleukin 7: molecular cloning and growth factor ac-tivity on human and murine B-lineage cells. *Proc. Natl. Acad. Sci. (USA)*, **86**, 302–6.

Grabstein, K.H., Namen, A.E., Shanebeck, K., Voice, R.F., Reed, S.G., and Widmer, M.B. (1990). Regulation of T cell proliferation by IL-7. *J. Immunol.*, **144**, 3015–20.

Grabstein, K.H., Waldschmidt, T.J., Finkelman, F.D., Hess, B.W., Al-pert, A.R., Boiani, N.E., Namen, A.E., and Morrissey, P.J. (1993). Inhibition of murine B and T lymphopoiesis *in vivo* by an anti-interleukin 7 monoclonal antibody. *J. Exp. Med.*, **178**, 257–64.

Henney, C.S. (1989). Interleukin 7: effects on early events in lympho-poiesis. *Immunol. Today*, **10**, 170–3.

Hock, H., Dorsch, M., Diamantstein, T., and Blankenstein, T. (1991). Interleukin 7 induces CD4+ T cell-dependent tumour rejection. *J. Exp. Med.*, **174**, 1291–8.

Jicha, D.L., Mule, J.J., and Rosenberg, S.A. (1991). Interleukin 7 generates antitumour cytotoxic T lymphocytes against murine sarcomas with efficacy in cellular adoptive immunotherapy. *J. Exp. Med.*, **174**, 1511–5.

Jicha, D.L., Schwarz, S., Mule, J.J., and Rosenberg, S.A. (1992). Inter-leukin-7 mediates the generation and expansion of murine allo-sensitized and antitumour CTL. *Cell. Immunol.*, **141**, 71–83.

Londei, M., Verhoef, A., Hawrylowicz, C., Groves, J., De Berardinis, P., and Feldmann, M. (1990). Interleukin 7 is a growth factor for mature human T cells. *Eur. J. Immunol.*, **20**, 425–8.

Lupton, S.D., Gimpel, S., Jerzy, R., Brunton, L.L., Hjerrild, K.A., Cosman, D., and Goodwin, R.G. (1990). Characterization of the human and murine IL-7 genes. *J. Immunol.*, **144**, 3592–601.

Lynch, D.H., and Miller, R.E. (1990). Induction of murine lymphokine-activated killer cells by recombinant IL-7. *J. Immunol.*, **145**, 1983–90.

Lynch, D.H., Namen, A.E., and Miller, R.E. (1991). *In vivo* evaluation of the effects of interleukins 2, 4 and 7 on enhancing the immunotherapeutic efficacy of anti-tumour cytotoxic T lymphocytes. *Eur. J. Immunol.*, **21**, 2977–85.

McBride, W.H., Thacker, J.D., Comora, S., Economou, J.S., Kelley, D., Hogge, D., Dubinett, S.M., and Dougherty, G.J. (1992). Genetic modification of a murine fibrosarcoma to produce interleukin 7 stimulates host cell infiltration and tumour immunity. *Cancer Res.*, **52**, 3931–7.

Namen, A.E., Lupton, S., Hjerrild, K., Wignall, J., Mochizuki, D.Y., Schmierer, A., Mosley, B., March, C.J., Urdal, D., and Gillis, S. (1988). Stimulation of B-cell progenitors by cloned murine interleukin-7. *Nature*, **333**, 571–3.

Samaridis, J., Casorati, G., Traunecker, A., Iglesias, A., Gutierrez, J.C., Muller, U., and Palacios, R. (1991). Development of lymphocytes in interleukin 7-transgenic mice. *Eur. J. Immunol.*, **21**, 453–60.

Suda, T., Murray, R., Guidos, C., and Zlotnik, A. (1990). Growth-promoting activity of IL-1 alpha, IL-6, and tumour necrosis factor-alpha in combination with IL-2, IL-4, or IL-7 on murine thymocytes. Differential effects on CD4/CD8 subsets and on CD3+/CD3⁻ double-negative thymocytes. *J. Immunol.*, **144**, 3039–45.

Touw, I., Pouwels, K., van Agthoven, T., van Gurp, R., Budel, L., Hoogerbrugge, H., Delwel, R., Goodwin, R., Namen, A., and Lowenberg, B. (1990). Interleukin-7 is a growth factor of precursor B and T acute lymphoblastic leukaemia. *Blood*, **75**, 2097–101.

Anthony E. Namen:
Immunex Corporation,
51 University Street,
Seattle, WA 98101, USA

Interleukin-7 Receptor

The interleukin-7 (IL-7) receptor is a member of a superfamily of cytokine receptors, which includes the growth hormone and prolactin receptors, that are homologous in their extracellular cytokine binding domains. IL-7 receptors are expressed on early B-lineage cells, thymocytes, mature T cells, and monocytes. Though the cloned receptor is capable of binding IL-7 with high affinity, an associated protein(s) is presumed to be required for its biological signalling. Engagement of the receptor by its ligand stimulates tyrosine phosphorylation and inositol phospholipid turnover.

■ The IL-7 receptor protein

The human IL-7 receptor is produced as a 459-amino-acid precursor with a predicted 20-amino-acid signal peptide (GenBank accession number M29696) (Fig. 1). It contains a single transmembrane domain of 25 amino acids, an extracellular domain of 219 amino acids, and a cytoplasmic domain of 195 amino acids. The mature protein has a predicted molecular weight of 49 500 but an observed molecular weight by SDS-PAGE analysis of about 75 000, presumably due to post-translational modification of one or more of the six potential N-linked glycosylation sites (Goodwin *et al.* 1990). The extracellular domain of the IL-7 receptor shows significant homology to a number of other cytokine receptors, forming what is often referred to as the haemopoietin receptor superfamily. Though the IL-7 receptor has conserved only two of the four cysteine residues found in other members of this family, it does contain the classic Trp-Ser-X-Trp-Ser domain common to other members of this receptor superfamily. Alternatively spliced human IL-7 receptor cDNAs have been identified which delete the transmembrane and cytoplasmic domains (Goodwin *et al.* 1990; Pleiman *et al.* 1991). This results in the production of a secreted, soluble form of the receptor which is still capable of binding IL-7 with high affinity. Several other members of this family of receptors have also been shown to produce soluble forms of their receptors, either encoded by specific transcripts or released *via* the action of proteases. The physiological role of these soluble receptors is unknown at present but they could function as both inhibitors or carriers of their respective ligands.

The murine IL-7 receptor (GenBank accession number M29697) is 459 amino acids long and shows 64 per cent sequence identity with the human protein. Attempts to identify alternatively spliced transcripts which would encode a soluble form of the murine receptor have thus far been unsuccessful (Pleiman *et al.* 1991).

■ IL-7 receptor gene and expression

The human IL-7 receptor gene is at chromosome 5p13 (Lynch *et al.* 1992) and the murine at the proximal end of chromosome 15. This places the IL-7 receptor gene in close proximity to the genes for several other members of this receptor superfamily including the growth hormone and prolactin receptors. Analysis of the 5′ flanking sequence of

```
         ↓
Human    MTILGTTFGMVFSLLQVVSGESGYAQNGDLEDAELDDYSFSCYSQLEVNG   30
         |  ||  |  || | |||||| || ||||| || || | ||||| |
Murine   MMALGRAFAIVFCLIQAVSGESGNAQDGDLEDADADDHSFWCHSQLEVDG   30

Human    SQHSLTCAFEDPDVNTTNLEFEICGALVEVKCLNFRKLQEIYFIETKKFL   80
         ||| |||||| | | ||||| ||||| |||| ||| ||||| |  ||
Murine   SQHLLTCAFNDSDINTANLEFQICGALLRVKCLTLNKLQDIYFIKTSEFL   80

Human    LIGKSNICVKVGEKSLTCKKIDLTTIVKPEAPFDLSVIYREGANDFVVTF   130
         ||| |||||| | | |||    |||| ||| || | | |  |||| ||
Murine   LIGSSNICVKLGQKNLTCKNMAINTIVKAEAPSDLKVVYRKEANDFLVTF   130

Human    NTSHLQKKYVKVLMHDVAYRQEKDENKWTHVNLSSTKLTLLQRKLQPAAM   180
         |  || ||| |    |||||| |     ||||| |  |  ||| | ||
Murine   NAPHLKKKYLKKVKHDVAYRPARGESNWTHVSLFHTRTTIPQRKLRPKAM   180

Human    YEIKVRSIP-DHYFKGFWSEWSPSYYFRTPEINNSSGEMD PILLTISILS   229
         ||||||||   |||||||||||||| ||  | |    |       |||
Murine   YEIKVRSIPHNDYFKGFWSEWSPSSTFETPEPKNQGG-WD PVLPSVTILS   229

Human    FFSVALLVILACVLWKKRIKPIVWPSLPDHKKTLEHLCKKPRKNLNVSFN   279
          ||| |||||| |||||||||| |||||||||||||||| |||||  |||||
Murine   LFSVFLLVILAHVLWKKRIKPVVWPSLPDHKKTLEQLCKKPKTSLNVSFI   279

Human    PESFLDCQIHRVDDIQARDEVEGFLQDTFPQQLEESEKQRLGGDVQSPNC   329
         ||  ||||||| |     ||||||  |    | | ||    |  |||
Murine   PEIFLDCQIHEVKGVEARDEVEIFLPNDLPAQPEELETQGHRAAVHSANR   329

Human    PSEDVVVTPESFGRDSSLTCLAGNVSACDAPILSSSRSLDCRESGKNGPH   379
          |  |   ||  |  | | | |||  | | | | || | |||| | |  | |
Murine   SPETSVSPPETVRRESPLRCLARNLSTCNAPPLLSSRSPDYRDGDRNRPP   379

Human    VYQDLLLSLGTTN—STLPPPFSLQSGILTLNPVAQGQPILTSLGSNQEE   427
         ||||||  | |    ||| | |    ||| |          |    ||||
Murine   VYQDLLPNSGNTNVPVPVPQPLPFQSGILI—PFSQRQPISTSSVLNQEE   427

Human    AYVTMSSFYQNQ   439
         ||||||||||||
Murine   AYVTMSSFYQNK   439
```

Figure 1. Alignment of the human and murine IL-7 receptor amino acid sequences. Vertical lines indicate identical amino acids while horizontal lines mark gaps that were introduced to provide maximum alignment. The arrow indicates the predicted amino-terminus of the mature protein, which is designated amino acid number 1. The predicted transmembrane domains are underlined.

the murine gene has demonstrated a functional interferon regulatory element, to which the interferon-induced nuclear factors IRF-1 and IRF-2 are capable of binding and inducing expression (Pleiman *et al.* 1991).

High affinity receptors for IL-7 have been found on progenitor cells of both B and T lymphocytes, as well as mature T cells (Park *et al.* 1990; Chazen *et al.* 1989). Binding has also been detected in bone-marrow-derived macrophages, and peripheral blood monocytes. IL-7 binding has recently been demonstrated on some leukaemic cells in patients with chronic lymphocytic leukaemia (Frishman *et al.* 1993), and functional receptors are expressed on precursor B and T acute lymphoblastic leukaemia cells (Touw *et al.* 1990; Eder *et al.* 1990; Dibirdik *et al.* 1991). Thus, the possibility exists that the inappropriate expression of IL-7 and/or its receptor may play a role in the growth of leukaemic cells.

■ Binding characteristics of the IL-7 receptor

Binding studies using radiolabelled IL-7 have demonstrated that a variety of human and murine primary cells and cell lines exhibit biphasic binding curves with K_a values in the range of 5×10^9 to $5 \times 10^{10} M^{-1}$ for the high affinity component and 5×10^7 to $1 \times 10^9 M^{-1}$ for the low affinity component. Both the cloned human and murine IL-7 receptor subunits also exhibit biphasic binding curves, with similar high and low affinity components, when expressed transiently in COS-7 cells (Goodwin *et al.* 1990). These results, in addition to results from cross-linking and kinetic experiments, indicate that the low- and high-affinity components result from binding of IL-7 to IL-7 receptor monomers and dimers, respectively. Structural studies of the growth hormone receptor, which also belongs to the haemopoietin receptor superfamily and maps near the IL-7 receptor on chromosome 5p13, have shown that the high affinity binding complex consists of a single molecule of growth hormone binding to a dimer of growth hormone receptor (de Vos *et al.* 1992). It appears likely that the IL-7/IL-7 receptor complex has similar characteristics.

In addition, studies using IL-7 labelled with biotin and flow cytometry have demonstrated the presence of a novel low-affinity receptor for IL-7 that appears to be distinct from the cloned molecule. The estimated affinity of binding of IL-7 to this receptor is 1 to $3 \times 10^6 \, M^{-1}$. The function of this low-affinity receptor is at present unknown, though there is some evidence it may be capable of signalling (Armitage *et al.* 1992).

■ Signalling mechanism

Analysis of the intracellular domain of the IL-7 receptor failed to reveal any structural features common to protein kinases or any other catalytic function. Thus, signal transduction mediated through the IL-7 receptor presumably involves associated proteins. Such proteins have indeed been identified for several members of this family of receptors, and these are described in their respective chapters. Several recent studies have demonstrated that engagement of the IL-7 receptor stimulates tyrosine phosphorylation and inositol phospholipid turnover (Dibirdik *et al.* 1991; Uckun *et al.* 1991a,b). The specific kinase involved in this reaction as well as the proteins which are phosphorylated have yet to be identified.

■ References

Armitage, R.J., Ziegler, S.F., Friend, D.J., Park, L.S., and Fanslow, W.C. (1992). Identification of a novel low-affinity receptor for human interleukin-7. *Blood*, **79**, 1738–45.

Chazen, G.D., Pereira, G.M., Legros, G., Gillis, S., and Shevach, E.M. (1989). Interleukin 7 is a T-cell growth factor. *Proc. Natl. Acad. Sci. (USA)*, **86**, 5923–7.

de Vos, A.M., Ultsch, M., and Kossiakoff, A.A. (1992). Human growth hormone and extracellular domain of its receptor: crystal structure of the complex. *Science*, **255**, 306–12.

Dibirdik, I., Langlie, M.-C., Ledbetter, J.A., Tuel-Ahlgren, L., Obuz, V., Waddick, K.G., Gajl-Peczalska, K., Schieven, G.L., and Uckun, F.M. (1991). Engagement of interleukin-7 receptor stimulates tyrosine phosphorylation, phosphoinositide turnover, and clonal proliferation of human T-lineage acute lymphoblastic leukaemia cells. *Blood*, **78**, 564–70.

Eder, M., Ottmann, O.G., Hansen-Hagge, T.E., Bartram, C.R., Gillis, S., Hoelzer, D., and Ganser, A. (1990). Effects of recombinant human IL-7 on blast cell proliferation in acute lymphoblastic leukaemia. *Leukemia*, **4**, 533–40.

Frishman, J., Long, B., Knospe, W., Gregory, S., and Plate, J. (1993). Genes for interleukin 7 are transcribed in leukemic cell subsets of individuals with chronic lymphocytic leukaemia. *J. Exp. Med.*, **177**, 955–64.

Goodwin, R.G., Friend, D., Ziegler, S.F., Jerzy, R., Falk, B.A., Gimpel, S., Cosman, D., Dower, S.K., March, C.J., Namen, A.E., and Park, L.S. (1990). Cloning of the human and murine interleukin-7 receptors; demonstration of a soluble form and homology to a new receptor superfamily. *Cell*, **60**, 941–51.

Lynch, M., Baker, E., Park, L.S., Sutherland, G.R., and Goodwin, R.G. (1992). The interleukin-7 receptor gene is at 5p13. *Hum. Genet.*, **89**, 566–8.

Park, L.S., Friend, D.J., Schmierer, A.E., Dower, S.K., and Namen, A.E. (1990). Murine interleukin 7 (IL-7) receptor. Characterization on an IL-7-dependent cell line. *J. Exp. Med.*, **171**, 1073–89.

Pleiman, C.M., Gimpel, S.D., Park, L.S., Harada, H., Taniguchi, T., and Ziegler, S.F. (1991). Organization of the murine and human interleukin-7 receptor genes: two mRNAs generated by differential splicing and presence of a type I-interferon-inducible promoter. *Mol. Cell. Biol.*, **11**, 3052–9.

Touw, I., Pouwels, K., van Agthoven, T., van Gurp, R., Budel, L., Hoogerbrugge, H., Delwel, R., Goodwin, R., Namen, A., and Lowenberg, B. (1990). Interleukin-7 is a growth factor of precursor B and T acute lymphoblastic leukaemia. *Blood*, **75**, 2097–101.

Uckun, F.M., Dibirdik, I., Smith, R., Tuel-Ahlgren, L., Chandan-Langlie, M., Schieven, G.L., Waddick, K.G., Hanson, M., and Ledbetter, J.A. (1991a). Interleukin 7 receptor ligation stimulates tyrosine phosphorylation, inositol phospholipid turnover, and clonal proliferation of human B-cell precursors. *Proc. Natl. Acad. Sci. (USA)*, **88**, 3589–93.

Uckun, F.M., Tuel-Ahlgren, L., Obuz, V., Smith, R., Dibirdik, I., Hanson, M., Langlie, M.-C., and Ledbetter, J.A. (1991b). Interleukin 7 receptor engagement stimulates tyrosine phosphorylation, inositol phospholipid turnover, proliferation, and selective differentiation to the CD4 lineage by human fetal thymocytes. *Proc. Natl. Acad. Sci. (USA)*, **88**, 6323–7.

Raymond G. Goodwin and Linda S. Park:
Immunex Research and Development Corporation,
Seattle, WA, USA

Interleukin-8 and Related Chemokine α Family Members

A group of at least 14 structurally related 8–10 kDa cytokines has recently been designated as the chemokine superfamily (Taub and Oppenheim 1993). These cytokines exhibit between 20 and 45 per cent similarity in their amino acid sequences, are all basic heparin-binding proteins, and act on various inflammatory cell types (Oppenheim et al. 1991). Members of the chemokine superfamily are also related by having four conserved cysteine residues: interleukin-8 (IL-8), β-thromboglobulin (β-TG) which is cleaved to produce neutrophil activating peptide-2 (NAP-2), platelet factor-4 (PF-4), melanoma growth stimulating activity (MGSA) otherwise known as growth related cytokine (GRO), ENA-78, granulocyte chemotactic peptide-1 (GCP-1), and interferon-inducible protein-10 (IP-10) are members of the α (or C–X–C) chemokine family. These α chemokines induce the directional migration of various cell types including neutrophils, monocytes, T lymphocytes, basophils or fibroblasts. Four of them (e.g. IL-8, GRO, NAP-2, and ENA-78) primarily attract and activate neutrophils. These proteins are active at nanomolar concentrations and are produced by a wide variety of cell types in response to exogenous irritants and endogenous mediators such as IL-1, TNF, PDGF, and IFNγ (Table 1) (Oppenheim et al. 1991). IL-8 and several other α chemokines have been shown to be produced in various disease states, including rheumatoid arthritis, inflammatory bowel disease, atherosclerosis, asthma, leprosy, psoriasis, and a variety of respiratory syndromes. The generation of antagonists specific for IL-8 and other α chemokines or chemokine receptor-specific antibodies may provide a means of controlling both acute and chronic inflammatory disease states.

■ Chemokine α family proteins

The chemokine α family members from mouse and man exhibit a great deal of homology in their amino acid sequences (Fig. 1) (Oppenheim et al. 1991; Schall 1991). The cDNAs code for a precursor protein which contains a leader sequence. This presumably enables the chemokines to be secreted and enzymatically cleaved to yield a mature form. The mature form of all chemokine α proteins contains four cysteine residues which form two disulphide bridges. The first two cysteines in this subfamily are separated by only one amino acid. Members of this family are typically secreted as non-glycosylated dimers and tetramers containing between 66 and 80 amino acids in their mature monomeric protein form (Oppenheim et al. 1991; Schall 1991; Matsushima et al. 1992).

Table 1. Stimulants and cellular sources of α chemokines (Taub and Oppenheim 1993; Oppenheim et al. 1991; Schall 1991; Matsushima et al. 1992)

Chemokine	Stimulants	Cell source(s)
hIL-8	PHA, ConA, LPS, silica, IL-1α, IL-1β, TNFα, IL-3, IFNγ, anti-CD3 mAb, phorobol esters, viruses, bacteria	Monocytes/macrophages, T cells, fibroblasts, neutrophils, keratinocytes, endothelial cells, hepatocytes, chondrocytes, glioblastoma cells, astrocytoma cells, NK cells
hGROα/MGSA	IL-1, TNFα	Monocytes/macrophages, fibroblasts, endothelial cells, synovial cells, tumour cell lines
hPF-4	Platelet activators	Platelets
hIP-10	IFNγ, LPS, anti-CD3 mAb	Monocytes/macrophages, fibroblasts, endothelial cells, keratinocytes, T cells
hβ-TG/CTAP III/NAP-2	Platelet activators	Monocytes/macrophages, platelets
hENA-78	LPS/IL-1β/TNFα	Epithelial cell line
hGCP-2	PBMC mitogen-activated supernatants	Osteosarcoma
mMIP-2	LPS	Monocytes/macrophages

```
                1       10        20        30        40        50        60        70        80
                .   .    .    .    .    .    .    .    .    .    .    .    .    .    .    .    .

hIL-8/NAP-1/    AVLPRSAKELRCQCIKTYSKPFHPKFIKELRVIESGPHCANT-EIIVKLSD-GRELCLDPKENWVQRVVEKFLKRAENS
hGCP-1

hGROα/MGSA      ASVATELRCQCLQTLQGI-HPKNIQSVNVKSPGPHCAQT-EVIATLKN-GRKACLNPASPIVKKIIEKMLNSDKSN

hGROß/hMIP-2α   APLATELRCQCLQTLQGI-HLKNIQSVKVKSPGPHCAQT-EVIATLKN-GQKACLNPASPMVKKIIEKMLKNGKSN

hGROγ/hMIP-2ß   ASVVTELRCQCLQTLQGI-HLKNIQSVNVRSPGPHCAQT-EVIATLKN-GKKACLNPASPMVQKIIEKILNKGSTN

mMIP-2          VVASELRCQCLKTLPRV-DFKNIQSLSVTPPGPHCAQT-EVIATLKG-GQKVCLDPEAPLVQKIIQKILNKGKAN

mKC             APIANELRCVCLQTMAGI-HLKNIQSLKVLPSGPHCTQT-EVIATLKN-GREACLDPEAPLVQKIVQKMLKGVPK

hIP-10          VPLSRTVRCTCISISNQPVNPRSLEKLEIIPASQFCPRV-EIIATMKKKGEKRCLNPESKAIKNLL-KAVSKEMSKRSP

mIP-10/mC7/     IPLARTVRCNCIHIDDGPVRMRAIGKLEIIPASLSCPRV-EIIATMKKNDEQRCLNPESKTIKNLM-KAFSQKRSKRAP
mCRG-2

mMIG            TLVIRNARCSCISTSRGTIHYKSLKDLKQFAPSPNCNKT-EIIATLKN-GDQTCLDPDDSANVKKLMKEWEKKI..

hPF-4           EAEEDGDLQCLCVKTTSQV-RPRHITSLEVIKAGPHCPTA-QLIATLKN-GRKICLDLQAPLYKKII-KKLLES

hCTAP/ß-TG      SLDSDLYAELRCMCIKTTSGI-HPKNIQSLEVIGKGTHCNQV-EVIATLKD-GRKICLDPDAPRIKKIVQKKLAGDESAD

hNAP-2          AELRCMCIKTTSGI-HPKNIQSLEVIGKGTHCNQV-EVIATLKD-GRKICLDPDAPRIKKIVQKKLAGDESAD

hENA-78         GPAAAVLRELRCVCLQTTQGV-HPKMISNLQVFAIGPQCSKV-EVVASLKN-GKEICLDPEAPFLKKVIQKILDGGNKEN

mMIP-2          AVVASELRCQCLKTLPRV-DFKNIQSLSVTPPGPHCAQJ-EVIATLKG-GQKVCLDPEAPLVQKIIQKILNGKAN

mGCP-2          GPVSAVLTELRCTCLRVTLR.....
```

Figure 1. Amino acid sequence alignment of the α chemokine subfamily (Taub and Oppenheim 1993; Oppenheim *et al.* 1991; Schall 1991; Matsushima *et al.* 1992; Walz *et al.* 1989; Walz *et al.* 1991; Proost *et al.* 1993).

Of the α chemokines, IL-8 has been the most extensively studied, at both a molecular and biochemical level. The precursor form of human IL-8 consists of 99 amino acids with a putative signal sequence that is cleaved to generate a non-glycosylated mature protein of approximately 8 kDa. The amino terminal end of the mature IL-8 protein starts with serine at residue 28. At least four N-terminal variants of human IL-8 containing 79, 77, 72, or 69 amino acids have been described (Yoshimura *et al.* 1989). The 72-amino-acid (monocyte) form is most active while the 77-amino-acid (endothelial) form and other forms of IL-8 have lower biological activity. Similar truncated forms of chemokines have been observed for both IP-10 and PF-4. The roles of these intermediate forms remains to be determined. There are 14 basic amino acids (lysine and arginine) in the 72-amino-acid form of mature IL-8, contributing to the basic characteristic of the protein. There are no potential N-glycosylation sites in the amino acid sequences of any of the α chemokines nor do the mature proteins appear to undergo any post-translational modifications, such as phosphorylation, sulphation, or glycosylation. The four α family members that chemoattract neutrophils all share the common amino acid sequence -E-L-R- before the first two cysteine residues. Recent studies have demonstrated that conservation of these residues is essential for the neutrophil effects of these α chemokines. The absence of the ELR sequence in IP-10 and PF-4 may account for their failure to chemoattract neutrophils (Clark-Lewis *et al.* 1993).

GRO has three variants (α, β, and γ) which exhibit about 95 per cent amino acid similarity (Schall 1991; Haskill *et al.* 1990). They are probably homologues of murine macrophage derived KC, macrophage inflammatory peptide-2α (MIP-2α) and MIP-2β, respectively. In addition, several of the listed α chemokine family members are cleavage products of platelet basic protein (PBP), a protein released upon degranulation from the α granules of platelets. Proteolytic removal of the first nine amino acids of PBP yields connective tissue activating protein III (CTAP III), reported to stimulate synovial cells to replicate and to secrete glycosaminoglycans. Loss of four more amino acids from CTAP III yields β-thromboglobulin (β-TG), a potent chemotactic agent for fibroblasts. Proteolytic removal of 15 amino acids from CTAP III yields the 70 amino acid neutrophil activating peptide-2 (NAP-2), which possesses many of the activities of IL-8 (Walz and Baggiolini 1990).

IL-8 exerts its biological activities by binding to specific cell receptors and causing a rapid rise in cytosolic free Ca^{2+} concentration. Recently, three groups have independently cloned and sequenced the IL-8 receptors from rabbit neutrophils, human neutrophils, and the human cell-line HL-60 (Holmes *et al.* 1991; Murphy and Tiffany 1991). The cDNA sequences of all of the IL-8 receptors encode proteins with seven hydrophobic transmembrane domains. The two cloned human IL-8 receptor isotypes exhibit 77 per cent homology in their amino acid sequence and have different binding affinity for IL-8. IL-8 can bind to both the IL-8R-A (IL-8R type I) and IL-8R-B (IL-8R type II). However, the IL-8R-B also binds GRO, muMIP-2, ENA-78, and NAP-2 with high affinity similar to that of IL-8, enabling these chemokines to act as neutrophil activators and chemoattractants (Taub and Oppenheim 1993; Oppenheim *et al.* 1991; Lee *et al.* 1992; Gayle *et al.* 1993). CTAP-III/β-TG, IP-10, and PF-4 fail to compete for IL-8 binding strongly supporting the existence of distinct receptors for the chemokines. In addition, GRO, unlike IL-8, binds well to melanoma cells and fibroblasts suggesting the existence of unique receptors for GRO (Horuk *et al.* 1993). Signal transduction through the α chemokine receptors is believed to operate through G proteins, phosphoinositol hydrolysis, and via rapid elevation of diacylglycerol and cytosolic Ca^{2+} levels (Taub and Oppenheim 1993). Many of the α chemokines have been shown to induce a rapid rise in cytosolic free Ca^{2+} concentration in human neutrophils, which is inhibitable by pertussis toxin. While intracellular calcium measurements have been an effective method for characterizing and classifying chemokine receptors, the relevance of calcium mobilization in initiating a subsequent cell response remains to be determined. Studies using a variety of kinase inhibitors have also revealed that serine/threonine kinase and protein kinase C activation is essential for the action of α chemokines on neutrophils and that a pertussis toxin-sensitive GTP binding protein may be associated with these receptors (Taub and Oppenheim 1993; Oppenheim *et al.* 1991; Schall 1991; Matsushima *et al.* 1992).

All of these proteins are highly stable, not easily denatured by heat, extreme pH changes, or proteases. IL-8 and the other α chemokines can be purified using conventional chromatography steps using heparin–Sepharose columns and HPLC as well as affinity chromatography using monoclonal antibodies and acid elution. The structure of IL-8 and PF-4 has been determined by X-ray crystallography. IL-8 typically exists as a hydrogen bonded dimer, whereas PF-4 preferentially forms tetramers. The structure of crystalline IL-8 consists of three antiparallel β-strands connected with loops, and one long α helix made of carboxyl-terminal residues 57–72 (Clore *et al.* 1990). Monomeric PF-4 is quite similar in its crystalline structure to IL-8 (St. Charles *et al.* 1989). The IL-8 dimer is probably stabilized by hydrogen bonds between the first β-strand in each molecule and additional side chain interactions. A variety of structure–function studies have determined that the helices are important for binding of IL-8 to its receptor(s). The crystalline structure of the other α chemokine family members remains to be determined.

■ α chemokine genes and transcription

Many of the human α chemokine genes, including IL-8, GRO/MGSA, PF-4, β-thromboglobulin, and IP-10 have been localized and assigned to human chromosome 4q12–21 (Oppenheim *et al.* 1991; Schall 1991; Matsushima *et al.* 1992). While the precise distance between each of these genes has not yet been determined, it seems likely that this family arose through gene duplication and subsequent divergence. The IL-8, GRO, and IP-10 genes consist of four exons and three introns, while PF-4 contains three exons and two introns. The first and second introns are conserved within this family and the first intron separates the leader sequence from the mature protein. The IL-8 gene encompasses approximately 5.2 kb and typically generates a single mRNA transcript of 1.8 kb (Mukaida *et al.* 1989). The IP-10 gene encompasses 5.25 kb and also yields a single 1.8 kb mRNA transcript (Luster and Ravetch 1987a). The cDNA for these chemokines have been recognized by their characteristic conserved single open reading frames, typical signal sequences in the 5′ region, AT rich sequences in their 3′ untranslated regions, and rapidly inducible mRNA expression (Oppenheim *et al.* 1991; Matsushima *et al.* 1992). Expression of mRNA for α chemokines in both macrophages and T cells increases rapidly after stimulation, reaches a maximum in 3–4 h, then declines. The transient expression of both IL-8 and IP-10 mRNA has been attributed to the AT rich sequence in the 3′ untranslated region of the transcript that leads to message destabilization. The instability conferred by the AT sequence in the mRNA was partially alleviated by treatment of the cells with cycloheximide. Interestingly, IP-10 mRNA is inducible with cycloheximide alone. Also, the fourth exon of the IP-10 gene almost exclusively contains this long 3′ untranslated region containing the sequence TTATTTAAT and ATTTA and may play a role in the inducible transient expression of the IP-10 mRNA (Luster and Ravetch 1987a,b). Both the IL-8 and PF-4 genes contain a short 3′ untranslated region and PF-4 mRNA is constitutively expressed in platelets where the protein is stored in α granules (Oppenheim et al. 1991).

A single TATA- and CAT-like structure can be found within the IL-8 genomic DNA sequences. The flanking region of the IL-8 gene contains several potential binding sites for known nuclear factors including AP-1, AP-2, AP-3, octamer motif, NF-κb, hepatocyte nuclear factor-1, glucocorticoid receptor, and interferon regulatory protein-1 (Matsushima et al. 1992; Mukaida et al. 1989). Similarly, the 5′ flanking region within the IP-10 gene also contain potential binding sites for AP-2, NF-kb, and an interferon-stimulated response element (ISRE) (Luster and Ravetch 1987a). The 5′-flanking region of the IL-8 and IP-10 genes shows no overall sequence homology with the flanking regions of known cytokines, including those also induced by IL-1, TNFα, and IFN. Cycloheximide studies have revealed that IL-1, TNF, and phorbol esters rapidly activate the transcription of the IL-8 gene in the absence of new protein synthesis (Oppenheim et al. 1991; Matsushima et al. 1988). Analysis of the 5′ flanking region of the IL-8 gene revealed that the sequences from 94 to 71 bp are important for inducible expression of the IL-8 gene by IL-1, TNFα, and phorbol esters. This site contains a number of potential binding sites for nuclear factors. However, two cis-acting factors, an NF-κB-like and an NF-IL-6-like element, were responsible for the transcriptional activation of IL-8 by these stimuli in a human fibrosarcoma cell line. Similarly, 235 to 172 of the 5′ regulatory region of the IP-10 gene contains one or more IFNγ-inducible positive cis regulatory elements, with sequence similarity to the reported ISREs found within a number of genes whose transcription is activated by IFNα/β and IFNγ (Luster and Ravetch 1987a,b). Additionally, while the IP-10 gene contains an NF-κb-like site, suggesting the possibility that IP-10 may be regulated

Table 2. *In vitro* effects of α chemokine family members (Taub and Oppenheim 1993; Oppenheim et al. 1991; Schall 1991; Matsushima et al. 1992)

Chemokine	Target	Effect
IL-8	Neutrophil	Chemotaxis; shape change; degranulation; respiratory burst; increased cytosolic Ca^{2+}; increased adhesion to endothelial cells, fibrinogen, and extracellular matrix proteins; increased *Candida albicans* growth inhibition; increased expression of CD11a, CD11b, CD11c, and CD18; lysosomal enzyme release
	T cells	Chemotaxis
	B cells	Inhibits IL-4-induced IgE production
	Basophils	Chemotaxis; histamine release; increased leukotriene release
	Monocytes	Increased adherence to endothelial cells
	Keratinocytes	Increased proliferation
	Endothelial cells	Increased blood vessel proliferation
	Melanoma cells	Increased adhesiveness and haptotactic response
	Smooth muscle	Chemotaxis
	Stem cells	Suppresses colony formation of immature myeloid progenitors
GRO/MGSA	Neutrophils	Chemotaxis; lysosomal enzyme release
	Fibroblasts	Growth stimulatory activity
	Melanoma cells	Growth factor
PF-4	Fibroblasts	Chemotaxis; increased proliferation
	T cells	Immunostimulant; decrease bone resorption
	Endothelial cells	Increased ICAM-1 expression; decreased blood vessel proliferation
ENA-78	Neutrophils	Chemotaxis
CTAP/β-TG/NAP-2	Fibroblasts	Chemotaxis; increased proliferation; increased activator
	Chondrocytes	Increased glycosaminoglycan synthesis
	Neutrophils	Chemotaxis; superoxide release
IP-10	T cells	Chemotaxis; increased adhesion to endothelial cells and extracellular matrix proteins
	Monocytes	Chemotaxis
MIP-2	Neutrophils	Chemotaxis; degranulation; suppresses colony formation of immature myeloid progenotors

Table 3. *In vivo* effects of α chemokine family members (Taub and Oppenheim 1993; Oppenheim *et al.* 1991; Schall 1991; Matsushima *et al.* 1992; Taub *et al.* 1993*b*)

Chemokine	Effects
IL-8	Neutrophil infiltration in human, primates, mouse, rat, rabbit, and dig
	T cell infiltration in rat
	Human T cell infiltration in SCID mouse
	Neutrophilia
	Angiogensis
	Synovial inflammation
	Plasma leakage in rabbits
GROα	Neutrophil infiltration in rabbits
GCP-2	Granulocyte infiltration in rabbits
IP-10	Human T cell infiltration in SCID mice
PF-4	Fibrosis
	Neutrophil and mononuclear cell infiltration
	Inhibits angiogensis
MIP-2	Neutrophil infiltration

by PMA or TNFα, neither of these factors were able to stimulate IP-10 expression. Although little is known about the regulatory regions of the other α chemokines, the similarity in mRNA expression mediated by inflammatory cytokines and LPS would suggest that many common DNA binding proteins are involved.

■ Biological activities

Most of the members of the chemokine α family induce the directional migration of various cell types, including neutrophils, monocytes, T lymphocytes, basophils, and fibroblasts. Four of the α chemokine family members namely IL-8, GRO, NAP-2, and ENA-78 chemoattract neutrophils (Oppenheim *et al.* 1991; Haskill *et al.* 1990; Walz and Baggiolini 1990; Walz *et al.* 1991). IP-10 is the only α chemokine member that chemoattracts monocytes rather than neutrophils (Taub *et al.* 1993a). IL-8 and IP-10 have also been reported to chemoattract T lymphocytes (Taub *et al.* 1993a,b; Larsen *et al.* 1989). In addition to being potent chemoattractants for neutrophils, many chemokines also have a wide range of other pro-inflammatory effects (Table 2). In *in vitro* experiments, some of these effects include neutrophil degranulation with release of myeloperoxidase, elastase, and β-glucuronidase in the presence of cytochalasin; increased expression of cell adhesion molecules and enhanced adhesion to endothelial cells and extracellular matrix proteins; and enhanced cytostatic effects of neutrophils on *Candida albicans* (Taub and Oppenheim 1993; Oppenheim *et al.* 1991; Schall 1991; Matsushima *et al.* 1992). In addition, many of these α chemokines have been shown to recruit selectively leucocytes to sites of injection, supporting an *in vivo* role for these peptides (Table 3) (Taub and Oppenheim 1993; Oppenheim *et al.* 1991; Schall 1991; Matsushima *et al.* 1992).

■ Pathology

IL-8 and other α chemokines play a key role in the accumulation of leucocytes at sites of inflammation. A number of respiratory disease states exhibit increased IL-8 expression at both the protein and RNA levels in both histological sections and interstitial lung fluids as well as associated with increased neutrophils present within the lung compared to normal lavage controls. These disease states include idiopathic pulmonary fibrosis, active sacoidosis, and bronchial carcinoma (Taub and Oppenheim 1993; Lynch *et al.* 1992; Rolfe *et al.* 1991; Kunkel *et al.* 1991). Increased levels of both IL-8 and ENA-78 have also been found in the lavage fluid of cystic fibrosis patients (Nakamura *et al.* 1992). Synovial fluids from rheumatoid arthritis, but not osteoarthritis, exhibit increased levels of IL-8 which corresponds to increased numbers of neutrophils in collected fluids (Koch *et al.* 1992a). Elevated levels of IL-8 and GRO can also be detected in biopsies and extracted scales of psoriatic lesions (Schroder *et al.* 1990). IL-8 has been detected in inflammatory bowel disease, gastritis, alcoholic hepatitis, and in the circulation of patients with β-thalassaemia, post major surgery, septic shock, acute meningococcal infections, and in various peritonitis models (Taub and Oppenheim 1993; Matsushima *et al.* 1992). PF-4 has been reported to induce sequential neutrophil and monocyte as well as fibroblast infiltration at sites of injury, suggesting a role in tissue repair.

Several studies have demonstrated the hyperexpression of IL-8 and GRO in neoplastic cells and IL-8 in cell lines of both diffuse and intestinal gastric cancers (Taub and Oppenheim 1993). IL-8 has also been reported to be a potent angiogenic factor, promoting vascularization of tumours and traumatized tissues (Koch *et al.* 1992b). In contrast, PF-4 has been shown to be an inhibitor of angiogenesis in chicken chorioallantoic membranes and has been promulgated as an inhibitor of neovascularization of tumours (Maione *et al.* 1990). PF-4 is at present being evaluated in phase II studies in man.

IP-10 can readily be detected on cells participating in delayed-type hypersensitivity responses and in lesions of lepromatous leprosy patients (Kaplan *et al.* 1987; Enk and Katz 1992). In addition, psoriatic plaques which are characterized by neutrophil infiltration are also a rich source of IP-10 (Gottlieb *et al.* 1988). While the precise role of IP-10 in these disease states remains to be defined, the recent finding that IP-10 is a monocyte and T cell chemoattractant suggests that this protein may play a role in development of chronic inflammatory responses (Taub *et al.* 1993a).

Thus, α chemokines are implicated as major participants in acute as well as more prolonged inflammatory reactions, modulation of angiogenesis, and fibroplasia. The possiblity that they may also contribute to the normal homing and distribution of leucocytes needs to be evaluated. In addition to their obvious direct effects on cell function, the possibility that they may act as costimulants of cell growth needs more study.

■ References

Clark-Lewis, I., Dewald, B., Geiser, T., Moser, B., and Baggiolini, M. (1993). Platelet factor 4 binds to interleukin 8 receptors and activates neutrophils when its N terminus is modified with Glu–Leu–Arg. *Proc. Natl. Acad. Sci. (USA)*, **90**, 3574–7.

Clore, G.M., Appella, E., Yamada, M., Matsushima, K., and Gronenborn, A.M. (1990). Three-dimensional structure of interleukin 8 in solution. *Biochemistry*, **29**, 1689–96.

Enk, A.H., and Katz, S.I. (1992). Early molecular events in the induction phase of contact sensitivity. *Proc. Natl. Acad. Sci. (USA)*, **89**, 1398–1402.

Gayle III, R.B., Sleath, P.R., Srinivason, S., Birks, C.W., Weerawarna, K.S., Cerretti, D.P., Kozlosky, C.J., Nelson, N., Vanden Bos, T., and Beckmann, M.P. (1993). Importance of the amino terminus of the interleukin-8 receptor in ligand interactions. *J. Biol. Chem.*, **268**, 7283–9.

Gottlieb, A.B., Luster, A.D., Posnett, D.N., and Carter, D.M. (1988). Detection of a gamma interferon-induced protein IP-10 in psoriatic plaques. *J. Exp. Med.*, **168**, 941–8.

Haskill, S., Peace, A., Morris, J., Sporn, S.A., Anisowicz, A., Lee, S.W., Smith, T., Martin, G., Ralph, P., and Sager, R. (1990). Identification of three related human GRO genes encoding cytokine functions. *Proc. Natl. Acad. Sci. (USA)*, **87**, 7732–6.

Holmes, W.E., Lee, J., Kuang, W.-J., Rice, G.C., and Wood, W.I. (1991). Structure and functional expression of a human interleukin-8 receptor. *Science*, **253**, 1278–80.

Horuk, R., Yansura, D.G., Reilly, D., Spencer, S., Bourell, J., Henzel, W., Rice, G., and Unemori, E. (1993). Purification, receptor binding analysis, and biological characterization of human melanoma growth stimulating activity (MGSA). Evidence for a novel MGSA receptor. *J. Biol. Chem.*, **268**, 541–6.

Kaplan, G., Luster, A.D., Hancock, G., and Cohn, Z.A. (1987). The expression of a gamma interferon-induced protein (IP-10) in delayed immune responses in human skin. *J. Exp. Med.*, **166**, 1098–108.

Koch, A.E., Kunkel, S.L., Harlow, L.A., Johnson, B., Evanoff, H.L., Haines, G.K., Burdick, M.D., Pope, R.M., and Strieter, R.M. (1992a). Enhanced production of monocyte chemoattractant protein-1 in rheumatoid arthritis. *J. Clin. Invest.*, **90**, 772–9.

Koch, A.E., Polverini, P.J., Kunkel, S.L., Harlow, L.A., DiPietro, L.A., Elner, V.M., Elner, S.G., and Strieter, R.M. (1992b). Interleukin-8 as a macrophage-derived mediator of angiogenesis. *Science*, **258**, 1798–801.

Kunkel, S.L., Standiford, T., Kasahara, K., and Strieter, R.M. (1991). Interleukin-8 (IL-8): the major neutrophil chemotactic factor in the lung. *Exp. Lung Res.*, **17**, 17–23.

Larsen, C.G., Anderson, A.O., Appella, E., Oppenheim, J.J., and Matsushima, K. (1989). The neutrophil-activating protein (NAP-1) is also chemotactic for T lymphocytes. *Science*, **243**, 1464–6.

Lee, J., Kuang, W.J., Rice, G.C., and Wood, W.I. (1992). Characterization of complementary DNA clones encoding the rabbit IL-8 receptor. *J. Immunol.*, **148**, 1261–4.

Luster, A.D., and Ravetch, J.V. (1987a). Genomic characterization of a gamma-interferon-inducible gene (IP-10) and identification of an interferon-inducible gene (IP-10) and identification of an interferon-inducible hypersensitive site. *Mol. Cell. Biol.*, **7**, 3723–31.

Luster, A.D., and Ravetch, J.V. (1987b). Biochemical characterization of a gamma interferon-inducible cytokine (IP-10). *J. Exp. Med.*, **166**, 1084–97.

Lynch III, J.P., Standiford, T.J., Rolfe, M.W., Kunkel, S.L., and Strieter, R.M. (1992). Neutrophilic alveolitis in idiopathic pulmonary fibrosis. The role of interleukin-8. *Am. Rev. Respir. Dis.*, **145**, 1433–9.

Maione, T.E., Gray, G.S., Petro, J., Hunt, A.J., Donner, A.L., Bauer, S.I., Carson, H.F., and Sharpe, R.J. (1990). Inhibition of angiogenesis by recombinant human platelet factor-4 and related peptides. *Science*, **247**, 77–9.

Matsushima, K., Morishita, K., Yoshimura, T., Lavu, S., Kobayashi, Y., Lew, W., Appella, E., Kung, H.F., Leonard, E.J., and Oppenheim, J.J. (1988). Molecular cloning of a human monocyte-derived neutrophil chemotactic factor (MDNCF) and the induction of MDNCF mRNA by interleukin 1 and tumour necrosis factor. *J. Exp. Med.*, **167**, 1883–93.

Matsushima, K., Baldwin, E.T., and Mukaida, N. (1992). In *Interleukins: molecular biology and immunology* (ed. T. Kishimoto), Vol. 51, pp. 236–65. Karger, Basel, Switzerland.

Mukaida, N., Shiroo, M., and Matsushima, K. (1989). Genomic structure of the human monocyte-derived neutrophil chemotactic factor IL-8. *J. Immunol.*, **143**, 1366–71.

Murphy, P.M., and Tiffany, H.L. (1991). Cloning of complementary DNA encoding a functional human interleukin-8 receptor. *Science*, **253**, 1280–3.

Nakamura, H., Yoshimura, K., McElvaney, N.G., and Crystal, R.G. (1992). Neutrophil elastase in respiratory epithelial lining fluid of individuals with cystic fibrosis induces interleukin-8 gene expression in a human bronchial epithelial cell line. *J. Clin. Invest.*, **89**, 1478–84.

Oppenheim, J.J., Zachariae, C.O.C., Mukaida, N., and Matsushima, K. (1991). Properties of the novel proinflammatory supergene 'intercrine' cytokine family. *Annu. Rev. Immunol.*, **9**, 617–48.

Proost, P., De Wolf-Peeters, C., Conings, R., Opdenakker, G., Billiau, A., and Van Damme, J. (1993). Identification of a novel granulocyte chemotactic protein (GCP-2) from human tumour cells. *In vitro* and *in vivo* comparison with natural forms of GRO, IP-10, and IL-8 *J. Immunol.*, **150**, 1000–10.

Rolfe, M.W., Kunkel, S.L., Standiford, T.J., Chensue, S.W., Allen, R.M., Evanoff, H.L., Phan, S.H., and Strieter, R.M. (1991). Pulmonary fibroblast expression of interleukin-8: a model for alveolar macrophage-derived cytokine networking. *Am. J. Respir. Cell Mol. Biol.*, **5**, 493–501.

Schall, T.J. (1991). Biology of the RANTES/SIS cytokine family. *Cytokine*, **3**, 165–83.

Schroder, J.-M., Persoon, N.L.M., and Christophers, E. (1990). Lipopolysaccharide-stimulated human monocytes secrete, apart from neutrophil-activating peptide 1/interleukin 8, a second neutrophil-activating protein. NH2-terminal amino acid sequence identity with melanoma growth stimulatory activity. *J. Exp. Med.*, **171**, 1091–100.

St. Charles, R., Walz, D.A., and Edwards, B.F. (1989). The three-dimensional structure of bovine platelet factor 4 at 3.0-A resolution. *J. Biol. Chem.*, **264**, 2092–9.

Taub, D.D., and Oppenheim, J.J. (1993). Review of the chemokine meeting. The Third International Symposium of Chemotactic Cytokines. *Cytokine*, **5**, 175.

Taub, D.D., Lloyd, A.R., Conlon, K., Wang, J.M., Ortaldo, J.R., Harada, A., Matsushima, K., Kelvin, D.J., and Oppenheim, J.J. (1993a). Recombinant human interferon-inducible protein 10 is a chemoattractant for human monocytes and T lymphocytes and promotes T cell adhesion to endothelial cells. *J. Exp. Med.*, **177**, 1809–14.

Taub, D.D., Conlon, K., Lloyd, A.R., Oppenheim, J.J., and Kelvin, D.J. (1993b). Preferential migration of activated CD4+ and CD8+ T cells in response to MIP-1 alpha and MIP-1 beta. *Science*, **260**, 355–8.

Walz, A., and Baggiolini, M. (1990). Generation of the neutrophil-activating peptide NAP-2 from platelet basic protein or connective tissue-activating peptide III through monocyte proteases. *J. Exp. Med.*, **171**, 449–54.

Walz, A., Burgener, R., Car, B., Baggiolini, M., Kunkel, S.L., and Strieter, R.M. (1991). Structure and neutrophil-activating properties of a novel inflammatory peptide (ENA78) with homology to interleukin 8. *J. Exp. Med.*, **174**, 1355–62.

Walz, A., Dewald, B., von Tscharner, V., and Baggiolini, M. (1989). Effects of the neutrophil-activating peptide NAP-2, platelet basic protein, connective tissue-activating peptide III and platelet factor 4 on human neutrophils. *J. Exp. Med.*, **170**, 1745–50.

Yoshimura, T., Robinson, E.A., Appella, E., Matsushima, K., Showalter, S.D., Skeel, A., and Leonard, E.J. (1989). Three forms of monocyte-derived neutrophil chemotactic factor (MDNCF) dis-tinguished by different lengths of the amino-terminal sequence. *Mol. Immunol.*, **26**, 87–93.

Dennis D. Taub and Joost J. Oppenheim:
NCI-Frederick Cancer Research and Development Center,
National Institute of Health,
Frederick, MD 21702–1201, USA

Monocyte Chemoattractant Protein-1 (MCP-1)

Human MCP-1 is a 76-amino-acid basic protein that has chemotactic activity for blood monocytes (Yoshimura et al. 1989; Yoshimura and Leonard 1992). MCP-1 can be produced by a variety of cells including leukocytes, endothelial cells, fibroblasts, and tumour cells with or without addition of stimulus such as LPS, IL-1, TNFα, or PDGF. MCP-1 may mediate the infiltration of blood monocytes into immune or non-immune sites of inflammation.

■ Alternative names

MCP-1 was first described as lymphocyte-derived chemotactic factor (LDCF). Other names are tumour cell-derived chemotactic factor (TDCF), glioma-derived monocyte chemotactic factor (GDCF), smooth muscle cell-derived chemotactic factor (SMC-CF), monocyte chemotactic and activating factor (MCAF). The gene of the mouse homologue of MCP-1 was named JE.

■ MCP-1 sequence

The open reading frame of human MCP-1 cDNA encodes a 99-amino-acid protein, the first 23 amino acids of which form a signal peptide (GenBank accession number X14768) (Yoshimura *et al.* 1989; Yoshimura and Leonard 1992) (Fig. 1). MCP-1 and NAP-1/IL-8 are members of the β-thromboglobulin superfamily. The superfamily comprises 2 subfamilies, according to the relative position of the first two cysteines in the proteins, which is either C–C or C–X–C. MCP-1, MCP-2, MCP-3, RANTES, LD-78/PAT464/GOS19–1/MIP-1α, ACT-2/PAT744/G25/MIP-1β, and I-309/TCA-3 are in the C–C subfamily (Yoshimura and Leonard 1992; Oppenheim *et al.* 1991; Van Damme *et al.* 1992). MCP-1 homologues have been purified or cloned from baboon (Yoshimura and Leonard 1992), bovine (Wempe *et al.* 1991), rabbit (M57440) (Yoshimura and Yuhki 1991), guinea pig (L04985) (Yoshimura 1993), mouse (J04467) (Rollins *et al.* 1988; Kawahara and Deuel 1989), and rat (M57441) (Yoshimura *et al.* 1991) (Fig. 1). The amino acid sequence similarities between human MCP-1 and bovine, rabbit, guinea pig, mouse, and rat MCP-1s are 71, 74, 56, 55, and 52 per cent, respectively. Rabbit, guinea pig, mouse, and rat MCP-1s have extra amino acids at the C-termini.

The amino acid sequence of human MCP-1β was completely analysed. The N-terminus is a blocked Gln (pyroglutamic acid) at position 24 of the protein encoded by the open reading frame of human MCP-1 cDNA. Since the sequence in this region is conserved among species, the same N-terminus is expected for MCP-1s in all species noted above.

■ MCP-1 protein

There are two major forms of human MCP-1; MCP-1α (15 kDa) and MCP-1β (13 kDa). The 8700 Da core protein is modified to higher molecular mass forms by O-linked glycosylation and addition of terminal sialic acid residues, which probably accounts for the migration positions of MCP-1 on SDS-PAGE. Heterogeneity of MCP-1 appears to be due to variations in O-linked carbohydrate processing. Recent immunoprecipitation experiments have revealed small amounts of two additional forms of MCP-1 in locations by SDS-PAGE above MCP-1α. Glycosylation is not required for *in vitro* chemotactic activity. A truncated form of MCP-1, with five fewer amino acids at the N-terminus, has been reported. MCP-1 has two disulphide bridges (Cys11–Cys36 and Cys12–Cys52). In the case of NAP-1/IL-8, reductive cleavage of the paired disulphide loops causes loss of biological activity. It thus appears that the oxidized state is the functionally active structure, and the paired disulphide loops are required for biological activity; the same requirements probably apply to MCP-1.

MCP-1 can be purified from serum-free U-105MG human malignant glioma cell-line or PHA-stimulated peripheral blood mononuclear leukocyte culture fluid in three steps: dye–ligand chromatography on orange-A Agarose, CM-HPLC, and RP-HPLC. The cation exchange column separates MCP-1 into two well separated peaks. MCP-1 can be also

```
                10                20                30                40                50
Human   M K V S A A L L C L L L I A A T F I P Q G L A Q P D A I N A P V T C C Y N F T N R K I S V Q R L A S
Bovine  - - - - - - - - - - - - T V - A - S T E V E - - - - - - S Q - A - - - T - N S K - - - M - - - M N
Rabbit  - - - - - T - - - - - - - V A - S S H V - - - - V - S - - - - - T - - - K T - - - K - - M - -
G.P.    - Q R - S V - - - V - E - - C S L L M - - - - G V - T - - - - - T - - - K Q - P L K - V K G
Mouse   - Q - P V M - - - G - - F T V - G W S I H V - - - - - V - - - L - - - - S - - S K M - P M S - - E -
Rat     - Q - - V T - - G - - F T V - A C S I H V - S - - - - V - - - L - - - - S - - G K M - P M S - - E N
```

```
                60                70                80                90                100
Y R R I T S S K C P K E A V I F K T I V A K E I C A D P K Q K W V Q D S M D H L D K Q T Q T P K T
- - - V - - - - - - - - - - - - - - - - - L G - - L - - - - - - - - - - - I N Y - N - K N - - - - P
- - - - N - T - - - - - - - - - - M - K L - - G - - - T - - - - - - - - A I A N - - - K M - - - - - L T S Y S
- E - - - - - R - - Q - - - - R - L K N - - V - - - - T - - - - - - ` Y I A K - - Q R - - Q K Q N S T A P Q
- K - - - - - R - - - - - - V - V - L L K R - V - - - - - K E - - - T Y I K N - - R N Q M R S E P T T L F K
- K - - - - - R - - - K L - V - V - K L K R - - - - - - N K E - - - K Y I R L K - Q N Q V R S E T T V F Y K
```

```
     110               120               130               140               150
T T Q E H T T N L S S T R T P S T T T S L
T S K P L N I R F T T Q D P K N R S
T A S A L R S S A P L N V K L T R K S E A N A S T T F S T T T S S T S V G V T S V T V N
I A S T L R T S A P L N V N L T H K S E A N A S T L F S T T T S S T S V E V T S M T E N
```

Figure 1. Amino acid sequences of MCP-1s from different species.

purified by affinity chromatography with monoclonal anti-MCP-1 and HPLC.

Most of the MCP-1 protein obtained from rabbit, guinea pig, mouse and rat is also glycosylated. The proteins appear as broad bands on SDS-polyacrylamide gel at about 28, 25, 30, and 30 kDa, respectively.

■ MCP-1 gene

Human genomic MCP-1 DNA has been cloned (Shyy *et al.* 1990). The gene comprises three 45, 118, and 478 base-pair exons, and two introns of 800 and 385 base pairs in length. Phorbol-ester-responsive elements were found 129 and 157 base pairs upstream from the translation site. The MCP-1 gene (SCYA2) has been mapped to chromosome 17q11.2–q21.1 along with another family member, RANTES. In mice, the MCP-1/JE gene has been mapped on chromosome 11 between Evi-2 and Hox-2 loci, forming a tightly linked cluster with other members of the family, such as TCA-3, MIP-1α, and MIP-1β (Wilson *et al.* 1990). The region of the mouse chromosome is evolutionarily conserved and corresponds to the q11.2–q21.1 region of human chromosome 17.

■ Biological activities

MCP-1 was named for its ability to attract blood monocytes in chemotaxis chambers. The optimal concentration of MCP-1 in the assay system is approximately 1 nM and about 30–40 per cent of input monocytes are induced to migrate. The dose–response curve of MCP-1 is bell-shaped, which is typical for chemotactic factors. Human MCP-1 attracts blood monocytes, basophils, eosinophils, and lymphocytes but not neutrophils. Animal MCP-1s are as good

chemoattractants as human MCP-1 for human monocytes. In studies on the responses of macrophages to MCP-1, it was found that guinea pig MCP-1 did not attract guinea pig resident peritoneal macrophages, and it attracted only one per cent of exudate macrophages. This suggests that MCP-1 functions to recruit blood monocytes, but has little effect on tissue macrophages.

Intradermal injection of guinea pig MCP-1 (1 μg/site) into guinea pig skin caused monocyte migration into the injected sites after 4–6 hours but no significant basophil or eosinophil infiltration was seen. The infiltrate induced by intradermal injection of human MCP-1 into guinea-pigs was smaller than the response to guinea pig MCP-1.

In addition to monocyte chemotactic activity, other activities have been reported. MCP-1 augments the growth inhibitory effects of monocytes on tumour cell lines. MCP-1 causes increased expression of monocyte integrins (CD11c and 11a) and secretion of IL-1 and IL-6 (Jiang *et al.* 1992). MCP-1 causes an increase of calcium influx in monocytes and histamine release by basophils.

■ Pathology

Monocyte/macrophage infiltration is observed in many kinds of inflammatory diseases or tumours. Therefore, the expression or production of MCP-1 has been studied in diseased human tissue. MCP-1 has been detected by *in situ* hybridization or immunohistochemistry in the lesions of atherosclerosis (Yoshimura and Leonard 1992; Takeya *et al.* 1993), rheumatoid arthritis (Villiger *et al.* 1992), idiopathic pulmonary fibrosis (Antoniades *et al.* 1992), nephritis, and tumours such as malignant fibrous histiocytoma (Takeya *et al.* 1991), malignant glioma, or melanoma (Graves *et al.* 1992). It appears that MCP-1 accounts for monocyte/macrophage infiltration in these diseases or tu-mours. Transplantation of MCP-1-transfected cells induced marked monocyte accumulation and regression of the transplanted cells (Rollins and Sunday 1991; Walter *et al.* 1991). In contrast to the elimination of transfected tumour inocula in these experimental models, spontaneously aris-ing tumours in man have a malignant course despite infil-tration by macrophages. The differences may relate to the size of the tumour burden, which is relatively small in the transplantation model, and to the capacity of MCP-1 to stimulate the cytotoxic potential of monocytes. Evidence for the latter is limited at present.

■ References

Antoniades, H. N., Neville-Golden, J., Galanopoulos, T., Kradin, R.L., Valente, A.J., and Graves, D.T. (1992). Expression of monocyte chemoattractant protein 1 mRNA in human idiopathic pulmonary fibrosis. *Proc. Natl. Acad. Sci. (USA)*, **89**, 5371–5.

Graves, D.T., Barnhill, R., Galanopoulos, T., and Antoniades, H.N. (1992). Expression of monocyte chemotactic protein-1 in human melanoma in vivo. *Am. J. Pathol.*, **140**, 9–14.

Jiang, Y., Beller, D.I., Frendl, G., and Graves, D.T. (1992). Monocyte chemoattractant protein-1 regulates adhesion molecule ex-pression and cytokine production in human monocytes. *J. Immu-nol.*, **148**, 2423–8.

Kawahara, R.S., and Deuel, T.F. (1989). Platelet-derived growth fac-tor-inducible gene JE is a member of a family of small inducible genes related to platelet factor 4. *J. Biol. Chem.*, **264**, 679–82.

Oppenheim, J.J., Zachariae, C.O., Mukaida, N., and Matsushima, K. (1991). Properties of the novel proinflammatory supergene 'inter-crine' cytokine family. *Annu. Rev. Immunol.*, **9**, 617–48.

Rollins, B. J., and Sunday, M.E. (1991). Suppression of tumour forma-tion in vivo by expression of the JE gene in malignant cells. *Mol. Cell. Biol.*, **11**, 3125–31.

Rollins, B.J., Morrison, E.D., and Stiles, C.D. (1988). Cloning and expression of JE, a gene inducible by platelet-derived growth factor and whose product has cytokine-like properties. *Proc. Natl. Acad. Sci. (USA)*, **85**, 3738–42.

Shyy, Y.-J., Li, Y.-S., and Kolattukudy, P.E. (1990). Structure of human monocyte chemotactic protein gene and its regulation by TPA. *Biochem. Biophys. Res. Commun.*, **169**, 346–51.

Takeya, M., Yoshimura, T., Leonard, E.J., Kato, T., Okabe, H., and Takahashi, K. (1991). Production of monocyte chemoattractant protein-1 by malignant fibrous histiocytoma: relation to the ori-gin of histiocyte-like cells. *Exp. Mol. Pathol.*, **54**, 61–71.

Takeya, M., Yoshimura, T., Leonard, E.J., and Takahashi, K. (1993). Detection of monocyte chemoattractant protein-1 in human ath-erosclerotic lesions by an anti-monocyte chemoattractant pro-tein-1 monoclonal antibody. *Hum. Pathol.*, **24**, 534–9.

Van Damme, J., Proost, P., Lenaerts, J.-P., and Opdenakker, G. (1992). Structural and functional identification of two human, tumour-derived monocyte chemotactic proteins (MCP-2 and MCP-3) belonging to the chemokine family. *J. Exp. Med.*, **176**, 59–65.

Villiger, P. M., Terkeltaub, R., and Lotz, M. (1992). Production of monocyte chemoattractant protein-1 by inflamed synovial tissue and cultured synoviocytes. *J. Immunol.*, **149**, 722–7.

Walter, S., Bottazzi, B., Govoni, D., Colotta, F., and Mantovani, A. (1991). Macrophage infiltration and growth of sarcoma clones expressing different amounts of monocyte chemotactic pro-tein/JE. *Int. J. Cancer*, **49**, 431–5.

Wempe, F., Henschen, A., and Scheit, K.H. (1991). Gene expression and cDNA cloning identified a major basic protein constituent of bovine seminal plasma as bovine monocyte-chemoattractant pro-tein-1 (MCP-1). *DNA Cell Biol.*, **10**, 671–9.

Wilson, S.D., Billings, P.R., D'Eustachio, P., Fournier, R.E., Geissler, E., Lalley, P.A., Burd, P.R., Housman, D.E., Taylor, B.A., and Dorf, M.E. (1990). Clustering of cytokine genes on mouse chromosome 11. *J. Exp. Med.*, **171**, 1301–14.

Yoshimura, T. (1993). cDNA cloning of guinea pig monocyte chemo-attractant protein-1 and expression of the recombinant protein. *J. Immunol.*, **150**, 5025–32.

Yoshimura, T., and Leonard, E. J. (1992). Interleukin 8 (NAP-1) and related chemotactic cytokines. In *Cytokine* (ed. M. Baggiolini and C. Sorg), Vol. 4, pp.131–52. Karger, Basel.

Yoshimura, T., and Yuhki, N. (1991). Neutrophil attractant/activa-tion protein-1 and monocyte chemoattractant protein-1 in rabbit. cDNA cloning and their expression in spleen cells. *J. Immunol.*, **146**, 3483–8.

Yoshimura, T., Yuhki, N., Moore, S.K., Appella, E., Lerman, M.I., and Leonard, E.J. (1989). Human monocyte chemoattractant protein-1 (MCP-1). Full-length cDNA cloning, expression in mitogen-stimu-lated blood mononuclear leukocytes, and sequence similarity to mouse competence gene JE. *FEBS Lett.*, **244**, 487–93.

Yoshimura, T., Takeya, M., and Takahashi, K. (1991). Molecular cloning of rat monocyte chemoattractant protein-1 (MCP-1) and its expression in rat spleen cells and tumour cell lines. *Biochem. Biophys. Res. Commun.*, **174**, 504–9.

Teizo Yoshimura and Edward J. Leonard:
Immunopathology Section,
Laboratory of Immunobiology,
National Cancer Institute-Frederick Cancer Research and Development Center

Receptors for Interleukin-8 (IL-8)

The IL-8 receptor belongs to a family of receptors that can be classified into two distinct groups based on ligand-binding specificity. The first group comprises the Type A and B IL-8 receptors which both bind IL-8 with high affinity (K_D 1 nM). The type B receptor also binds the related chemokine MGSA with high affinity (K_D 1 nM). Both of these receptors have been cloned and belong to a superfamily of seven transmembrane domain-containing molecules that are coupled to GTP-binding proteins. The cloned receptors signal in response to IL-8 and are expressed on a variety of cells including neutrophils, monocytes, fibroblasts and lymphocytes. In contrast to the cloned IL-8 receptors, which have a fairly narrow ligand specificity, a promiscuous IL-8 binding protein on human erythrocytes—which binds the chemokines IL-8, MGSA, MCP-1 and RANTES with high affinity—has been characterized. This protein, known as the chemokine receptor, does not appear to signal in response to IL-8. It has, however, been postulated to play a major role in inflammation by limiting the concentration of IL-8 and of the other chemokines in the circulation. The chemokine receptor has so far only been identified on erythrocytes and recently it has been shown to be a receptor for the malarial parasite Plasmodium vivax.

■ Human type A and B IL-8 receptors

Recently, cDNA clones encoding two human IL-8 receptors were isolated and characterized (Holmes *et al*. 1991; Murphy and Tiffany 1991). One receptor, termed the IL-8RA, was isolated (Holmes *et al*. 1991) from a human neutrophil cDNA library by expression cloning, using binding to ^{125}I-IL-8 as the method of detection. The IL-8RA is a 350-amino-acid protein (GenBank accession number M68932) with five potential N-linked glycosylation sites and seven hydrophobic membrane-spanning stretches characteristic of the G-protein-linked family of molecules (Strosberg 1991). The receptor is glycosylated since the translated molecular size of the receptor is around 40 kDa, which is considerably smaller than the reported size of 55 to 69 kDa based on cross-linking of the neutrophil receptor (Holmes *et al*. 1991; Samanta *et al*. 1989). The amino acid sequence of the IL-8RA is around 77 per cent identical with that of a second, IL-8 receptor, termed the IL-8RB (Murphy and Tiffany 1991) (Fig. 1). The IL-8RB was isolated from a neutrophil-like cell line, HL-60, and is a 355-amino-acid protein (GenBank accession number M73969) with a single potential N-linked glycosylation site.

■ Mouse IL-8 receptors

Screening of a mouse library with a cDNA probe for the type A human IL-8 receptor has identified a single nucleotide sequence that has around 70 per cent homology with both type A and type B human IL-8 receptors (Lee, J., Wood, W.I., and Moore, M. unpublished observations). Whether this sequence codes for a putative mouse IL-8 receptor remains to be determined, but the failure to detect mouse versions of the human type A and type B receptors is somewhat surprising. These preliminary findings suggest that mice may have a different type of IL-8 receptor than those cloned from human cells. This is especially interesting in light of a

recent communication claiming that the IL-8 gene is absent in mice and rats (Yoshimura 1992).

■ Rabbit IL-8 receptors

Recently, a protein that was initially identified as the rabbit fMLP receptor has been shown by two separate groups to be the rabbit homologue of the human IL-8 receptor (Beckmann *et al*. 1991; Lee *et al*. 1992a). The rabbit IL-8 receptor is a 355-amino-acid protein (GenBank accession number M82873) and its sequence shows 84 per cent and 73 per cent homology to the human types A and B IL-8 receptors, respectively (Fig. 1). Cells transfected with the rabbit IL-8 receptor bind human IL-8 with high affinity (K_D 1.9–3.6 nM) (Lee *et al*. 1992a) and induce a rapid increase in the mobilization of intracellular free Ca^{2+}. Interestingly, cells transfected with the rabbit IL-8 receptor do not bind radiolabelled MGSA nor does unlabelled MGSA cross-compete for IL-8 binding (Beckmann *et al*. 1991). It is unlikely that the failure of MGSA to bind to the rabbit IL-8 receptor reflects species differences since experiments with rabbit neutrophils demonstrate that these cells respond chemotactically to MGSA (Beckmann *et al*. 1991). The failure of the rabbit IL-8 receptor to bind to MGSA suggests that it more closely resembles the human type A IL-8 receptor than the type B receptor. Indeed, based on sequence homologies (Fig. 1), the rabbit IL-8 receptor is more closely related to the type A receptor than to the type B receptor.

■ Human erythrocyte chemokine receptor

In contrast to the cloned IL-8 receptors described above, human erythrocytes express a receptor with a broad specificity for chemokine binding (Darbonne *et al*. 1991; Neote *et al*. 1993; Horuk *et al*. 1993a). This multispecific receptor which has been designated as the erythrocyte chemokine

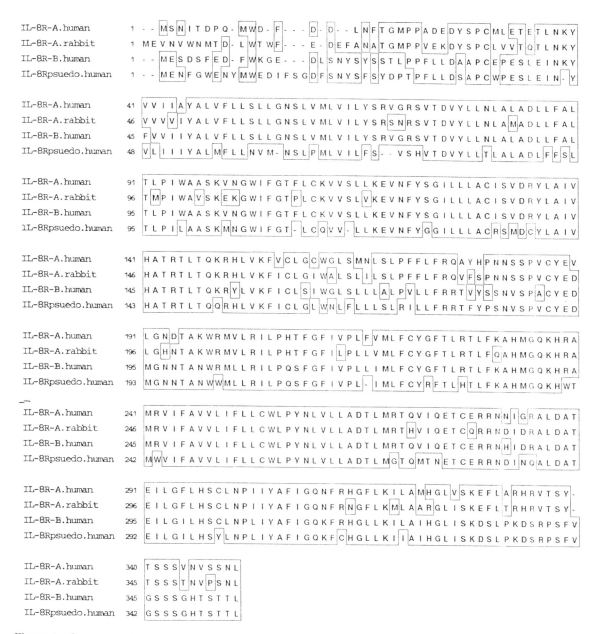

Figure 1. Sequence alignments.

(CK) receptor, binds the C–X–C chemokines IL-8 and MGSA and the C–C chemokines MCP-1 and RANTES but not the C–C chemokines MIP-1α and MIP-1β (Neote et al. 1993). In addition to differences in ligand binding profiles there are also fundamental differences in the biochemical properties of the CK receptor compared to the cloned IL-8 receptors (Horuk et al. 1993a). For example, chemical cross-linking experiments have demonstrated that the CK receptor is a 39 kDa glycosylated protein that does not appear to be regulated by G-proteins (Horuk et al. 1993a). By contrast the

cloned IL-8 receptors have a much larger molecular mass and are coupled to G-proteins (Samanta et al. 1989; Kupper et al. 1992).

Recently the erythrocyte chemokine receptor was identified as a receptor for the malarial parasite *Plasmodium vivax* (Horuk et al. 1993b). This observation was based on three observations. First, erythrocytes from a majority of African Americans who failed to bind IL-8 also lacked the Duffy blood-group antigen. The Duffy blood-group antigen has previously been shown to be required for the invasion of human erythrocytes by the human malarial parasite *Plasmodium vivax* (Miller *et al.* 1976). Second, a monoclonal antibody to the Duffy blood-group antigen blocked binding of IL-8 and other chemokines to Duffy-positive erythrocytes. Third, both MGSA and IL-8 blocked the binding of the parasite ligand and the invasion of human erythrocytes by *P. knowlesi*, a related monkey malarial parasite.

■ IL-8 receptor gene and expression

The genes for the type A and type B IL-8 receptors as well as a pseudogene for the type B IL-8 receptor (Fig. 1) have been mapped to human chromosome 2q35 (Morris *et al.* 1992). The facts that these three genes are clustered so close together and their high degree of sequence homology suggest that they may have arisen from a duplication of a common ancestral gene. Although the erythrocyte chemokine receptor has not been cloned, the precise chromosomal location of its *alter ego*, the Duffy blood group antigen, has been determined. This protein has been mapped to human chromosome 1 q21–q25 where it is flanked by genes for spectrin and Na/K ATPase (O'Connell *et al.* 1989).

The type A and B IL-8 receptor genes are expressed in neutrophils, HL-60, and THP-1 myeloid precursor cell lines but is absent in B cell and T cell lineages (Holmes *et al.* 1991; Murphy and Tiffany 1991). Small levels of message for the type A receptor have also been detected in synovial fibroblasts (Unemori *et al.* 1993). Little is known regarding the transcriptional regulation of these genes.

■ Receptor binding and structure/function analysis of the IL-8 receptors

Binding assays with [125]I- IL-8 and ^{125}I-MGSA in mammalian cells transfected with the type A and type B IL-8 receptor have recently been reported (Lee *et al.* 1992*b*). In direct contrast to the studies of Murphy and Tiffany (1991), which suggested that the IL-8RB expressed in *Xenopus* oocytes is a low affinity receptor for IL-8 and MGSA, analysis of the binding properties and intracellular Ca^{2+} response of the two human IL-8 receptors have demonstrated that the affinity of IL-8 for the A and B IL-8 receptors is similar (K_D = 1.2–3.6 nM). However, these receptors differ considerably in their affinities for the related ligand MGSA. IL-8RB binds both IL-8 and MGSA with similar affinities (K_D 2 nM), while IL-8RA binds MGSA with a 200-fold reduced affinity (K_D = 450 nM). The transfected receptors are also able to signal in response to ligand binding. A transient increase in the intracellular Ca^{2+} concentration is found for IL-8RB with both IL-8 and MGSA, while IL-8RA responds only to its high affinity ligand, IL-8.

The ligand specificity of the IL-8 receptor appears to reside in the amino terminal region of the molecule. In a series of domain swapping experiments LaRosa *et al.* (1992) constructed hybrid IL-8 receptors in which the entire N-terminal domain of one receptor was replaced by that of

the other. The hybrid type A receptor which had the type B receptor amino terminus exhibited high-affinity binding for both IL-8 and MGSA. Conversely the hybrid type B receptor which had the type A receptor amino terminus bound IL-8 with high affinity but lost the ability to bind MGSA. Additional information on the ligand-binding domain of the IL-8 receptors has come from alanine scanning mutagenesis of the receptor (Hébert *et al.* 1993). A series of type A IL-8 receptor mutants was created by replacing extracellular acidic residues of the receptor with alanine. These studies not only confirmed that the N-terminal region is an important functional binding determinant of IL-8 receptors but also that Glu 275 and Arg 280, two residues found in the third extracellular loop of the receptor, are critical for ligand binding.

■ Signalling mechanisms

Neutrophils are the major target cell for IL-8 and the chemokine triggers three main responses in these cells including shape change and transendothelial migration, degranulation and a respiratory burst (Oppenheim *et al.* 1991). Degranulation leads to the release of enzymes such as elastase and myeloperoxidase and the upregulation of certain adhesion molecules while the respiratory burst leads to superoxide and hydrogen peroxide formation. The molecular events that govern these processes are initiated by IL-8 binding to specific cell-surface receptors coupled to pertussis toxin sensitive G-proteins. Receptor binding leads to a dissociation of the heterotrimeric G-proteins into α and $\beta\gamma$ subunits. The G-protein α subunits then activate phospholipase C which in turn raises intracellular Ca^{2+} and leads to an activation of protein kinase C (Oppenheim *et al.* 1991; Moser *et al.* 1991). Recently both types of IL-8 receptor have been shown to signal through this pathway in response to both IL-8 and MGSA (Dianqing *et al.* 1993). In addition, the specific $G\alpha$ subunits that couple to the IL-8 receptor have been identified and were shown to activate a particular isoform of phospolipase C, PLC-β2 (Dianqing *et al.* 1993).

■ References

Beckmann, M.P., Munger, W.E., Kozlosky, C., Vanden Bos, T., Price, V., Lyman, S., Gerard, N.P., Gerard, C., and Cerretti, D.P. (1991). Molecular characterization of the interleukin-8 receptor. *Biochem. Biophys. Res. Commun.*, **179**, 784–9.

Darbonne, W.C., Rice, G.C., Mohler, M.A., Apple, T., Hebert, C.A., Valente, A.J., and Baker, J.B. (1991). Red blood cells are a sink for interleukin 8, a leukocyte chemotaxin. *J. Clin. Invest.*, **88**, 1362–9.

Dianqing, W., LaRosa, G.J., and Simon, M.I. (1993). G-protein-coupled signal transduction pathways for interleukin-8. *Science*, **261**, 101–3.

Hébert, C.A., Chuntharapai, A., Smith, M., Colby, T.J., Kim, J., and Horuk, R. (1993). Partial functional mapping of the human interleukin-8 type A receptor. Identification of a major ligand binding domain. *J. Biol. Chem.*, **268**, 18549–53.

Holmes, W.E., Lee, J., Kuang, W-J, Rice, G.C., and Wood, W.I. (1991). Structure and functional expression of a human interleukin-8 receptor. *Science*, **253**, 1278–80.

Horuk, R., Colby, T.J., Darbonne, W.C., Schall, T.J., and Neote, K. (1993a). The human erythrocyte inflammatory peptide (chemokine) receptor. Biochemical characterization, solubilization, and development of a binding assay for the soluble receptor. *Biochemistry*, **32**, 5733–8.

Horuk, R., Chitnis, C.E., Darbonne, W.C., Colby, T.J., Rybicki, A., Hadley, T.J., and Miller, L.H. (1993b). A receptor for the malarial parasite *plasmodium vivax*: the erythrocyte chemokine receptor. *Science*, **261**, 1182–4.

Kupper, R.W., Dewald, B., Jacobs, K.H., Baggiolini, M., and Gierschik, P. (1992). G-protein activation by interleukin 8 and related cytokines in human neutrophil plasma membranes. *Biochem. J.*, **282**, 429–34.

LaRosa, G.J., Thomas, K.M., Kaufmann, M.E., Mark, R., White, M., Taylor, L., Gray, G., Witt, D., and Navarro, J. (1992). Amino terminus of the interleukin-8 receptor is a major determinant of receptor subtype specificity. *J. Biol. Chem.*, **267**, 25402–6.

Lee, J., Kuang, W.-J., Rice, G. C., and Wood, W.I. (1992a). Characterization of complementary DNA clones encoding the rabbit IL-8 receptor. *J. Immunol.*, **148**, 1261–4.

Lee, J., Horuk, R., Rice, G.C., Bennet, G.L., Camerato, T., and Wood, W.I. (1992b). Characterization of two high affinity human interleukin-8 receptors. *J. Biol. Chem.*, **267**, 16283–7.

Miller, L.H., Mason, S.J., Clyde, D.F., and McGinniss, M.H. (1976). The resistance factor to *Plasmodium vivax* in blacks. The Duffy-blood-group genotype, FyFy. *New Eng. J. Med.*, **295**, 302–4.

Morris, S.W., Nelson, N., Valentine, M.B., Shapiro, D.N., Look, A.T., Kozlosky, C.J., Beckmann, M.P., and Cerretti, D.P. (1992). Assignment of the genes encoding human interleukin-8 receptor types 1 and 2 and an interleukin-8 receptor pseudogene to chromosome 2q35. *Genomics*, **14**, 685–91.

Moser, B., Schumacher, C., von Tscharner, V., Clark-Lewis, I., and Baggiolini, M. (1991). Neutrophil-activating peptide 2 and gro/melanoma growth-stimulatory activity interact with neutrophil-activating peptide 1/interleukin 8 receptors on human neutrophils. *J. Biol. Chem.*, **266**, 10666–71.

Murphy, P.M., and Tiffany, H.L. (1991). Cloning of complementary DNA encoding a functional human interleukin-8 receptor. *Science*, **253**, 1280–3.

Neote, K., Darbonne, W. C., Ogez, J., Horuk, R., and Schall, T.J. (1993). Identification of a promiscuous inflammatory peptide receptor on the surface of red blood cells. *J. Biol. Chem.*, **268**, 12247–9.

O'Connell, P., Lathrop, G.M., Nakamura, Y., Leppert, M.L., Ardinger, R.H., Murray, J.L., Lalouel, J.-M., and White, R. (1989). Twenty-eight loci form a continuous linkage map of markers for human chromosome 1. *Genetics*, **4**, 12–20.

Oppenheim, J.J., Zachariae, C.O., Mukaida, N., and Matsushima, K. (1991). Properties of the novel proinflammatory supergene 'intercrine' cytokine family. *Ann. Rev. Immunol.*, **9**, 617–48.

Samanta, A.K., Oppenheim, J.J., and Matsushima, K. (1989). Identification and characterization of specific receptors for monocyte-derived neutrophil chemotactic factor (MDNCF) on human neutrophils. *J. Exp. Med.*, **169**, 1185–9.

Strosberg, A.D. (1991). Structure/function relationship of proteins belonging to the family of receptors coupled to GTP-binding proteins. *Eur. J. Biochem.*, **196**, 1–10.

Unemori, E.N., Amento, E.P., Bauer, E.A., and Horuk, R. (1993). Melanoma growth-stimulatory activity/GRO decreases collagen expression by human fibroblasts. Regulation by C–X–C but not C–C cytokines. *J. Biol. Chem.*, **268**, 1338–42.

Yoshimura, T. (1992). Neutrophil attractant protein-1 (NAP-1) is highly conserved in guinea pig but not in mouse or rat. *FASEB J.*, **6**, A1340.

Richard Horuk:
The Department of Protein Chemistry,
Genentech Inc.,
South San Francisco, California 94080, USA

Interleukin-9 (IL-9)

IL-9 is a 32–39 kDa glycoprotein that was initially identified in the mouse as a growth factor for certain T cell clones. Subsequently, it has been found to potentiate the proliferative response of mast cell lines to IL-3, of fetal thymocytes to IL-2, to be mitogenic for megakaryoblastic leukaemia cells and to support erythroid colony formation. Its production and secretion is induced in peripheral blood mononuclear cells or lymphocytes by lectins, phorbol ester and calcium ionophore and in CD4+ T-helper cells by immune activation. Pathologically it might be involved in the development of T-cell tumours.

■ Alternative names

P40, mast cell growth enhancing activity (MEA), T cell growth factor III (TCGF-III).

■ IL-9 sequence

The nucleotide sequences of IL-9 cDNAs cloned from mouse and man (GenBank accession numbers X14045 and X17543), in both cases, predict a protein of 144 amino acids, each containing an 18-amino-acid leader sequence (Van Snick *et al.* 1989; Renauld *et al.* 1990a) The complete amino acid sequence of mouse IL-9 has been determined at the protein level (PIR accession number, S05531) (Simpson *et al.* 1989). These proteins are 55 per cent identical at the amino acid level (Fig. 1), with perfect conservation of the 10 cysteine residues in the mature protein. Human IL-9 has an additional cysteine in the presumptive signal peptide. The homology between mouse and human IL-9 is 69 per cent at the nucleotide level. Murine IL-9 has been shown to be active on human T cell lines, whereas human IL-9 is not active on mouse IL-9 dependent cell lines (Renauld *et al.* 1990*a*).

■ IL-9 protein

IL-9 is a secreted glycoprotein monomer containing 126 amino acid residues (mouse) in the mature protein. The complete amino acid sequence of mouse IL-9 has been determined at the protein level (the calculated molecular mass is 14150 Da) (Simpson *et al.* 1989). The amino terminal residue of the mature protein (glutamine) was found to occur in its cyclized form (pyroglutamic acid). Murine IL-9 contains N-linked glycosylation giving rise to several molecular weight species ranging from 32 to 39 kDa. Positions 32, 60, 83, and 96 in mouse IL-9 are N-glycosylated (three of these positions (32, 60, and 96) are conserved in the deduced amino acid sequence for human IL-9, while human IL-9 has one additional site at position 45) (Simpson *et al.* 1989). Treatment of native mouse IL-9 with N-glycanase F reduces the molecular mass to 15–16 kDa. Mouse IL-9 is a basic glycoprotein (pI ~10) due to the high incidence of basic amino acids. IL-9 can be purified by sequential chromatography on a TSK-phenyl column, Mono Q column, and RP-HPLC (Uyttenhove *et al.* 1988).

■ IL-9 gene and transcription

The human IL-9 gene (GenBank accession number M30135) is localized to chromosome 5q31–q35 (Modi *et al.* 1991) in a region containing the genes for various growth factors and has been mapped between the IL-3 and EGR-1 (early growth response-1) genes (Warrington *et al.* 1992). In the mouse, the IL−9 gene (GenBank accession number M30136) does not appear to be linked to the same gene cluster as it has been localized to chromosome 13 while the IL-3, IL-4, IL-5, and GM-CSF genes are located on chromosome 11. The human and mouse IL-9 genes share a similar structure with five exons and four introns encompassing about 4 kbp. Numerous ATTTA motives are found in the 3′ untranslated

region of the messages and could be involved in the regulation of the mRNA stability. Both genes contain a classical TATA box preceeded by several consensus sequences for transcription factors. Particularly, potential recognition sites for AP-1, AP-2, and IRF-1 (interferon responsive element-1) are conserved in the human and mouse promoter (Renauld *et al.* 1990*b*).

The IL-9 gene is expressed in freshly isolated T cells upon stimulation by lectins or anti-CD3 antibodies and is enhanced by PMA which does not induce IL-9 expression by itself. The kinetics of IL-9 mRNA expression and its inhibition by cycloheximide indicate the existence of secondary signals. IL-2 plays a key role in this process as an anti-IL-2 receptor antibody completely blocks the IL-9 induction by PMA and anti-CD3 (Houssiau *et al.* 1992*a*).

■ Biological functions

IL-9 was initially described on the basis of its T cell growth factor activity for some T helper clones but not freshly isolated T cells (Uyttenhove *et al.* 1988). More recently, studies performed with human T cells have shown that IL-9 responsiveness is not restricted to a particular T cell subset but is strictly dependent on the stage of activation of the cells (Houssiau *et al.* 1992*b*). The only activity of IL-9 described so far on fresh T cells is a synergistic response of mouse fetal thymocytes to the combination of IL-9 and IL-2 (Suda *et al.* 1990).

IL-9 also enhances IL-4 induced IgE and IgG production by human or mouse B cells (Petit-Frère *et al.* 1993; Dugas *et al.* 1993).

IL-9 induces the proliferation of mouse bone marrow-derived mast cell lines, alone or in synergy with IL−3 (Hültner *et al.* 1990). IL-9 also induces mast cell differentiation by stimulating IL-6 production, the synthesis of serine proteases, and the expression of the high-affinity IgE receptor (unpublished data).

In the haemopoietic system, IL-9 appears to have a restricted specificity for erythroid progenitors in the adult (Donahue *et al.* 1990) and a broader spectrum of activity for fetal progenitors, including myeloid precursors (Holbrook *et al.* 1991). IL-9 also stimulates the *in vitro* proliferation of MO7e, a human megakaryoblastic leukaemia cell line (Yang *et al.* 1989). IL-9 induces morphological maturation

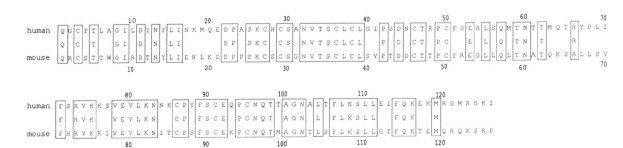

Figure 1. Alignment of human and mouse IL-9 amino acid sequence.

and cellular excitability of murine neuronal progenitor cells (Mehler et al. 1993).

■ Pathology

Contrasting with the absence of activity of IL-9 on fresh T cells, IL-9 efficiently stimulates in vitro proliferation of mouse T cell lymphomas (Vink et al. 1993). Moreover, constitutive IL-9 expression in transgenic mice results in the occurence of thymic lymphomas in about five per cent of the animals and renders them particularly susceptible to subliminal doses of a chemical mutagen, thereby suggesting that dysregulated IL-9 expression could be involved in T cell oncogenesis (Renauld et al. 1994). In the human, IL-9 mRNA expression has been found in lymph nodes from patients with Hodgkin's disease or large cell anaplastic lymphomas (Merz et al. 1991) and an autocrine IL-9 loop has been demonstrated for the in vitro proliferation of one Hodgkin-derived cell line (Gruss et al. 1992).

■ References

Donahue, R.E., Yang, Y.C., and Clark, S.C. (1990). Human P40 T-cell growth factor (interleukin-9) supports erythroid colony formation. Blood, 75, 2271–5.

Dugas, B., Renauld, J-C., Pène, J., Bonnefoy, J.Y., Petit-Frère, C., Braquet, P., Bousquet, J., Van Snick, J., and Mencia-Huerta, J.M. (1993). Interleukin-9 potentiates the interleukin-4-induced immunoglobulin (IgG, IgM and IgE) production by normal human B lymphocytes. Eur. J. Immunol., 23, 1687–92.

Gruss, H.J., Brach, M.A., Drexler, H.G., Bross, K.J., and Herman, F. (1992). Interleukin 9 is expressed by primary and cultured Hodgkin and Reed–Sternberg cells. Cancer Res., 52, 1026–31.

Holbrook, S.T., Ohls, R.K., Schibler, K.R., Yang, Y.C., and Christensen, R.D. (1991). Effect of interleukin-9 on clonogenic maturation and cell-cycle status of fetal and adult hematopoietic progenitors. Blood, 77, 2129–34.

Houssiau, F.A., Renauld, J.-C., Fibbe, W., and Van Snick, J. (1992a) IL-2 dependence of IL-9 expression in human T lymphocytes. J. Immunol., 148, 3147–51.

Houssiau, F.A., Renauld, J.-C., Stevens, M., Lehmann, F., Lethe, B., Coulie, P.G., and Van Snick, J. (1992b) Human T cell lines and clones respond to IL-9. J. Immunol., 150, 2634–40.

Hültner, L., Druez, C., Moeller, J., Uyttenhove, C., Schmitt, E., Rüde, E., Dörmer, P., and Van Snick, J. (1990). Mast cell growth-enhancing activity (MEA) is structurally related and functionally identical to the novel mouse T cell growth factor P40/TCGFIII (interleukin 9). Eur. J. Immunol., 20, 1413–6.

Mehler, M.F., Rozental, R., Dougherty, M., Spray, D.C., and Kessler, J.A. (1993). Cytokine regulation of neuroneeal differentiation of hippocampal progenitor cells. Nature, 362, 62–5.

Merz, H., Houssiau, F.A., Orscheschek, K., Renauld, J.-C., Fliedner, A., Herin, M., Noel, H., Kadin, M., Mueller-Hermelink, H.K., Van Snick, J., and Feller, A.C. (1991). Interleukin-9 expression in human malignant lymphomas: unique association with Hodgkin's disease and large cell anaplastic lymphoma. Blood, 78, 1311–7.

Modi, W.S., Pollock, D.D., Mock, B.A., Banner, C., Renauld, J.-C., and Van Snick, J. (1991). Regional localization of the human glutaminase (GLS) and interleukin-9 (IL9) genes by in situ hybridization. Cytogenet. Cell Genet., 57, 114–6.

Petit-Frère, C., Dugas, B., Braquet, P., and Mencia-Huerta, J.M. (1993). Interleukin-9 potentiates the interleukin-4-induced IgE and IgG1 release from murine B lymphocytes. Immunology, 79, 146–51.

Renauld, J.-C., Goethals, A., Houssiau, F., Van Roost, E., and Van Snick, J. (1990a). Cloning and expression of a cDNA for the human homolog of mouse T cell and mast cell growth factor P40. Cytokine, 2, 9–12.

Renauld, J.-C., Goethals, A., Houssiau, F., Merz, H., Van Roost, E., and Van Snick, J. (1990b). Human P40/IL-9. Expression in activated CD4+ T cells, genomic organization, and comparison with the mouse gene. J. Immunol., 144, 4235–41.

Renauld, J.C., van der Lugt, N., Vink, A., van Roon, M., Godfraind, C., Warnier, G., Merz, H., Feller, A., Berns, A., and Van Snick, J. (1994). Thymic lymphomas in interleukin α transgenic mice. Oncogene, 9, 1327–32.

Simpson, R.J., Moritz, R.L., Rubira, M.R., Gorman, J.J., and Van Snick, J. (1989). Complete amino acid sequence of a new murine T-cell growth factor P40. Eur. J. Biochem., 183, 715–22.

Suda, T., Murray, R., Fischer, M., Yokota, T., and Zlotnik, A. (1990). Tumor necrosis factor-alpha and P40 induce day 15 murine fetal thymocyte proliferation in combination with IL-2. J. Immunol., 144, 1783–7.

Uyttenhove, C., Simpson, R.J., and Van Snick, J. (1988). Functional and structural characterization of P40, a mouse glycoprotein with T-cell growth factor activity. Proc. Natl. Acad. Sci. (USA), 85, 6934–8.

Van Snick, J., Goethals, A., Renauld, J.-C., Van Roost, E., Uyttenhove, C., Rubira, M.R., Moritz, R.L., and Simpson, R.J. (1989). Cloning and characterization of a cDNA for a new mouse T cell growth factor (P40). J. Exp. Med., 169, 363–8.

Vink, A., Renauld, J.-C., Warnier, G., and Van Snick, J. (1993). Interleukin-9 stimulates in vitro growth of mouse thymic lymphomas. Eur. J. Immunol., 23, 1134–8.

Warrington, J.A., Bailey, S.K., Armstrong, E., Aprelikova, O., Alitalo, K., Dolganov, G.M., Wilcox, A.S. Sikela, J.M., Wolfe, S.F., Lovett, M., and Vasmuth, J. (1992). A radiation hybrid map of 18 growth factor, growth factor receptor, hormone receptor, or neurotransmitter receptor genes on the distal region of the long arm of chromosome 5. Genomics, 13, 803–8.

Yang, Y.C., Ricciardi, S., Ciarletta, A., Calvetti, J., Kelleher, K., and Clark, S.C. (1989). Expression cloning of cDNA encoding a novel human hematopoietic growth factor: human homologue of murine T-cell growth factor P40. Blood, 74, 1880–4.

Richard J. Simpson:
Joint Protein Structure Laboratory,
Ludwig Institute for Cancer Research (Melbourne Branch) and The Walter and Eliza Hall Institute for Medical Research, Melbourne, Australia

Jean-Christophe Renauld:
Ludwig Institute for Cancer Research, University of Leuven, Brussels, Belgium

The Interleukin-9 Receptor

High affinity binding of IL-9 to various cell types including mouse T helper clones, T cell lymphomas, mast cells, and macrophage cell lines is mediated by a single chain receptor belonging to the haemopoietin receptor superfamily. The mechanisms involved in signal transduction are still to be unravelled. However, in the cytoplasmic domain, the presence of a P–X–P motif conserved among many cytokine receptors as well as a significant homology with the erythropoietin receptor and the β-chain of the IL-2 receptor suggest the existence of similarities in the signal transduction pathways used by the receptors of this superfamily. Messages encoding potential isoforms of the IL-9 receptors are generated by alternative splicing, including a putative soluble form.

■ Binding characteristics and expression of the IL-9 receptor

The mouse IL-9 receptor has been characterized using T cell clones and mast cell lines that express a single class of high affinity binding site for IL-9 (K_D ~100 pM) (Druez *et al.* 1990). Cross-linking analysis indicated that the receptor consists of a ~64 kDa glycoprotein containing ~10 kDa N-linked sugar.

IL-9 binding sites were found on T helper cell clones and T cell tumours such as EL4 or LBRM-33. By contrast, freshly isolated lymphoid cells did not bind IL-9. These data are in accordance with the observation that T cell clones and T cell lymphomas respond to IL-9 while freshly isolated T cells do not. More surprisingly, IL-9 receptors were detected on macrophage cell lines while no activity of IL-9 has been described so far on this cell type. On the other hand, IL-9 has been recently reported to enhance IL-4-induced immunoglobulin production (Petit-Frère *et al.* 1993) although no IL-9 binding was detected on resting B cells.

■ The IL-9 receptor protein

A cDNA encoding the mouse IL-9 receptor has been isolated by expression cloning in COS cells (GenBank accession number M84746) (Renauld *et al.* 1992). The deduced protein contains 468 amino acids and its expression in COS cells resulted in high affinity binding sites with an affinity similar to that of the natural receptor. The extracellular domain of the receptor is composed of 233 amino acids and shows some features characteristic of the haemopoietin receptor superfamily, namely the conservation of cysteine residues and a Trp-Ser-X-Trp-Ser motif located 26 amino acids before the transmembrane domain.

A human IL-9 receptor cDNA has been identified by cross-hybridization with the mouse probe (GenBank accession number M84747). The human protein, which confers to murine cells the responsiveness to human IL-9, is a 522 amino acid molecule showing 53 per cent identity with its murine homologue. In addition, other cDNAs have been identified which result from alternative splicing. These cDNAs predict the existence of truncated isoforms of the

receptor, including a putative soluble form. Such an alternative splicing, as well as the use of alternative polyadenylation signals, probably accounts for the six different bands detectable by northern blot analysis from various human IL-9 responsive cells.

A schematic representation of the human IL-9 receptor is shown in Fig. 1.

■ Signalling mechanisms

Analysis of the intracellular sequence of the IL-9 receptor does not provide useful information regarding the signal transduction pathway. As already observed for many cytokine receptors, this region is particularly rich in serine and proline residues with an intriguing stretch of nine successive serines in the human molecule. Moreover, proximal to the transmembrane domain, a short region of homology was found with other haemopoietic receptors. This region includes a sequence characterized by a Pro-X-Pro motif previously described as 'box 1' for the gp130 chain of

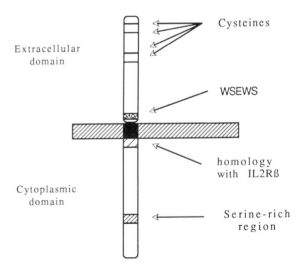

Figure 1. Schematic representation of the human IL-9 receptor.

the IL-6 receptor and potentially involved in signal transduction by this receptor (Murakami *et al.* 1991). Downstream of this motif a significant homology was observed specifically between the IL-9 receptor and the erythropoietin receptor and the B chain of the IL-2 receptor.

Analysis of the effect of IL-9 on protein tyrosine kinase activity in the human megakaryoblastic leukaemia cell line M07e has shown that IL-9 induced or enhanced at least four specific tyrosine phosphorylated bands in western blots. Moreover, IL-9 did not stimulate phosphorylation of the 42 kDa mitogen activated protein (MAP) kinase or Raf-1 which are serine/threonine kinases that are phosphorylated and activated by many growth factors, thereby suggesting that the IL-9 receptor activates a specific signal transduction pathway (Miyazawa *et al.* 1992).

■ References

Druez, C., Coulie, P., Uyttenhove, C., and Van Snick, J. (1990). Functional and biochemical characterization of mouse p40/IL-9 receptors. *J. Immunol.*, **145**, 2494–9.

Miyazawa, K., Hendrie, P.C., Kim, Y.J., Mantel, C., Yang, Y.C., Kwon, B.S., and Broxmeyer, H.E. (1992). Recombinant human interleukin-9 induces protein tyrosine phosphorylation and synergizes with steel factor to stimulate proliferation of the human factor-dependent cell line, M07e. *Blood*, **80**, 1685–92.

Murakami, M., Narazaki, M., Hibi, M., Yawata, H., Yasukawa, K., Hamaguchi, M., Taga, T., and Kishimoto, T. (1991). Critical cytoplasmic region of the interleukin 6 signal transducer gp130 is conserved in the cytokine receptor family. *Proc. Natl. Acad. Sci. (USA)*, **88**, 11349–53.

Petit-Frère, C., Dugas, B., Braquet, P., and Mencia-Huerta, J.M. (1993). Interleukin-9 potentiates the interleukin-4-induced IgE and IgG1 release from murine B lymphocytes. *Immunology*, **79**, 146–51.

Renauld, J.-C., Druez, C., Kermouni, A., Houssiau, F., Uyttenhove, C., Van Roost, E., and Van Snick, J. (1992). Expression cloning of the murine and human interleukin 9 receptor cDNAs. *Proc. Natl. Acad. Sci. (USA)*, **89**, 5690–4.

Jean-Christophe Renauld:
Ludwig Institute for Cancer Research, University of Louvain, Brussels, Belgium

Interleukin-10 (IL-10)

Interleukin-10 is a secreted 17–21 kDa protein which in humans is not glycosylated but exists as a non-covalent homodimer. It is produced by some subclasses of T-lymphocytes (primarily of the Th2 type), B-lymphocytes, and macrophages late after activation by antigen or bacterial products and its production is inhibited by interleukin-4 and interferon-γ IL-10 acts on macrophages to downregulate class II major histocompatibility complex (MHC) expression, inhibit the production of reactive intermediates, and, in turn, results in inhibition of cytokine synthesis by activated Th1 T-cells and natural killer (NK) cells. It could result in a switch in the immune response from a delayed type hypersensitivity to an antibody type. IL-10 can also act as a co-stimulator of the growth of several haemopoietic cell lineages including B-cells, T-cells, and mast cells. A viral analogue of IL-10 exists in some viruses including the Epstein–Barr (EB) virus and may be involved in evasion of host immune responses. IL-10 may find use in the amelioration of inflammatory and hyperimmune biological responses.

■ Alternative names

Cytokine synthesis inhibitory factor (CSIF); mast cell growth factor (MCGF); thymocyte growth promoting factor.

■ IL-10 sequence

The nucleotide sequences of interleukin-10 cDNAs cloned from mouse and man (GenBank accession numbers M37897 and M57627, respectively) predict proteins of 178 (human) or 175 (mouse) amino acids each including an 18-amino-acid leader sequence (Moore *et al.* 1990; Vieira *et al.* 1991). These proteins are 73 per cent identical at the amino acid level but, while human IL-10 acts on both human and mouse responsive cells, mouse IL-10 is inactive on human cells. Human IL-10 is highly homologous to proteins predicted for open reading frames present in Epstein–Barr virus (BCRFI) and equine herpes virus (GenBank accession numbers M11924 and S59624) (84 per cent and 75 per cent amino acid identity, respectively) (Moore *et al.* 1990; Vieira *et al.* 1991; Rode *et al.* 1993). These sequences vary most from human at the N-terminal end.

```
                10          20          30          40          50          60
HUMAN   MHSSALLCCLVLLTGVRASPGQGTQSENSCTHFPGNLPNMLRDLRDAFSRVKTFFQMKDQ
MOUSE   MPGSALLCCLLLLTGMRISRGQYSREDNNCTHFPVGQSHMLLELRTAFSQVKTFFQTKDQ
RAT     MPGSALLCCLLLLAGVKTSKGHSIRGDNNCTHFPVSQTHMLRELRAAFSQVKTFFQKKDQ


                70          80          90          100         110         120
HUMAN   LDNLLLKESLLEDFKGYLGCQALSEMIQFYLEEVMPQAENQDPDIKAHVNSLGENLKTLR
MOUSE   LDNILLTDSLMQDFKGYLGCQALSEMIQFYLVEVMPQAEKHGPEIKEHLNSLGEKLKTLR
RAT     LDNILLTDSLLQDFKGYLGCQALSEMIKFYLVEVMPQAENHGPEIKEHLNSLGEKLKTLW


                130         140         150         160         170
HUMAN   LRLRRCHRFLPCENKSKAVEQVKNAFNKLQEKGIYKAMSEFDIFINYIEAYMTMKIRN
MOUSE   MRLRRCHRFLKCENKSKAVEQVKSDFNKLEDQGVYKAMNEFDIFINCIEAYMMIKMKS
RAT     IQLRRCHRFLPCENKSKAVEQVKNDFNKLQDKGVYKAMNEFDIFINCIEAYVTLKMKN
```

Figure 1. Amino acid sequence alignments for human, mouse, and rat IL-10. The first amino acid of the mature protein begins at position 19 in each case.

■ IL-10 protein

Human IL-10 is an 18 kDa protein with no evidence of attached carbohydrate but mouse IL-10 is variably glycosylated at asparagine 11 of the mature protein to yield 17, 19, and 21 kDa species. Nevertheless, glycosylation is not required for activity of mouse IL-10. The BCRFI protein (viral IL-10) is also not glycosylated and is a 17 kDa protein. Rat anti-human IL-10 antibodies have been described and these recognize human IL-10 and viral IL-10 but not mouse IL-10 (Moore *et al*. 1993). Mouse IL-10 contains five cysteine residues, four of which are conserved in human and viral IL-10 and form two intramolecular disulphide bonds. Both human and murine IL-10 form non covalent homodimers in solution (39 kDa and 33 kDa, respectively) but it is unknown if monomers are biologically active. Mouse and human IL-10 are inactivated at low pH ($<$5) but are stable at neutral and alkaline pH (Moore *et al*. 1993).

■ IL-10 gene and transcription

The mouse and human IL-10 genes are located on syntenic regions of mouse and human chromosome 1. The mouse IL-10 gene (GenBank accession number M84340) consists of five exons that span approximately 5.1 kb of genomic DNA and contains several possible transcriptional control sequences including glucocorticoid and interleukin-6 response elements, CBP, PEA3, NFKB, and AP-1 binding sites as well as interferon and acute phase stimuli response elements. These motifs are similar to those found in the 5′-flanking region of the mouse and human IL-6 genes (Kim *et al*. 1992).

Mouse IL-10 is produced by Th2 T-lymphocytes activated by anti-CD3 antibodies and phorbol ester, but unlike the production of other cytokines, this is not inhibitable by cyclosporin. It is also expressed late in the activation process of B-cells, monocytes/macrophages, mast cells, and by keratinocytes (Moore *et al*. 1993; Rennick *et al*. 1992; de Waal Malefyt *et al*. 1992). Expression of human IL-10 is similar to that of the mouse except that it is not restricted to Th2 cells but occurs in a wider range of T cells. Eptein–Barr virus-transformed cells are also induced to produce IL-10 early in infection, apparently preceding viral IL-10 production. Ly-1 or CD5+ B-cells also express IL-10 which may act as an autocrine growth factor. IL-10 production by macrophages is inhibited by IL-4 and interferon-γ.

■ Biological functions

IL-10 inhibits cytokine production by T cells indirectly by affecting macrophages (Fiorentino *et al*. 1991). The inhibition is at the transcriptional level and in the mouse affects only cytokines produced late in the activation of Th-1 cells (interferon-γ and interleukin-3). In the human system, inhibition is not restricted to Th-1 cells so that synthesis of GM-CSF, TNFα, and TNFβ are also affected (Vieira *et al*. 1991). The switch from Th-1 to Th-2 type immune responses suggests that IL-10 may have a role in switching from delayed-type hypersensitivity (Th-1) to antibody (Th-2) type immune responses. Cytokine synthesis by IL-2 stimulated natural killer (NK) cells (IFNγ and TNFα) was also inhibited by IL-10 in a macrophage-dependent manner (Hsu *et al*. 1992).

The mechanism of IL-10 action in inhibiting antigen-dependent cytokine production and T cell proliferation appears to involve, in part, downregulation of constitutive and interferon-γ-induced expression of class II MHC molecules on the surface of monocytes and macrophages (de Waal Malefyt *et al.* 1991a). However, other macrophage surface molecules that directly stimulate Th-1 and NK cells may also be involved.

IL-10 also inhibits bacterial lipopolysaccharide or interferon-γ induced cytokine synthesis by macrophages (IL-1α, β, IL-6, IL-8, TNFα, GM-CSF, and G-CSF (de Waal Malefyt *et al.* 1991b), inhibits the production of reactive oxygen intermediates and nitric oxide by macrophages induced by interferon-γ, and inhibits adherence of macrophages suggesting that it has anti-inflammatory activity (Moore *et al.* 1993; Bogdan *et al.* 1991).

In contrast to its T-cell and macrophage inhibitory activities, IL-10 can also act directly as a co-stimulator of the proliferation of thymocytes and T cells in the presence of IL-2, IL-4, or IL-7 and of mucosal-like mast cells in the presence of IL-3 or IL-4, at least in the mouse. Similarly, IL-10 may be a co-stimulator of B cell growth (especially of the CD5 lineage) and can upregulate class II MHC and IgM, IgG, and IgA levels in some classes of B cells. Finally, IL-10 also acts as a co-stimulator of the proliferation of megakaryocyte, erythroid, and primitive haemopoietic cells with IL-3, IL-6, and stem cell factor (Moore *et al.* 1993; Rennick *et al.* 1992; de Waal Malefyt *et al.* 1992).

Viral IL-10 (product of the BCRFI gene of the EB virus) shares the cytokine synthesis inhibitory activity of human and mouse IL-10, although it has a lower specific activity, and this may be important in suppressing the antiviral activities of interferon-γ and macrophages. It also shares their activities on the survival and co-stimulation of B cells which also may be important, but viral IL-10 lacks the activity of mammalian IL-10s in co-stimulating thymocyte and mast cell proliferation and in inducing class II MHC expression on B cells (Vieira *et al.* 1991; Moore *et al.* 1993; de Waal Malefyt *et al.* 1991b, 1992). These observations must reflect some complexity in IL-10 receptors which has not yet been resolved.

■ Pathology

Because of its capacity to inhibit the production and release of tumour necrosis factor by macrophages in response to endotoxin (LPS), IL-10 reduces the lethality of experimental endotoxaemia (septic shock) in mice (Howard *et al.* 1993; Gerard *et al.* 1993). Conversely, mice treated with anti-IL-10 antibodies showed increased mortality in response to LPS-induced shock, decreased serum IgM and IgA levels and an increase in IgG2 levels, a decreased responsiveness to some types of thymus independent antigens, and a specific depletion of Ly1 (CD5) B-cells (Ishida *et al.* 1993). Since CD5+ B cells produce IL-10, IL-10 may be important for the expansion of this cell subpopulation. Viral IL-10 produced by EB-virus-transformed B-cells inhibits cytokine synthesis by T cells, NK cells, and macrophages and stimulates the survival and co-stimulation of proliferation of B cells, processes which may aid the infectivity and evasion of host immune responses by the virus (O'Garra *et al.* 1992).

■ References

Bogdan, C., Vodovotz, Y., and Nathan, C. (1991). Macrophage deactivation by interleukin 10. *J. Exp. Med.*, **174**, 1549–55.

de Waal Malefyt, R., Haanen, J., Spits, H., Roncarolo, M.G., te Velde, A., Figdor, C., Johnson, K., Kastelein, R., Yssel, H., and de Vries, J.E. (1991a). Interleukin 10 (IL-10) and viral IL-10 strongly reduce antigen-specific human T cell proliferation by diminishing the antigen-presenting capacity of monocytes via downregulation of class II major histocompatibility complex expression. *J. Exp. Med.*, **174**, 915–24.

de Waal Malefyt, R., Abrams, J., Bennett, B., Figdor, C.G., and de Vries, J.E. (1991b). Interleukin 10(IL-10) inhibits cytokine synthesis by human monocytes: an autoregulatory role of IL-10 produced by monocytes. *J. Exp. Med.*, **174**, 1209–20.

de Waal Malefyt, R., Yssel, H., Roncarolo, M.G., Spits, H., and de Vries, J.E. (1992). Interleukin-10. *Curr. Opin. Immunol.*, **4**, 314–20.

Fiorentino, D.F., Zlotnik, A., Vieira, P., Mosmann, T.R., Howard, M., Moore, K.W., and O'Garra, A. (1991). IL-10 acts on the antigen-presenting cell to inhibit cytokine production by Th1 cells. *J. Immunol.*, **146**, 3444–51.

Gerard, C., Bruyns, C., Marchant, A., Abramowicz, D., Vandenabeele, P., Delvaux, A., Fiers, W., Goldman, M., and Velu, T. (1993). Interleukin 10 reduces the release of tumour necrosis factor and prevents lethality in experimental endotoxemia. *J. Exp. Med.*, **177**, 547–50.

Howard, M., Muchamuel, T., Andrade, S., and Menon, S. (1993). Interleukin 10 protects mice from lethal endotoxemia. *J. Exp. Med.*, **177**, 1205–8.

Hsu, D.H., Moore, K.W., and Spits, H. (1992). Differential effects of IL-4 and IL-10 on IL-2-induced IFN-gamma synthesis and lymphokine-activated killer activity. *Int. Immunol.*, **4**, 563–9.

Ishida, H., Hastings, R., Thompson-Snipes, L., and Howard, M. (1993). Modified immunological status of anti-IL-10 treated mice. *Cell. Immunol.*, **148**, 371–84.

Kim, J.M., Brannan, C.I., Copeland, N.G., Jenkins, N.A., Khan, T.A., and Moore, K.W. (1992). Structure of the mouse IL-10 gene and chromosomal localization of the mouse and human genes. *J. Immunol.*, **148**, 3618–23.

Moore, K.W., Vieira, P., Fiorentino, D.F., Trounstine, M.L., Khan, T.A., and Mosmann, T.R. (1990). Homology of cytokine synthesis inhibitory factor (IL-10) to the Epstein–Barr virus gene BCRFI [published erratum appears in *Science* (1990) **250. 494**]. *Science*, **248**, 1230–4.

Moore, K.W., O'Garra, A., de Waal Malefyt, R., Vieira, P., and Mosmann, T.R. (1993). Interleukin-10. *Annu. Rev. Immunol.*, **11**, 165–90.

O'Garra, A., Chang, R., Go, N., Hastings, R., Haughton, G., and Howard, M. (1992). Ly-1 B (B-1) cells are the main source of B cell-derived interleukin 10. *Eur. J. Immunol.*, **22**, 711–7.

Rennick, D., Berg, D., and Holland, G. (1992). Interleukin 10: an overview. *Prog. Growth Factors Res.*, **4**, 207–27.

Rode, H.J., Janssen, W., Rosen-Wolff, A., Bugert, J.J., Thein, P., Becker, Y., and Darai, G. (1993). The genome of equine herpesvirus type 2 harbors an interleukin 10 (IL10)-like gene. *Virus Genes*, **7**, 111–6.

Vieira, P., de Waal Malefyt, R., Dang, M.N., Johnson, K.E., Kastelein, R., Fiorentino, D.F., de Vries, J.E., Roncarolo, M.G., Mosmann, T.R., and Moore, K.W. (1991). Isolation and expression of human cytokine synthesis inhibitory factor cDNA clones: homology to Epstein–Barr virus open reading frame BCRFI. *Proc. Natl. Acad. Sci. (USA)*, **88**, 1172–6.

Nicos A. Nicola:
Walter and Eliza Hall Institute of Medical Research,
Royal Melbourne Hospital,
Victoria 3050, Australia

Interleukin-11 (IL-11)

IL-11 is a 19–21 kDa protein that is a member of the cytokine family which includes interleukin-6 (IL-6), leukaemia inhibitory factor (LIF), oncostatin-M (OSM), and ciliary neurotrophic factor (CNTF). It mediates signal transduction through gp130, a protein originally identified as a component of the IL-6 receptor. Like other members of this family, IL-11 is a pleiotropic cytokine with significant activities in the haemopoietic system (primitive, erythroid, and megakaryocytic progenitors), hepatic system (inducer of acute phase protein response), stromal cell system (inhibition of adipocyte differentiation), and intestinal epithelial system (protection/regeneration). In contrast to IL-6, IL-11 has little if any direct effect on either T or B lymphocytes. IL-11 in animal models has shown considerable potential in myelorestoration, particularly relating to reconstitution of platelets and in protection against or recovery from radiation or drug-induced damage to intestinal epithelium.

■ Alternative names

Adipogenesis inhibitory factor (Kawashima *et al.* 1991).

■ IL-11 sequence

The nucleotide sequences of IL-11 cDNAs cloned from mouse and man (mouse sequence submitted to GenBank, human sequence GenBank accession number M37006) encode proteins with 199 amino acids (Fig. 1) (Paul *et al.* 1990). At the amino acid level, the mouse and human proteins are highly conserved (88 per cent identity). The human protein has been shown to include a 21-amino-acid signal sequence which is likely to be identical given the conservation of amino acids around the processing site. That the human and murine proteins are very similar is also supported by the observation that both molecules are active with target cells from the alternate species (C. Wood, unpublished results). Careful analysis of the amino acid sequences and gene structures has revealed that IL-11, IL-6, granulocyte colony-stimulating factor (G-CSF), OSM, LIF, CNTF, and growth hormone are all distantly related in evolution (Bruce *et al.* 1992). However, IL-11 and IL-6 do not directly compete in receptor binding (Yin *et al.* 1992a,b).

■ IL-11 protein

IL-11 is a secreted monomeric protein of 178 amino acids (core protein molecular weight of 19 000 Da) which has no asparagine-linked and little, if any, serine- or threonine-linked carbohydrate (Paul *et al.* 1990). It has no cysteine residues and therefore no disulphide bridges. Despite the lack of disulphides for tertiary structure stabilization, IL-11 is a highly stable molecule. In addition to the lack of cysteine residues, IL-11 has an unusual amino acid composition which includes 12 per cent proline, 23 per cent leucine, and 14 per cent arginine plus lysine residues. The high proportion of positively charged amino acids results in an unusually basic isoelectric point calculated to be 11.7. This very high p*I* greatly facilitates separation of IL-11 from other proteins by cation exchange chromatography. IL-11 is also readily fractionated using reverse-phase HPLC (K. Turner, unpublished results). The three-dimensional structure of IL-11 has yet to be determined but, given its interaction with gp130, it is likely to adopt a four-helix bundle structure common to many of the other cytokines.

■ IL-11 gene and transcription

The human IL-11 gene has been localized to the long arm of chromosome 19 at l9ql3.3–ql13.4 (McKinley *et al.* 1992). The gene, encompassing approximately 7 kbp, consists of five exons and four introns. Two distinct mRNAs with sizes estimated to be 1.5 and 2.5 kb are generated from this gene by alternative utilization of poly-A addition sequences in the 3′ non-coding region (Paul *et al.* 1990). Both transcripts have AU-rich regions found in other cytokine mRNAs which render them unstable (Shaw and Kamen 1986).

IL-11 expression was originally identified in several fibroblast cell lines derived from bone marrow or lung (Paul *et al.* 1990). With these cell lines, IL-11 expression was substantially enhanced by treatment with a variety of stimuli including IL-1, phorbol esters and TGFβ (Paul *et al.* 1990; Elias *et al.* 1994). IL-1 and TGFβ have been found to interact synergistically in activating IL-11 gene expression in lung fibroblasts. IL-11 expression has also been observed in a human trophoblast cell line (Paul *et al.* 1990) and cells from a thyroid malignancy (Tohyama *et al.* 1992). In contrast to many of the cytokines, IL-11 mRNA can not be detected in lectin-stimulated peripheral blood cells.

The human IL-11 gene has been found to contain a variety of consensus sequences for transcription factor binding (McKinley *et al.* 1992). These include those for AP1, CTF/NF1, and EF/C in the 5′ flanking region. The gene also contains several copies of interferon-inducible elements in the 5′ flanking region and a copy of the IL-1-inducible element (ACATGGCACAATCT) found in the 5′ flanking region of the IL-6 gene located within the 3′ flanking region of the IL-11 gene.

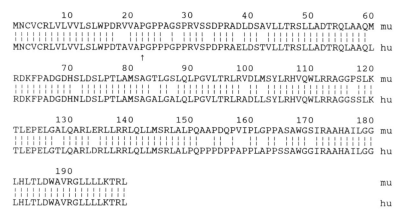

```
           10        20        30        40        50        60
MNCVCRLVLVVLSLWPDRVVAPGPPAGSPRVSSDPRADLDSAVLLTRSLLADTRQLAAQM  mu
||||||||||||||||| |||||| | |||| |||| ||| |||||||||||||||||||
MNCVCRLVLVVLSLWPDTAVAPGPPPGPPRVSPDPRAELDSTVLLTRSLLADTRQLAAQL  hu
                ↑
           70        80        90        100       110       120
RDKFPADGDHSLDSLPTLAMSAGTLGSLQLPGVLTRLRVDLMSYLRHVQWLRRAGGPSLK  mu
||||||||| ||||||||||||||| || |||||||||||| ||| ||||||||||| |||
RDKFPADGDHNLDSLPTLAMSAGALGALQLPGVLTRLRADLLSYLRHVQWLRRAGGSSLK  hu
           130       140       150       160       170       180
TLEPELGALQARLERLLRRLQLLMSRLALPQAAPDQPVIPLGPPASAWGSIRAAHAILGG  mu
||||||| ||||| |||||||||||||||||||||| ||  || || |||||||||||||
TLEPELGTLQARLDRLLRRLQLLMSRLALPQPPPDPPAPPLAPPSSAWGGIRAAHAILGG  hu
           190
LHLTLDWAVRGLLLLKTRL                                           mu
||||||||||||||||||||
LHLTLDWAVRGLLLLKTRL                                           hu
```

Figure 1. Sequence comparison between human and murine IL-11.

■ Biological functions

IL-11 was originally identified by its mitogenic activity with the IL-6-dependent plasmacytoma, T1165 (Paul *et al.* 1990). However, in comparison to IL-6, IL-11 has proved to be a relatively poor mitogen for plasmacytomas and myelomas and seems to only indirectly regulate B cell development (Paul *et al.* 1992; Yin *et al.* 1992c). IL-11 has proved to share with IL-6 the capacity for acting synergistically with other cytokines, notably steel factor and IL-3 in supporting the growth of primitive haemopoietic progenitors (Musashi *et al.* 1991). In the mouse, these progenitors have been shown to yield both lymphoid and myeloid progeny indicative of their early status within the hierarchy of haemopoietic cells (Hirayama *et al.* 1992). IL-11 directly promotes megakaryocyte maturation (Burstein *et al.* 1992) and acts in concert with IL-3 or steel factor in promoting megakaryo-

cyte colony formation (Yonemura *et al.* 1992). IL-11, also in combination with IL-3 or steel factor, supports the growth of erythroid colonies and, in the mouse, bipotent erythroid/megakaryocytic colonies (Quesniaux *et al.* 1992). As with other members of the gpl30 signal transducing cytokines (Baumann and Schendel 1991), IL-11 promotes hepatic acute phase protein synthesis and inhibits adipocyte development (Kawashima *et al.* 1991). Although IL-11, like IL-6 (in contrast to LIF and OSM), does not prevent embryonal stem cell differentiation, it has recently been found that IL-11 in combination with steel factor promotes haemopoietic differentiation by developing embryoid bodies in culture (Keller *et al.* 1993). In contrast to IL-6, IL-11 has not been found to interact with T cells. A comparison of the biological activities of IL-6 and IL-11 is shown in Table 1.

Administration of IL-11 to normal mice and primates typically results in elevation in the numbers of circulating platelets but otherwise has little effect on normal haemopoiesis (Neben *et al.* 1992; Bree *et al.* 1991). In primates, IL-11 altered the levels of acute phase proteins as predicted from its effects on hepatocytes in culture. In myelosuppressed mice rescued by bone marrow transplantation, IL-11 was found to accelerate both platelet and neutrophil recovery (Du *et al.* 1993). In a murine model of myelosuppression with sublethal radiation followed by sublethal administration of 5-fluorouracil, IL-11 strikingly promoted survival, an effect that appeared to be in large part due to protection of small intestinal mucosa from the severe damage induced by this regimen (Du *et al.* 1994). All of these findings are consistent with the pleiotropic nature of IL-11, on the one hand suggesting that IL-11 may have important clinical utility as an adjunct to cancer therapy and, on the other, that there is still much to be learned about the cells that respond to IL-11 and what other functions this cytokine may have.

Table 1. Comparison of the biological activities of IL-11 and IL-6

Activity	IL-11	Il-6
Enhancement of IL-3- or SF-supported colony formation by		
Primitive progenitors	+	+
Megakaryocyte progenitors	+	+/–
Erythroid progenitors	+	–
Enhancement of megakaryocyte maturation	+	+
Hybridoma growth factor	+/–	+
Augmentation of antibody responses	+	+
T cell activation	–	+
Cytoytic T cell differentiation	–	+
Hepatic acute-phase protein induction	+	+
Inhibition of adipocyte differentiation	+	+
Protection of intestinal mucosa from 5-FU	+	?
Promotion of embryoid body haemopoiesis	+	–

■ Pathology

Administration of IL-11 in primates is associated with a modest drop in hematocrit despite the demonstrated stimulatory effects of this cytokine on erythroid progenitors (Bree *et al*. 1991). Although the mechanism of this effect is not known, it may represent dilution from expanded plasma volume due to fluid retention. Long term expression of high levels of IL-11 in mice has been associated with some weight loss, consistent with the known ability of IL-11 to inhibit adipocyte differentiation (Hawley *et al*. 1993).

■ References

Baumann, H., and Schendel, P. (1991). Interleukin-11 regulates the hepatic expression of the same plasma protein genes as interleukin-6. *J. Biol. Chem*., **266**, 20424–7.

Bree, A., Schlerman, F., Timony, G., McCarthy, K., Stoudimire, J., and Garnick, M. (1991). Pharmacokinetics and thrombopoietic effects of recombinant human interleukin-11 (rhIL-11) in nonhuman primates and rodents. *Blood*, **78** (Suppl. 1), 132a.

Bruce, A.G., Linsley, P.S., and Rose, T.M. (1992). Oncostatin M. *Prog. Growth Factor Res*., **4**, 157–70.

Burstein, S.A., Mei, R.-L., Henthorn, J., Friese, P., and Turner, K. (1992). Leukemia inhibitory factor and interleukin-11 promote maturation of murine and human megakaryocytes *in vitro*. *J. Cell. Physiol*., **153**, 305–12.

Du, X.X., Neben, T., Goldman, S., and Williams, D.A. (1993). Effects of recombinant human interleukin-11 on hematopoietic reconstitution in transplant mice; acceleration of recovery of peripheral blood neutrophils and platelets. *Blood*, **81**, 27–34.

Du, X. X., Doerschuk, C.M., Orazi, A., and Williams, D.A. (1994) A bone marrow stromal-derived growth factor, interleukin-11 stimulates recovery of small intestinal mucosal cells after cytoablative therapy. *Blood*, **83**, 33–7.

Elias, J.A., Zheng, T., Whiting, N.L., Trow, T.K., Merrill, W.W., Zitnik, R., Ray, P., and Alderman, E.M. (1994). Il-1 and TGF-β regulation of fibroblast IL-11. *J. Immunol*., **152**, 2421–9.

Hawley, R.G., Fong, A.Z.C., Ngan, B.Y., de Lanux, V.M., Clark, S.C., and Hawley, T.S. (1993). Progenitor cell hyperplasia with rare development of myeloid leukaemia in interleukin 11 bone marrow chimeras. *J. Exp. Med*., **178**, 1175–88.

Hirayama, F., Shih, J.-P., Awgulewitsch, A., Warr, G.W., Clark, S.C., and Ogawa, M. (1992). Clonal proliferation of murine lymphohematopoietic progenitors in culture. *Proc. Natl. Acad. Sci. (USA)*, **89**, 5907–11.

Kawashima, I., Ohsumi, J., Mita-Honjo, K., Shimoda-Takano, H., Ishikawa, S., Sakakibara, K., Miyadai, K., and Takiguchi, Y. (1991). Molecular cloning of cDNA encoding adipogenesis inhibitory factor and identity with interleukin-11. *FEBS Lett*., **283**, 199–202.

Keller, G., Kennedy, M., Papayannopoulou, T., and Wiles, M.V. (1993). Hematopoietic commitment during embryonic stem cell differentiation in culture. *Mol. Cell. Biol*., **13**, 473–86.

McKinley, D., Wu, Q., Yang-Feng, T., and Yang, Y.-C. (1992). Genomic sequence and chromosomal location of human interleukin-11 gene (IL11). *Genomics*, **13**, 814–9.

Musashi, M., Yang, Y.-C., Paul, S.R., Clark, S.C., Sudo, T., and Ogawa, M. (1991). Direct and synergistic effects of interleukin 11 on murine hemopoiesis in culture. *Proc. Natl. Acad. Sci. (USA)*, **88**, 765–9.

Neben, T.Y., Loebelenz, J., Hayes, L., McCarthy, K., Stoudemire, J., Schaub, R., and Goldman, S.J. (1992). Recombinant human interleukin-11 stimulates megakaryocytopoiesis and increases peripheral platelets in normal and splenectomized mice. *Blood*, **80**, 901–8.

Paul, S.R., Bennett, F., Calvetti, J.A., Kelleher, K., Wood, C.R., O'Hara, M., Leary, A.C., Sibley, B., Clark, S.C., Williams, D.A., and Yang, Y.-C. (1990). Molecular cloning of a cDNA encoding interleukin 11, a stromal cell-derived lymphopoietic and hematopoietic cytokine. *Proc. Natl. Acad. Sci. (USA)*, **87**, 7512–6.

Paul, S.R., Barut, B.A., Bennett, F., Cochran, M.A., and Anderson, K.C. (1992). Lack of a role of interleukin 11 in the growth of multiple myeloma. *Leuk. Res*., **16**, 247–52.

Quesniaux, V.F., Clark, S.C., Turner, K., and Fagg, B. (1992). Interleukin-11 stimulates multiple phases of erythropoiesis *in vitro*. *Blood*, **80**, 1218–23.

Shaw, G., and Kamen, R. (1986). A conserved AU sequence from the 3' untranslated region of GM-CSF mRNA mediates selective mRNA degradation. *Cell*, **46**, 659–67.

Tohyama, K., Yoshida, Y., Ohashi, K., Sano, E., Kobayashi, H., Endo, K., Naruto, M., and Nakamura, T. (1992). Production of multiple growth factors by a newly established human thyroid carcinoma cell line. *Jpn. J. Cancer Res*., **83**, 153–8.

Yin, T., Miyazawa, M., and Yang, Y.-C. (1992a). Characterization of interleukin-11 receptor and protein tyrosine phosphorylation induced by interleukin-11 in mouse 3T3-L1 cells. *J. Biol. Chem*., **267**, 8347–51.

Yin, T., Taga, T., Tsang, M.L.-S, Yasukawa, K., Kishimoto, T., and Yang, Y.-C. (1992b). Interleukin (IL) -6 signal transducer, gp130, is involved in IL-11 mediated signal transduction. *Blood*, **80** (Suppl. 1), 151a.

Yin, T., Schendel, P., and Yang, Y.-C. (1992c) Enhancement of *in vitro* and *in vivo* antigen-specific antibody responses by interleukin 11. *J. Exp. Med*., **175**, 211–6.

Yonemura, Y., Kawakita, M., Masuda, T., Fugimoto, K., Kato, K., and Takatsuki, K. (1992). Synergistic effects of interleukin 3 and interleukin 11 on murine megakaryopoiesis in serum-free culture. *Exp. Hematol*., **20**, 1011–6.

Steven C. Clark:
Genetics Institute Inc.,
Cambridge, MA, USA

Interleukin 12 (IL-12)

IL-12 is a heterodimeric glycoprotein composed of unrelated subunits of 35 kDa (p35) and 40 kDa (p40). Co-expression of both chains is necessary for biological activity. Il-12 was originally cloned from B cell lines but the major source of IL-12 in human peripheral blood mononuclear cells (PBMC) is monocytes stimulated by bacteria, intracellular parasites, or their products. Both human and murine IL-12 have been cloned and while murine IL-12 is active on responsive human and murine cells, human IL-12 is not active on murine cells. Cells expressing high-affinity receptors for IL-12 appear to be limited to NK and activated T cells. Biological responses include induction of IFNγ from T and NK cells, stimulation of T cell proliferation as a co-stimulant, and enhancement of cytolytic activity of T and NK cells. Recent evidence has shown that IL-12 is a key factor in the differentiation of Th1 cells from progenitors and in suppressing IgE synthesis. This array of activities suggests clinical utility as an anti-tumour and anti-viral agent, in restoring immune function in immune suppressed patients, as a vaccine adjuvant in promoting cellular immune response, and in combating opportunistic infections in immune suppressed patients. Conversely, these same activities may prove to exacerbate autoimmune activity.

■ Alternative names

Natural killer cell stimulatory factor (NKSF), cytotoxic lymphocyte maturation factor (CLMF), T cell stimulatory factor (TSF).

■ IL-12 sequence

cDNAs for human and murine IL-12 subunits have been cloned (Wolf *et al*. 1991; Gubler *et al*. 1991; Schoenhaut *et al*. 1992). The nucleotide sequences of the p40 subunit from man and mouse (GenBank accession numbers M65292, M65272, and M86671) predict polypeptides of 328 and 335 amino acids, respectively. Both include amino terminal sequences, 22 amino acids in length, which are characteristic of hydrophobic signal peptides. At the amino acid level, the p40 subunits are 70 per cent identical and 79 per cent similar.

The p35 subunit nucleotide sequences from man and mouse (GenBank accession numbers M65291, M65271, and M86672) predict polypeptides of 253 and 215 amino acids, respectively. The human p35 cDNA contains two potential initiation codons 5′ of the amino terminus of the mature protein. Initiation at the second methionine codon encodes a characteristic hydrophobic signal peptide. The amino acid sequence from the first methionine is less hydrophobic. Comparable sequence patterns are found in some membrane-associated proteins, suggesting a potential membrane-bound form for IL-12 or p35. Murine p35 cDNA lacks a methionine codon comparable to the first methionine codon in the human cDNA but does encode a characteristic signal peptide sequence. Murine p35 is 67 per cent identical to human p35 at the amino acid sequence level. There is no sequence homology between p35 and p40, and they do not appear to be structurally related to each other. The p40 subunit is most closely related by sequence homology to the extracellular domain of the IL-6 receptor (Gearing and Cosman 1991). The p35 subunit is more distantly related to IL-6 itself (Merberg *et al*. 1992).

■ IL-12 protein

The predicted molecular weight of the human IL-12 heterodimer is 57 200, but the observed molecular mass is 75–80 kDa on SDS-PAGE gels under non-reducing conditions. Upon reduction, the 75–80 kDa species is converted to two sets of bands corresponding to the p35 and p40 subunits. The p35 subunit has a predicted protein molecular mass of 22.5 kDa but migrates on SDS-PAGE as a series of bands at about 35 kDa. The p40 subunit has a predicted molecular mass of 34.7 kDa but migrates on SDS-PAGE as a group of at least three species at about 40 kDa. On p35, both Asn-71 and Asn-85 are glycosylated. Of the four potential N-linked glycosylation sites on p40, only the site at Asn-200 is glycosylated to a significant degree (E. Nickbarg, manuscript in preparation). Murine IL-12 retains only one potential N-linked site on p35 but has one additional potential N-linked site on p40. Human p35 contains seven cysteines that are all conserved in murine p35, while human p40 has ten cysteines, nine of which are conserved in murine p40. Murine p40 has an additional three cysteines clustered near the C-terminus. The two subunits in human IL-12 are covalently linked by a single intermolecular disulphide bond. The p35 subunit contains three intramolecular disulphides, while p40 contains four intramolecular disulphides and one free cysteine (E. Nickbarg, manuscript in preparation). Disulphide bonding is necessary for the maintenance of biological activity as shown by the fact that IL-12 rapidly loses activity upon reduction with dithiothreitol (Podlaski *et al*. 1992). Chemical modification and antibody binding studies of human IL-12 showed that p40 is involved in mediating the IL-12 biological response (Podlaski *et al*.

1992; Chizzonite et al. 1991). However, p35 appears to determine the species specificity (Schoenhaut et al. 1992).

Human and murine IL-12 have been purified by conventional chromatographic procedures, including ion-exchange, heparin–sepharose, lentil–lectin, and hydroxyl-apatite steps, and by HPLC reversed-phase and ion-exchange steps (Kobayashi et al. 1989; Stern et al. 1990). Human IL-12 has also been purified by immunoaffinity chromatography using a monoclonal antibody specific for the 75 kDa heterodimer (Chizzonite et al. 1992).

■ IL-12 genes and transcription

The gene for p40 (IL12B) maps to human chromosome 5q31–33 and that for p35 (IL12A) to 3p12–3q13.2 (Sieburth et al. 1992). The mouse homologues have not been mapped. The p40 gene spans approximately 20kb and contains at least six exons (S. Wolf, unpublished). The gene for p35 has not been as well characterized but Southern blot analysis suggests a size of approximately 6kb (Wolf et al. 1991). Expression of the p40 gene is limited in human peripheral blood to monocytes, B cell lines and presently undefined non-adherent cells, perhaps B cells. Monocytes are the major source in human peripheral blood mononuclear cells (D'Andrea et al. 1992). The p40 gene is not constitutively expressed but is induced by bacterial products or infection with microorganisms (D'Andrea et al. 1992). Expression by monocytes is enhanced by co-stimulation with IFNγ and suppressed by IL-4 and IL-10 (D'Andrea et al. 1993). Although p40 is related to the haemopoietic receptor family, alternate transcripts of p40 encoding a membrane form have not been detected (S. Wolf, unpublished).

P35 transcription is constitutive. Transcripts are detectable by PCR analysis in essentially all cells and lineages tested and by northern analyses in several cell lines which have been analyzed (S. Wolf, unpublished). In contrast to p40 transcription, levels of p35 mRNA are not dramatically changed by stimulation with a wide variety of agents.

■ Biological functions

Murine IL-12 is active on human and murine cells (Schoenhaut et al. 1992). Although reported to weakly induce IFNγ production from murine spleen cells, human IL-12 does not induce proliferation of activated mouse splenocytes (Schoenhaut et al. 1992). Activities of IL-12 include induction of IFNγ production by resting and activated T and NK cells (Kobayashi et al. 1989; Chan et al. 1991, 1992); enhancement of T cell proliferation in combination with co-stimulation (Kobayashi et al. 1989; Gately et al. 1991; Perussia et al. 1992); some enhancement of NK cell proliferation (Robertson et al. 1992; Naume et al. 1992; Naume et al. 1993); and enhancement of NK and T cell cytolytic activity (Kobayashi et al. 1989; Gately et al. 1991). For induction of IFNγ, IL-12 synergizes strongly with IL-2 (Kobayashi et al. 1989; Chan et al. 1991, 1992), an effect largely due to stabilization of the induced mRNA (Chan et al. 1992). Cells induced to proliferate include CD4[+] and CD8[+] T cells

(Gately et al. 1991) and, to a lesser degree, activated NK cells (Robertson et al. 1992; Naume et al. 1992, 1993). Proliferation is not IL-2 dependent nor as pronounced as for IL-2 and is not sustained (Perussia et al. 1992). In contrast, IL-12 inhibits proliferation induced by IL-2 of TCR$\gamma\delta^+$ cells and NK blasts (Perussia et al. 1992). Direct cytolytic T cell (CTL) activation has been demonstrated for alloreactive T cells and as measured by anti-CD3 redirected lysis (Gately et al. 1992; Chehimi et al. 1993). Although maximal levels of NK cytolytic activity induced by IL-12 are lower than those induced by IL-2, only picomolar concentrations of IL-12 are necessary (Kobayashi et al. 1989). Enhancement of NK cytolytic activity extends to the suppressed activity found with NK cells from HIV infected patients and advanced cancer patients (Chehimi et al. 1992; Soiffer et al. 1993). Consistent with cells shown to be responsive to IL-12, high affinity receptors have only been detected on T and NK cells (Desai et al. 1992).

Recent studies have shown that IL-12 is a key cytokine in the differentiation of Th1 T lymphocytes from progenitors and in promoting natural immunity. Hsieh et al. (1993) used IL-12 to induce differentiation in vitro of naive T cells expressing a transgenic ovalbumin-specific T cell receptor into Th1 cells. Sypek et al. (1993) and Heinzel et al. (1993) have demonstrated, in murine models of leishmaniasis, the role of IL-12 in inducing a Th1 response in vivo. A role for IL-12 in initiating natural immunity was demonstrated in studies by Tripp et al. (1993) and Gazzinelli et al. (1993). IL-12 induced by treatment of SCID spleen cells with Listeria monocytogenes or Toxoplasma gondii was shown to be responsible for the production of IFNγ in these models and to prolong the survival of infected mice (Gazzinelli et al. 1993). Thus, in vivo, IL-12 appears to be critical both for initiating and sustaining a cellular immune response.

Perhaps related to its activity in promoting Th1 development and IFNγ production, IL-12 has been shown to affect IgE synthesis (Kiniwa et al. 1992). In addition, IL-12 appears to suppress IgE synthesis by IFNγ independent mechanisms. Umbilical cord blood lymphocytes cultured in the presence of IL-12 and hydrocortisone plus IL-4 were suppressed in the synthesis of IgE. Under these conditions, IFNγ is not detected nor did added neutralizing antibody to IFNγ reverse the suppression. In vivo, mice treated with goat anti-mouse IgD produce increased IL-4, IgE, and IgG1. Treatment of anti-IgD treated mice with IL-12 suppressed all three. However, addition of neutralizing anti-IFNγ antibody to the anti-IgD protocol reversed the effect on IL-4 and IgG1, but not on IgE levels (Morris et al. 1993).

Activities of IL-12 in normal and tumour bearing mice reflect those detected in vitro. Administration of 1 μg per day of IL-12 to C57BL/6 mice induced high serum level of IFN-γ. NK activity in liver, spleen, lung, and peritoneal cavity was also increased by IL-12 with peak NK activity in liver and spleen reached after two days of dosing. A dose-dependent increase in NK cytolytic activity could be measured from 1 ng (IP, qd) (Gately et al. 1994). Co-administration of IL-12 and allogeneic cells also enhanced development of an allo-specific cytolytic T-lymphocyte response (Gately et al. 1994). In studies in tumour-bearing mice, IL-12 suppressed metastasis formation and intradermal and subcutaneous tumour growth in all tumours tested (Brunda *et al.* 1993). In a renal cell sarcoma (Renca) tumour model, intratumour injection of IL-12 reproducibly cured 70–80 per cent of mice and a majority were resistant to re-challenge (Brunda *et al.* 1993). Depletion of CD4+, CD8+, or NK cells by administration of antibodies demonstrated that CD8+ cells were necessary for the anti-tumour effect (Brunda *et al.* 1993).

■ Pathology

Daily injection of IL-12 (0.1–1μg, IP, qd) results in splenomegaly due to extramedullary haemopoiesis. Despite enhanced extramedullary haemopoiesis, mice developed a reversible anaemia, lymphopenia, and neutropenia (Gately *et al.* 1994).

■ References

Brunda, M.J., Luistro, L., Warier, R.R., Wright, R.B., Hubbard, B.R., Murphy, M., Wolf, S.F., and Gately, M.K. (1993). Antitumour and antimetastatic activity of interleukin 12 against murine tumours. *J. Exp. Med.*, **178**, 1223–30

Chan, S.H., Perussia, B., Gupta, J.W., Kobayashi, M., Pospisil, M., Young, H.A., Wolf, S.F., Young, D., Clark, S.C., and Trinchieri, G. (1991). Induction of interferon gamma production by natural killer cell stimulatory factor: characterization of the responder cells and synergy with other inducers. *J. Exp. Med.*, **173**, 869–79.

Chan, S.H., Kobayashi, M., Santoli, D., Perussia, B., and Trinchieri, G. (1992). Mechanisms of IFN-gamma induction by natural killer cell stimulatory factor (NKSF/IL-12). Role of transcription and mRNA stability in the synergistic interaction between NKSF and IL-2. *J. Immunol.*, **148**, 92–8.

Chehimi, J., Starr, S.E., Frank, I., Rengaraju, M., Jackson, S.J., Llanes, C., Kobayashi, M., Perussia, B., Young, D., Nickbarg, E., Wolf, S.F., and Trinchieri, G. (1992). Natural killer (NK) cell stimulatory factor increases the cytotoxic activity of NK cells from both healthy donors and human immunodeficiency virus-infected patients. *J. Exp. Med.*, **175**, 789–96.

Chehimi, J., Valiante, M.N., D'Andrea, A., Rengaraju, M., Rosado, Z., Kobayashi, M., Perussia, B., Wolf, S.F., Starr, S.E., and Trinchieri, G. (1993). Enhancing effect of natural killer cell stimulatory factor (NKSF/interleukin-12) on cell-mediated cytotoxicity against tumour-derived and virus-infected cells. *Eur. J. Immunol.*, **23**, 1826–30.

Chizzonite, R., Truitt, T., Podlaski, F.J., Wolitzky, A.G., Quinn, P.M., Nunes, P., Stern, A.S., and Gately, M.K. (1991). IL-12: monoclonal antibodies specific for the 40-kDa subunit block receptor binding and biologic activity on activated human lymphoblasts. *J. Immunol.*, **147**, 1548–56.

Chizzonite, R., Truitt, T., Desai, B.B., Nunes, P., Podlaski, F.J., Stern, A.S., and Gately, M.K. (1992). IL-12 receptor. I. Characterization of the receptor on phytohemagglutinin-activated human lymphoblasts. *J. Immunol.*, **148**, 3117–24.

D'Andrea, A., Rengaraju, M., Valiante, N.M., Chehimi, J., Kubin, M., Aste, M., Chan, S.H., Kobayashi, M., Young, D., Nickbarg, E., Chizzonite, R., Wolf, S.F., and Trinchieri, G. (1992). Production of natural killer cell stimulatory factor (interleukin 12) by peripheral blood mononuclear cells. *J. Exp. Med.*, **176**, 1387–98.

D'Andrea, A., Aste-Amezaga, M., Valiante, M.N., Ma, X., Kubin, M., and Trinchieri, G. (1993). Interleukin 10 (IL-10) inhibits human lymphocyte interferon g-production by suppressing natural killer cell stimulatory factor/IL-12 synthesis in accessory cells. *J. Exp. Med.*, **178**, 1041–8

Desai, B.B., Quinn, P.M., Wolitzky, A.G., Mongini, P.K., Chizzonite, R., and Gately, M.K. (1992). IL-12 receptor. II. Distribution and regulation of receptor expression. *J. Immunol.*, **148**, 3125–32.

Gately, M.K., Desai, B.B., Wolitzky, A.G., Quinn, P.M., Dwyer, C.M., Podlaski, F.J., Familletti, P.C., Sinigaglia, F., Chizzonnite, R., Gubler, U., and Stern, A.S. (1991). Regulation of human lymphocyte proliferation by a heterodimeric cytokine, IL-12 (cytotoxic lymphocyte maturation factor). *J. Immunol.*, **147**, 874–82.

Gately, M.K., Wolitzky, A.G., Quinn, P.M., and Chizzonite, R. (1992). Regulation of human cytolytic lymphocyte responses by interleukin-12. *Cell. Immunol.*, **143**, 127–42.

Gately, M.K., Warrier, R.R., Honasoge, S., Faherty, D.A., Connaughton, S.E., Anderson, T.D., Sarmiento, U., Hubbard, B., and Murphy, M. (1994). Administration of recombinant IL-12 to normal mice enhances cytolytic lymphocyte activity and induces production of IFN-γ *in vivo*. *Int. Immunol.*, **6**, 157–67.

Gazzinelli, R.T., Heiny, S., Wynn, T.A., Wolf, S.F., and Sher, A. (1993). Interleukin 12 is required for the T-lymphocyte-independent induction of interferon gamma by an intracellular parasite and induces resistance in T-cell-deficient hosts. *Proc. Natl. Acad. Sci. (USA)*, **90**, 6115–9.

Gearing, D.P., and Cosman, D. (1991). Homology of the p40 subunit of natural killer cell stimulatory factor (NKSF) with the extracellular domain of the interleukin-6 receptor. *Cell*, **66**, 9–10.

Gubler, U., Chua, A.O., Schoenhaut, D.S., Dwyer, C.M., McComas, W., Motyka, R., Nabavi, N., Wolitzky, A.G., Quinn, P.M., Familletti, P.C., and Gately, M.K. (1991). Coexpression of two distinct genes is required to generate secreted bioactive cytotoxic lymphocyte maturation factor. *Proc. Natl. Acad. Sci. (USA)*, **88**, 4143–7.

Heinzel, F.P., Schoenhaut, D.S., Rerko, R.M., Rosser, L.E., and Gately, M.K. (1993). Recombinant interleukin 12 cures mice infected with Leishmania major. *J. Exp. Med.*, **177**, 1505–9.

Hsieh, C.S., Macatonia, S.E., Tripp, C.S., Wolf, S.F., O'Garra, A., and Murphy, K.M. (1993). Development of TH1 CD4+ T cells through IL-12 produced by Listeria-induced macrophages. *Science*, **260**, 547–9.

Kiniwa, M., Gately, M., Gubler, U., Chizzonite, R., Fargeas, C., and Delespesse, G. (1992). Recombinant interleukin-12 suppresses the synthesis of immunoglobulin E by interleukin-4 stimulated human lymphocytes. *J. Clin. Invest.*, **90**, 262–6.

Kobayashi, M., Fitz, L., Ryan, M., Hewick, R.M., Clark, S.C., Chan, S., Loudon, R., Sherman, F., Perussia, B., and Trinchieri, G. (1989). Identification and purification of natural killer cell stimulatory factor (NKSF) a cytokine with multiple biologic effects on human lymphocytes. *J. Exp. Med.*, **70**, 827–45.

Merberg, D., Wolf, S.F., and Clark, S. (1992). Sequence similarity between NKS and IL-6/G-CSF family. *Immunology Today*, **13**, 7778.

Morris, S.C., Hubbard, B., Gately, M., Gause, W.C., and Finkelman, F.D. (1993). Effects of interleukin-12 (IL-12) on an *in vivo*, T cell dependent, humoral immune response. *J. Immunol.*, **150**, 45A.

Naume, B., Gately, M., and Espevik, T. (1992). A comparative study of IL-12 (cytotoxic lymphocyte maturation factor)-, IL-2-, and IL-7-induced effects on immunomagnetically purified CD56+ NK cells. *J. Immunol.*, **148**, 2429–36.

Naume, B., Gately, M.K., Desai, B.B., Sundan, A., and Espevik, T. (1993). Synergistic effects of interleukin 4 and interleukin 12 on NK cell proliferation. *Cytokine*, **5**, 38–46.

Perussia, B., Chan, S.H., D'Andrea, A., Tsuji, K., Santoli, D., Pospisil, M., Young, D., Wolf, S.F., and Trinchieri, G. (1992). Natural killer (NK) cell stimulatory factor or IL-12 has differential effects on the proliferation of TCR-alpha beta+, TCR-gamma delta+ T lymphocytes, and NK cells. *J. Immunol.*, **149**, 3495–502.

Podlaski, F.J., Nanduri, V.B., Hulmes, J.D., Pan, Y.C., Levin, W., Danho, W., Chizzonite, R., Gately, M.K., and Stern, A.S. (1992). Molecular characterization of interleukin 12. *Arch. Biochem. Biophys.*, **294**, 230–7.

Robertson, M.J., Soiffer, R.J., Wolf, S.F., Manley, T.J., Donahue, C., Young, D., Herrmann, S.H., and Ritz, J. (1992). Response of human natural killer (NK) cells to NK cell stimulatory factor (NKSF): cytolytic activity and proliferation of NK cells are differentially regulated by NKSF. *J. Exp. Med.*, **175**, 779–88.

Schoenhaut, D.S., Chua, A.O., Wolitzky, A.G., Quinn, P.M., Dwyer, C.M., McComas, W., Familletti, P.C., Gately, M.K., and Gubler, U. (1992). Cloning and expression of murine IL-12. *J. Immunol.*, **148**, 3433–40.

Sieburth, D., Jabs, E.W., Warrington, J.A., Li, X., Lasota, J., LaForgia, S., Kelleher, K., Huebner, K., Wasmuth, J.J., and Wolf, S.F. (1992). Assignment of genes encoding a unique cytokine (IL12) composed of two unrelated subunits to chromosomes 3 and 5. *Genomics*, **14**, 59–62.

Soiffer, R.J., Robertson, M.J., Murray, C., Cochran, K., and Ritz, J. (1993). Interleukin-12 augments cytolytic activity of peripheral blood lymphocytes from patients with hematologic and solid malignancies. *Blood*, **82**, 2790–6.

Stern, A.S., Podlaski, F.J., Hulmes, J.D., Pan, Y.E., Quinn, P.M., Wolitzky, A.G., Familletti, P.C., Stremlo, D.L., Truitt, T., Chizzonite, R., and Gately, M.K. (1990). Purification to homogeneity and partial characterization of cytotoxic lymphocyte maturation factor from human B-lymphoblastoid cells. *Proc. Natl. Acad. Sci. (USA)*, **87**, 6808–12.

Sypek, J.P., Chung, C.L., Mayor, S.E., Subramanyam, J.M., Goldman, S.J., Sieburth, D.S., Wolf, S.F., and Schaub, R.G. (1993). Resolution of cutaneous leishmaniasis: interleukin 12 initiates a protective T helper type 1 immune response. *J. Exp. Med.*, **177**, 1797–802.

Tripp, C.S., Wolf, S.F., and Unanue, E.R. (1993). Interleukin 12 and tumour necrosis factor alpha are costimulators of interferon gamma production by natural killer cells in severe combined immunodeficiency mice with listeriosis, and interleukin 10 is a physiologic antagonist. *Proc. Natl. Acad. Sci. (USA)*, **90**, 3725–9.

Wolf, S.F., Temple, P.A., Kobayashi, M., Young, D., Dicig, M., Lowe, L., Dzialo, R., Fitz, L., Ferenz, C., Hewick, R.M., Kelleher, K., Herrmann, S.H., Clark, S.C., Azzoni, L., Chan, S.H., Trinchieri, G., and Perussia, B. (1991). Cloning of cDNA for natural killer cell stimulatory factor, a heterodimeric cytokine with multiple biologic effects on T and natural killer cells. *J. Immunol.*, **146**, 3074–81.

Stanley F. Wolf and Elliott Nickbarg:
Genetics Institute,
87 CambridgePark Drive,
Cambridge, MA 02140, USA

Interleukin-13 (IL-13)

IL-13 is a T cell-derived cytokine that regulates the function of monocytes and human B cells. It is rapidly synthesized and secreted in response to a number of immunological stimuli. IL-13 shares approximately 25–30 per cent amino acid homology with interleukin-4 (another T cell-derived cytokine the gene for which is closely linked to that encoding IL-13), and it is noteworthy that these cytokines share many biological activities including the induction of IgE production by human B cells. It is also apparent that the receptors for these two cytokines are distinct but share at least one common subunit.

■ Alternative names

P600 referring to the mouse cDNA clone. The human cDNA has also been referred to as NC30.

■ IL-13 sequence

The mouse IL-13 cDNA (GenBank accession number M23504) encodes a protein of 131 amino acids (Brown *et al.* 1989). The human IL-13 (hIL-13) cDNA (GenBank accession No L06801) encodes a protein of 132 amino acids (McKenzie *et al.* 1993a; Minty *et al.* 1993). An alternative splicing event results in the inclusion/exclusion of a GLN residue at position 98 (McKenzie *et al.* 1993b). Cleavage of the leader peptide results in the mature protein starting at SER19 or GLY21 (Minty *et al.* 1993). The mouse and human cDNAs have 66 per cent sequence identity over the coding region and the proteins are 58 per cent identical (Fig. 1). Although mIL-13 shows no species specificity, hIL-13 is ~100-fold less active on the mouse cell line B9. There is between 25–30 per cent sequence homology between IL-13 and IL-4 (Fig. 1) (Minty *et al.* 1993; Zurawski *et al.* 1993).

■ IL-13 protein

IL-13 is secreted as a monomeric protein that has heterogeneous glycosylation, although the predominant form appears to be unglycosylated (McKenzie *et al.* 1993a). In the absence of carbohydrate the protein has a M_r of ~10 000. The four cysteines present in the mature proteins are

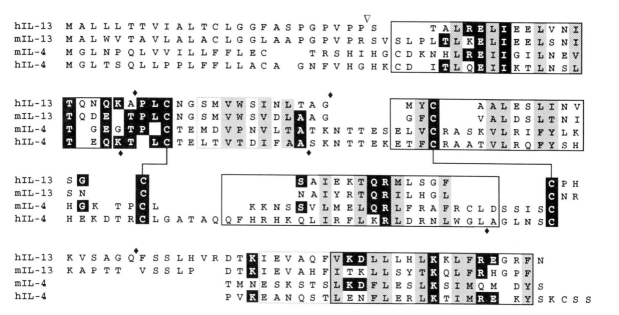

Figure 1. Sequence relatedness of IL-4 and IL-13. Alignment of the mature protein sequences of human and mouse IL-4 and IL-13. Residues in black backgrounds are common to at least one IL-4 and one IL-13 sequence. Residues in grey backgrounds have hydrophobic side chains in hIL-4 that are buried in the structural core. Black boxes delineate the four α-helical regions of hIL-4 and grey boxes delineate the two β-strands of hIL-4. The inverted open triangle indicates the hIL-4 leader peptide processing site and the filled diamonds indicate the gene exon/intron junctions. Lines joining cystines indicate the disulphide linkages that are known for hIL-4.

conserved between mIL-13 and hIL-13. In addition to three conserved potential N-linked glycosylation sites, hIL-13 contains one extra site. N-linked glycosylation results in the production of a range of protein species with M_r of 14 000 to 40 000. Although the crystal structure of IL-13 has not been determined, it is likely to be similar to that of IL-4 which is an anti-parallel four α-helical bundle protein (Zurawski et al. 1993; Walter et al. 1992).

■ IL-13 gene and transcription

The human and mouse IL-13 genes have been mapped to chromosome 5q31, and the middle of chromosome 11, respectively (Fig. 2) (a) (McKenzie et al. 1993b; Morgan et al. 1992). In both species these chromosomal regions also contain the genes for IL-3, IL-4, IL-5, and GM-CSF, and it would appear that hIL-13 is probably situated within ~50 kbp of IL-4 (Fig. 2(b)) (Morgan et al. 1992). The four-exon structures of the IL-13 genes are contained within ~3 kbp. The structural similarity of the cytokine genes in this cluster suggests that they arose by gene duplication. A single 1.3 kbp mRNA transcript has been described for both mouse

and human IL-13 (Brown et al. 1989; McKenzie et al. 1993a; Minty et al. 1993). An alternatively spliced form of hIL-13 incorporates an extra GLN at position 98 in the cDNA, this corresponds to the intron/exon boundary of exon 4 (McKenzie et al. 1993b) (no such alternative splicing occurs in the mIL-13 gene). The 3'-untranslated region of the IL-13 mRNA contains a number of copies of a sequence that has been reported to confer message instability (Minty et al. 1993).

IL-13 expression appears to be T cell specific and requires cellular activation (McKenzie et al. 1993a; Minty et al. 1993). In mice, IL-13 is predominantly expressed by activated Th2 cells (Cherwinski et al. 1987). A variety of inducers has been shown to upregulate IL-13 expression, including LPS, calcium ionophore, phorbol ester, and concanavalin A (McKenzie et al. 1993a; Minty et al. 1993; de Waal Malefyt et al. 1993).

The 5'-flanking regions of the mouse and human IL-13 genes contain potential recognition sequences for various transcription factors. Such sequences which are conserved between the human and mouse genes include interferon-responsive elements, binding sites for AP-1, AP-2 and AP-3, an NF-IL6 site, and a TATA-like sequence (McKenzie et al. 1993b). No significant sequence similarity is observed between the IL-13 genes and other cytokine genes (McKenzie et al. 1993b).

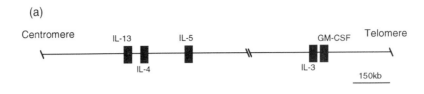

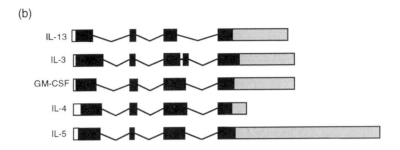

Figure 2. Diagrammatic representation of the chromosomal localization (a) and structure (b) of the genes encoding IL-13, IL-3, IL-4, IL-5, and GM-CSF. In (b) the white boxes indicate the 5'-untranslated sequences, the black boxes are exon sequences, and the grey boxes are the 3'-untranslated sequences. The intron sequences are not shown to scale.

■ Biological functions

To date, IL-13 has been shown to regulate monocyte and B cell function in the human system, but only appears to affect the macrophage lineage in the mouse. The cells from the adherent fraction of human PBMC (predominantly monocytes) show profound morphological changes upon incubation with IL-13, becoming flattened and developing extensive processes (McKenzie *et al.* 1993*a*). These monocytes also modulate their cell surface markers, upregulating CD23, MHC classII, and a number of integrin family members (de Waal Malefyt *et al.* 1993). In addition, IL-13 also regulates the synthesis of cytokines by these monocytes. For example, IL-13 inhibits synthesis of IL-1α, IL-1β, IL-6, IL-8, G-CSF, and TNFα by LPS-activated monocytes (de Waal Malefyt *et al.* 1993). Similar morphological changes and inhibition of cytokine synthesis have been observed in mouse macrophages (Doherty *et al.* 1993). IL-13 also stimulates the production of macrophage–like cells from bone marrow precursors (Heslan *et al.* 1994). Thus, it would appear that IL-13, like IL-4, is capable of activating cells of the monocyte lineage and inhibiting their production of inflammatory cytokines.

In the presence of a co-stimulatory signal (anti-CD40 or anti-IgM) human B cells proliferate in response to IL-13, and this activity is comparable to that of IL-4 (McKenzie *et al.* 1993*a*; Briere *et al.* 1993). Interestingly, IL-13, in the presence of the CD40 ligand or T cell membranes also causes immunoglobulin isotype switching, resulting in the synthesis of IgE (Briere *et al.* 1993; Punnonen *et al.* 1993; Cocks *et al.* 1993). The induction of this immunoglobulin had previously only been attributable to IL-4. In contrast to these results on human B cells, mouse B cells do not appear to respond to IL-13 (B. Coffman, unpublished data).

A common feature of the biological functions characteristic of IL-13 is that they are often also elicited by IL-4. A key difference is that IL-13, unlike IL-4, does not appear to stimulate T cells (H. Yssel, unpublished data). The overlapping of biological functions of these two lymphokines has been explained in part by recent evidence demonstrating that IL-13 and IL-4 share a receptor subunit (Zurawski *et al.* 1993).

■ References

Briere, F., Bridon, J.-M., Servet, C., Rousset, F., Zurawski, G., and Banchereau, J. (1993). IL-10 and IL-13 as B cell growth and differentiation factors. *Nouv. Rev. Fr. Hematol.*, **35**, 233–5.

Brown, K.D., Zurawski, S.M., Mosmann, T.R., and Zurawski, G. (1989). A family of small inducible proteins secreted by leukocytes are members of a new superfamily that includes leukocyte and fibroblast-derived inflammatory agents, growth factors, and indicators of various activation processes. *J. Immunol.*, **142**, 679–87.

Cherwinski, H. M., Schumacher, J. H., Brown, K. D., and Mosmann, T. R. (1987). Two types of mouse helper T cell clone. III. Further differences in lymphokine synthesis between Th1 and Th2 clones revealed by RNA hybridization, functionally mono-specific bioassays, and monoclonal antibodies. *J. Exp. Med.*, **166**, 1229–44.

Cocks, B.G., de Waal Malefyt, R., Galizzi, J.-P., de Vries, J.E., and Aversa, G. (1993). IL-13 induces proliferation and differentiation of human B cells activated by the CD40 ligand. *Int. Immunol.*, **5**, 657–63.

de Waal Malefyt, R., Figdor, C., Huijbens, R., Mohan-Peterson, S., Bennett, B., Culpepper, J., Dang, W., Zurawski, G., and de Vries, J.E. (1993). Effects of IL-13 on phenotype, cytokine production, and cytotoxic function of human monocytes. *J. Immunol.* **151**, 6370–81.

Doherty, T.M., Kastelein, R., Menon, S., Andrade, S., and Coffman, R.L. (1993). Modulation of murine macrophage function by interleukin-13. *J. Immunol.* **151**, 7151–60.

Heslan, J.-M., Guilbert, L., Kastelein, R., Elliott, J.F., and Mosmann, T.M. (1994). TH2 cells secrete a cytokine, P600 (IL-13), that stimulates production of macrophage-like cells from bone marrow precursors. (Submitted for publication.)

McKenzie, A.N., Culpepper, J.A., de Waal Malefyt, R., Briere, F., Punnonen, J., Aversa, G., Sato, A., Dang, W., Cocks, B.G., Menon, S., de Vries, J.E., Banchereau, J., and Zurawski, G. (1993a). Interleukin 13, a T-cell-derived cytokine that regulates human monocyte and B-cell function. *Proc. Natl. Acad. Sci. (USA)*, **90**, 3735–9.

McKenzie, A.N., Li, X., Largaespada, D.A., Sato, A., Kaneda, A., Zurawski, S.M., Doyle, E.L., Milatovich, A., Francke, U., Copeland, N.G., Jenkins, N.A., and Zurawski, G. (1993b). Structural comparison and chromosomal localization of the human and mouse IL-13 genes. *J. Immunol.*, **150**, 5436–44.

Minty, A., Chalon, P., Derocq, J.M., Dumont, X., Guillemot, J.C., Kaghad, M., Labit, C., Leplatois, P., Liauzun, P., Miloux, B., Minty, C., Casellas, P., Loison, G., Lupker, J., Shire, D., Ferrara, P., and Caput, D. (1993). Interleukin-13 is a new human lymphokine regulating inflammatory and immune responses. *Nature*, **362**, 248–50.

Morgan, J.G., Dolganov, G.M., Robbins, S.E., Hinton, L.M., and Lovett, M. (1992). The selective isolation of novel cDNAs encoded by the regions surrounding the human interleukin 4 and 5 genes. *Nucleic Acids Res.*, **20**, 5173–9.

Punnonen, J., Aversa, G., Cocks, B.G., McKenzie, A.N., Menon, S., Zurawski, G., de Waal Malefyt, R., and de Vries, J.E. (1993). Interleukin 13 induces interleukin 4-independent IgG4 and IgE synthesis and CD23 expression by human B cells. *Proc. Natl. Acad. Sci. (USA)*, **90**, 3730–4.

Walter, M.R., Cook, W.J., Zhao, B.G., Cameron, R.P. Jr., Ealick, S.E., Walter, R.L. Jr., Reichert, P., Nagabhushan, T.L., Trotta, P.P., and Bugg, C.E. (1992). Crystal structure of recombinant human interleukin-4. *J. Biol. Chem.*, **267**, 20371–6.

Zurawski, S.M., Vega, Jr. F., Huyghe, B., and Zurawski, G. (1993). Receptors for interleukin-13 and interleukin-4 are complex and share a novel component that functions in signal transduction. *EMBO J.*, **12**, 3899–905.

Andrew N. J. McKenzie and Gerard Zurawski†:
†Department of Molecular Biology,
DNAX Research Institute for Cellular and Molecular Biology
*Laboratory of Molecular Biology, Cambridge

High Molecular Weight B Cell Growth Factor (Interleukin-14)

IL-14 is a 50–60 kDa cytokine that induces proliferation of activated B cells. It probably has importance in the generation/maintenance of B cell memory, and may have relevance to particular B cell malignancies and immunodeficiencies.

■ IL-14 protein

50–60 kDa forms of IL-14 have been identified in supernatants of normal human T cells, T-ALL cells, B cell lines, and B cell lymphomas after stimulation with phytohemagglutinin (PHA), and in ascites and pleural fluid from patients with B cell lymphomas (Ambrus and Fauci 1985; Ford *et al.* 1992). Purified 60 kDa IL-14 from PHA-stimulated Namalva cells or T-ALL cells is glycosylated, and its apparent molecular weight does not change under reducing conditions. Isoelectric focusing reveals that the 60 kDa band can be resolved into two distinct proteins at pI 6.7 and 7.8, both of which demonstrate B cell growth factor activity (Ambrus and Fauci 1985; Ambrus *et al.* 1985).

■ IL-14 cDNA

A cDNA encoding pI 7.8 HMW-BCGF has been isolated from a PHA-stimulated Namalva cell library (GenBank accession number L15344) and designated IL-14 (Ambrus *et al.* 1993). The nucleotide sequence predicts a protein of 483 amino acids with a MW of 53.1 kDa with three N-linked glycosylation sites. The predicted amino acid sequence shows no significant homology to any other growth factors or cytokines, except pI 6.7 HMW-BCGF (25 per cent sequence homology, unpublished data). There is eight per cent sequence homology with the complement fragment Bb, a protein that binds to the IL-14 receptor and has antigenic similarities with IL-14. Recombinant protein made from the IL-14 cDNA induces proliferation of activated human B cells, and inhibits immunoglobulin secretion, like native HMW-BCGF (Ambrus *et al.* 1993).

Biology

IL-14 is a product of normal activated T lymphocytes (Ambrus *et al.* 1987) and follicular dendritic cells (unpublished data) which induces activated B lymphocytes to proliferate and inhibits their differentiation into antibody secreting cells (Ambrus *et al.* 1990). In normal human T cells IL-14 mRNA appears at about 8 hours, is still detectable at 24 hours, but is no longer found at 36–48 hours after PHA stimulation (Ambrus *et al.* 1993). IL-14 protein appears at 8–12 hrs after PHA stimulation, and is no longer detectable after 36 hours (Ambrus *et al.* 1987). Resting mature human B cells express 50–350 IL-14R sites/cell, and do not proliferate in response to IL-14. B cells activated with *Staphylococcus aureus* Cowan I (SAC) or antibodies to surface IgM express 10 000–15 000 copies of the IL-14R/cell, and proliferate in response to IL-14 (Ambrus *et al.* 1988). To date, the IL-14R has been found only on cells of B lymphocyte lineage (Ambrus *et al.* 1988; Uckun *et al.* 1987, 1989), and is specifically absent on neutrophils, monocytes/macrophages, and mast cells. B cells which are surface IgD-negative exhibit the greatest proliferative response to IL-14 *in vitro* (Ambrus *et al.* 1990). If activated B lymphocytes are exposed to differentiation factors in the presence of IL-14, immunoglobulin secretion is inhibited. However, if cells expanded by IL-14 are allowed to differentiate in its absence, IgG is the predominant antibody made (Ambrus *et al.* 1990).

From the above observations, one can hypothesize that IL-14 plays a role in the secondary immune response. In this model, a sIgD-negative B cell circulating through a lymph node becomes activated by encountering its antigen and other activation signals supplied by the lymph node microenvironment. Lymph node T cells and follicular dendritic cells release IL-14, apoptosis of the B cells is inhibited, and multiple rounds of proliferation occur. As the B cell migrates through the lymph node, it is exposed to other soluble and cell surface signals and becomes either a memory B cell or an IgG-secreting plasma cell. This model is supported by the recent unpublished observations that germinal centre T cells and follicular dendritic cells express IL-14 mRNA, germinal centre B cells express IL-14R protein, and IL-14 upregulates bcl-2 in IL-14R-positive cells.

After IL-14 binds to its receptor, several intracellular signalling events have been observed. There is a gradual increase in intracellular calcium derived from extracellular sources (Ambrus *et al.* 1991a). Diacylglycerol and glycaninositols are released (Ambrus *et al.* 1991b) and a delayed increase in cAMP, dependent upon the release of arachidonic acid and formation of PGE, is noted (Ambrus *et al.* 1991b). Cyclic AMP is important in both the IL-14 induced inhibition of antibody secretion and the upregulation of IL-14 receptors (Ambrus *et al.* 1991b).

Pathology

Proliferation to IL-14 has been noted in freshly isolated malignant B cells from patients with chronic lymphocytic leukaemia, hairy cell leukaemia, and high-grade follicular and Burkitt's lymphomas (Ford *et al.* 1992; Uckun *et al.* 1989). A subpopulation of patients with acute lymphocytic leukaemia of the B cell precursor type have malignant B cells which proliferate in response to IL-14 (Uckun *et al.* 1987). Production of IL-14 *in vivo* by high grade B cell lymphomas suggests an autostimulatory role for IL-14 in these tumours (Ford *et al.* 1992).

Increased production of IL-14 *in vivo* and responsiveness to IL-14 *in vitro* has been noted in B cells from patients with SLE (unpublished data). Failure to respond to IL-14, but not other human BCGF, characterizes B cells from patients with common variable immunodeficiency (Ambrus *et al.* 1991c).

Purification

IL-14 is most efficiently purified from culture supernatants by hydroxyapatite chromatography and fluid phase isoelectric focusing (Ambrus *et al.* 1993).

Activity assays

Proliferative activity is measured on B cells purified from peripheral blood, tonsil or spleen which have been activated with SAC for 72 h. IL-14 is added for an additional 72 h, and ^{3}H-thymidine incorporation is determined over the final 16 h (Ambrus *et al.* 1993). Alternatively, B cells can be activated with anti-μ or pre-activated B cells isolated by elutriation for these assays (Ambrus and Fauci 1985; Ambrus *et al.* 1990). The effects of IL-14 on antibody secretion can be determined in assays using mononuclear cells stimulated with pokeweed mitogen or purified B cells stimulated with SAC and IL-2/IL-6. Secretion of antibody is determined by ELISA (Ambrus *et al.* 1990).

Antibodies

A polyclonal serum recognizing recombinant IL-14 and both forms of native HMW-BCGF is available. Monoclonal antibodies to these proteins are currently being produced. A mAb BA5 recognizes the IL-14R (Ambrus *et al.* 1988; Uckun *et al.* 1989). All of these antibodies can be used to inhibit IL-14 induced B cell proliferation. BA5 can be used for Western blots, immunoprecipitations, and flow cytometry (Ambrus *et al.* 1988; Ambrus *et al.* 1991c).

References

Ambrus, J.L., Jr., and Fauci, A.S. (1985). Human B lymphoma cell line producing B cell growth factor. *J. Clin. Invest.*, **75**, 732–9.

Ambrus, J.L., Jr., Jurgensen, C.H., Brown, E.J., and Fauci, A.S. (1985). Purification to homogeneity of a high molecular weight human B cell growth factor; demonstration of specific binding to activated

B cells; and development of a monoclonal antibody to the factor. *J. Exp. Med.*, **162**, 1319–35.

Ambrus, J.L., Jr., Jurgensen, C.H., Bowen, D.L., Tomita, S., Nakagawa, T., Nakagawa, N., Goldstein, H., Witzel, N.L., Mostowski, H.S., and Fauci, A.S. (1987). In *Mechanisms of lymphocyte activation and immune regulation* (eds S. Gupta, W. Paul, and A. Fauci), pp. 163–75. Plenum, San Francisco.

Ambrus, J.L., Jr., Jurgensen, C.H., Brown, E.J., McFarland, P., and Fauci, A.S. (1988). Identification of a receptor for high molecular weight human B cell growth factor. *J.Immunol.*, **141**, 861–9.

Ambrus, J.L., Jr., Chesky, L., Stephany, D., McFarland, P., Mostowski, H., and Fauci, A.S. (1990). Functional studies examining the subpopulation of human B lymphocytes responding to high molecular weight B cell growth factor. *J.Immunol.*, **145**, 3949–55.

Ambrus, J.L., Jr., Chesky, L., McFarland, P., Young, K.R., Jr., Mostowski, H., August, A., and Chused, T.M. (1991a). Induction of proliferation by high molecular weight B cell growth factor or low molecular weight B cell growth factor is associated with increases in intracellular calcium in different subpopulations of human B lymphocytes. *Cell.Immunol.*, **131**, 314–24.

Ambrus, J.L., Jr., Chesky, L., Chused, T., Young, K.R., Jr., McFarland, P., August, A., and Brown, E.J. (1991b). Intracellular signaling events associated with the induction of proliferation of normal human B lymphocytes by two different antigenically related human B cell growth factors (high molecular weight B cell growth factor (HMW-BCGF) and the complement factor Bb). *J.Biol.Chem.*, **266**, 3702–8.

Ambrus, J.L., Jr., Haneiwich, S., Chesky, L., McFarland, P., Peters, M.G., and Engler, R.J. (1991c). Abnormal response to a human B cell growth factor in patients with common variable immunodeficiency (CVI). *J. Allerg. Clin. Immunol.*, **87**, 1138–49.

Ambrus, J.L., Jr., Pippin, J., Joseph, A., Xu, C., Blumenthal, D., Tamayo, A., Claypool, K., McCourt, D. Srikiatchatochorn, A., and Ford, R.J. (1993). Identification of a cDNA for a human high-molecular-weight B-cell growth factor. *Proc. Natl. Acad. Sci. (USA)*, **90**, 6330–4.

Ford, R.J., Tamayo, A., and Ambrus, J.L., Jr. (1992). The role of growth factors in human lymphomas. *Curr. Top. Microbiol. Immunol.*, **182**, 341–7.

Uckun, F.M., Fauci, A.S., Heerema, N.A., Song, C.W., Mehta, S.R., Gajl-Peczalska, K., and Ambrus, J.L., Jr. (1987). B-cell growth factor receptor expression and B-cell growth factor response of leukemic B cell precursors and B lineage lymphoid progenitor cells. *Blood*, **70**, 1020–34.

Uckun, F.M., Fauci, A.S., Chandan-Langlie, M., Myers, D.E., and Ambrus, J.L., Jr. (1989). Detection and characterization of human high molecular weight B cell growth factor receptors on leukemic B cells in chronic lymphocytic leukaemia. *J.Clin.Invest.*, **84**, 1595–608.

David E. Blumenthal:
Wahington University School of Medicine, St Louis, MO, USA

Richard J. Ford:
M.D. Anderson Cancer Center, Houston, TX, USA

Julian L. Ambrus Jr.:
Washington University School of Medicine, St Louis, MO, USA

Human B Cell Growth Factor-12 kDa (BCGF-12 kDa)

The cDNA and the genomic domain encoding BCGF-12 kDa have recently been isolated and characterized. Immunoprecipitated recombinant BCGF-12 kDa is a 14kDa glycoprotein that primarily functions as a potent growth factor in human B cells. Its expression can be induced in T cells by mitogen activation. However, it is constitutively expressed in Jurkat T cells and Epstein–Barr virus (EBV)-positive as well as EBV-negative lymphoma B cells. Our recent results clearly suggest that it might be involved in AIDS and non-AIDS-associated lymphomagenesis.

■ Alternative name

Low molecular weight B cell growth factor (LMW-BCGF).

■ BCGF-12 kDa cDNA

The BCGF-12 kDa cDNA was first cloned using size fractionated poly (A+) mRNA from mitogen activated T cells (Sharma *et al*. 1987). BCGF-12 kDa mRNA was selected on the basis of its ability to encode B cell growth factor activity after expression in Xenopus oocytes. The open reading frame in the BCGF-12 kDa cDNA predicts a protein of 120 amino acids (Fig. 1). The hydrophobicity analysis suggests a 13-amino-acid putative leader sequence. An unusual feature of the cDNA is the presence of an Alu repeat unit at the 3′ end. Even more intriguing, this Alu repeat unit contributes 40 amino acids to the carboxy terminus of the BCGF-12 kDa protein. Other cDNAs that have been shown to contain Alu repeat element in the coding region encode the delay accelerating factor and the A_4 amyloid precursor protein respectively (Caras et al. 1987; de Sauvage and Octave 1989). There is no sequence homology between BCGF-12 kDa and other cytokines, thus making it a distinct B cell growth factor moiety.

■ BCGF-12 kDa gene and its transcription

The presence of an Alu sequence in the BCGF-12 kDa open reading frame has made it imperative to further characterize the BCGF-12 kDa gene for its genetic make-up and expression so that it could be established as a functional gene. We have recently isolated a genomic fragment encoding the BCGF-12 kDa gene. A sequence homology search shows that the previously identified cDNA sequence (Sharma et al. 1987) is colinear with the genomic sequence, suggesting that BCGF-12 kDa is an intronless gene (Fig. 1). The BCGF-12 kDa genomic sequence is also found to contain additional Alu repeat units, arbitrarily designated Alu-2, Alu-3, and Alu-4, both within and flanking the open reading frame (Fig. 1). The polyadenylation signal (AA TAAA) is located 700 bp downstream from the originally identified cDNA domain (Fig. 1) suggesting that the BCGF-12 kDa gene may be alternatively transcribed to generate multiple mRNA species. Recently, in collaboration with Dr Thomas Shows of Roswell Park Institute, Buffalo, we have mapped the BCGF-12 kDa gene to the long arm of chromosome 16a (manuscript in preparation). This specific chromosomal localization further supports the notion that the BCGF-12 kDa gene is a unique locus distinct from other cytokines.

To study BCGF-12 kDa mRNA expression, we have employed RNA-based PCR amplification. Our results suggest that expression of the BCGF-12 kDa gene is mitogen inducible in normal human T cells and constitutive in Jurkat T cells (manuscript in preparation). Furthermore, it is constitutively expressed in both EBV-positive as well as EBV-negative malignant B cells including AIDS lymphoma cells. BCGF-12 kDa mRNA expression in chronic lymphocytic leukaemia (CLL) B cells may be of importance since CLL-B cells exhibit a significant response to BCGF-12 kDa (Fournier et al. 1992). A computer search of the BCGF-12 kDa promoter region reveals the presence of binding sites for a number of known transcription factors such as API, C/EBP, and OCT-1 as well as an NFAT like sequence required for cyclosporin A sensitivity (Yaseen et al. 1993; Jain et al. 1992; Clipstone and Crabtree 1992). Experiments are currently in progress to study these consensus sequences for their regulatory effects on BCGF-12 kDa expression.

■ BCGF-12 kDa protein and its biological activity

It is fair to say that purification of large quantities of natural or recombinant BCGF-12 kDa has been the most difficult part of our studies on the biological characterization of this lymphokine. Several unexpected molecular and biochemical complexities have considerably slowed the progress on this molecule. However, we have recently been able to express recombinant BCGF-12 kDa in a baculovirus expression system (Invitrogen). Rabbit polyclonal antibodies have been raised against two synthetic BCGF-12 kDa peptides. Baculovirus expressed BCGF-12 kDa has been purified through an antibody affinity column. Highly purified recombinant BCGF-12 kDa has an apparent molecular weight of 14 kDa. On a non-denaturing SDS gel, BCGF-12 kDa migrates as multimers. The mature protein contains one N-glycosylation site and three cysteine residues. Our preliminary data suggest that BCGF-12 kDa may exist as a membrane-bound molecule.

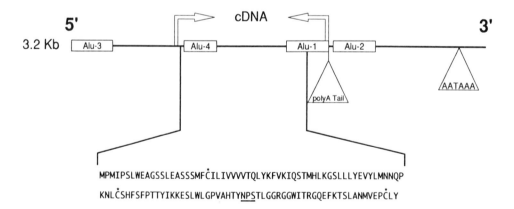

Figure 1. Gene structure of BCGF-12 kDa.

Highly purified recombinant BCGF-12 kDa has been studied for its ability to induce proliferation in human B cells. Activated normal B cells as well as established EBV-positive and EBV-negative malignant cell lines have been used for assessing the biological function of BCGF-12 kDa. Of significance is the observation that BCGF-12 kDa stimulates DNA synthesis in both EBV-positive and EBV-negative AIDS and non-AIDS lymphoma B cells (manuscript in preparation). This is important in light of the fact that other growth factors including IL-2, IL-4, IL-5, IL-10, and IL-14 fail to induce S-phase entry in immortalized B cells under serum-free conditions, although immortalized and malignantly transformed B cells have been shown to produce some of these interleukins (Blay et al. 1993; Ambrus and Fauci 1985). We are in the process of characterizing the pleiotropic effects of BCGF-12 kDa in other immune cell types.

■ Pathologic implications

Cytokines are rapidly becoming a central focus in molecular medicine, since molecular and/or immunological reagents related to them may be utilized therapeutically to specifically counteract a pathogenetic event within a malignant cell. Targeting a single activated gene which might be associated with the transformed phenotype may be useful since this approach may not induce cellular reistance. In this respect, a rapid development of antisense therapeutics represents a revolution in molecular medicine (Agrawal 1992; Rapaport et al. 1992; Crooke 1993; Stein and Cheng 1993). Our recent in vitro studies using BCGF-12 kDa antisense oligonucleotides clearly suggest that such a treatment inhibits the autocrine growth observed in both EBV-positive and EBV-negative lymphoma B cells. Further experimentation is needed to assess the involvement of BCGF-12 kDa and its receptors in the pathogenesis of other B cell malignancies (Ford et al. 1988). Taken together, our recent studies are exciting and provide us and others with an opportunity to exploit this cytokine for novel therapeutic modalities.

■ References

Agrawal, S. (1992). Trends in Biotechnology, 10, 152–8.

Ambrus, J.L., Jr., and Fauci, A.S. (1985). Human B lymphoma cell line producing B cell growth factor. J. Clin. Invest., 75, 732–9.

Blay, J.-Y., Bardin, N., Rousset, F., Lenoir, G., Biron, P., Phillip, T., Banchereau, J., and Favrot, M.C. (1993). Serum interleukin-10 in non-Hodgkin's lymphoma: a prognostic factor. Blood, 82, 2169–74.

Caras, I.W., Davitz, M.A., Rhee, L., Weddell, G., Martin, D.W. Jr., and Nussenzweig, V. (1987). Cloning of decay-accelerating factor suggests novel use of splicing to generate two proteins. Nature, 325, 545–9.

Clipstone, N.A., and Crabtree, G.R. (1992). Identification of calcineurin as a key signalling enzyme in T-lymphocyte activation. Nature, 357, 695–7.

Crooke, S.T. (1993). Progress toward oligonucleotide therapeutics: pharmacodynamic properties. FASEB J., 7, 533–9.

Ford, R.J., Mehta, S.R., and Sharma, S. (1988). In Hodgkin's disease and non-Hodgkin's lymphomas (ed. L. Fuller et al.) pp. 47–88. Raven, New York.

Fournier, S., Jackson, J., Kumar, A., King, T., Sharma, S., Biron, G., Rubio, M., Delespesse, G., and Sarfati, M. (1992). Low-molecular weight B cell growth factor (BCGF-12KDa) as an autocrine growth factor in B cell chronic lymphocytic leukaemia. Eur. J. Immunol., 22, 1927–30.

Jain, J., McCaffrey, P.G., Valge-Archer, V.E., and Rao, A. (1992). Nuclear factor of activated T cells contains Fos and Jun. Nature, 356, 801–4.

Rapaport, E., Misiura, K., Agrawal, S., and Zamecnik, P. (1992). Antimalarial activities of oligodeoxynucleotide phosphorothioates in chloroquine-resistant Plasmodium falciparum. Proc. Natl. Acad. Sci. (USA), 89, 8577–80.

de Sauvage, F., and Octave, J.N. (1989). A novel mRNA of the A4 amyloid precursor gene coding for a possibly secreted protein. Science, 245, 651–3.

Sharma, S., Mehta, S., Morgan, J., and Maizel, A. (1987). Molecular cloning and expression of a human B-cell growth factor gene in Escherichia coli. Science, 235, 1489–92.

Stein, C.A., and Cheng, Y.-C. (1993). Antisense oligonucleotides as therapeutic agents–is the bullet really magical?. Science, 261, 1004–12.

Yaseen, N.R., Maizel, A.L., Wang, F., and Sharma, S. (1993). Comparative analysis of NFAT (nuclear factor of activated T cells) complex in human T and B lymphocytes. J. Biol. Chem., 268, 14285–93.

Surendra Sharma:
Section of Experimental Pathology,
Department of Pathology,
Roger Williams Medical Center–Brown University,
Providence, RI 02908 USA

CD40 Ligand (CD40L)

CD40L is a 33 kDa type II glycoprotein, expressed on activated T cells, which exhibits significant homology to other ligands in the TNF family of molecules. Stimulation of B cells with CD40L enhances expression of activation-associated surface antigens and leads to proliferation. In the presence of soluble cytokines, CD40L induces the secretion of multiple Ig isotypes. The CD40L gene is located on the X chromosome, and mutations within this gene are responsible for the inherited human condition of hyper-IgM syndrome, which is characterized by a virtual absence of Ig isotypes other than IgM and an inability to mount an antigen-specific antibody response resulting in susceptibility to opportunistic infections. Thus, CD40L provides an essential signal to B cells to undergo Ig isotype switching and to secrete mature Ig.

■ CD40

CD40 is a 50 kDa type I glycoprotein which is a member of the TNF receptor (TNFR) family. This group of related molecules includes both p60 and p80 forms of TNFR, the T cell activation antigens OX40, 4–1BB, CD27 and CD30, Fas antigen and a Shope fibroma virus-encoded TNFR homologue (Mallett and Barclay 1991). CD40 is expressed on B cells, follicular dendritic cells, thymic epithelium, mast cells, monocytes and some carcinomas (Clark 1990; Galy and Spits 1992; Alderson *et al.* 1993). Monoclonal antibodies (mAb) specific for CD40 induce a range of biological activities on human B cells including the induction of homotypic adhesion, short- and long-term proliferation in concert with various costimuli, production of IgE in the presence of IL-4 and production of other Ig isotypes with IL-2 or IL-10 (Armitage *et al.* 1993a). In addition, ligation with CD40 mAb prevents the spontaneous apoptosis of germinal centre B cells *in vitro* (Liu *et al.* 1989). Murine and human ligands for CD40 have been identified and cloned.

■ CD40L sequence

The nucleotide sequence of murine and human CD40L (GenBank accession numbers X65453 and X67878) predict proteins of 260 and 261 amino acids respectively (Armitage *et al.* 1992; Spriggs *et al.* 1992). These molecules contain a 22-amino-acid N-terminal cytoplasmic domain, a 24-amino-acid transmembrane region, and 215 or 216 amino acids in the extracellular domain which contains a single potential N-linked glycosylation site. Murine and human CD40L exhibit 75 per cent identity at the amino acid level and cross-react both in terms of receptor binding and biological activity. CD40L exhibits significant homology to other ligands for members of the TNFR family (Smith *et al.* 1993) (Table 1). The strongest homology is in the carboxy-terminal portion of the extracellular domains. Since in mature TNF the homologous region forms a β-sandwich which trimerizes, it has been predicted that CD40L may have a similar oligomeric tertiary structure (Peitsch and Jongeneel 1993).

■ CD40L protein

CD40L is a type II membrane glycoprotein with a M_r of approximately 33 kDa. As the predicted M_r based on amino acid sequence is 29.4 kDa, it appears that the single glycosylation site in the extracellular domain is utilized. To date, expression of CD40L seems to be largely restricted to activated T cells, predominantly of a CD4+ phenotype (Spriggs *et al.* 1992; Lane *et al.* 1992), although a recent study has demonstrated that CD40L can also be expressed by basophils and mast cells (Gauchat *et al.* 1993). Binding studies have revealed a single class of high affinity interaction (K_a $\sim$1–2$\times$10^9 M^{-1}) of CD40L with its receptor (Armitage *et al.* 1992).

Northern blot analysis performed on activated T cells indicates that CD40L is transcribed into a single mRNA species of approximately 2000 nucleotides in the mouse (Armitage *et al.* 1992), while human T cells appear to contain two forms of CD40L mRNA, a predominant species of 2000 nucleotides and a less abundant, smaller form (Spriggs *et al.* 1992). The exact nature of this smaller band is unclear although it appears to contain the complete coding region and may arise as a result of the use of an alternative polyadenylation site in the 3' untranslated region.

■ Biological activities

CD40L has various activities on murine and human B cells including the enhancement of expression of activation-associated molecules CD23, HLA class II and BB1/B7, and the induction of proliferation (Armitage *et al.* 1993a). Membrane-bound CD40L is directly mitogenic for resting B cells in the absence of costimuli (Armitage *et al.* 1992; Spriggs *et al.* 1992), while enhanced proliferation occurs in the presence of IL-2, IL-4, or IL-10 (Armitage *et al.* 1993b). In addition, soluble oligomeric constructs of the extracellular CD40L domain costimulate proliferation of B cells activated with anti-IgM, phorbol ester or CD20 mAb (Hollenbaugh *et al.* 1992; Lane *et al.* 1993). CD40L and IL-4 provide the

Table 1. Percentage of identity between homologous regions of TNF family ligands based on amino acid sequence

	Hu TNFα	Mu TNFα	Hu TNFβ	Mu TNFβ	Hu CD40L	Mu CD40L	Hu CD27L	Hu CD30L	Mu CD30L
HuTNFα	100	80	36	39	27	25	18	16	18
MuTNFα		100	37	41	25	24	21	15	18
HuTNFβ			100	79	24	22	18	14	18
MuTNFβ				100	25	25	18	17	18
HuCD40L					100	75	22	12	17
MuCD40L						100	18	15	16
HuCD27L							100	12	13
HuCD30L								100	78
MuCD30L									100

essential signals required for the production of murine and human IgE (Armitage et al. 1992; Spriggs et al. 1992), while a combination of CD40L and IL-2 or IL-10 induces IgM, IgG, and IgA secretion from human B cells (Armitage et al. 1993b). In the murine system, stimulation by CD40L together with IL-4 and IL-5 is optimal for polyclonal IgM and IgG production (Maliszewski et al. 1993), while antigen-specific antibody secretion mediated by CD40L appears to require IL-2 as the costimulus (Grabstein et al. 1993). Recently it was shown that in vivo administration of a blocking antibody specific for murine CD40L inhibited the development of collagen-induced arthritis (Durie et al. 1993), and both primary and secondary humoral responses to T-dependent antigen (Foy et al. 1993). Furthermore, CD40L has been shown to prevent germinal centre B cells from undergoing spontaneous apoptosis in vitro (Holder et al. 1993).

In addition to its activities on B cells, CD40L costimulates TNFα, IL-6, and IL-8 production and induces tumouricidal activity in monocytes (Alderson et al. 1993). Furthermore, CD40L costimulates proliferation and IL-2 and TNFα production from murine and human T cells, and the expression of CD25 (IL-2R α chain), CD69, and CD40L itself on the T cell surface (Armitage et al. 1993a).

■ CD40L gene and pathology

The murine CD40L gene is located in the proximal region of the X chromosome linked to hprt. The corresponding position in the human is $X_q26.3–27.1$ (Graf et al. 1992; Allen et

al. 1993). The human disorder X-linked hyper-IgM (HIGM) syndrome maps to the X_q26 region (Padayachee et al. 1992) and it has been demonstrated that mutations in the CD40L gene are the underlying cause of this disease (Callard et al. 1993). HIGM is characterized by normal or elevated levels of serum IgM with a virtual absence of other Ig isotypes, susceptibility to opportunistic infections and a generalized failure to form germinal centres in lymphoid tissue. All but one of the CD40L mutations reported to date (Callard et al. 1993) (Fig. 1) lie in the TNF family homologous region of the carboxy-terminal domain (amino acids 123–261). Studies comparing the binding of soluble CD40 constructs with that of an antibody specific for human CD40L indicate that, upon activation, T cells from a proportion of HIGM patients can express mutated forms of CD40L on the cell surface which are unable to bind CD40 (Callard et al. 1993). However, B cells isolated from HIGM patients can be induced to secrete IgE and IgG when cultured with recombinant CD40L and either IL-4 or IL-10, suggesting that CD40L provides an essential signal for the process of Ig isotype switching and secretion in vivo.

■ References

Alderson, M.R., Armitage, R.J., Tough, T.W., Strockbine, L., Fanslow, W.C., and Spriggs, M.K. (1993). CD40 expression by human monocytes: Regulation by cytokines and activation of monocytes by the ligand for CD40. J. Exp. Med., **178**, 669–74.

Allen, R.C., Armitage, R.J., Conley, M.E., Rosenblatt, H., Jenkins,

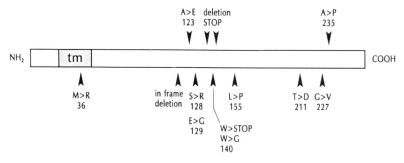

Figure 1. CD40L mutations identified to date in HIGM.

N.A., Copeland, N.G., Bedell, M.A., Edelhoff, S., Disteche, C.M., Simoneaux, D.K., Fanslow, W.C., Belmont, J., and Spriggs, M.K. (1993). CD40 ligand gene defects responsible for X-linked hyper-IgM syndrome. *Science*, **259**, 990–3.

Armitage, R.J., Fanslow, W.C., Strockbine, L., Sato, T.A., Clifford, K.N., Macduff, B.M., Anderson, D.M., Gimpel, S.D., Davis-Smith, T., Maliszewski, C.R., Clark, E.A., Smith, C.A., Grabstein, K.H., Cosman, D., and Spriggs, M.K. (1992). Molecular and biological characterization of a murine ligand for CD40. *Nature*, **357**, 80–2.

Armitage, R.J., Maliszewski, C.R., Alderson, M.R., Grabstein, K.H., Spriggs, M.K., and Fanslow, W.C. (1993a). CD40L: a multi-functional ligand. *Seminars in Immunology*, **5**, 401–12.

Armitage, R.J., Macduff, B.M., Spriggs, M.K., and Fanslow, W.C. (1993b). Human B cell proliferation and Ig secretion induced by recombinant CD40 ligand are modulated by soluble cytokines. *J. Immunol.*, **150**, 3671–80.

Callard, R.E., Armitage, R.J., Fanslow, W.C., and Spriggs, M.K. (1993). CD40 ligand and its role in X-linked hyper-IgM and syndrome. *Immunol. Today*, **14**, 559–64.

Clark, E.A. (1990). CD40: a cytokine receptor in search of a ligand. *Tissue Antigens*, **36**, 33–6.

Durie, F.H., Fava, R.A., Foy, T.M., Aruffo, A., Ledbetter, J.A., and Noelle, R.J. (1993). Prevention of collagen-induced arthritis with an antibody to gp39, the ligand for CD40. *Science*, **261**, 1328–30.

Foy, T.M., Shepherd, D.M., Durie, F.H., Aruffo, A., Ledbetter, J.A., and Noelle, R.J. (1993). *In vivo* CD40–gp39 interactions are essential for thymus dependent humoral immunity. II. Prolonged suppression of the humoral immune response by an antibody to the ligand for CD40, gp39. *J. Exp. Med.*, **178**, 1567–76.

Galy, A.H., and Spits, H. (1992). CD40 is functionally expressed on human thymic epithelial cells. *J. Immunol.*, **149**, 775–82.

Gauchat, J.-F., Henchoz, S., Mazzei, G., Aubry, J.-P., Brunner, T., Blasey, H., Life, P., Talabot, D., Flores-Romo, L., Thompson J., Kishi K., Butterfield J., Dahinden C. and Bonnefoy, J.-Y. (1993). Induction of human IgE synthesis in B cells by mast cells and basophils. *Nature*, **365**, 340–3.

Grabstein, K.H., Maliszewski, C.R., Shanebeck, K., Sato, T.A., Spriggs, M.K., Fanslow, W.C., and Armitage, R.J. (1993). The regulation of T cell-dependent antibody formation in vitro by CD40 ligand and IL-2. *J. Immunol.*, **150**, 3141–7.

Graf, D., Korthauer, U., Mages, H.W., Senger, G., and Kroczek, R.A. (1992). Cloning of TRAP, a ligand for CD40 on human T cells. *Eur. J. Immunol.*, **22**, 3191–4.

Holder, M.J., Wang, H., Milner, A.E., Casamayor, M., Armitage, R.J, Spriggs, M.K., Fanslow, W.C., MacLennan, I.C.M., Gregory, C.D., and Gordon, J. (1993). Suppression of apoptosis in normal and neoplastic human B lymphocytes by CD40 ligand is independent of Bcl–2 induction. *Eur. J. Immunol.*, **23**, 2368–71.

Hollenbaugh, D., Grosmaire, L.S., Kullas, C.D., Chalupny, N.J., Braesch-Andersen, S., Noelle, R.J., Stamenkovic, I., Ledbetter, J.A., and Aruffo, A. (1992). The human T cell antigen gp39, a member of the TNF gene family, is a ligand for the CD40 receptor: expression of a soluble form of gp39 with B cell co-stimulatory activity. *EMBO J.*, **11**, 4313–21.

Lane, P., Traunecker, A., Hubele, S., Inui, S., Lanzavecchia, A., and Gray, D. (1992). Activated human T cells express a ligand for the human B cell-associated antigen CD40 which participates in T cell-dependent activation of B lymphocytes. *Eur. J. Immunol.*, **22**, 2573–8.

Lane, P., Brocker, T., Hubele, S., Padovan, E., Lanzavecchia, A., and McConnell, F. (1993). Soluble CD40 ligand can replace the normal T cell-derived CD40 ligand signal to B cells in T cell-dependent activation. *J. Exp. Med.*, **177**, 1209–13.

Liu, Y.J., Joshua, D.E., Williams, G.T., Smith, C.A., Gordon, J., and MacLennan, I.C. (1989). Mechanism of antigen-driven selection in germinal centres. *Nature*, **342**, 929–31.

Maliszewski, C.R., Grabstein, K., Fanslow, W.C., Armitage, R., Spriggs, M.K., and Sato, T.A. (1993). Recombinant CD40 ligand stimulation of murine B cell growth and differentiation: cooperative effects of cytokines. *Eur. J. Immunol.*, **23**, 1044–9.

Mallett, S., and Barclay, A.N. (1991). A new superfamily of cell surface proteins related to the nerve growth factor receptor. *Immunol. Today*, **12**, 220–3.

Padayachee, M., Feighery, C., Finn, A., McKeown, C., Levinsky, R.J., Kinnon, C., and Malcolm, S. (1992). Mapping of the X-linked form of hyper-IgM syndrome (HIGM1) to Xq26 by close linkage to HPRT. *Genomics*, **14**, 551–3.

Peitsch, M.C., and Jongeneel, C.V. (1993). A 3-D model for the CD40 ligand predicts that it is a compact trimer similar to the tumour necrosis factors. *Int. Immunol.*, **5**, 233–8.

Smith, C.A., Gruss, H-J., Davis, T., Anderson, D., Farrah, T., Baker, E., Sutherland, G.R., Brannan, C.I., Copeland, N.G., Jenkins, N.A., Grabstein, K.H., Gliniak, B., McAlister, I.B., Fanslow, W., Alderson, M., Falk, B., Gimpel, S., Gillis, S., Din, W.S., Goodwin, R.G., and Armitage, R.J. (1993). CD30 antigen, a marker for Hodgkin's lymphoma, is a receptor whose ligand defines an emerging family of cytokines with homology to TNF. *Cell*, **73**, 1349–60.

Spriggs, M.K., Armitage, R.J., Strockbine, L., Clifford, K.N., Macduff, B.M., Sato, T.A., Maliszewski, C.R., and Fanslow, W.C. (1992). Recombinant human CD40 ligand stimulates B cell proliferation and immunoglobulin E secretion. *J. Exp. Med.*, **176**, 1543–50.

Richard J. Armitage:
Immunex Research and Development Corporation,
Seattle, USA

Tumour Necrosis Factor (TNF)

Tumour necrosis factor (TNF) is a 17 kDa protein produced primarily by macrophages in response to a wide variety of stimuli including mitogens, cytokines, bacteria, viruses, and parasites. Besides having antitumour activity, TNF plays a role in immune modulation, viral replication, haemopoiesis, inflammation, anorexia, cachexia, septic shock, infection, and immunity.

■ Alternative names

Cachectin, macrophage cytotoxin, necrosin, cytotoxin, tumour necrosis factor-alpha, haemorrhagic factor, macrophage cytotoxic factor, differentiation inducing factor (Aggarwal 1992).

■ Sources

Besides macrophages, various other cell types known to express TNF include natural cytotoxic cells, T and B lymphocytes, granulocytes, fibroblasts, mast cells, smooth muscle cells, breast, ovarian, and glial tumour cells, astrocytes, Kupffer's cells, epidermal cells (A431,KB), adipocytes, and granulosa cells (Aggarwal 1992; Spriggs et al. 1992).

■ Induction

TNF is produced in response to both Gram-negative and Gram-positive bacteria and their products (e.g. lipopolysaccharide, lipid A analogue muramyl peptide) , viruses (e.g. HIV-1), mycoplasma, immune complexes, cytokines (e.g. GM-CSF, IL-1, IL-2, IFNγ), tumour cells, complement (C5a), glucose-modified proteins, neuropeptides, myelin P2 protein, protein kinase C activators (phorbol esters), protein phosphatase inhibitors (okadaic acid), reactive oxygen species, cyclooxygenase inhibitors, and platelet activating factor (PAF) (Aggarwal 1992; Spriggs et al. 1992).

■ Suppression

The production of TNF is suppressed by dexamethasone, prostaglandin E_2, transforming growth factor-beta, cyclosporin A, phosphodiesterase inhibitors (e.g. pentoxifylline), lipooxygenase inhibitors, vitamin D_3, retinoic acid, thalidomide, PAF receptor antagonists, IL-4, and IL-6 (Aggarwal 1992; Spriggs et al. 1992).

■ Bioassay

A typical bioassay for TNF involves L-929 mouse cells (40 000 cells per well in a 96-well plate) plated overnight and then incubated at 37 °C for 16 hours with a test sample serially diluted in medium containing 1 μg/ml actinomycin D. The test sample is then removed and TNF-induced cytotoxicity determined by staining viable cells with crystal violet and measuring absorbance at 570 nm in a multiscan autoplate reader. One unit of TNF is defined as the amount of cytokine required to kill 50 per cent of the cells (Aggarwal and Kohr 1985).

■ Purification

Cell conditioned medium is filtered through 3 μm Pall–Sealkleen filter, chromatographed on controlled-pore glass beads, DEAE-Sepharose, Mono Q fast protein liquid chromatography, and then subjected to gel permeation chromatography (Aggarwal and Kohr 1985; Aggarwal et al. 1985).

■ Physicochemical properties

TNF is a 157-amino-acid long acidic (pI around 5.8) polypeptide with an apparent M_r of 17 000 Da under denaturing conditions but is a trimer of $\approx$ 45–55 000 Da under native conditions (Aggarwal et al. 1985). TNF contains two cysteines at positions 69 and 101 involved in a disulphide bridge (Aggarwal et al. 1985). TNF belongs to a family of proteins which includes lymphotoxin, CD40 ligand, CD30 ligand, CD29 ligand, and lymphotoxin-binding protein with 20–30 per cent sequence homolgy (Farrah and Smith 1992). The amino acid sequences of human, murine, rat, rabbit, feline, goat, bovine, and porcine TNF are 70–90 per cent identical to each other (Sprang and Eck 1992) and the proteins do not display species-specificity but rather species preference in their biological action. Unlike human TNF, the murine form is a glycoprotein. Both murine and human forms of TNF are highly sensitive to proteolysis with trypsin, chymotrypsin, and staphylococcal V8 protease.

Gel filtration, ultracentrifugation, and X-ray crystallographic studies have revealed that human TNF is a trimeric molecule with an antiparallel β-sheet sandwich and an edge-to-edge packing 'jelly-roll' structural motif (for references see Aggarwal and Vilcek 1992; Beutler 1992; Sprang and Eck 1992). The protein is stable at 4 °C for up to six months, but only for 10 min at 56 °C and pH above 4.

■ Gene structure

The size of the TNF gene is $\sim$ 3.6 kbp, contains three introns, and is localized in the major histocompatability region of the short arm of human chromosome 6 between p21.1 and p21.3. The 5′ flanking region of the TNF gene, which affects its transcriptional regulation, contains several regulatory sites including an Igk chain kB enhancer homologue, a cytokine-1-type kB enhancer variant homologue, a homologue of an MHC-II promoter element, a cyclic AMP responsive element, a c-jun/AP-1 binding site homologue, a GC box/SP-1 binding site, an AP-2 binding site and a TATAA box. The 3′ untranslated region, which controls post-transcriptional regulation of TNF gene expression contains an octanucleotide TTATTTAT which is associated with degradation of mRNA and interferes with translation. The mRNA for TNF is approximately 1.7 kbp long (Spriggs et al. 1992).

■ Biological effects

In vitro, TNF is a multipotential cytokine with a wide variety of biological effects. Some of the in vitro biological activities include inhibition of proliferation of certain tumour cells, stimulation of proliferation of normal diploid fibroblasts, antiviral effects against both DNA and RNA viruses; induction of viral replication; induction of endothelial cell adhesion molecules; induction of plasminogen activator and its inhibitor; activation of neutrophils; myeloid cell differentiation; and induction of other cytokines (Aggarwal and Vilcek 1992; Beutler 1992).

In vivo, TNF displays anti-tumour activities against certain solid tumours, induces tumorigenesis, metastasis; acts as an immunomodulator; induces septic shock, fever; promotes cereberal malaria; plays a role in multiple sclerosis; and is involved in autoimmunity and in bacterial, parasitic, and viral infections (Aggarwal and Vilcek 1992; Beutler 1992).

■ Clinical role

TNF alone and in combinations has been used as an anti-tumour agent. High doses of TNF, adminstered by limb perfusion to patients with melanoma and other solid tumours, have produced a cure rate as high as 90 per cent (Lienard et al. 1993). Some of the major side-effects of TNF encountered during its systemic administration in cancer patients include nephrotoxicity, hepatotoxicity, pulmonary toxicity and neurotoxicity, while the minor side effects include hypertension, fever, rigors, acute dyspnea, cyanosis with hypoxia, and myalgia (Saks and Rosenblum 1992).

■ References

Aggarwal, B.B., (1992). Tumor necrosis factor. In Human cytokines: handbook for basic and clinical researchers, (ed. J.U. Gutterman and B.B. Aggarwal), pp. 270–85. Blackwell, New York.

Aggarwal, B.B., and Kohr, W.H. (1985). Human tumour necrosis factor. In Methods in enzymology (ed. G. DiSabato), Vol. 116 pp. 448–56.

Aggarwal, B.B., and Vilcek, J. (eds.) (1992). Tumor necrosis factor: structure, function and mechanism of action., pp. 1–624. Marcel Dekker, New York.

Aggarwal, B.B., Kohr, W.J., Hass, P.E., Moffat, B., Spencer, S.A., Henzel, W.J., Bringman, T.S., Nedwin, G.E., Goeddel, D.V., and Harkins, R.N. (1985). Human tumour necrosis factor: Production, purification and characterization. J. Biol. Chem., 260, 2345–54.

Beutler, B. (ed.) (1992). Tumor necrosis factor: the molecules and their emerging role in medicine, pp. 1–590. Raven, New York.

Farrah, T., and Smith, C.A. (1992). Emerging cytokine family. Nature, 358, 26.

Lienard, D., Eggermont, A.M.M., Koops, H.S., and Lejeune, F.J. (1993). High dose of rTNFα, rIFNγ and melphalan in isolation perfusion produce 90 per cent complete response in melanoma in transit metastases. In Tumor necrosis factor: molecular and cellular biology and clinical relevance, (ed. W. Fiers and W.A. Buurman), pp. 233–8. Karger, Basel.

Saks, S., and Rosenblum, M. (1992). Recombinant human TNF-α: preclinical studies and results from early clinical trials. In Tumor necrosis factor: structure, function and mechanism of action, (ed. B.B. Aggarwal and J. Vilcek), pp. 567–88. Marcel Dekker, New York.

Sprang, S.R., and Eck, M.J. (1992). The 3-D structure of TNF. In Tumor necrosis factors: the molecules and their emerging role in medicine, (ed. B. Beutler), pp. 11–32. Raven, New York.

Spriggs, D.R., Deutsch, S., and Kufe, D.W. (1992). Genomic structure, induction, and production of TNF-α. In Tumor necrosis factor: structure, function and mechanism of action, (ed. B.B. Aggarwal and J. Vilcek), pp. 3–34. Marcel Dekker, New York.

Bharat B. Aggarwal and Shrikanth Reddy:
Cytokine Research Section,
Department of Clinical Immunology & Biological Therapy,
University of Texas,
M.D. Anderson Cancer Center,
Houston, Texas 77030, USA

TNFβ is a tightly regulated product of lymphocytes that can be induced by a wide variety of antigenic and viral stimuli. It is a 25 kDa secreted protein that can also associate with a transmembrane protein, LTβ, to form a membrane heteromeric complex. TNFβ binds as a homotrimer to two TNFR, p55 and p75. Because of the wide cellular distribution of one or both TNFR, TNFβ exerts a broad spectrum of biological activities ranging from killing to inducing gene expression (cytokines, adhesion molecules, viruses) and promoting cell growth and differentiation. TNFβ is closely related to TNFα and apparently to LTβ. The TNFβ gene is tightly linked to TNFα and LTβ, within the MHC. Its expression is regulated transcriptionally and post-transcriptionally. The contribution of TNFβ to the pathogenicity of numerous diseases (experimental allergic encephalomyelitis, insulin-dependent diabetes mellitus) is well documented. Most TNFβ activities are also carried out by TNFα. A clear understanding of their different biologic roles and the extent of their redundancy will come from a delineation of differences in their molecular regulation and from a comparison of animal models deficient in one or the other cytokine.

■ Alternative names and related family members

Tumor necrosis factor-β is also known as lymphotoxin, lymphotoxin-α, and cytotoxic factor. TNFβ is closely related and genetically linked to TNFα (Turetskaya *et al.* 1992) and LTβ (Browning *et al.* 1993). It has many of the same activities as TNFα. It is also related to CD40 ligand, CD27 ligand, and CD30 ligand. In contrast to the other family members it does not possess a transmembrane leader peptide and it is secreted. However, it can form a heteromeric membrane associated complex with LTβ (Fig. 1). TNFβ binds with high affinity to two different receptors, p55 TNF receptor and p75 TNF receptor. These receptors are members of a family that includes: the low affinity nerve growth factor receptor, CD40, CD27, CD30, and fas (Loetscher *et al.* 1992).

■ Assays

TNFβ is measured as a cytotoxic activity against WEHI 164 or L929 cells. Because the same assay is used to detect TNFα, the molecules are frequently confused. However, human TNFα and TNFβ can be resolved with specific antibodies. ELISA assays for human TNFβ are also commercially available. No available antibody distinguishes between murine TNFα and TNFβ. The most widely used 'anti-TNF' antibody neutralizes and western blots both (Sheehan *et al.* 1989)

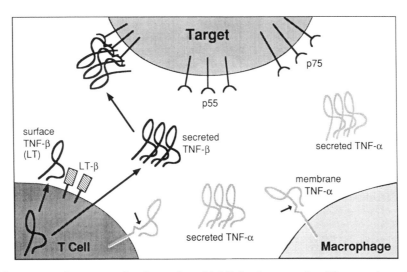

Figure 1. TNFβ forms a membrane-associated complex with LT-β or is secreted and forms an homotrimer which binds to p55 and p75 TNFR.

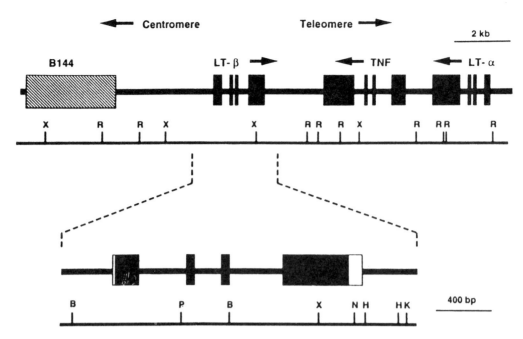

Figure 2. Organization of the human TNF gene complex on chromosome 6 (Browning *et al.* 1993).

and the commercially available 'TNFα' ELISA kits detect both.

■ TNFβ sequence

The TNFβ genes have been sequenced and their cDNAs have been identified in human, cow, rabbit, and mouse (GenBank accession number Y00467) (Turetskaya *et al.* 1992). The cDNA sequences predict proteins of 205 (human), 204 (cow), 197 (rabbit), or 202 (mouse) amino acids. Each contains a leader peptide ranging from 26 to 34 amino acids. The mature secreted proteins are highly homologous with 75 per cent identity in the 169 terminal amino acids. Nevertheless, antibodies against TNFβs of different species do not generally cross react. As noted above, TNFβ binds to two TNF receptors with affinities comparable to that of TNFα. There is a species preference for ligand binding to the p75 but not the p55 TNF receptor.

■ TNFβ protein

TNFβ is a glycoprotein that by SDS-PAGE has a molecular weight of 25 kDa (Aggarwal *et al.* 1984). By gel filtration, the model method of purification, its molecular weight is 60–70 kDa due to complex formation. It has been difficult to prepare large amounts of recombinant murine TNFβ. Studies with recombinant *E. coli* derived human TNFβ indicate that glycosylation is not essential for *in vitro* biologic activities. It has an isoelectric point of 5.8, it is only slightly sensitive to trypsin, resistant to chymotrypsin, acid labile, and unstable in detergents. The molecular structure of the

biologically active form of TNFβ is a homotrimer (Schoenfeld *et al.* 1991) which can bind to both TNF (p55 and p75) receptors (Eck *et al.* 1992). Analysis of human recombinant derived TNFβ indicates the importance of surface loops near the base of each monomer. Mutation of Asp 50 or Tyr 108 results in loss of L929 killing activity (Goh *et al.* 1991). These regions are homologous to regions on the TNFα molecule which are also crucial for function (Van Ostade *et al.* 1991; Yamagishi *et al.* 1990). The TNFβ ligand–p55 TNF receptor complex has been crystallized and analysed at 2.85 Å resolution (Banner *et al.* 1993). This complex is composed of three soluble receptor molecules bound symmetrically to one TNFβ trimer.

■ Molecular regulation

TNFβ is a tightly regulated product of lymphocytes. Transcription is induced after activation of certain T and B cell subsets by a variety of antigenic and viral stimuli. TNFβ can be induced by several viruses including HIV (Sastry *et al.* 1990), HTLV-I (Paul *et al.* 1990), and it is expressed by many Epstein–Barr virus transformed cells. The TNFβ gene maps within the major histocompatibility complex (MHC) on human chromosome 6 and mouse chromosome 17 (summarized in Turetskaya *et al.* 1992) between the complement and class I genes in close proximity to the TNFα locus (Fig. 2). The 3' end of the TNFβ gene is approximately 1 kbp from the transcription initiation site of the TNFα gene. Transcription through the TNF complex is from TNFβ to TNFα on the same coding strand. The LTβ gene is transcribed in the opposite direction. The TNFβ gene consists of 4 exons and 3 introns. This organization, which is conserved in mouse, human, rabbit, and cow, is similar to that of TNFα and LTβ. TNF complex linkage relationships are also conserved between the species. These observations suggest that TNFβ, LTβ, and TNFα have arisen through a process of gene duplication.

Extensive similarities are seen upon alignment of the human, rabbit, and mouse TNFβ genes. The most complete homology among the three species in the 5′ region is in the 300 bases upstream of the transcription initiation sites. Two functional cap sites within 15 bases have been identified. In the three species analysed, within 100 bp upstream of a TATA box, are consensus sites for SP1 transcription factor and near-consensus sites for Ap-2 (summarized in Turetskaya et al. 1992). An upstream NF-κB-like sequence has been identified in all species and its function has been demonstrated in T cells. Additional 5′ regulatory elements in murine TNFβ include positive and negative regulatory elements (reviewed in Turetskaya et al. 1992). One long poly (dA-dT) rich region binds a high mobility group I non-histone nuclear protein also known as HMG/Y and is crucial for TNFβ expression in pre-B cells (Fashena et al. 1992). The region between the TNFβ and TNFα genes is of considerable interest with regard to transcriptional regulation of both TNFβ and TNFα. This inter-genic region (IGR) consists of approximately 1 kbp of DNA and includes several potential NF-κB sites (Fig. 3). Some of these elements may act as shared enhancers, functioning in those situations in T cells in which both TNFα and TNFβ are transcribed.

TNFβ is also regulated post-transcriptionally. TNFβ mRNA isolated from activated T cells has a relatively long half-life, particularly when compared to that of TNFα (English et al. 1991; Millet and Ruddle 1994). It has also been suggested that the rate of splicing of precursor mRNA contributes to TNFβ regulation (Weil et al. 1990; Millet and Ruddle 1994). The regions of the gene which participate in these regulatory mechanisms have not been definitively identified, though a 3′ AT rich region has been suggested.

■ Biological roles

TNFβ has a broad spectrum of activities ranging from cytotoxicity to inducing cellular and viral gene expression and promoting cell proliferation and/or differentiation (Paul and Ruddle 1988). These multiple activities affecting many cell types (from lymphocytes to endothelial, bone, and nervous cells) are due in part to the almost ubiquitous expression of one or both of its receptors on most cell types and have implicated TNFβ in a wide variety of syndromes, including disease pathogenesis associated with HIV, HTLV-I, and EBV infections, autoimmune disease, and lymphoid development. TNFβ can kill a wide variety of transformed cells and some nontransformed cells such as oligodendrocytes, usually through DNA fragmentation (apoptosis). It can also synergize with many cytokines, particularly IFNγ, and can also induce and/or enhance expression of cytokines, including TNFα, and expression of adhesion molecules. TNFβ can induce expression of HIV in part through its activation of the transcription factor NF-κB (Poli and Fauci 1992), again implicating it in the pathogenesis of AIDS (Ruddle 1986).

TNFβ has been implicated in inflammatory autoimmune diseases. It has been detected in multiple sclerosis plaques and its inhibition by an antibody that also neutralizes TNFα prevents transfer of experimental allergic encephalomyelitis in mice (Ruddle et al. 1990). Mice transgenic for the insulin promoter driving expression of TNFβ develop an inflammatory mononuclear accumulation in and around islets (Picarella et al. 1992). These phenomena are most likely due in part to TNFβ's induction of adhesion molecule expression resulting in cellular recruitment.

Most TNFβ activities are also carried out by TNFα. Biologic activites for LTβ and the TNFβ/LTβ complex have not yet been identified. Definition of the different biological roles of the members of this family and an understanding of their apparent redundancy will derive from a comparison of animal models deficient in one or the other cytokines. So far, only TNFβ knockout mice have been developed. These animals exhibit marked derangement in splenic organization and lymphoid architecture, suggesting a crucial role for this cytokine in development of peripheral lymphoid organs (De Togni et al. 1994).

■ References

Aggarwal, B.B., Moffat, B., and Harkins, R.N. (1984). Human lymphotoxin. Production by a lymphoblastoid cell line, purification, and initial characterization. J. Biol. Chem., 259, 686–91.

Banner, D.W., D'Arcy, A., Janes, W., Gentz, R., Schoenfeld, H.J., Broger, C., Loetscher, H., and Lesslauer, W. (1993). Crystal structure of the soluble human 55 kd TNF receptor-human TNF beta complex: implications for TNF receptor activation. Cell, 73, 431–45.

Browning, J.L., Ngam-ek, A., Lawton, P., DeMarinis, J., Tizard, R., Chow, E.P., Hession, C., O'Brine-Greco, B., Foley, S F., and Ware,

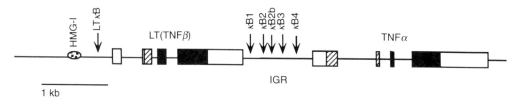

Figure 3. Molecular organization of the murine TNF gene complex with potentially important sites for gene regulation. Note the 5 NF-κB sites in the intergenic region (IGR).

C.F. (1993). Lymphotoxin beta, a novel member of the TNF family that forms a heteromeric complex with lymphotoxin on the cell surface. *Cell*, **72**, 847–56.

De Togni, P., Goeller, J., Ruddle, N., Streeter, P., Fick, A., Mariathasan, S., Strauss-Schoenberger, J., Shornick, L., Russell, J., Karr, R., and Caplin, D. (1994). Abnormal development of peripheral lymphoid organs in mice deficient in lymphotoxin. *Science*, **264**, 703–6.

Eck, M.J., Ultsch, M., Rinderknecht, E., de Vos, A.M., and Sprang, S.R. (1992). The structure of human lymphotoxin (tumour necrosis factor-beta) at 1.9-Å resolution. *J. Biol. Chem.*, **267**, 2119–22.

English, B.K., Weaver, W.M., and Wilson, C.B. (1991). Differential regulation of lymphotoxin and tumour necrosis factor genes in human T lymphocytes. *J. Biol. Chem.*, **266**, 7108–13.

Fashena, S.J., Reeves, R., and Ruddle, N.H. (1992). A poly (dA-dT) upstream activating sequence binds high-mobility group I protein and contributes to lymphotoxin (tumour necrosis factor-beta) gene regulation. *Mol. Cell. Biol.*, **12**, 894–903.

Goh, C.R., Loh, C.S., and Porter, A.G. (1991). Aspartic acid 50 and tyrosine 108 are essential for receptor binding and cytotoxic activity of tumour necrosis factor beta (lymphotoxin). *Protein Eng.*, **4**, 785–91.

Loetscher, H., Brockhaus, M., Dembic, Z., Gallati, H., Gentz, R., Gubler, U., Lahm, H-W., Lustig, A., Pan, Y-C.E., Schlaeger, E-J., Tabuchi, H., Zulauf, M., and Lesslauer, W. (1992). In *Tumor necrosis factor: structure–function relationship and clinical application* (ed. T. Osawa and B. Bonavida.), pp. 34–47. Karger, Basel.

Millet, I., and Ruddle, N. H. (1994). Differential regulation of lymphotoxin (LT), lymphotoxin-β (LT-β) and TNF-α in murine T cell clones activated through the TCR. *J. Immunol.*, **152**, 4336–46.

Paul, N.L., and Ruddle, N.H. (1988). Lymphotoxin. *Ann. Rev. Immunol.*, **6**, 407–38.

Paul, N.L., Lenardo, M.J., Novak, K.D., Sarr, T., Tang, W.L., and Ruddle, N.H. (1990). Lymphotoxin activation by human T-cell leukaemia virus type I-infected cell lines: role for NF-κB. *J. Virol.*, **64**, 5412–9.

Picarella, D.E., Kratz, A., Li, C.B., Ruddle, N.H., and Flavell, R.A. (1992). Insulitis in transgenic mice expressing tumour necrosis factor beta (lymphotoxin) in the pancreas. *Proc. Natl. Acad. Sci. (USA)*, **89**, 10036–40.

Poli, G., and Fauci, A.S. (1992). The effect of cytokines and pharmacologic agents on chronic HIV infection. *AIDS Research & Human Retroviruses*, **8**, 191–7.

Ruddle, N.H. (1986). Lymphotoxin production in AIDS. *Immunol. Today*, **7**, 8–9.

Ruddle, N.H., Bergman, C.M., McGrath, K.M., Lingenheld, E.G., Grunnet, M.L., Padula, S.J., and Clark, R.B. (1990). An antibody to lymphotoxin and tumour necrosis factor prevents transfer of experimental allergic encephalomyelitis. *J. Exp. Med.*, **172**, 1193–200.

Sastry, K.J., Reddy, H.R., Pandita, R., Totpal, K., and Aggarwal, B.B. (1990). HIV-1 tat gene induces tumour necrosis factor-beta (lymphotoxin) in a human B-lymphoblastoid cell line. *J. Biol. Chem.*, **265**, 20091–3.

Schoenfeld, H.J., Poeschl, B., Frey, J.R., Loetscher, H., Hunziker, W., Lustig, A., and Zulauf, M. (1991). Efficient purification of recombinant human tumour necrosis factor beta from *Escherichia coli* yields biologically active protein with a trimeric structure that binds to both tumour necrosis factor receptors. *J. Biol. Chem.*, **266**, 3863–9.

Sheehan, K.C.F., Ruddle, N.H., and Schreiber, R.D. (1989). Generation and characterization of hamster monoclonal antibodies that neutralize murine tumour necrosis factors. *J. Immunol.*, **142**, 3884–93.

Turetskaya, R., Fashena, S.J., Paul, N.L, and Ruddle, N.H. (1992). In *Tumor necrosis: structure, function and mechanism of action* (ed. B. Aggarwal and J. Vilcek). pp. 35–60. Marcel Dekker, New York.

Van Ostade, X., Tavernier, J., Prange, T., and Fiers, W. (1991). Localization of the active site of human tumour necrosis factor (hTNF) by mutational analysis [published erratum appears in EMBO J. 1992 Aug. 11(8): 3156]. *EMBO J.*, **10**, 827–36.

Weil, D., Brosset, S., and Dautry, F. (1990). RNA processing is a limiting step for murine tumour necrosis factor beta expression in response to interleukin-2. *Mol. Cell. Biol.*, **10**, 5865–75 .

Yamagishi, J., Kawashima, H., Matsuo, N., Ohue, M., Yamayoshi, M., Fukui, T., Kotani, H., Furuta, R., Nakano, K., and Yamada, M. (1990). Mutational analysis of structure–activity relationships in human tumour necrosis factor-alpha. *Protein. Eng.*, **3**, 713–9.

Isabelle Millet and Nancy H. Ruddle:
Yale University School of Medicine,
New Haven, CT USA

Tumor Necrosis Factor Receptor

Almost all cell types interact with TNF through a single class of specific high-affinity (K_D of 0.1–1.0 nM) receptors. Two different TNF receptors with molecular masses of 55–60 kDa and 75–80 kDa, referred to as p60 (type I or type B) and p80 (type II or type A), have been identified. The p60 form of the receptor is primarily expressed on epithelial cells, whereas p80 is expressed mainly on myeloid cells. Both TNF and lymphotoxin bind to both p60 and p80 with equal affinity. The amino acid sequence of the extracellular domain of the two receptors is rich in cysteine and is 25 per cent homologous but the cytoplasmic domains are homologous neither to each other nor to any other receptor. These receptors are members of the TNF receptor family, which includes nerve growth factor receptor, CD27, CD30, CD40, fas antigen, and proteins expressed by poxviruses (Dembic et al. 1990). Although both receptors can transduce a transcriptional factor NF-κB, most functions of TNF are mediated through p60 receptor. The p80 receptor of TNF is species-specific. Animals with a p60 receptor gene knockout are resistant to septic shock.

■ TNF receptor protein

The p80 form of the TNF receptor contains 461 amino acid residues consisting of an extracellular domain (ECD), a single transmembrane domain, and a cytoplasmic domain of 235, 30, and 174 amino acid residues, respectively. In contrast, the p60 form of the TNF receptor contains 455 amino acid residues consisting of extracellular, transmembrane, and cytoplasmic domains of 182, 21, and 223 amino acid residues, respectively. The ECD of the p80 receptor has two potential N-linked and several O-linked glycosylation sites, whereas the p60 receptor has only three N-linked sites. The ECD of the p60 and p80 receptors have 24 and 22 cysteine residues, respectively, organized into four cysteine-rich domains. The theoretical molecular masses of p60 and p80 receptor protein are 47 526 and 46 000 Da, respectively. Both receptors have one net positive charge. The ECD of both are highly susceptible to cleavage by proteases activated by kinases and phosphatase inhibitors. The soluble form of both the p60 and p80 receptors, consisting primarily of the ECD, has been detected in cell-conditioned media, urine, serum, amniotic fluids, and synovial fluids.

The cytoplasmic domain of the p60 receptor contains potential phosphorylation sites for c-AMP-dependent protein kinase, protein kinase C, and tyrosine kinases, whereas the p80 receptor does not. Neither intracellular domain shows any resemblance to tyrosine kinases. The amino acid sequence of the murine and human receptors are 64 per cent homologous for p60 and 62 per cent homologous for p80. The cytoplasmic domain of the p80 receptor has a serine-rich domain that is conserved between murine and human. This domain constitutively undergoes phosphorylation (Tartaglia and Goeddel 1992; Dembic et al. 1990; Sprang 1990). The X-ray crystal structure of the complex of the p60 ECD with TNFβ (lymphotoxin) indicates that three receptor molecules bind symmetrically to one TNFβ trimer (Banner et al. 1993).

■ TNF receptor gene and expression

The genes for human p60 and p80 receptors are located on chromosome 12p13 and 1, whereas those for murine are on chromosome 6 and 4, respectively. The chromosomal localization of p60 receptor gene in human is closely linked to CD4. The gene for human p60 receptor containing the entire coding region is 17 kb long. The gene is composed of 10 exons covering the protein coding region and the 3′ untranslated region spanning 13 kb. Each of the four repeats of the ECD of the p60 receptor is interrupted by an intron. The gene sequence excludes the possibility that alternative splicing generates soluble receptor protein. The murine p60 receptor promoter region does not respond to a variety of cytokines, and structural and functional analyses suggest that p60 receptor expression is directed by a non-inducible housekeeping-type promoter (Rothe et al. 1993; Fuchs et al. 1992).

■ TNF receptor mRNA

A leader sequence of 29 and 22 amino acids long has been found for p60 and p80 receptor. The p60 and p80 mRNAs are 3 kb and 4.5 kb long, respectively (Dembic et al. 1990). The p80 mRNA is regulated by phorbol ester, dibutyryl cAMP, dexamethasone, vitamin D3, interleukin-2, TNF, and IL-1. The p60 mRNA has been shown to be downmodulated on differentiation of HL-60 cells.

■ Cell surface expression and regulation of TNF receptors

All cell types so far studied express high-affinity (10^{-11} to 10^{-9} M) receptors for TNF. Both p60 and p80 receptors bind TNF with equally high affinity and transduce signals independent of each other. Epithelial cells mainly express p60, whereas myeloid cells express the p80 form of the TNF receptor. After TNF binds its receptor, the receptor–ligand complex is quickly internalized and degraded. On cross-linking of TNF to its cell surface receptor in most cells, a molecular mass of 80 kDa to 100 kDa has been found by sodium dodecyl sulphate polyacrylamide gel electrophoresis. The cell surface expression of the receptor is regulated by a variety of agents including cytokines (e.g. interferons, IL-1, IL-2, IL-4, IL-6, IL-8, TNF and GM-CSF), protein kinase (PK)-C and PK-A activators, microtubule depolymerizing agents, steroids and Ca^{+2} ionophores. Cell surface expression of TNF receptors appears to be a requirement for its biological action but it is not sufficient (Tsujimoto and Oku 1992; Smith and Baglioni 1992).

■ TNF signal transduction

TNF is a highly pleiotropic cytokine (Vilcek and Lee 1991). A role for several intermediates including transcriptional factors (e.g. NF-κB, c-jun/AP-1), phospholipases, protein kinases, tyrosine kinases, G-proteins, phosphatases, sphingomyelinase and free radicals in the action of TNF has been demonstrated. Which signals are mediated through which receptor is not entirely clear. By using receptor-specific agonistic and antagonistic antibodies and by transfection of cells with specific receptor cDNA, it has been shown that although both receptors can activate the transcriptional factor NF-κB, most biological effects of TNF are mediated through the p60 receptor. Similar to p60, the role of p80 receptor in growth modulation of certain cells has been demonstrated but whether it exhibits all the activities of p60 is not yet clear. Mice deficient in p60 receptor were found to be resistant to septic shock but susceptible to *Listeria monocytogenes* infection, suggesting a role for the p60 receptor in host defence against microorganisms and their pathogenic factors. Also, lymphocytes isolated from such mice showed loss of the ability to induce NF-κB by TNF (Pfeffer *et al.* 1993).

■ References

Banner, D.W., D'Arcy, A., Janes, W., Gentz, R., Schoenfeld, H.-J., Broger, C., Loetscher, H., and Lesslauer, W. (1993). Crystal structure of the soluble human 55 kd TNF receptor–human TNFβ complex: implications for TNF receptor activation. *Cell*, **73**, 431–45.

Dembic, Z., Loetscher, U., Gubler, U., Pan, Y.-C.E., Lahm, H.-W., Gentz, R., Brockhaus, M., and Lasslauer, W. (1990). Two human TNF receptors have similar extracellular, but distinct intracellular domain sequences. *Cytokine*, **2**, 231–7.

Fuchs, P., Strehl, S., Dworzak, M., Himmler, A., and Ambros, P.F. (1992). Structure of the human TNF receptor 1 (p60) gene (TNRF1) and localization to chromosome 12p13. *Genomics*, **13**, 219–24.

Pfeffer, K., Matsuyama, T., Kundig, T.M., Wakeham, A., Kishihara, K., Shaninian, A., Wiegmann, K., Ohashi, P.S., Kronke, M., and Mak, T.W. (1993). Mice deficient for the 55 kd tumour necrosis factor receptor are resistant to endotoxic shock, yet succumb to *L. monocytogenes* infection. *Cell*, **73**, 457–67.

Rothe, J., Bluethmann, H., Gentz, R., Lesslauer, W., and Steinmetz, M. (1993). Genomic organization and promoter function of the murine tumour necrosis factor receptor β gene. *Mol. Immunol.*, **30**, 165–75.

Smith, R.A., and Baglioni, C. (1992). Characterization of TNF receptors. In *Tumor necrosis factor: structure, function and mechanism of action*, (ed. B.B. Aggarwal and J. Vilcek), pp. 131–48. Marcel Dekker, New York.

Sprang, S.R. (1990). The divergent receptors for TNF. *Trends Biochem. Sci.*, **15**, 366–8.

Tartaglia, L.A., and Goeddel, D.V. (1992). Two TNF receptors. *Immunol. Today*, **13**, 151–3.

Tsujimoto, M., and Oku, N. (1992). Regulation of TNF receptors. In *Tumor necrosis factor: structure, function and mechanism of action*, (ed. B.B. Aggarwal and J. Vilcek), pp. 149–60. Marcel Dekker, New York.

Vilcek, J., and Lee, T.H. (1991). Tumor necrosis factor. *J. Biol. Chem.*, **266**, 7313–6.

Bharat B.Aggarwal and Shrikanth Reddy:
Cytokine Research Section,
Department of Clinical Immunology & Biological Therapy,
University of Texas,
M.D. Anderson Cancer Center,
Houston, Texas 77030, USA

Type I interferons (IFNs) are a closely related family of 165–172 amino acid proteins that are produced by leucocytes (α subtypes) and fibroblasts (β subtypes). IFNω and trophoblast IFN (IFNτ) are also members of the type I IFN family and are produced in lymphocytes and ruminant embryos, respectively. IFN induces a wide range of effects on target cells including anti-viral, antiproliferative and immunomodulatory activity. IFNs are produced in response to viral, bacterial, mycoplasma, and protozoan infection as well as in the presence of certain cytokines including colony-stimulating factor-1, interleukin-1 (IL-1), IL-2, and tumour necrosis factor. The secreted IFN binds to the Type I receptor in a species-specific manner on target cells which initiates the various target-cell responses. Type I IFNs are currently utilized in therapy for hairy cell leukaemia, chronic myelogenous leukaemia, condyloma acuminata, mycosis fungoides, chronic hepatitis B and C, and Kaposi's sarcoma and is currently in clinical trials for various other carcinomas and viral diseases (Dianzani 1992).

■ Nomenclature

IFNs were originally classified according to the producer cells as leucocyte, fibroblast, and immune IFN (Pestka 1983, 1986). Leucocyte IFNs were then designated as α and ω, fibroblast IFN as β, and immune IFN as γ. The α, β, and ω have since been grouped as type I due to the extensive sequence homologies in their amino acid sequences and the observation that they bind to a common cell-surface receptor

```
          1                                                                                              100
IFN- α1  (D)   ..SP.A...V LV...C..S. .......E.. ..D...T.M. ....S...S ..M........ .......P..... ..L..I... .T........
IFN- α2  (A)   ........V. L...C..S. .V........ ....S..T.M. ....R...L. .......... ...ET.P.... ..L...I....
IFN- α4B       .......... .......... .......... .......... .......H.. .......... ...ET.P.... ..I.......
IFN- α5  (G)   ...P.V.... LV..NC.... .......... ...S...T.MI M........ .......E...H. .T........ .......E....
IFN- α6  (K)   ...P.A.... LV...C..S. .D........ ...H..TMM. ....R...L. .......... ...E....... ..........T.
IFN- α7  (J1)  ..R.....V. .......... .......R.. .......... .......E.R .E...H.... ...E......V. ........V..
IFN- α8  (B)   ...T.Y..V. LV.....FS .......... .......... ....R..... .......E. ...DK......
IFN- α9        .......... .......... .......... .......G.. .......... ........... ..........L
IFN- α10 (L,C) ..........* .......... .T.R...... .....G.... .......R I .......... ..........E.....
IFN- α14 (M)   ...P.A.M.. LV...C..S. ....N.S... .N...T.M. M..R...... .......E.. ..........M....... .......
IFN- α16 (WA,N) .......... .......... .......... .......H.. ....Y.... ...V...... ...AF .........N.....
IFN- α17 (I)   .......... .......... .......... .......... .......P.. L........ ...T....... .......E......
IFN- α21 (F)   .......... .......... .......... .......... .......... .......... .......... .......E....T.
IFN- αCons.    MALSFSLLMA VLVLSYKSIC SLGCDLPQTH SLGNRRALIL LAQMGRISPF SCLKDRHDFG FPQEEFDGNQ FQKAQAISVL HEMIQQTFNL FSTKDSSAAW

          101                                                                                   189
IFN- α1  (D)   ..D...D..C. .......... ...M..ER.G ......A.. ...K..R.. .......... ...L.L.... ..E.....E
IFN- α2  (A)   ..T..D.... .......... ....G..T ....K..... .......... ...K...... ...L.L.... ..E.....E
IFN- α4B       EQ......S. .......... .......V.. .......... ...K...... ...L... ..ES..S.E
IFN- α5  (G)   ..T..D.... .......... ...MM..... D....V... T.......... .......L......
IFN- α6  (K)   ..R..D.L.. .......... ...M..W.G G.......... .......... ...L.A.. ..E.....E
IFN- α7  (J1)  EQ......S. .......... .......F... .......M.. .......... ...S.R.. ..E.....E
IFN- α8  (B)   ..T..DE..I ..D....... S..M....I .S...Y.... .......... ...K.G....
IFN- α9        EQ......S. .......... .......... .......S......... ...L.I.... ..KS.E
IFN- α10 (L,C) EQ......S. .I........ .......... ...I.R.... .......... ...L.......
IFN- α14 (M)   ..T......I .F.M...... .......... ...I.R.... .......... ...L.......
IFN- α16 (WA,N) ..T..D...I .F........ ...T.....IA .......K... ...M.......
IFN- α17 (I)   EQ......S. .......N... ....M..... .......MG.. .......... ...G.....
IFN- α21 (F)   EQ......S. ..N....... .......V...K... .......... ...L.KI F.E....E
IFN- αCons.    DESLLEKFYT ELYQQLNDLE ACVIQEVGVE ETPLMNEDSI LAVRKYFQRI TLYLTEKKYS PCAWEVVRAE IMRSFSFSTN LQKRLRRKD
```

Figure 1. Alignment of the human IFNα protein sequences. Representative amino acid sequences of each of the major human IFNα subtypes are shown as compared with the consensus sequence (bottom line, 'IFN-αCons') as determined by the GCG sequence analysis software (Devereux *et al.* 1984). Only the amino acids differing from the consensus sequences are shown. The single amino acid deletion in IFNα2 is indicated by a dash; the asterisk indicates a termination codon in IFNα10; position 109 in the consensus sequence can be either a tyrosine (Y) or a serine (S) as both are found equally in the sequences. The GenBank accession numbers of the sequences are: IFNα1 (D), V00537 (Mantei *et al.* 1980); IFNα2 (A), V00548 (Streuli *et al.* 1980); IFNα4b, X02955 (Henco *et al.* 1985); IFNα5 (G), X02956 (Henco *et al.* 1985; IFNα6 (K), X02958 (Henco *et al.* 1985); IFNα7 (J1), X02960 (Henco *et al.* 1985); IFNα8 (B), X03125 (Henco *et al.* 1985); IFNα9, V00551 (Goedel *et al.* 1981); IFNα10 (C,L), X02961 (Henco *et al.* 1985); IFNα14 (M), V00542 (Goedel *et al.* 1981); IFNα16 (WA,N), X02957 (Henco *et al.* 1985); IFNα17 (I), V00532 (Lawn *et al.* 1981); IFNα21 (F), X00145 (Gren *et al.* 1983).

```
                  1                                                                                              100
  Human IFN- β    ...K....I. .......... .M..N..G.L ....FQ... ..W....RLE Y..K...N.D I...I..L.. .........T. .....N..A. F.Q.S.....
  Mouse IFN- β    .N..WI.HA. F......... .IN.KQ.QL. E.TNIRK..E ..E....KIN --.TY.A..K I.M..TE--K M..SYT.FA. Q....NV.LV F.NN.....
  Horse IFN- β    ..Y.WI.P.. .......... .VN.D...S. L....S..LM .R....A.. R.P..T.N. ...IE..... .......... ....HTWR. F..N.A....
Bovine 1 IFN- β   ..Y......V .......... ....Q.LK.. ....PS.S. ...A..... M.....E... .....I..M .V..H..G. .T........
Bovine 2 IFN- β   ..H.......V .......... ....R.LAL. ...R..PS.. ...A..... M......... ......I... .......... .T........
Bovine 3 IFN- β   ..Y....P.V .......... ....R.A.V. ....HS... ..AK..... ...R....I .......... ....T.....
  Swine IFN- β    .A.K.I..I. ..M....... .M..DV..Y. ........L. ....P....N.E ...IM.PP. .....V.I. H........G. .....N.....
  Cons. IFN- β    MTNRCLLQMA LLLCFSTTAL SRSYSLLRFQ QRSSNEACQK LLGQLNGTPQ HCLEDRMDFQ VPEEMKQAQQ FQKEDAALVI YEMLQQIFNI LRRDFSSTGW

                  101                                                                                            187
  Human IFN- β    ....V.N..A NV.H.I.H.K .V...KLEK. D....KLMSS ....R..G.. .H..A...S H.....I.... ......Y.I. ........
  Mouse IFN- β    ....VVR..D ..HQ.TVF.K .V....K... RL.WEMSS.A ...S..W.V QR...LMK.. SY..M...A. .F...LIIR. ..RNFQ.
  Horse IFN- β    ....VKN... .VHL..D... .N.....E. SS.WN-..I .R....G.. S..A.K.S H.....QA. M...LA...G. .D..Q.
Bovine 1 IFN- β   S........ ...W.....Q P.QK....KQ .S.TE.-.I. P.G....NL M...E...D .....Q.Q .T.V...M. ....V.D
Bovine 2 IFN- β   S.......E ....H...Q P.QK....KQ .S.M.-... ....R....NL V......... .......Q .......T. ......E
Bovine 3 IFN- β   S......... .Q P.QK....Q ....M.-... ....NL V..... ...Q. .T...M. ..AS..D
  Swine IFN- β    ...V.KTI.. .D...DD... ....E...P...-M.I .......LS. ...R S......Q. .........D....
  Cons. IFN- β    NETIIEDLLV ELYGQMNRLE TILEEIMQEE NFTRGD-TTV LHLKKYYFRI LQYLKSKEYN RCAWTVVRVE ILRNFSFLNR LTGYLRN
```

Figure 2. Mammalian IFNβ amino acid sequence alignment. Mammalian IFNβ amino acid sequences are shown as compared to the consensus sequence as determined by the GCG sequence analysis package (Devereux *et al.* 1984). GenBank accession numbers of the sequences are: human J00218 (Taniguchi *et al.* 1980); mouse, K00020 (Higashi *et al.* 1983); horse, M14546 (Himmler *et al.* 1986); bovine 1, M15477 (Leung *et al.* 1984); bovine 2, M15478 (Leung *et al.* 1984); bovine 3, M15479 (Leung *et al.* 1984); swine, S109938 (Artursson *et al.* 1992).

(Mariano *et al.* 1992). The specific nomenclature employed for the several α subtypes has varied from laboratory to laboratory where the genes and proteins have been identified and has not been standardized.

■ Type I IFN genes and proteins

The type I IFN genes are clustered on human chromosome 9 (Diaz *et al.* 1991). The cluster consists of at least 14 functional IFNα genes, one IFNβ and one IFNω gene. In addition to these functional genes, at least five pseudogenes are found within this region. Type I IFN genes do not contain introns. The regulatory regions of the type I IFNs contain sequences shown to be both negative regulatory elements (NRDI and NRDII), which allow for low-level endogenous expression of IFNs, and positive regulatory elements (PRDI–PRDIV), which are capable of binding a transcriptional activator protein (IRF-1) in response to viral infection (Dron and Tovey 1992). The specific sequence in the promoter of each gene may allow for differential regulation of specific α-subtypes.

The nucleotide sequences of the human IFN α subtypes show coding sequences for a 23 amino acid signal peptide and a 165- to 166-amino-acid mature protein (Fig. 1). The nucleotide sequences of the Hu-IFNα genes show 68 per cent invariant nucleotides in the coding region resulting in 73 per cent similarity in the proteins (Zoon *et al.* 1992). Glycosylation of some Hu-IFNα proteins results in native molecular weights ranging from 16 000 to 27 500 Da and a family of over 25 proteins with similar structure and biological function (Pestka 1986). Each protein contains at least four cysteine residues allowing for the formation of two intramolecular disulphide bridges, though mutagenesis studies have shown that in some proteins the formation of the Cys1 to Cys98 disulphide bridge is not required for activity. The Hu-IFNα protein appears to bind to the receptor as a monomer despite the ability of some of the subtypes to form dimers in solution (Rehberg *et al.* 1982).

The single gene encoding human IFNβ produces a protein with a 23-amino-acid signal peptide and a 164-amino-acid mature protein (Fig. 2). IFNβ is glycosylated at Asn80 resulting in a native glycoprotein with an apparent molecular weight of 22 000 to 23 000 (Utsumi and Shimizu 1992). Though in humans and mice IFNβ is encoded by a single gene, cows, sheep and pigs possess multiple IFNβ genes (Dron and Tovey 1992). A comparison of the amino acid sequences of several mammalian IFNβ sequences shows a high degree of conservation between human and other species (Fig. 2).

IFNω genes are found in most mammalian species, excluding mouse, with functional genes coding for a protein of 195 amino acids including the 23-amino-acid signal peptide. The human IFNω protein shares approximately 60 per cent homology with the human IFNα proteins (Capon *et al.* 1985). No IFNω gene has been found in mouse, but similar to the IFNβ genes, several independent IFNω genes are found in cows, horses, and sheep (Himmler *et al.* 1986). IFNω proteins also contain potential glycosylation sites. Comparison of the amino acid sequences of several IFNωs is shown in Fig. 3 and reveals a sequence homology between species similar to that seen in the IFNβ proteins. In addition to the IFNω genes, a closely related group of multiple copy genes is found in domestic ruminants and are designated as the trophoblast interferons (IFNτ) due to the fact that they are produced in the embryonic trophectoderm (Leaman and Roberts 1992). The proteins encoded by the trophoblast IFN genes have a high degree of homology with the IFNω subgroup (as much as 77 per cent within a species) and contain 195 amino acids with a 23-amino-acid leader sequence (Fig. 4) which is the same as the IFNω subgroup. The trophoblast IFNs are identified as a separate gene family by the sequences surrounding the coding region which are homologous within the trophoblast IFNs but differ greatly from the IFNω subgroup. Also the expression

```
            1                                                                                                         100
Human  IFN- ω   ...L.P..A.  ...T..S.V.  ......P...  G.L...NT.V.  .H....I..F  ......R....  ......K....  ....HVM....  ......S...  ..........
Bovine 1 IFN- ω ..........  ..........  ......P...  ...G.Q.....  ..........  ....R...Q...  A.....V.  F.E........  ......S...  ..K.......
Bovine 2 IFN- ω ..........  ......R...  .Y..E..M.GA.....  .AR.N.....  P..Q....G  ......N...  .D........  ......C...  .Y..H.....
Bovine 3 IFN- ω ..........  ......R...  .Y..ED..M.GA.....  .AR.N.....  P..Q....G  ......N...  .D........  ......C...  .Y..H.....
Sheep  1 IFN- ω ..........  ..........  ......G.K.....  .......T...  ...Q....A  ......G...  .E........  ......S...  ..........
Sheep  2 IFN- ω ...QL..T..  .V...C.....  PH.S APL...ST.V.  .D...V..V  ......R..Q  .R.V.N...  F..N.T....  .........A.  ..........
Horse  1 IFN- ω ...SV.S...  .V..SS.VS  .S...PASL D.RKQ.T..V  .H..ETI..P  S...H.T...  .QL..R. FPE...T...  Q......VS.  ..........
Horse  2 IFN- ω ...LP...T.  ..VYELW.C.  A.....P...  ...K..V...  .S.I.SA I...  ....A.R. FPE..A.....  ......S...  ..........
Cons.  IFN- ω   MAFVLSLLMA  LVLVSYGPGG  SLGCDLSQNH  VLVSRENLRL  LGQMRRLSPH  LCLKDRKDFR  FPQEMVEGSQ  LQKAQAISVL  HEMLQQIFNL  FHTERSSAAW

            101                                                                                                       195
Human  IFN- ω   .M.....H. E.....QH. ...L..V.G ...GAISS.A .TLR...... R............  ..K..FL... ......S..R ....S
Bovine 1 IFN- ω D.......L. .....A...LLT.....T...M........Q..G......S........M...K..
Bovine 2 IFN- ω ..........Q.....A..P...K .D..M..I .T.K......E.....M.A....T ..K...KMG. ..N.L
Bovine 3 IFN- ω ..........Q.....A..P...K .D..M..I .T.K......E.....V. M.A....T ..K...KMG. ..N.L
Sheep  1 IFN- ω ...K..RN .LD..V...A...D.......T........G.T.V.......S.....M...VN..
Sheep  2 IFN- ω .N....E.H. A.....QG....V.A....V.TADS...ML....R. RL..D...H. G...L....R.A...TAD ..S.S.....A..
Horse  1 IFN- ω .....R.LA .......N ..DEQT.....T.....RR. RL..T........V.......A. ..G.G.....
Horse  2 IFN- ω .....E...LR.....E.E......T.R.......R...L........A..K....
Cons.  IFN- ω   NTTLLEQLCT GLHQQLEDLD TCLGQVMGEE DSALGRVGPT LAVKRYFQGI HVYLKEKKYS DCAWEIVRME IMRSLSSSTN LQERLRMKDG DLGSP
```

Figure 3. Mammalian IFNω amino acid sequence alignment. Mammalian IFNω amino acid sequences are shown as compared to the consensus sequence as determined by the GCG sequence analysis package (Devereux *et al.* 1984). Six positions in the consensus sequence are equally represented by two amino acids: position 27, S and P; position 54, K and Q; position 71, L and F; position 73, K and E; position 121, T and A; position 188, M and K. GenBank accession numbers of the sequences are: human, M11003 (Capon *et al.* 1985); bovine 1, M11002 (Capon *et al.* 1985); bovine 2, M60908 (Hansen *et al.* 1991); bovine 3, M60913 (Hansen *et al.* 1991); sheep 1, M73245 (Leaman and Roberts 1992); sheep 2, X59068 (Whaley *et al.* 1991); horse 1, M14544 (Himmler *et al.* 1986); horse 2, M14545 (Himmler *et al.* 1986).

of the trophoblast IFNs in development may suggest that they play a different role than the other Type I IFNs.

Biological functions

Type I interferons exert a variety of effects on target cells including anti-viral, antiproliferative, and immunomodulatory activities. All of the type I interferons compete for binding to a single cell-surface receptor complex which initiates the signalling mechanism for the cellular effects observed. The cellular response to the binding of interferons to the cell induces the activation of specific DNA-binding proteins that complex with the promoters of interferon stimulated genes (ISG) at the interferon stimulated response element (ISRE) resulting in the induction of several gene products including 2–5 A synthetase, MHC class I surface antigens, p68 kinase, and the Mx family of proteins, as well as several other proteins whose function is not yet known (Sen and Lengyel 1992). Trophoblast IFNs are expressed early in embryogenesis and therefore may play a key role in the implantation process at least in domestic ruminants.

References

Artursson, K., Gobl, A., Lindersson, M., Johansson, M., and Alm, G. (1992). Molecular cloning of a gene encoding porcine interferon-beta. *J. Interferon Res.*, **12**, 153–60.

Capon, D.J., Shepard, H.M., and Goeddel, D.V. (1985). Two distinct families of human and bovine interferon-alpha genes are coordinately expressed and encode functional polypeptides. *Mol. Cell. Biol.*, **5**, 768–79.

Devereux, J., Haeberli, P., and Smithies, O. (1984). A comprehensive set of sequence analysis programs for the VAX. *Nucleic Acids Res.*, **12**, 387–95.

Dianzani, F. (1992). Interferon treatments: how to use an endogenous system as a therapeutic agent. *J. Interferon Res.* (Special issue May 1992), 109–18.

Diaz, M.O., Pomukala, H., Bohlander, S., Maltepe, E., and Olopade, O. (1991). A complete physical map of the type-I interferon gene cluster. *J. Interferon Res.*, **11**, suppl. 1,S85.

Dron, M., and Tovey, M.G. (1992). Interferon *α/β*, gene structure and regulation. In *Interferon: principles and medical applications* (eds S. Baron *et al.*), pp. 33–45. The University of Texas Medical Branch at Galveston, Galveston, TX.

Goedel, D.V., Leung, D.W., Dull, T.J., Gross, M., Lawn, R.M., McCandliss, R., Seeburg, P.H., Ullruch, A., Yelverton, E., and Gray, P.W. (1981). The structure of eight distinct cloned human leukocyte interferon cDNAs. *Nature*, **290**, 20–6.

Gren, E.Y., Bersin, V.M., Tsimanis, A.Y., Apsalon, U.R., Vishnevskii, Y.I., Yansone, I.V., Dishler, A.V., Pudova, N.V., Smorodintsev, A.A., Iovlev, V.I., Stepanov, A.N., Feldmane, G.Y., Meldrais, Y.A., Lozha, V.P., Kavsan, V.M., Efimov, V.A., and Sverdlov, E.D. (1983). A new type of leukocytic interferon. *Dolk. Biochem.*, **269**, 91–5.

Hansen, T.R., Leaman, D.W., Cross, J.C., Mathialagan, N., Bixby, J.A., and Roberts, R.M. (1991). The genes for the trophoblast interferons and the related interferon-alpha II possess distinct 5′-promoter and 3′-flanking sequences. *J. Biol. Chem.*, **266**, 3060–7.

Henco, K., Brosius, J., Fujisawa, A., Fujisawa, J.-I., Haynes, J.R., Hochstadt, J., Kovacic, T., Pasek, M., Schambock, A., Schmid, J., Todokoro, K., Walchli, M., Nagata, S., and Weissmann, C. (1985). Structural relationship of human interferon alpha genes and pseudogenes. *J. Mol. Biol.*, **185**, 227–60.

Higashi, Y., Sokawa, Y., Watanabe, Y., Kawade, Y., Ohno, S., Takaoka, C., and Taniguchi, T. (1983). Structure and expression of a cloned cDNA for mouse interferon-beta. *J. Biol. Chem.*, **258**, 9522–9.

Himmler, A., Hauptmann, R., Adolf, G.R., and Swetly, P. (1986). Molecular cloning and expression in *Escherichia coli* of equine type I interferons. *DNA*, **5**, 345–56.

Lawn, R.M., Adelman, J., Dull, T.J., Gross, M., Goeddel, D., and

```
                       1                                                                                                                      100
Bovine 1 IFN-τ  .......... .......... .......R .......EDH ..G....... ..A....... ..P....... .......... .....N.... ....I... ..H.....C... ..........
Bovine 2 IFN-τ  .......... .......... .......R ......EDH ..G....... ..A....... ..P....... .......... .....N.... ...I... ..H.....C... ..........
Sheep 1 IFN-τ   .......... .......... .......... .......... .......... .......... .......... .......EA..C.. .......... .H..R.....
Sheep 2 IFN-τ   .......... .......... .......... .......... .......... .......... .......... .......EA..C.. .......... .H..R.....
Sheep 3 IFN-τ   .......... .......... .......... .......... ......K... .......... .......... .......... ......P... ..........
Goat    IFN-τ   .......... .......... ......R... .......... .......... ..Q....... .......... ....SC... ..........
Musk Ox IFN-τ   .......R.. .....C... ......R.P T..V..... .......... ..Q....... .......... .......... .....R.... .H....C...
Cons.IFN-τ      MAFVLSLLMA LVLVSYGPGG SLGCYLSQRL MLDARENLRL LDRMNRLSPH SCLQDRKDFG LPQEMVEGDQ LQKDQAFSVL YEMLQQSFNL FYTEHSSAAW

                       101                                                                                                                    195
Bovine 1 IFN-τ  .......... .......... A.L....... ....R.G.. L........ .V..K..E.. .......... .....S.... ...L
Bovine 2 IFN-τ  .......... .......... A.L....... ....R.G.. L........ .V..K..E.. .......M.. ....S.... ...L
Sheep 1 IFN-τ   .......... .......... .......... .......... .F..K..E.. .......... .....S...S .E.....
Sheep 2 IFN-τ   .......... .......... .......... .......... .......... ....T..... .......... .K.G*
Sheep 3 IFN-τ   D......... ....H... ....Q....E .....N.... .......... Y......... .......... ...V.... ...T.... ..........
Goat    IFN-τ   D......... ....H... ....Q.... .....N.... YY........ .......T... .......A... ...T.T.... .......
Musk Ox IFN-τ   .......R.. .H........ .......... .......... .......... Y......... .......... ......K.T.. ......
Cons.IFN-τ      NTTLLEQLCT GLQQQLEDLD TCRGPVMGEK DSELGKMDPI VTVKKYFQGI HDYLQEKGYS DCAWEIVRVE MMRALTSSTT LQKRLRKMGG DLNSP
```

Figure 4. Ruminant trophoblast IFN amino acid alignment. Ruminant trophoblast IFN amino acid sequences are shown as compared to the consensus sequence as determined by the GCG sequence analysis package (Devereux *et al.* 1984). GenBank accession numbers of the sequences are: bovine 1, M60903 (Hansen *et al.* 1991); bovine 2, X65539 (Stewart *et al.* 1990); sheep 1, M73241 (Leaman and Roberts 1992); sheep 2, M73242 (Leaman and Roberts 1992); sheep 3, X07920 (Stewart *et al.* 1990); goat, M73243 (Leaman and Roberts 1992); musk ox, M73244 (Leaman and Roberts 1992).

Ullrich, A. (1981). DNA sequence of two closely linked human leukocyte interferon genes. *Science*, **212**, 1159–62.

Leaman, D.W., and Roberts, R.M. (1992). Genes for the trophoblast interferons in sheep, goat, and musk ox and distribution of related genes among mammals. *J. Interferon Res.*, **12**, 1–11.

Leung, D.W., Capon, D.J., and Goeddel, D.V. (1984). The structure and bacterial expression of three distinct bovine interferon-beta genes. *Bio/Technology*, **2**, 458–64.

Mantei, N., Schwarzstein, M., Streuli, M., Panem, S., Nagata, S., and Weissmann, C. (1980). The nucleotide sequence of a cloned human leukocyte interferon cDNA. *Gene*, **10**, 1–10.

Mariano, T.M., Donnelly, R.J., Soh, J., and Pestka, S. (1992). Structure and function of the type I interferon receptor. In *Interferon: Principles and Medical Applications*, (eds S. Baron *et al.*), pp. 129–38. The University of Texas Medical Branch at Galveston, Galveston, TX.

Pestka, S. (1983). The human interferons—from protein purification and sequence to cloning and expression in bacteria: before, between, and beyond. *Arch. Biochem. Biophys.*, **221**, 1–37.

Pestka, S. (1986). Interferon from 1981 to 1986. *Meth. Enzymol.*, **119**, 3–14.

Rehberg, E., Kelder, B., Hoal, E.G., and Pestka, S. (1982). Specific molecular activities of recombinant and hybrid leukocyte interferons. *J. Biol. Chem.*, **257**, 11497–502.

Sen, G.C., and Lengyel, P. (1992). The interferon system. A bird's eye view of its biochemistry. *J. Biol. Chem.*, **267**, 5017–20.

Stewart, J.H., McCann, S.H., and Flint, A.P. (1990). Structure of an interferon-alpha 2 gene expressed in the bovine conceptus early in gestation. *J. Mol. Endocrinol.*, **4**, 275–82.

Streuli, M., Nagata, S., and Weissmann, C. (1980). At least three human type alpha interferons: structure of alpha 2. *Science*, **209**, 1343–7.

Taniguchi, T., Ohno, S., Fujii-Kuriyama, Y., and Muramatsu, M. (1980). The nucleotide sequence of human fibroblast interferon cDNA. *Gene*, **10**, 11–15.

Utsumi, J., and Shimizu, H. (1992). Human interferon β, protein structure and function. In *Interferon: Principles and Medical Applications*, (eds S. Baron *et al.*), pp. 107–16. The University of Texas Medical Branch at Galveston, Galveston, TX.

Whaley, A.E., Carroll, R.S., and Imakawa, K. (1991). Cloning and analysis of a gene encoding ovine interferon alpha-II. *Gene*, **106**, 281–2.

Zoon, K.C., Bekisz, J., and Miller, D. (1992). Human interferon alpha family: Protein structure and function. In *Interferon: Principles and Medical Applications*, (eds. S. Baron *et al.*), pp. 95–105. The University of Texas Medical Branch at Galveston, Galveston, TX.

Robert J. Donnelly:
Department of Molecular Genetics and Microbiology,
University of Medicine and Dentistry of New Jersey,
Robert Wood Johnson Medical School,
Piscataway, NJ USA

Receptors for the Alpha and Beta Interferon (IFN) Family

The alpha/beta IFN receptor functional unit is a multi-ligand system that binds the family of alpha and beta IFN subtypes. Strong evidence suggests that functional binding sites for the alpha/beta IFNs involve several proteins. The IFNAR (IFN alpha/beta receptor) chain has been cloned and character-ized as a component of the alpha/beta IFN receptor functional unit. The predicted structure of the extracellular part of IFNAR puts this protein in the class II of the cytokine receptor family.

■ Binding characteristics of the alpha/beta IFN receptor

Receptors are generally found on all human cells whatever their origin, even on cells poorly responsive to interferon. Cell lines showing no binding at all of one or other of the family are relatively rare or deliberate constructs. Binding studies with radiolabelled ligand have been reported for the following human recombinant interferon: alphas, 1(D), 2 (A), 8(B), 4 and for beta interferon. Competitive binding experiments with unlabelled ligand against a given radiolabelled interferon have been reported for a wider range of alpha interferons. Taking account of differences in specific activity (e.g. that of alpha-1 is markedly less on human cells than those of the other alpha beta interferons), these results all tend to the conclusion that the alpha beta interferons interact with the same receptor site (Branca 1988). However, direct binding studies with the radiolabelled ligand reveal differences that are not easily apparent from competitive studies. Thus, provisionally, binding falls into two classes giving either quasi-linear Scatchard plots over the pM to nM range (β and alpha-8) or descending concave plots (alphas: 1, D and presumably 4). Our experience is that binding to cells (or membranes from those cells), sensitive to interferon in the pM to nM range rarely shows sign of saturability (i.e. semi-log plots do not inflect in this range). Estimates of receptor density and binding affinity are therefore weighted by the range of concentrations chosen. Generally tangents to the Scatchard plot give apparent K_Ds of 100 to 500 pM and a density of $>10\,000$ to >1000 sites per cell. While interferon-insensitive cells often show low binding, there is no obvious correlation between the Scatchard coefficients and the sensitivity of the cells. Binding studies on the solubilized receptor show a wide range of avidities which depend on the detergent used. Stable complexes of high M_r (600kDa) are obtained with electrically neutral steroid detergents; complexes of M_r 100–150kDa can be obtained, with low stability, by treating with non-steroid detergents and by treating the *in situ* complex with cross-linking reagents followed by extraction into SDS (Branca 1988; Eid and Mogensen 1990). An isolated protein of M_r c. 90kDa can be recognized by ligand blotting (Schwabe *et al.* 1988); antibody studies suggest that this is

the protein encoded by the IFNAR gene (Uzé *et al.* 1991). Recent studies suggest that the soluble form of the putative extracellular part of the IFNAR protein binds interferon with only a low avidity. In sum, these results suggest that high affinity binding of the alpha beta interferons is to organised sites that contain the IFNAR protein. Biological activity correlates only with high affinity binding though the existence of high affinity sites is, in itself, not a sufficient condition for activity. For sensitive cells, a one-to-one correlation between receptor-mediated uptake of ligand and biological activity has been described for the recombinant interferons (Mogensen *et al.* 1989). For the naturally occurring mixtures, uptake of the individual components appears to occur independently and it is reasonable to assume that it is receptor activation that correlates with the effect.

■ Structure of the IFNAR chain of the alpha/beta IFN receptor

The IFNAR chain of the human alpha/beta interferon receptor (Uzé *et al.* 1990) is a 557 amino acid precursor with a predicted 27-amino-acid leader sequence and a single predicted transmembrane segment of 21 amino acids. The predicted molecular weight of the human IFNAR is 63 000 Da but there are 15 potential N-linked glycosylation sites. The 100 amino acid intracellular part of IFNAR does not carry any recognized enzymatic activities. It is rich in serine and negatively charged amino acids. The 409 amino acid extracellular part is organized in two domains of 200 amino acids (D200) showing an internal symmetry (Gaboriaud *et al.* 1990). This structure appears to be a duplication of D200 characteristic of the extracellular part of receptors that belong to the cytokine receptor family (Bazan 1990; Thoreau *et al.* 1991). Each D200 domain can be further subdivided into two sub domains of 100 amino acids (SD100A, SD100B, SD100A', and SD100B'). The predicted folding pattern of each SD100 is that of seven β strands (S1–S7) which has been solved for the growth hormone receptor (de Vos *et al.* 1992). The positions of the conserved cysteines in the D200s of the cytokine receptor family define two classes of D200 (Bazan 1990). The two D200 of IFNAR belong to class II, together with the D200 of the interferon gamma receptor, the tissue factor and the CRFB4 receptor for which no

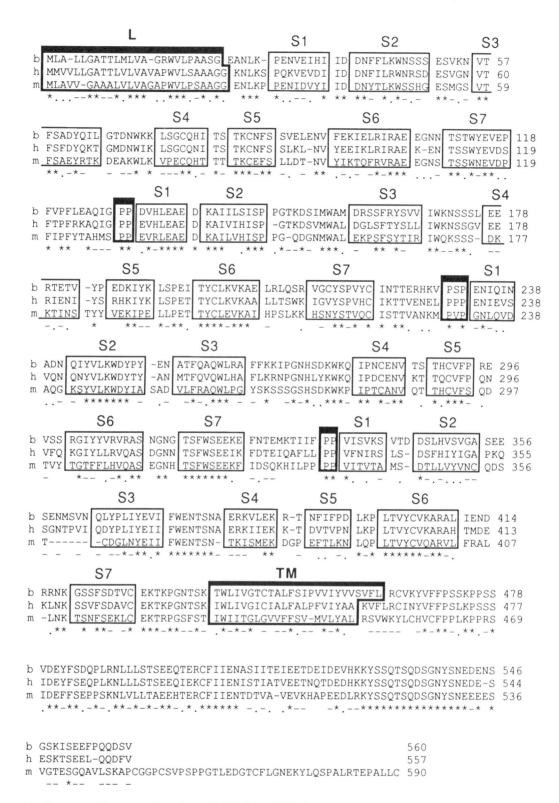

Figure 1. Sequence alignments (from Mouchel-Vielh *et al.* 1992).

ligand has yet been identified (Lutfalla et al. 1993). The alignment of the sequences of the human (Uzé et al. 1990) (GenBank accession number J03171), bovine (Mouchel-Vielh et al. 1992) (EMBL data library accession number X68443) and murine (Uzé et al. 1992) (GenBank accession number M89647) IFNARs is shown in Fig. 1. The positions of the predicted β strands are boxed as are the transmembrane, the leader regions and the proline stretches corresponding to the hinge regions between SD100s.

The alpha/beta IFN receptor functional unit

The IFNAR chain is likely implicated in the binding and activities of all alpha/beta IFN species (Uzé et al. 1992). However, in addition to IFNAR, the existence of other component(s) to form a fully functional receptor system for the alpha/beta IFNs is indicated by cross-species cDNA transfectants (the human receptor expressed in mouse cells [Uzé et al. 1990] and the mouse receptor in human cells [Uzé et al. 1992]) which suggest that a heterospecific background imposes constraints on the functioning of the receptor. There are alpha interferon subspecies which are only active when their specific receptors are expressed in cells of the same species. This implies a local structure for functional high-affinity receptors which involves more than a single protein; a conclusion that corresponds well with the estimated size of high affinity sites isolated by detergent extraction (Eid and Mogensen 1990). On human cells, the differences between the specific activities of the different human alpha IFNs lie in the structure of the IFNAR chain rather than in the other components of the functional receptor (Mouchel-Vielh et al. 1992).

Signalling mechanisms

The immediate early response of a cell following the binding of IFN to its receptor is the activation of transcription of a set of genes containing a conserved response element called ISRE (IFN stimulated response element) in their promoter (Kerr and Stark 1991). ISRE binds the ISGF-3 transcription factor which consists of four proteins (Schindler et al. 1992). The current model is that the ISGF-3 transcription factor is activated by phosphorylation on tyrosine in a set of reactions which are likely to occur in the proximity of the plasma membrane (David and Larner 1992). The activated ISGF-3 then translocates to the nucleus to bind the ISRE sequences.

The tyrosine kinase Tyk-2 has been identified as a component of the IFN alpha signalling pathway (Velazquez et al. 1992). Its role in the activation of ISGF-3 components is not yet clear but the interesting point is that a cell deficient in Tyk-2 does not express high affinity binding sites for IFNs. This suggests an interaction between the Tyk-2 activity and the binding activity of the alpha/beta IFN receptor functional unit.

The IFNAR genetic locus

The human IFNAR gene is localized on human chromosome 21 in a 400 kbp Mlu I fragment containing also GART and D21S58 (Lutfalla et al. 1992). The D21S58 probe corresponds to the exon 4 of the CRFB4 gene which is localized at less than 35 kbp from IFNAR (Lutfalla et al. 1993). A similar organization is found on mouse chromosome 16 (Cheng et al. 1993).

The 32 906 bp of the IFNAR gene (GenBank accession number X60459) includes 11 exons. Each sub-domain of 100 amino acids (SD100) of the extracellular part of IFNAR is encoded by two separate exons, the junction of which falls at the end of the third predicted β strand for SD100A and at the beginning of the fourth β strand for SD100B. This splice site pattern with regards to the predicted structure of the extracellular part of other receptors belonging to the class I and class II of the cytokine receptor family is conserved (Lutfalla et al. 1992).

References

Bazan, J.F. (1990). Structural design and molecular evolution of a cytokine receptor superfamily. Proc. Natl. Acad. Sci. (USA), **87**, 6934–8.

Branca, A.A. (1988). Interferon receptors. In Vitro Cell. Devel. Biol., **24**, 155–65.

Cheng, S., Lutfalla, G., Uzé, G., Chumakov, I.M., and Gardiner, K. (1993). GART, SON, IFNAR and CRF2–4 genes cluster on human chromosome 21 and mouse chromosome 16. Mammalian Genome, **4**, 338–42.

David, M., and Larner, A.C. (1992). Activation of transcription factors by interferon-alpha in a cell-free system. Science, **257**, 813–5.

de Vos, A.M., Ultsch, M., and Kossiakoff, A.A. (1992). Human growth hormone and extracellular domain of its receptor: crystal structure of the complex. Science, **255**, 306–12.

Eid, P., and Mogensen, K. (1990). Detergent extraction of the human alpha-beta interferon receptor: a soluble form capable of binding interferon. Biochim. Biophys. Acta., **1034**, 114–7.

Gaboriaud, C., Uzé, G., Lutfalla, G., and Mogensen, K. (1990). Hydrophobic cluster analysis reveals duplication in the external structure of human alpha-interferon receptor and homology with gamma-interferon receptor external domain. FEBS Lett., **269**, 1–3.

Kerr, I.M., and Stark, G.R. (1991). The control of interferon-inducible gene expression. FEBS Lett., **285**, 194–8.

Lutfalla, G., Gardiner, K., Proudhon, D., Vielh, E., and Uzé G. (1992). The structure of the human interferon alpha/beta receptor gene. J. Biol. Chem., **267**, 2802–9.

Lutfalla, G., Gardiner, K., and Uzé, G. (1993). A new member of the cytokine receptor gene family maps on chromosome 21 at less than 35 kb from IFNAR. Genomics, **16**, 366–73.

Mogensen, K.E., Uzé, G., and Eid, P. (1989). The cellular receptor of the alpha-beta interferons. Experientia, **45**, 500–8.

Mouchel-Vielh, E., Lutfalla, G., Mogensen, K.E., and Uzé, G. (1992). Specific antiviral activities of the human alpha interferons are determined at the level of receptor (IFNAR) structure. FEBS Lett., **313**, 255–9.

Schindler, C., Shuai, K., Prezioso, V.R., and Darnell, J.E. Jr. (1992). Interferon-dependent tyrosine phosphorylation of a latent cytoplasmic transcription factor. Science, **257**, 809–13.

Schwabe, M., Princler, G.L., and Faltynek, C.R. (1988). Characterization of the human type I interferon receptor by ligand blotting. *Eur. J. Immunol.*, **18**, 2009–14.

Thoreau, E., Petridou, B., Kelly, P.A. Djiane, J., and Mornon, J.P. (1991). Structural symmetry of the extracellular domain of the cytokine/growth hormone/prolactin receptor family and interferon receptors revealed by hydrophobic cluster analysis. *FEBS Lett.*, **282**, 26–31.

Uzé, G., Lutfalla, G., and Gresser, I. (1990). Genetic transfer of a functional human interferon alpha receptor into mouse cells: cloning and expression of its cDNA. *Cell*, **60**, 225–34.

Uzé, G., Lutfalla, G., Eid, P., Maury, C., Bandu, M.-T., Gresser, I., and Mogensen, K.E. (1991). Murine tumour cells expressing the gene for the human interferon alpha beta receptor elicit antibodies in syngeneic mice to the active form of the receptor. *Eur. J. Immunol.*, **21**, 447–51.

Uzé, G., Lutfalla, G., Bandu, M.-T., Proudhon, D., and Mogensen, K.E. (1992). Behavior of a cloned murine interferon alpha/beta receptor expressed in homospecific or heterospecific background. *Proc. Natl. Acad. Sci. (USA)*, **89**, 4774–8.

Velazquez, L., Fellous, M., Stark, G.R., and Pellegrini, S. (1992). A protein tyrosine kinase in the interferon alpha/beta signaling pathway. *Cell*, **70**, 313–22.

G. Uzé, G. Lutfalla, and K.E. Mogensen:
Institut de Génétique Moléculaire de Montpellier, CNRS UMR 9942,
BP. 5051. 34033 Montpellier Cedex 1, France

Interferon-Gamma (IFNγ)

IFNγ is a glycoprotein of monomer molecular weight 20–25 kDa which is secreted from T cells and natural killer (NK) cells. This molecule exhibits many different biological activities in vitro, *including anti-viral, anti-proliferation, macrophage activation, and induction of MHC class I and II antigens. IFNγ plays a key role in host defence and in inflammatory processes. Clinically, IFNγ has been shown to alleviate the symptoms of chronic granulomatous disease and exhibits promising potential in the treatment of infectious disease and neoplasia.*

■ Alternative names

Type II interferon, immune interferon, macrophage activation factor.

■ IFNγ sequence

The human IFNγ cDNA (GenBank accession number X13274) encodes a 143-amino-acid mature protein with a 23 residue signal sequence (Gray et al. 1982). The activity of IFNγ is species specific due to the low level of homology between different species sequences. Mouse IFNγ only contains 40 per cent identical residues (Gray and Goeddel 1983). The mouse cDNA (GenBank K00083) encodes a 133 residue mature protein with a 22 amino acid signal sequence (Gray and Goeddel 1983).

■ IFNγ protein

Human IFNγ is a secreted glycoprotein which is extensively modified (Rinderknecht et al. 1984). There are two potential N-linked glycosylation sites; position 25 is almost always glycosylated while position 125 is glycosylated in half of purified preparations. The amino terminal glutamine cyclizes to form pyroglutamic acid. The carboxyl terminal end is of variable length due to proteolytic degradation (at least nine residues may be removed without alteration of bio-

logical activity). Purified natural preparations of IFNγ exhibit monomer molecular weights of 25 000 and 20 000 Da on SDS-PAGE (with a minor 15 500 Da band representing a non-glycosylated species). However, the active moiety is a dimer, held together by strong non-covalent interactions (Farrar and Schreiber 1993). The X-ray crystallographic structure has recently been determined (Ealick et al. 1991) and suggests that the monomer subunits associate by the helical intertwining of subunits. The monomers interact in an antiparallel association, such that amino and carboxyl regions of opposite monomers are adjacent. The overall dimer structure is compact and globular. Experimental data suggests that each IFNγ dimer is able to bind two IFNγ receptors and that receptor clustering may be important for signal transduction (Farrar and Schreiber 1993).

IFNγ can be purified by chromatography on lectin–sepharose, size exclusion gels, and HPLC (Rinderknecht et al. 1984).

■ IFNγ gene and transcription

The human IFNγ gene (GenBank accession number V00536) is located on human chromosome 12 at q24.1 (the murine IFNγ gene is on chromosome 10) (Naylor et al. 1983, 1984). The human gene is approximately 6 kbp in length and the coding sequence is interrupted by three introns (Gray and Goeddel 1982). The transcript is 1200 bases in length. IFNγ is produced by T cells and NK cells, specifically by CD8+ T cells and subsets of CD4[+] T cells (T$_H$1 and T$_H$0) (Farrar and Schreiber 1993). The synthesis of IFNγ is induced by antigen stimulus of T cells or by mitogens such as concanavalin A. This expression may be enhanced by IL-2 or leukotrienes. IFNγ synthesis is inhibited by IL-10, dexamethasone, and cyclosporin A. Maximal synthesis occurs at 12–14 hours following stimulation.

IFNγ biology

IFNγ was originally identified by its *in vitro* anti-viral activity nearly 30 years ago (Wheelock 1965). IFNγ is also a potent antiviral agent *in vivo*; however, the immunomodulatory activities of IFNγ are probably more physiologically relevant (Farrar and Schreiber 1993; Vilcek *et al*. 1985). IFNγ induces the expression of many key molecules, including class I and class II MHC antigens (Basham and Merrigan 1983; King and Jones 1983), nitric oxide synthase (Farrar and Schreiber 1993; Corbett *et al*. 1991), and cytokines such as IL-1 (Farrar and Schreiber 1993; Vilcek *et al*. 1985). The activity of IFNγ is synergistic with other cytokines; for example, IFNγ enhances the cytotoxic activity of TNF and the antiviral activities of IFNα or IFNβ. These molecular events account for the broad array of observed biological activities *in vitro*: activation of macrophages to become tumouricidal and kill intracellular parasites (Pace *et al*. 1983; Buchmeier and Schreiber 1985), anti-proliferative activity of tumour cells (Farrar and Schreiber 1993; Vilcek *et al*. 1985), enhancement of B cell maturation and immunoglobulin secretion (Vilcek *et al*. 1985; O'Garra *et al*. 1988) and NK cell activation (Farrar and Schreiber 1993). IFNγ is responsible for inducing non-specific cell mediated mechanisms of host defence through its macrophage activation ability. IFNγ is required for the resolution of microbial infections, such as *Listeria*, *Toxoplasma*, and *Leishmania* (Farrar and Schreiber 1993; Buchmeier and Schreiber 1985). In the inflammatory response, IFNγ plays an indirect role by causing the induction of TNF and synergizing with its activities. IFNγ also appears to have a participatory role in the development of autoimmune disease; for example, the expression of IFNγ in the pancreas of transgenic mice has been shown to cause an autoimmune diabetes (Sarvetnick *et al*. 1988).

References

Basham, T.Y., and Merrigan, T.C. (1983). Recombinant interferon-gamma increases HLA-DR synthesis and expression. *J. Immunol*., **130**, 1492–4.

Buchmeier, N.A., and Schreiber, R.D. (1985). Requirement of endogenous interferon-gamma production for resolution of *Listeria monocytogenes* infection. *Proc. Natl. Acad. Sci. (USA)*, **82**, 7404–8.

Corbett, J.A., Lancaster, J.R., Sweetland, M.A., and McDaniel, M.L. (1991). Interleukin-1 beta-induced formation of EPR-detectable iron-nitrosyl complexes in islets of Langerhans. Role of nitric oxide in interleukin-1 beta-induced inhibition of insulin secretion. *J. Biol. Chem*., **266**, 21351–4.

Ealick, S.E., Cook, W.J., Vijay-Kumar, S., Carson, M., Nagabhushan, T.L., Trotta, P.P., and Bugg, C.E. (1991). Three-dimensional structure of recombinant human interferon-gamma. *Science*, **252**, 698–702.

Farrar, M.A., and Schreiber, R.D. (1993). The molecular cell biology of interferon-gamma and its receptor. *Ann. Rev. Immunol*., **11**, 571–611.

Gray, P.W., and Goeddel, D.V. (1982). Structure of the human immune interferon gene. *Nature*, **298**, 859–63.

Gray, P.W., and Goeddel, D.V. (1983). Cloning and expression of murine immune interferon cDNA. *Proc. Natl. Acad. Sci. (USA)*, **80**, 5842–6.

Gray, P.W., Leung, D.W., Pennica, D., Yelverton, E., Najarian, R., Simonsen, C.C., Derynck, R., Sherwood, P.J., Wallace, D.M., Berger, S.L., Levinson, A.D., and Goeddel, D.V. (1982). Expression of human immune interferon cDNA in E. coli and monkey cells. *Nature*, **295**, 503–8.

King, D.P., and Jones, P.P. (1983). Induction of Ia and H-2 antigens on a macrophage cell line by immune interferon. *J. Immunol*., **131**, 315–8.

Naylor, S.L., Sakaguchi, A.Y., Shows, T.B., Law, M.L., Goeddel, D.V., and Gray, P.W. (1983). Human immune interferon gene is located on chromosome 12. *J. Exp. Med*., **157**, 1020–7.

Naylor, S.L., Gray, P.W., and Lalley, P.A. (1984). Mouse immune interferon (IFN-gamma) gene is on chromosome 10. *Somat. Cell. Mol. Genet*., **10**, 531–4.

O'Garra, A., Umland, S., DeFrance, T., and Christiansen, J. (1988). 'B-cell factors' are pleiotropic. *Immunol. Today*, **9**, 45–54.

Pace, J.L., Russell, S.W., Schreiber, R.D., Altman, A., and Katz, D.H. (1983). Macrophage activation: priming activity from a T-cell hybridoma is attributable to interferon-gamma. *Proc. Natl. Acad. Sci. (USA)*, **80**, 3782–6.

Rinderknecht, E., O'Connor, B.H., and Rodriguez, H. (1984). Natural human interferon-gamma. Complete amino acid sequence and determination of sites of glycosylation. *J. Biol. Chem*., **259**, 6790–7.

Sarvetnick, N., Liggitt, D., Pitts, S.L., Hansen, S.E., and Stewart, T.A. (1988). Insulin-dependent diabetes mellitus induced in transgenic mice by ectopic expression of class II MHC and interferon-gamma. *Cell*, **52**, 773–82.

Vilcek, J., Gray, P.W., Rinderknecht, E., and Sevastopoulus, C.G. (1985). Interferon γ: A lymphokine for all seasons. *Lymphokines*, **11**, 1–32.

Wheelock, E.F. (1965). Interferon-like virus-inhibitor induced in human leukocytes by phytohemagglutinin. *Science*, **149**, 310–1.

Patrick W. Gray:
ICOS Corporation,
22021 20th Avenue S.E.,
Bothell, WA 98021, USA

Table 1. Biological actions of IFNγ

1 Anti-viral activity *in vitro* and provides *in vivo* protective effect.
2 Anti-proliferative activity on tumour cells *in vitro*; anti-tumour activity in animal model systems.
3 Induction of MHC class I and class II antigens, TNF, IL-1, chemokines, and cell surface markers such as ICAM-1.
4 Activates macrophages to become tumouricidal and kill intracellular parasites; required for the natural resolution of microbial infections.
5 Induces immunoglobulin secretion by B cells and enhances B cell proliferation and maturation.
6 Enhances natural killer cell activity.
7 Plays a key role in the regulation of cytokine activity by both inducing the synthesis of some cytokines and synergizing with their activities.
8 Participates in pathology of inflammation and autoimmune disease.

The IFNγ receptor consists of two integral membrane polypeptides that are members of the type II cytokine receptor family: a 90 kDa α subunit that is necessary and sufficient for ligand binding and receptor trafficking and necessary but not sufficient for signal transduction and a β subunit that participates in signal transduction only. Human and murine IFNγ receptors bind their respective ligands with high affinity in a strictly species specific manner. The mechanism of signal transduction used by the IFNγ receptor involves the sequential activation of receptor associated protein tyrosine kinases which in turn phosphorylate and activate latent cytosolic transcription factors leading to the assembly of an active multimolecular transcription factor complex that translocates to the nucleus and initiates transcription of IFNγ inducible genes.

■ The IFNγ receptor α subunit protein

The α subunit of the human IFNγ receptor (CDw119) is produced as a 489-amino-acid precursor that contains a 17-amino-acid signal sequence (GenBank accession number J03143) (Aguet et al. 1988) (Fig. 1). Although the mature, nonglycosylated 472 amino acid polypeptide has a predicted molecular mass of 50.5 kDa, it displays an Mr of 65 kDa when analysed by SDS-PAGE (Farrar and Schreiber 1993). The molecule is symmetrically oriented around a single 23-amino-acid transmembrane domain. The extracellular domain consists of 228 amino acids and includes ten cysteine residues and five potential N-linked glycosylation sites. In the fully mature molecule, all five sites are occupied by carbohydrate which contribute 20–35 kDa to the apparent molecular mass of the protein (Farrar and Schreiber 1993). IFNγ receptor α subunits from different cells display Mr that vary between 80–95 kDa due to cell-specific differences in carbohydrates (Farrar and Schreiber 1993). The intracellular domain consists of 220 amino acids, and contains five tyrosine residues and a high percentage (24 per cent) of serine and threonine residues. The α subunit is devoid of intrinsic kinase or phosphatase activities. Based on the general structural features of the extracellular domain, the IFNγ receptor α chain has been classified as a member of the type II cytokine receptor family that also includes the receptors for IFNα/β, IL-10, and tissue factor (Bazan 1990; Ho et al. 1993).

The murine IFNγ receptor α subunit (GenBank accession numbers M26711, M28233, M25764 and M28995) (Gray et al. 1989; Hemmi et al. 1989; Kumar et al. 1989; Munro and Maniatis 1989, respectively) is produced as a 477-amino-acid polypeptide containing a 26-amino-acid signal peptide. The mature 451-amino-acid protein shows only 52.5 per cent amino acid sequence identity with its human counterpart. Nevertheless, the murine and human proteins are organized in a similar manner. Both contain identically sized extracellular and transmembrane domains and relatively large serine and threonine rich intracellular domains (although the intracellular domain of the murine α subunit is 200 amino acids and therefore slightly smaller than that of the human protein).

■ The IFNγ receptor β subunit

The human and murine IFNγ receptor β subunits (also referred to as AF-1) have recently been identified at the molecular level. The human β subunit (GenBank accession number Y05875) (Soh et al. 1994) is a 337-amino-acid polypeptide which contains a 21-amino-acid signal sequence (Fig. 1). Like the α subunit, the β chain is also a member of

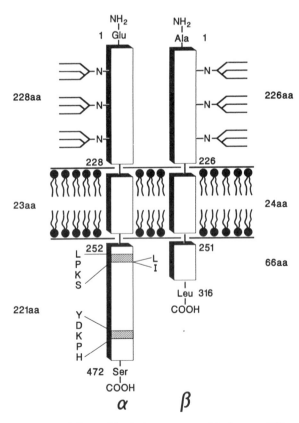

Figure 1. Polypeptide chain structure of the human IFNγ receptor.

the type II cytokine receptor family. The mature protein consists of 316 amino acids and has a predicted molecular mass of 34.8 kDa. It contains an extracellular domain of 226 amino acids, a single 24-amino-acid transmembrane domain, and a relatively short intracellular domain of only 66 amino acids.

The mature murine β chain contains a 224-amino-acid extracellular domain, a 24-amino-acid transmembrane domain and a 66-amino-acid intracellular domain (EMBL GenBank accession number X77133) (Hemmi et al. 1994). It has a predicted molecular mass of 34.5 kDa. No information is available at this time concerning postsynthetic modifications of the mature human or murine proteins or the structure–function relationships that are operative within them.

■ IFNγ receptor subunit genes and gene expression

The gene encoding the IFNγ receptor α subunit has been localized to human chromosome 6 (q16–q22) (Pfizenmaier et al. 1988) and murine chromosome 10 (Mariano et al. 1987). The human and murine genes are approximately 30 kb in size and contain seven exons (Merlin et al. 1989). Gene activation results in the generation of a single 2.3 kb transcript. Expression of the receptor α chain is constitutive and ubiquitous. Except for mature erythrocytes, all normal somatic and haemopoietic cells express α subunit message and protein. Expression of the protein ranges from 200–10 000 copies per cell in normal tissue and is not significantly modulated by external stimuli.

The gene encoding the human IFNγ receptor β chain is located on human chromosome 21 in the distal portion of band 21q22.1 between the D21S58 marker and the GART locus (Soh et al. 1993). The murine gene has been localized to a syntenic region on the distal fifth of murine chromosome 16 (Hibino et al. 1991). No information is currently available about the structure of the IFNγ receptor β chain gene. Activation of the IFNγ receptor β chain gene results in the generation of a 2.0 kb transcript (Hemmi et al. 1994)

■ Functional roles of IFNγ receptor subunits

Binding of human and murine IFNγ to their respective IFNγ receptors is strictly species-specific. The extracellular domain of the α subunit is both necessary and sufficient to confer high-affinity, species-specific ligand binding to cells (Farrar and Schreiber 1993). Binding of ligand to receptor α subunits either in the absence or presence of the β subunit is described by a single K_A of 1×10^{10} M^{-1} at 4°C. The molecular basis for the collaboration between α and β receptor subunits remains undefined. However, this interaction shows strict species specificity which is mediated at least in part by the extracellular domain of the α subunit (Gibbs et al. 1991; Hemmi et al. 1992; Hibino et al. 1992). Binding of IFNγ, a homodimeric protein, to the receptor α chain leads

to α subunit dimerization (Fountoulakis et al. 1992; Greenlund et al. 1993).

The intracellular domain of the α chain is also required in an obligatory manner for the development of IFNγ dependent biologic responses (Farrar et al. 1991; Farrar et al. 1992; Greenlund et al. 1994). However, the action of this portion of the α subunit is not species restricted. Three functionally important sequences within the intracellular domain of the α subunit have been defined which play distinct roles in ligand processing and signal transduction (Fig. 1). A membrane proximal leucine–isoleucine (LI) sequence (residues 270–271) has been identified that does not appear to participate in signal transduction events, but is required for efficient trafficking of receptor through the cell. Immediately upstream of this sequence is a leucine–proline–lysine–serine (LPKS) sequence (positions 266–269) that is required for the ligand-dependent activation of a receptor-associated tyrosine kinase and/or the association of the kinase with the receptor α subunit. The specific enzyme that associates with this sequence has not yet been identified. However, current evidence suggests that the α subunit may interact with one of two Janus family tyrosine kinases (JAK-1 or JAK-2) since both enzymes are (a) obligatorily required for the development of IFNγ induced cellular responses (Müller et al. 1993; Watling et al. 1993) and (b) known to bind to similar sequences (termed box 1 sequences) in other cytokine receptors (Murakami et al. 1991). The third sequence is located near the carboxy terminus of the α chain and contains a functionally critical tyrosine residue at position 440. Mutation of Y440 to alanine, phenylalanine or serine completely ablates the ability of the α subunit to participate in signal transduction processes. Y440 has recently been shown to be a physiologically important target of one of the IFNγ activated tyrosine kinases and phosphorylation of Y440 leads to the generation of a specific binding site for the latent cytosolic transcription factor p91 thereby effecting the coupling of the IFNγ receptor to its signal transduction system. The specific α subunit amino acids that form the p91 binding site have been defined and include: phosphorylated Y-440, the adjacent aspartic acid (D-441) and a closely spaced histidine residue (H-444).

No definitive structure-function information is currently available for the IFNγ receptor β subunit. However, generation of functionally active IFNγ receptors in heterologous cells shows an obligatory requirement for a species matched set of α and β chains (Jung et al. 1987, 1990; Farrar et al. 1992). Interestingly the β chain contains an intracellular domain sequence element that is similar to so-called box 2 elements found in the intracellular domains of other cytokine receptors that associate with JAK family tyrosine kinases (Murakami et al. 1991).

■ IFNγ receptor mediated signal transduction

The past year has seen an explosive growth in the understanding of the molecular events that underlie IFNγ depen-

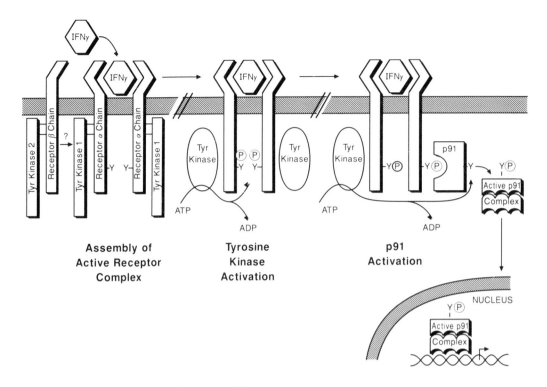

Figure 2. Proposed signalling mechanism of the IFNγ receptor.

dent signal transduction. Figure 2 represents a proposed model of the IFNγ signal transduction pathway. IFNγ, a homodimeric ligand, binds to IFNγ receptor α chains expressed at the cell surface effecting rapid subunit dimerization (Fountoulakis et al. 1992; Greenlund et al. 1993). The IFNγ receptor β chain then associates with this complex bringing into close juxtaposition at least two inactive intracellular tyrosine kinases, one associated with the LPKS box-1-like sequence of the receptor α chain and the other associated with the box-2-like sequence of the receptor β chain. It is possible that these kinases may in fact be JAK1 and JAK2 (Müller et al. 1993; Watling et al. 1993). These enzymes cross-activate one another and this event leads to the phosphorylation of Y440 in the intracellular domain of the IFNγ receptor α chain, thereby generating a binding site for latent p91 (Greenlund et al. 1994). Docking of p91 at the IFNγ receptor α chain brings p91 into close proximity with the activated receptor associated tyrosine kinase(s), leading to the tyrosine phosphorylation and activation of p91 (Schindler et al. 1992). Phosphorylated p91 then dissociates from the receptor, associates with either another phosphorylated p91 molecule or other activated transcription factors, and translocates to the nucleus where it interacts with specific sequences in IFNγ inducible genes, thereby initiating gene transcription (Shuai et al. 1992). Perhaps one of the most striking features of this model is that it comprises signal transduction effector molecules that are shared by a number of different receptors. It will therefore be important in the future to define how this system functions to

specifically effect the development of IFNγ inducible biologic responses.

■ References

Aguet, M., Dembic, Z., and Merlin, G. (1988). Molecular cloning and expression of the human interferon-γ receptor. *Cell*, **55**, 273–80.

Bazan, J.F. (1990). Structural design and molecular evolution of a cytokine receptor superfamily. *Proc. Natl. Acad. Sci. (USA)*, **87**, 6934–8.

Farrar, M.A., Fernandez-Luna, J., and Schreiber, R.D. (1991). Identification of two regions within the cytoplasmic domain of the human interferon-gamma receptor required for function. *J. Biol. Chem.*, **266**, 19626–35.

Farrar, M.A., Campbell, J.D., and Schreiber, R.D. (1992). Identification of a functionally important sequence motif in the carboxy terminus of the interferon-γ receptor. *Proc. Natl. Acad. Sci. (USA)*, **89**, 11706–10.

Farrar, M.A., and Schreiber, R.D. (1993). The molecular cell biology of interferon-γ and its receptor. *Annu. Rev. Immunol.*, **11**, 571–611.

Fountoulakis, M., Zulauf, M., Lustig, A., and Garotta, G. (1992). Stoichiometry of interaction between interferon-γ and its receptor. *Eur. J. Biochem.*, **208**, 781–7.

Gibbs, V.C., Williams, S.R., Gray, P.W., Schreiber, R.D., Pennica, D., Rice, G., and Goeddel, D.V. (1991). The extracellular domain of the human interferon gamma receptor interacts with a species-specific signal transducer. *Mol. Cell. Biol.*, **11**, 5860–6.

Gray, P.W., Leong, S., Fennie, E.H., Farrar, M.A., Pingel, J.T., Fernandez-Luna, J., and Schreiber, R.D. (1989). Cloning and expression of the cDNA for the murine interferon gamma receptor. *Proc. Natl. Acad. Sci. (USA)*, **86**, 8497–501.

Greenlund, A.C., Schreiber, R.D., Goeddel, D.V., and Pennica, D. (1993). Interferon-γ induces receptor dimerization in solution and on cells . *J. Biol. Chem.*, **268**, 18103–10.

Greenlund, A.C., Farrar, M.A., Viviano, B.L., and Schreiber, R.D. (1994). Ligand induced IFNγ receptor phosphorylation couples the receptor to its signal transduction system (p91). *EMBO J.*, **13**, 1591–600.

Hemmi, S., Peghini, P., Metzler, M., Merlin, G., Dembic, Z., and Aguet, M. (1989). Cloning of murine interferon gamma receptor cDNA: expression in human cells mediates high-affinity binding but is not sufficient to confer sensitivity to murine interferon gamma. *Proc. Natl. Acad. Sci. (USA)*, **86**, 9901–5.

Hemmi, S., Merlin, G., and Aguet, M. (1992). Functional characterization of a hybrid human-mouse interferon gamma receptor: Evidence for species-specific interaction of the extracellular receptor domain with a putative signal transducer. *Proc. Natl. Acad. Sci. (USA)*, **89**, 2737–41.

Hemmi, S., Bohni, R., Stark, G., DiMarco, F., and Aguet, M. (1994). A novel member of the interferon receptor family complements functionality of the murine interferon-gamma receptor in human cells. *Cell*, **76**, 803–10.

Hibino, Y., Mariano, T.M., Kumar, C.S., Kozak, C.A., and Pestka, S. (1991). Expression and reconstitution of a biologically active mouse interferon gamma receptor in hamster cells. *J. Biol. Chem.*, **266**, 6948–51.

Hibino, Y., Kumar, C.S., Mariano, T.M., Lai, D., and Pestka, S. (1992). Chimeric interferon-gamma receptors demonstrate that an accessory factor required for activity interacts with the extracellular domain. *J. Biol. Chem.*, **267**, 3741–9.

Ho, A. S.-Y., Liu, Y., Khan, T.A., Hsu, D.-H., Bazan, J.F., and Moore, K.W. (1993). A receptor for interleukin-10 is related to interferon receptors. *Proc. Natl. Acad. Sci. (USA)*, **90**, 11267–71.

Jung, V., Rashidbaigi, A., Jones, C., Tischfield, J.A., Shows, T.B., and Pestka, S. (1987). Human chromosomes 6 and 21 are required for sensitivity to human interferon gamma. *Proc. Natl. Acad. Sci. (USA)*, **84**, 4151–5.

Jung, V., Jones, C., Kumar, C.S., Stefanos, S., O'Connell, S., and Pestka, S. (1990). Expression and reconstitution of a biologically active human interferon-gamma receptor in hamster cells. *J. Biol. Chem.*, **265**, 1827–30.

Kumar, C.S., Muthukumaran, G., Frost, L.J., Noe, M., Ahn, Y.H., Mariano, T.M., and Pestka, S. (1989). Molecular characterization of the murine interferon gamma receptor cDNA. *J. Biol. Chem.*, **264**, 17939–46.

Mariano, T.M., Kozak, C.A., Langer, J.A., and Pestka, S. (1987). The mouse immune interferon receptor gene is located on chromosome 10. *J. Biol. Chem.*, **262**, 5812–4.

Merlin, G., Van der Leede, B.-J., Aguet, M., and Dembic, Z. (1989). The human Interferon gamma receptor gene. *J. Interferon Res.*, **9**, 89.

Müller, M., Briscoe, J., Laxton, C., Guschin, D., Ziemleckl, A., Silvennoinen, O., Harpur, A.G., Barbier, G., Witthuhn, B.A., Schindler, C., Pellegrini, S., Wilks, A.F., Ihle, J.N., Stark, G.R., and Kerr, I.M. (1993). The protein tyrosine kinase JAK1 complements a mutant cell line defective in the interferon-alpha/beta and -gamma signal transduction pathways. *Nature*, **366**, 129–35.

Munro, S., and Maniatis, T. (1989). Expression and cloning of the murine interferon-γ receptor cDNA. *Proc. Natl. Acad. Sci. (USA)*, **86**, 9248–52.

Murakami, M., Narazaki, M., Hibi, M., Yawata, H., Yasukawa, K., Hamaguchi, M., Taga, T., and Kishimoto, T. (1991). Critical cytoplasmic region of the interleukin 6 signal transducer gp130 is conserved in the cytokine receptor family. *Proc. Natl. Acad. Sci. (USA)*, **88**, 11349–53.

Pfizenmaier, K., Wiegmann, K., Scheurich, P., Krönke, M., Merlin, G., Aguet, M., Knowles, B.B., and Ucer, U. (1988). High affinity human IFN-gamma-binding capacity is encoded by a single receptor gene located in proximity to c-ros on human chromosome region 6q16 to 6q22. *J. Immunol.*, **141**, 856–60.

Schindler, C., Shuai, K., Prezioso, V.R., and Darnell, J.E., Jr. (1992). Interferon-dependent tyrosine phosphorylation of a latent cytoplasmic transcription factor. *Science*, **257**, 809–13.

Shuai, K., Schindler, C., Prezioso, V.R., and Darnell, J.E., Jr. (1992). Activation of transcription by IFNγ: tyrosine phosphorylation of a 91-kD DNA binding protein. *Science*, **258**, 1808–12.

Soh, J., Donnely, R.O., Kotenko, S., Mariano, T.M., Cook, J.R., Wang, N., Emanuel, S., Schwartz, B., Miki, T., and Pestka, S. (1994). Identification and sequence of an accessory factor required for activation of the human IFN-gamma receptor. *Cell*, **76**, 793–802.

Soh, J., Donnelly, R.J., Mariano, T.M., Cook, J.R., Schwartz, B., and Pestka, S. (1993). Identification of a yeast artificial chromosome encoding an accessory factor for the human interferon γ receptor: Evidence for multiple accessory factors. *Proc. Natl. Acad. Sci. (USA)*, **90**, 8737–41.

Watling, D., Guschin, D., Müller, M., Silvennoinen, O., Witthuhn, B.A., Quelle, F.W., Rogers, N.C., Schindler, C., Stark, G.R., Ihle, J.N., and Kerr, I.M. (1993). Complementation by the protein tyrosine kinase JAK2 of a mutant cell line defective in the interferon-gamma signal transduction pathway. *Nature*, **366**, 166–70.

Robert D. Schreiber and Michael Aguet
Dept. of Pathology,
Washington University School of Medicine,
St. Louis, MO USA
and
Institut fur Molekularbiologie,
Der-Universität Źrich-Hönggerberg Zürich, CH

Leukaemia Inhibitory Factor (LIF)

LIF is a basic and heavily glycosylated monomeric protein. LIF is known by a variety of alternative names and exerts diverse effects upon haemopoietic cells, embryonal stem cells, primitive germ cells, hepatocytes, neurones, adipocytes, myoblasts and osteoblasts. Consistent with its broad biological effects in vitro, elevation of LIF levels in vivo results in a complex pathology including wasting, elevated megakaryocyte and platelet numbers, and aberrant bone deposition. Although LIF is pleiotropic, mice that fail to produce this cytokine as a consequence of targeted disruption of the gene appear to mature normally. Females that fail to produce LIF, although capable of becoming pregnant, do not carry litters to term. The defect in this case is independent of the genotype of the embryo and resides in the inability of embryos to implant correctly in the uterine wall.

■ Alternative names

Differentiation inducing factor (DIF), differentiation stimulating factor (D-factor), differentiation inhibitory activity (DIA), differentiation retarding factor (DRF), cholinergic neuronal differentiating factor (CNDF), human interleukin for DA cells (HILDA), hepatocyte stimulating factor-III (HSF-III), melanocyte derived lipoprotein lipase inhibitor (MLPLI). (Hilton 1992.)

■ LIF protein

LIF is a secreted protein that is heavily glycosylated and quite basic (pl > 9.0) (Hilton 1992). As a native protein produced by mammalian cells, LIF exhibits an apparent molecular weight of 32 000 to 62 000 Da, depending on the source, but deglycosylation reduces this to 20–25 000 Da, similar to the molecular weight predicted from the cDNA sequence (Hilton 1992). Recombinant LIF produced in yeast is hyperglycosylated while that produced in E. coli is not glycosylated (Hilton 1992). The glycosylation of LIF does not appear to affect its biological activity either in vitro or in vivo. LIF contains six cysteine residues each of which are conserved across species. The pattern of disulphide-bond formation has been determined for murine LIF (Nicola et al. 1993): C13–C135, C19–C132, and C61–C164. A similar arrangement is observed for the two disulphide bonds of human oncostatin-M (Rose and Bruce 1991). Although LIF exhibits only a weak primary sequence similarity to oncostatin-M and other cytokines, these proteins are thought to form a family composed of four alpha-helical bundle structures (Bazan 1991).

Native LIF has been purified from a number of sources using conventional chromatography (Hilton 1992). The various purification schemes take advantage of LIF's high isoelectric point and use anion exchange chromatography as an early step (Hilton 1992). More recently recombinant LIF has been produced as a fusion product with glutathione-S transferase and can be readily purified using immobilized glutathione (Gearing et al. 1989).

■ LIF sequence and gene

cDNA molecules encoding human, murine, rat, porcine and ovine LIF have been cloned and sequenced (Hilton 1992). The primary mRNA species is 4.0–5.0 kb in length (Hilton 1992). LIFs from different species exhibit between 78 and 92 per cent amino acid sequence identity (Fig. 1), but while human LIF seems able to act on both human and murine cells, murine LIF is not active on human cells (Layton et al. 1992).

The genes encoding human and murine LIF are located on chromosomes 22q12 and 11A1, respectively (Hilton 1992). They contain three exons and two introns, are similar in structure, and are approximately 6 kbp in length. mRNA species utilizing alternative first exons have been described (Rathjen et al. 1990). These mRNA species give rise to proteins that differ in the first few amino acids of the signal peptide. This difference has been claimed to influence the extracellular location of secreted LIF (Rathjen et al. 1990).

The GenBank database accession numbers for human and mouse LIF cDNA and genomic sequences are X13987, X06381, M63419, S05435, M63420, and S05436.

■ LIF production

LIF is normally undetectable in both mouse and human serum (Hilton 1992). LIF is, however, detectable upon injection of mice with lipopolysaccharide and is found in the serum of patients with florid infections (Hilton 1992; Metcalf 1992). In these situations LIF appears to be produced by a wide range of tissues (Gough et al. 1992). Of particular interest in normal mice is the spike of production observed in the endometrial glands of the uterus at the time of embryo implantation (Bhatt et al. 1991).

Consistent with the multiple primary sources of LIF, many cell lines are capable of producing LIF (Hilton 1992). These include the human bladder carcinoma cell line 5637, human melanoma line SEKI, and the colon carcinoma cell line COLO-16. The following rodent cell lines also produce LIF:

```
              10        20          30  #       40       50        #  60
M   MKVLAAGIVPLLLLVLHWKHGAGSPLP ITPVNATCAIRHPCHGNLMNQIKNQLAQLNGSA
R   MKVLAAGIVP-LLLILHWKHGAGSPLP ITPVNATCAIRHPCHGNLMNQIKSQLAQLNGSA
H   MKVLAAGVVP-LLLVLHWKHGAGSPLP ITPVNATCAIRHPCHNNLMNQIRSQLAQLNGSA
O   MKILAAGVVP-LLLVLHWKPGAGSPLP INPVNATCNTHHPCPSNLMSQIRSQLAQLNGTA
P   MKVLAAGVVP-LLLVLHWKHGAGSPLS ITPVNATCATRHPCHSNLMNQIKNQLAHVNSSA


          70        80      #  90     #  100       110           #
M   NALFISYYTAQGEPFPNNVEKLCAPNMTDFPSFHGNGTEKTKLVELYRMVAYLSASLTNI
R   NALFISYYTAQGEPFPNNVDKLCAPNMTDFPPFHANGTEKTKLVELYRMVTYLGASLTNI
H   NALFILYYTAQGEPFPNNLDKLCGPNVTDFPPFHANGTEKAKLVELYRIVVYLGTSLGNI
O   NALFILYYTAQGEPFPNNLDKLCGPNVTDFPPFQPNGTEKVRLVELYRIVAYLGTALGNI
P   NALFILYYTAQGEPFPNNLDKLCGPNVTNFPPFHANGTEKARLVELYRIIAYLGASLGNI


        #          #          150       160       170        180
M   TRDQKVLNPTAVSLQVKLNATIDVMRGLLSNVLCRLCNKYRVGHVDVPPVPDHSDKEAFQ
R   TWDQKNLNPTAVSLQIKLNATTDVMRGLLSSVLCRLCNKYHVGHVDVPCVPDNSSKEAFQ
H   TRDQKILNPSALSLHSKLNATADILRGLLSNVLCRLCSKYHVGHVDVTYGPDTSGKDVFQ
O   TRDQKTLNPTAHSLHSKLNATADTLRGLLSNVLCRLCSKYHVAHVDVAYGPDTSGKDVFQ
P   TRDQRSLNPGAVNLHSKLNATADSMRGLLSNVLCRLCNKYHVAHVDVAYGPDTSGKDVFQ


        190       200
M   RKKLGCQLLGTYKQVISVVVQAF
R   RKKLGCQLLGTYKQVISVVVQAF
H   KKKLGCQLLGKYKQIIAVLAQAF
O   KKKLGCQLLGKYKQVMAVLAQAF
P   KKKLGCQLLGKYKQVISVLARAF
```

Figure 1. Alignment of the predicted amino acid sequence of human, murine, rat, porcine, and ovine LIF. The predicted signal sequence is shown by a solid line above the sequence. Conserved cysteine residues are shown in bold type. Potential N-linked glycosylation sites are indicated by #.

Krebs and Ehrlich's ascites cells, L929, T-lymphoid lines (eg. LB-3), and STO fibroblasts (Hilton 1992; Gough *et al.* 1992).

■ *In vitro* biological effects of LIF

The biological effects described for LIF *in vitro* are bewildering in their number and diversity (Hilton 1992; Metcalf 1992). LIF has been shown to induce the differentiation and suppress the clonogenicity of the mouse monocytic leukaemia cell-line M1. LIF also synergistically suppresses the clonogenicity of the human leukaemic cell-lines HL-60 and U937. The induction of M1 differentiation is the assay by which LIF activity is usually quantitated. The concentration of LIF that induces 50 per cent of M1 colonies to differentiate in a semi-solid agar culture is defined as 50 U/ml. Purified native and recombinant LIF have a specific activity of 1 to 2×10^8 U/mg.

A most widespread use of LIF is in the suppression of embryonic stem cell differentiation (Hilton 1992; Metcalf 1992). In the presence of LIF these cells may be manipulated genetically, for example by targeted disruption of genes by homologous recombination, and subsequently re-introduced into the blastocysts of recipient embryos where they contribute to all tissues, including the germline. LIF enhances the survival and proliferation of primitive germ cells (Matsui *et al.* 1991). LIF stimulates acute phase protein synthesis by HepG2 cells but inhibits lipoprotein lipase activity in 3T3-L1 cells (Hilton 1992; Metcalf 1992). It induces a cholinergic phenotype in adrenergic neurones and enhances the survival and proliferation of embryonic sensory neurones and myoblasts (Hilton 1992; Metcalf 1992). LIF stimulates the proliferation of the factor-dependent haemopoietic cell lines DA-1a and THP-1, enhances the generation of megakaryocytes by interleukin-3 and has been suggested to increase the frequency of infection of haemopoietic stem cells with retrovirus—perhaps by stimulating these cells to enter the cell cycle (Hilton 1992; Metcalf 1992).

■ *In vivo* actions of LIF

The pleiotropy and redundancy of LIF's biological effects are well-illustrated by comparing mice in which LIF levels have been artificially elevated (Metcalf 1992) to those that are unable to produce LIF because of disruption of the LIF gene by homologous recombination (Stewart *et al.* 1992).

LIF levels have been elevated in mice by injection of purified recombinant LIF and by engrafting mice with LIF-producing haemopoietic cell (Metcalf 1992). The effects in both cases are similar. There is an almost complete loss of subcutaneous and abdominal fat—resulting in a 30 per cent reduction in weight within three days (Metcalf 1992). Mice have elevated serum calcium levels, and in some cases excessive deposition of new bone and calcification of skeletal muscle, heart, and liver (Metcalf 1992). Symptoms of an ongoing acute-phase response are observed with a decrease in serum albumin concentration and an increase in erythrocyte sedimentation rate (Metcalf 1992). Platelet levels are increased in mice, and also primates, with increased circulating LIF levels, as are the numbers of megakaryocytes and megakaryocyte progenitors in the spleen (Metcalf 1992).

In contrast to the elevation of LIF levels, there appear to be very few detrimental effects of the failure to produce LIF (Stewart *et al.* 1992). Mice lacking a functional LIF gene appear to develop normally and have a normal lifespan. Mice also appear to be histologically normal. The most profound effect, however, is on the implantation of embryos in the uteri. Embryos, irrespective of their capacity to produce LIF, are unable to implant in the uteri of female mice that do not produce this cytokine (Stewart *et al.* 1992). This defect can be corrected by injection of purified LIF into the uterus at the time of implantation (Stewart *et al.* 1992).

■ References

Bazan, J.F. (1991). Neuropoietic cytokines in the hematopoietic fold. *Neuron*, **7**, 197–208.

Bhatt, H., Brunet, L.J., and Stewart, C.L. (1991). Uterine expression of leukaemia inhibitory factor coincides with the onset of blastocyst implantation. *Proc. Natl. Acad. Sci. (USA)*, **88**, 11408–12.

Gearing, D.P., Nicola, N.A., Metcalf, D., Foote, S., Willson, T.A., Gough, N.M., and Williams, R.L. (1989). Production of leukaemia inhibitory factor in *Escherichia coli* by a novel procedure and its use in maintaining embryonic stem cells in culture. *Biotechnology*, **7**, 1157–61.

Gough, N.M., Willson, T.A., Stahl, J., and Brown, M.A. (1992). Molecular biology of the leukaemia inhibitory factor gene. *CIBA Foundation Symposium*, **167**, 24–46.

Hilton, D.J. (1992). LIF: lots of interesting functions. *TIBS*, **17**, 72–6.

Layton, M.J., Cross, B.A., Metcalf, D., Ward, L.D., Simpson, R.J., and Nicola, N.A. (1992). A major binding protein for leukaemia inhibitory factor in normal mouse serum: identification as a soluble form of the cellular receptor. *Proc. Natl. Acad. Sci. (USA)*, **89**, 8616–20.

Matsui, Y., Toksoz, D., Nishikawa, S., Nishikawa, S., Williams, D., Zsebo, K., and Hogan, B.L. (1991). Effect of steel factor and leukaemia inhibitory factor on murine primordial germ cells in culture. *Nature*, **353**, 750–2.

Metcalf, D. (1992). Leukemia inhibitory factor–a puzzling polyfunctional regulator. *Growth Factors*, **7**, 169–73.

Nicola, N.A., Cross, B., and Simpson, R.J. (1993). The disulphide bond arrangement of leukaemia inhibitory factor: homology to oncostatin M and structural implications. *Biochem. Biophy. Res. Comm.*, **190**, 20–6.

Rathjen, P.D., Toth, S., Willis, A., Heath, J.K., and Smith, A.G. (1990). Differentiation inhibiting activity is produced in matrix-associated and diffusible forms that are generated by alternate promoter usage. *Cell*, **62**, 1105–14.

Rose, T.M., and Bruce, A.G. (1991). Oncostatin M is a member of a cytokine family that includes leukaemia-inhibitory factor, granulocyte colony-stimulating factor, and interleukin 6. *Proc. Natl. Acad. Sci. (USA)*, **88**, 8641–5.

Stewart, C.L., Kasoar, P., Brunet, L.J., Bhatt, H., Gadi, I., Kontgen, F., and Abbondanzo, S.J. (1992). Blastocyst implantation depends on maternal expression of leukaemia inhibitory factor. *Nature*, **359**, 76–9.

Douglas J. Hilton:
The Walter and Eliza Hall Institute of Medical Research and The Cooperative Research Centre for Cellular Growth Factors, Victoria, Australia

Oncostatin M (OSM)

OSM is a glycoprotein of approximately 28 kDa which is produced and secreted by activated monocytes and T lymphocytes. It is a member of a structurally and functionally related family of cytokines which includes leukaemia inhibitory factor (LIF), ciliary neurotrophic factor (CNTF), interleukin-6 (IL-6), interleukin-11 (IL-11), and granulocyte-colony stimulating factor (G-CSF). OSM exhibits diverse biological effects on the proliferation and differentiation of a variety of different cell types, and many of these functions are shared with the other family members.

■ OSM protein structure

OSM is a secreted glycoprotein monomer with a M_r of 28 000 Da (Zarling *et al.* 1986). Nucleotide sequence analysis of OSM cDNA and genomic clones (human sequence—GenBank accession numbers M27286, M27287, M27288, M26966; Simian sequence—unpublished), determined that OSM was initially expressed as a 252-amino-acid precursor polypeptide (Malik *et al.* 1989) (Fig.1). The mature form, containing 196 amino acids, results from the processing of the precursor by removal of a 25 N-terminal amino acid leader sequence and a 31-amino-acid C-terminal peptide, and is the predominant form detected in PMA-treated

```
                    -20                  -10                 -1
           M G V L L T Q R T L L S L V L A L L F P S M A S M
                               .  .   .  .   .  .   .  .  .  .

+1                    10                  20                  30
   A A I G S C S K E Y R V L L G Q L Q K Q T D L M Q D T S R L
   .  . M  .  .  .  .  .  .  . M  .  .  .  .  .  .  .  .  .  .  .

                     40                  50                  60
   L D P Y I R I Q G L D V P K L R E H C R E R P G A F P S E E
   .  .  .  .  .  .  . I  .  .  .  .  .  .  .  .  .  .  .  .  .

                     70                  80                  90
   T L R G L G R R G F L Q T L N A T L G C V L H R L A D L E Q
   .  .  .  .  .  .  .  .  .  . D  .  .  .  .  .  .  .  .  .  .

                    100                 110                 120
   R L P K A Q D L E R S G L N I E D L E K L Q M A R P N I L G
   H  .  .  .  .  .  .  .  .  .  .  .  .  .  .  .  .  . V  .  .

                    130                 140                 150
   L R N N I Y C M A Q L L D N S D T A E P T K A G R G A S Q P
   .  .  .  .  .  .  .  .  .  . M T  .  .  .  .  .  .  .  .  .

                    160                 170                 180
   P T P T P A S D A F Q R K L E G C R F L H G Y H R F M H S V
   .  .  .  . T  .  . V  .  .  .  .  . S  .  .  .  .  .  .  .

                    190       196       200                 210
   G R V F S K W G E S P N R S R R H S P H Q A L R K G V R R T
   . Q  .  .  .  .  .  .  .  .  .  .  .  .  .  .  .  .  .  .  .

                    220                 225
   R P S R K G K R L M T R G Q L P R

   .  .  .  .  .  . N  .  .  .  .  .  .
```

Figure 1. Alignment of the amino acid sequences of human (*top*) and simian (*bottom*) OSM. In the simian sequence, residues in common are denoted by a dot and unknown residues at the ends are unmarked. The N-terminal leader signal and the C-terminal peptide which are removed during processing are singly and doubly underlined, respectively.

human U937 histiocytic cells (Malik *et al.* 1989) and CHO cells transfected with OSM human cDNA (Malik *et al.* 1992). Comparison of amino acid sequences encoded by the human and simian OSM cDNAs shows that the two are 94 per cent identical (Fig.1). Within the encoded sequences are two potential *N*-glycosylation sites and sites for *O*-linked glycosylation. In addition, four of the five encoded cysteine residues are involved in two disulphide bonds (Cys6–Cys127 and Cys49–Cys167, the second of which is essential for biological activity (Kallestad *et al.* 1991). OSM is predicted to contain four amphipathic helical domains within a bundle structure similar to that determined for growth hormone (Rose and Bruce 1991). The exceptionally amphipathic C-terminal helix, along with other discontinuous regions within the molecule, have been shown by mutational analyses to be involved in receptor binding (Kallestad *et al.* 1991). There is significant amino acid sequence and structural homology between OSM and the cytokines LIF, CNTF, G-CSF, and IL-6 (Rose and Bruce 1991; Bruce *et al.* 1992).

Monoclonal and polyclonal antibodies have been produced against recombinant human OSM (Radka *et al.* 1992) and synthetic peptides (unpublished), respectively. One monoclonal antibody (OM2) neutralizes the biological activity of OSM. OSM can be purified by affinity chromatography with specific antibodies (Radka *et al.* 1992) or by gel permeation chromatography of the acid-soluble fraction of conditioned medium followed by reverse phase HPLC (Zarling *et al.* 1986). OSM is a highly stable protein and retains its biological activity after heating (56 °C, 1h) and in extremes of pH (2–11) (Zarling *et al.* 1986).

■ OSM gene structure and transcription

The entire human OSM gene, which spans approximately 5 kb, has been cloned and sequenced (GenBank accession numbers M27286, M27287, M27288; unpublished results). OSM is encoded by a mRNA of approximately 2 kb which is generated from three exons (Malik *et al.* 1989). As with many other cytokines, the 3' non-coding region of the OSM mRNA contains several AU rich regions which have been implicated in conferring mRNA instability (Shaw and Kamen 1986). No consensus polyadenylation site (AATAAA) was detected in either the cDNA clones or the downstream genomic sequences. The gene for OSM has been localized to human chromosome 22q12 within approximately 10 kb of the closely related LIF gene (Rose *et al.* 1993a). The OSM and LIF genes are tandemly arranged in the same transcriptional orientation with OSM being the most telomeric. These findings raise the possibility that the expression of OSM and LIF may be co-regulated. The gene structures of OSM and LIF are identical, and coupled with their similarities in amino acid sequence and their proximity in the genome, this suggests evolutionary descent from a common ancestral gene.

Expression of the OSM gene has been detected in monocytes after activation with various inducers, such as phorbol esters and bacterial endotoxin, and in T cells after antigen activation (Bruce *et al.* 1992).

■ Biological functions

OSM exhibits a variety of biological activities depending upon the cell type affected (Bruce *et al.* 1992) (Table 1). With cells derived from a wide variety of solid tissue tumours, OSM inhibits proliferation and induces changes in cellular morphology (Horn *et al.* 1990). Bovine aortic endothelial cells are similarly affected (Brown *et al.* 1990). However, with normal fibroblasts or cells derived from AIDS-related Kaposi's sarcoma, OSM acts as a mitogen to stimulate growth and proliferation (Bruce *et al.* 1992). OSM also has potent effects on cellular differentiation. OSM induces the differentiation of leukaemic cells to macrophage-like cells, and causes the synthesis of acute phase proteins in hepatic cells and vasoactive intestinal peptide in neuroneeal cells (Bruce *et al.* 1992). In both fibroblasts and endothelial cells, OSM also induces the expression of plasminogen activator (Bruce *et al.* 1992). In contrast, OSM inhibits the differentiation of totipotent embryonic stem cells (Rose *et al.* 1993b). The majority of the biological functions of OSM are shared with a number of cytokines, including LIF, IL-6, G-CSF, CNTF, and IL-11. Table 2 summarizes structural and functional similarities between OSM and other cytokines which appear to have evolved from the same ancestral gene (Bruce *et al.* 1992). The similarities in biological function of these cytokines can be explained, in most cases, by the utilization of common receptor subunits (Bruce *et al.* 1992). Since the extent of the overlap of activities between these cytokines has not been fully defined, OSM may share additional biological functions with other members of this cytokine family. The biological activites attributed to OSM *in vitro* suggest potential roles in haemopoiesis, inflammation, cholesterol regulation, and embryonic development *in vivo* (Bruce *et al.* 1992). Although no studies concerning the pathology of OSM are as yet available, the similarities with LIF would suggest, at least in some instances, common pathologies.

Table 1. *In vitro* activities of OSM

Inhibition of growth
Solid tissue tumour cells
Endothelial cells

Stimulation of growth
AIDS-related Kaposi's sarcoma cells
Fibroblasts

Induction of differentiation
Haematopoietic tumour cells
Hepatic cells
Fibroblasts
Endothelial cells
Neuronal cells

Inhibition of differentiation
Embryonal stem cells

Table 2. Similarities with OSM

Factor	Gene structure	Protein sequence	Secondary structure	Receptor subunit	Biological activity
LIF	++	+++	++	+	++
CNTF	+	++	+	+	++
IL-6	+	+	+	+	++
G-CSF	+	+++	+	−	+
IL-11	+	+/−	+	+	++
IL-7	+	+	+	−	?
GH	+	+/−	++	−	?
PRL	+	+/−	+	−	?
EP	+	+/−	+	−	?

Interleukin-11 (IL-11), interleukin-7 (IL-7), growth hormone (GH), prolactin (PRL), erythropoietin (EP) (Bruce *et al.* 1992).

■ References

Brown, T.J., Rowe, J.M., Shoyab, M., and Gladstone, P. (1990). Oncostatin M: a novel regulator of endothelial cell properties. In *Molecular biology of cardiovascular system*. UCLA Symposia on Molecular and Cellular Biology (new series) (ed. R. Roberts and M.D. Schneider), pp. 195–206. Wiley-Liss, New York.

Bruce, A.G., Linsley, P.S., and Rose, T.M. (1992). Oncostatin M. *Prog. Growth Factor Res.*, **4**, 157–70.

Horn, D., Fitzpatrick, W.C., Gompper, P.T., Ochs, V., Bolton-Hansen, M., Zarling, J., Malik, N., Todaro, G.J., and Linsley, P.S. (1990). Regulation of cell growth by recombinant oncostatin M. *Growth Factors*, **22**, 157–65.

Kallestad, J.C., Shoyab, M., and Linsley, P.S. (1991). Disulphide bond assignment and identification of regions required for functional activity of oncostatin M. *J. Biol. Chem.*, **266**, 8940–5.

Malik, N., Kallestad, J.C., Gunderson, N.L., Austin, S.C., Neubauer, M.G., Ochs, V., Marquardt, H., Zarling, J.M., Shoyab, M., Wei, C.-M., Linsley, P.S., and Rose T.M. (1989). Molecular cloning, sequence analysis, and functional expression of a novel growth regulator, oncostatin M. *Mol. Cell. Biol.*, **9**, 2847–53.

Malik, N., Graves, D., Shoyab, M., and Purchio, A.F. (1992). Amplification and expression of heterologous oncostatin M in Chinese hamster ovary cells. *DNA Cell Biol.*, **11**, 453–9.

Radka, S.F., Naemura, J.R., and Shoyab, M. (1992). Abrogation of the antiproliferative activity of oncostatin M by a monoclonal antibody. *Cytokine*, **4**, 221–6.

Rose, T.M., and Bruce, A.G. (1991). Oncostatin M is a member of a cytokine family that includes leukaemia-inhibitory factor, granulocyte colony-stimulating factor, and interleukin 6. *Proc. Natl. Acad. Sci. (USA)*, **88**, 8641–5.

Rose, T.M., Lagrou, M.J., Fransson, I., Werelius, B., Delattre, O., Thomas, G., de Jong, P.J., Todaro, G.J., and Dumanski, J.P. (1993*a*). The genes for oncostatin M (OSM) and leukaemia inhibitory factor (LIF) are tightly linked on human chromosome 22. *Genomics*, **17**, 136–40.

Rose, T.M., Weiford, D., Gunderson, N., and Bruce, A.G. (1993*b*). Oncostatin M (OSM) and leukaemia inhibitory factor (LIF) are tightly linked on human chromosome 22. *Cytokine*, (In press.)

Shaw, G., and Kamen, R. (1986). A conserved AU sequence from the 3′ untranslated region of GM-CSF mRNA mediates selective mRNA degradation. *Cell*, **46**, 659–67.

Zarling, J.M, Shoyab, M., Marquardt, H., Hanson, M.B., Lioubin, M.N., and Todaro, G.J. (1986). Oncostatin M: a growth regulator produced by differentiated histiocytic lymphoma cells. *Proc. Natl. Acad. Sci. (USA)*, **83**, 9739–43.

Timothy M. Rose:
Fred Hutchinson Cancer Research Center, and
Department of Pathobiology,
School of Public Health,
University of Washington,
Seattle, WA, USA

A. Gregory Bruce:
Bristol-Myers Squibb,
Pharmaceutical Research Institute,
Seattle, WA, USA

Receptors for Leukaemia Inhibitory Factor (LIF) and Oncostatin M (OSM)

The leukaemia inhibitory factor (LIF) receptor consists of a ligand-specific α-subunit that confers low-affinity binding and a β-subunit, gp130, that by itself does not bind LIF but endows the receptor complex with the capacity to bind LIF with higher affinity. Gp130 by itself binds oncostatin M (OSM) with low affinity. The combination of gp130 and LIFR-α also binds OSM with high affinity and CNTF with low affinity and signals in response to LIF, OSM and CNTF. A high-affinity OSM-specific receptor complex that uses gp130 but does not bind LIF has been identified on several human cell lines. Gp130 is also a component of the signalling complexes for interleukin-6 (IL-6), ciliary neurotrophic factor (CNTF) and interleukin-11 (IL-11). The LIF receptor-α/gp130 complex interacts with CNTF and the CNTF receptor α-subunit to form a functional high-affinity CNTF receptor complex. LIF receptor-α and gp130 are very similar glycoproteins that are members of the haemopoietin receptor family. Neither chain of the LIF receptor complex contains recognizable elements associated with signalling and the complex is presumed to associate with other cytoplasmic signalling molecules. Distinct domains of the cytoplasmic portions are required for proliferation in haemopoietic cells and gene activation in hepatic and neuronal cells. The LIF receptor is expressed on haemopoietic cells of the monocytic and lymphoid lineages and non-haemopoietic cells including embryonic stem cells, hepatocytes, neurones and osteoblasts.

■ The LIF receptor α-subunit protein

The α-subunit of the human LIF receptor is produced as a 1097 amino acid precursor with a predicted 44-amino-acid leader sequence (Gearing *et al.* 1991). There is a single predicted transmembrane segment of 26 amino acids, an extracellular domain of 789 amino acids and a relatively long cytoplasmic domain of 238 amino acids. The predicted molecular weight of the protein is 111 000 Da but the observed molecular weight on SDS-PAGE gels is 200 000 Da due to extensive N-glycosylation at several of the predicted 20 N-glycosylation sites. Distal to the transmembrane region the extracellular region comprises two haemapoietin receptor domains separated by an immunoglobulin-like domain and three membrane-proximal fibronectin type III domains. The haemopoietin domains consist of sequence elements conserved amongst members of the haemopoietin receptor family including several cysteine residues and the element Trp-Ser-X-Trp-Ser (Bazan 1990). The structure of this domain has been solved for the growth hormone receptor and shown to form two β-barrels each consisting of seven antiparallel β-sheets (deVos *et al.* 1992). The mouse LIF receptor α-chain precursor is 1093 amino acids long and shows 80 per cent identity with the human protein (Gearing *et al.* 1991). The cytoplasmic domain is the most conserved portion of the human and mouse alpha chains. The transmembrane domains of the human, mouse and rat LIF receptor α-chains are identical and contain a single cysteine residue.

■ Associated receptor subunits

The β-subunit of the human LIF receptor (Gearing *et al.* 1992) (also known as gp130) was first described as a component of the IL-6 receptor (Taga *et al.* 1989; Hibi *et al.* 1990). The combination of α and β subunits, but not either subunit alone, confers proliferative responsiveness upon transfected BAF-B03 haemopoietic cells to LIF, OSM, and, more weakly, to CNTF (Gearing *et al.* 1994). The β-subunit is a 918-amino-acid glycoprotein that contains a single haemopoietin domain in the extracellular region, a leader sequence of 22 amino acids, a single 22-amino-acid transmembrane region and a 277-amino-acid cytoplasmic tail (Hibi *et al.* 1990). The predicted molecular weight of the protein is 101 000 Da and on SDS-PAGE gels has an apparent molecular weight of 130 000 Da, the difference being due to glycosylation. Neither the α-subunit nor β-subunit have hallmark sequences associated with tyrosine kinase activity but instead contain at least three sequence motifs associated with signalling. Two membrane proximal motifs, which are conserved in several cytokine receptors, are minimally required for proliferative signalling in haemopoietic cells and a third motif, more centrally located, is required in addition to the membrane proximal motifs for gene activation in hepatocytes and neurones (Murakami *et al.* 1991; Baumann *et al.* 1994). Like the α-chain, the β-chain (gp130) contains cysteine residues (two) in its transmembrane region although there is no evidence yet for disulphide linked dimers between the LIF receptor α and β chains. The αβ-heterodimer forms a high-affinity CNTF receptor complex in combination with CNTF and the CNTF receptor α-subunit (Gearing *et al.* 1994; Baumann *et al.* 1993; Davis *et al.* 1991, 1993a; Ip *et al.* 1992; Stahl *et al.*

1993). The β-subunit is a component of functional IL-6 and IL-11 receptors (Taga et al. 1989; Hibi et al. 1990; Yin et al. 1993). In the case of the IL-6 receptor the β-subunit forms a disulphide-linked homodimer when complexed with IL-6 and the IL-6 receptor α-subunit and converts the α-subunit to higher affinity binding (Gearing et al. 1992; Hibi et al. 1990; Murakami et al. 1993). Mouse gp130 is 77 per cent homologous to human gp130 (Saito et al. 1992).

■ LIF receptor gene and expression

The LIF receptor α-subunit is located on human chromosome 5p12–13 and mouse chromosome 15, clustered with the genes for three other cytokine receptors, the IL-7 receptor, growth hormone receptor and prolactin receptor (Gearing et al. 1993a). There does not appear to be physiological relevance to this cluster. The gene for the β-chain, gp130, has been localized to two human chromosomes, 5 and 17, suggesting that at least one gene is close to the LIF receptor α-chain gene (Kidd et al. 1992). Multiple transcripts for the α-chain gene have been described in human placenta (Gearing et al. 1991). The receptors described above constitute one of the transcripts in each system. In mouse, an mRNA encoding a soluble form of the α-chain is expressed in liver cells in which an alternatively spliced exon between the second and third fibronectin type III domains results in premature termination (Gearing et al. 1991). The soluble receptor binds LIF with the same affinity as the membrane-bound version (Layton et al. 1992).

The LIF receptor α-chain is expressed at low levels in many tissues, including placenta, liver, bone marrow, monocytic cell lines and ubiquitously throughout the brain (Gearing et al. 1991, 1993b; Ip et al. 1993). Expression of the β-chain is similarly ubiquitous (Saito et al. 1992; Ip et al. 1993). The regulation of the α- and β-chain genes does not appear to be coordinated. High- and low-affinity LIF receptors have been detected on a wide variety of cell types, including haemopoietic, hepatic, neuronal, osteoblastic, placental and embryonic stem cells (Hilton and Nicola 1992; Hilton et al. 1988, 1991; Godard et al. 1992).

■ Binding characteristics of LIF receptors

LIF binds to isolated α-chains with high specificity but low affinity ($K_D \sim 1$nM) (Gearing et al. 1991). The isolated β-chain shows very low binding capacity for LIF ($K_D < 100$nM), binds OSM with low affinity ($K_D \sim 10$nM) and, in combination with the α-subunit, binds LIF and OSM with higher affinity ($K_D \sim 0.3$nM and ~ 1nM respectively) (Gearing et al. 1992) and CNTF with very low-affinity ($K_D < 10$nM) (Gearing et al. 1994). The α-β complex, but not either subunit alone, is internalized rapidly at 37 °C. The α-β complex cooperates with the low-affinity CNTF receptor (Davis et al. 1991) to form a high-affinity CNTF receptor complex (Gearing et al. 1994; Baumann et al. 1993; Ip et al. 1992; Davis et

al. 1993a). The LIF receptor α-chain forms a high-affinity CNTF receptor complex with the CNTF receptor α-chain in the absence of the β-chain (Gearing et al. 1994). As a consequence of the sharing of both the α- and β-receptor subunits, LIF, OSM, IL-6, and CNTF compete for binding of high-affinity receptors in human and murine cells where appropriate combinations of α- and β-chains are expressed (Gearing et al. 1992; Gearing et al. 1994). Cross-regulation of the various receptor types may also occur at the transcriptional level (Yamaguchi et al. 1992).

Soluble forms of each of the receptor components have been described (Gearing et al. 1991; Taga et al. 1989; Hibi et al. 1990; Layton et al. 1992; Mereau et al. 1993; Narazaki et al. 1993; Mllberg et al. 1993; Honda et al. 1992; Davis et al. 1993b). In the case of the LIF receptor α-chain the soluble receptor is antagonistic of LIF activity. Similarly, soluble gp130 molecules are antagonistic of IL-6, LIF, OSM and CNTF action. In each case, the soluble receptor would prevent the formation of cytoplasmic domain dimers and hence signalling would be blocked. By contrast, soluble versions of IL-6 and CNTF receptor alpha chains are agonists of their respective cognate's action (Fig. 1), acting more like cytokine presentation molecules. The action of the soluble versions of the IL-6 and CNTF receptor α-chains is analogous to the function performed by the p40 subunit of IL-12 (Gearing and Cosman 1991).

■ Signalling mechanisms

The α-chain, or the β-chain, in isolation is unable to signal in response to LIF, OSM, or CNTF. The α–β complex is required for signalling in hepatic and haemopoietic cells (Gearing et al. 1994; Baumann et al. 1993). The concentration of cytokine required for stimulation of the α–β complex on hematopoietic cells corresponds to the affinity of binding (Gearing et al. 1994). The LIFR α-chain/CNTF receptor α-chain complex binds CNTF with high affinity yet does not signal. In hepatic cells expressing endogenous gp130 the majority of the cytoplasmic domain of the LIF receptor α-chain is required for signalling (Baumann et al. 1994). Using chimeric receptor complexes, the regions of the cytoplasmic domains required for signalling in hepatic and neuronal cells have been localized to similar sequences mid-way between the transmembrane and C-termini of both α- and β-chains (α-chain amino acids 140–150, β-chain, 109–133) (Baumann et al. 1994). These domains are required in addition to the membrane-proximal domains needed for proliferation (amino acids ~ 1–30). Although tyrosine phosphorylation of both chains and several cytoplasmic proteins follows receptor activation, specific activation of a particular tyrosine kinase by the high-affinity LIF receptor complex has not been reported. Stimulation of the LIF receptor complex has variously been associated with weak activation of mitogen-activated protein (MAP) kinases, phosphorylation of heat shock protein 27, and activation of protein kinase C (Thoma et al. 1994; Selvakumaran et al. 1992; Michishita et al. 1991; Hoffmann-Liebermann and Liebermann 1991; Kalberg et al. 1993; Lord et al. 1991).

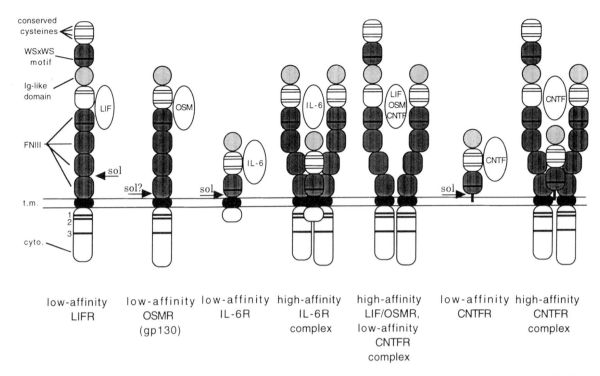

Figure 1. Cartoon structure of the cloned LIF receptor complex and its relationship to the cloned receptors for LIF, OSM, CNTF, and IL-6.

■ OSM-specific receptors

OSM binds to a separate class of receptors to those that are shared with LIF (Thoma *et al.* 1994; Linsley *et al.* 1989; Gearing and Bruce 1992). The OSM specific receptor complex binds OSM with high affinity, does not bind LIF, and involves the common β-subunit, gp130 (Liu *et al.* 1992). Specific low-affinity receptors for OSM have been described but it is not clear whether they are identical to gp130 or result from the expression of another binding component. OSM-specific receptors are expressed on a wide variety of cell types, including endothelial cells, hepatic cells and many human tumour cell lines. Activation of the OSM-specific receptor complex results in phosphorylation of p62[yes] (Liu *et al.* 1992; Schieven *et al.* 1992; Amaral *et al.* 1993) and mitogen-activated protein (MAP) kinases (Thoma *et al.* 1994).

■ References

Amaral, M.C., Miles, S., Kumar, G., and Nel, A.E. (1993). Oncostatin-M stimulates tyrosine protein phosphorylation in parallel with the activation of p42MAPK/ERK-2 in Kaposi's cells. Evidence that this pathway is important in Kaposi cell growth. *J. Clin. Invest.*, **92**, 848–57.

Baumann, H., Symes, A.J., Comeau, M.R., Morella, K.K., Wang, Y., Corpus, L., Friend, D., Ziegler, S.F., Fink, J.S., and Gearing, D.P. (1994). Multiple regions within the cytoplasmic domains of the leukaemia inhibitory factor receptor and gp130 cooperate in signal tansduction in hepatic and neuronal cells. *Mol. Cell. Biol.*, **14**, 138–46.

Baumann, H., Ziegler, S.F., Mosley, B., Morella, K.K., Pajovic, S., and Gearing, D.P. (1993). Reconstitution of the response to leukaemia inhibitory factor, oncostatin M, and ciliary neurotrophic factor in hepatoma cells. *J. Biol. Chem.*, **268**, 8414–7.

Bazan, J.F. (1990). Structural design and molecular evolution of a cytokine receptor superfamily. *Proc. Natl. Acad. Sci. (USA)*, **87**, 6934–8.

Davis, S., Aldrich, T.H., Valenzuela, D.M., Wong, V., Furth, M.E., Squinto, S.P., and Yancopoulos, G.D. (1991). The receptor for ciliary neurotrophic factor. *Science*, **235**, 59–63.

Davis, S., Aldrich, T.H., Stahl, N., Pan, L., Taga, T., Kishimoto, T., Ip, N.Y., and Yancopoulos, G.D. (1993a). LIFR beta and gp130 as heterodimerizing signal transducers of the tripartite CNTF receptor. *Science*, **260**, 1805–8.

Davis, S., Aldrich, T.H., Ip, N.Y., Stahl, N., Scherer, S., Farrugella, T., DiStefano, P.S., Curtis, R., Panayotatos, N., Gascan, H., Chevalier, S., and Yancopoulos, G.D. (1993b). Released form of CNTF receptor alpha component as a soluble mediator of CNTF responses. *Science*, **259**, 1736–9.

deVos, A.M., Ultsch, M., and Kossiakoff, A.A. (1992). Human growth hormone and extracellular domain of its receptor: crystal structure of the complex. *Science*, **255**, 306–12.

Gearing, D.P., and Bruce, G.A. (1992). Oncostatin M binds the high-affinity leukaemia inhibitory factor receptor. *New Biol.*, **4**, 61–5.

Gearing, D.P., and Cosman, D. (1991). Homology of the p40 subunit of natural killer cell stimulatory factor (NKSF) with the extracellular domain of the interleukin-6 receptor [letter]. *Cell*, **66**, 9–10.

Gearing, D.P., Thut, C.J., VandeBos, T., Gimpel, S.D., Delaney, P.B., King, J., Price, V., Cosman, D., and Beckmann, M.P. (1991). Leukaemia inhibitory factor receptor is structurally related to the IL-6 signal transducer, gp130. *EMBO J.*, **10**, 2839–48.

Gearing, D.P., Comeau, M.R., Friend, D.J., Gimpel, S.D., Thut, C.J., McGourty, J., Brasher, K.K., King, J.A., Gillis, S., Mosley, B., Ziegler, S.F., and Cosman, D. (1992). The IL-6 signal transducer, gp130: an oncostatin M receptor and affinity converter for the LIF receptor. *Science*, **255**, 1434–7.

Gearing, D.P., Druck, T., Huebner, K., Overhauser, J., Gilbert, D.J., Copeland, N.G., and Jenkins, N.A. (1993). The leukaemia inhibitory factor receptor [LIFR] gene is located within a cluster of cytokine receptor loci on the mouse chromosome 15 and human chromosome 5p12–13. *Genomics*, **18**, 148–50.

Gearing, D.P., Ziegler, S.F., Comeau, M.R., Friend, D.J., Thoma, B., Cosman, D., Park, L., and Mosley, B. (1994). Proliferative responses and binding properties of hemopoietic cells transfected with low affinity receptors for leukaemia inhibitory factor, oncostatin M and ciliary neurotrophic factor. *Proc. Natl. Acad. Sci. (USA)*, **91**, 1119–23.

Gearing, D.P., Comeau, M.R., Friend, D., and Park, L.S. (1993*b*). Unpublished results.

Godard, A., Heymann, D., Raher, S., Anegon, I, Peyrat, M.A., Le-Mauff, B., Mouray, E., Gregoire, M., Virdee, K., Soulillou, J.-P., Moreau, J.-F., and Jacques, Y. (1992). High and low affinity receptors for human interleukin for DA cells/leukaemia inhibitory factor on human cells. Molecular characterization and cellular distribution. *J. Biol. Chem.*, **267**, 3214–22.

Hibi, M., Murakami, M., Saito, M., Hirano, T., Taga, T., and Kishimoto, T. (1990). Molecular cloning and expression of an IL-6 signal transducer, gp130. *Cell*, **63**, 1149–57.

Hilton, D.J., and Nicola, N.A. (1992). Kinetic analyses of the binding of leukaemia inhibitory factor to receptor on cells and membranes and in detergent solution. *J. Biol. Chem.*, **267**, 10238–47.

Hilton, D.J., Nicola, N.A., and Metcalf, D. (1988). Specific binding of murine leukaemia inhibitory factor to normal and leukemic monocytic cells. *Proc. Natl. Acad. Sci. (USA)*, **85**, 5971–5.

Hilton, D.J., Nicola, N.A., and Metcalf, D. (1991). Distribution and comparison of receptors for leukaemia inhibitory factor on murine haemopoietic and hepatic cells. *J. Cell. Physiol.*, **146**, 207–15.

Hoffmann-Liebermann, B., and Liebermann, D.A. (1991). Interleukin-6- and leukaemia inhibitory factor-induced terminal differentiation of myeloid leukaemia cells is blocked at an intermediate stage by constitutive c-myc. *Mol. Cell. Biol.*, **11**, 2375–81.

Honda, M., Yamamoto, S., Cheng, M., Yasukawa, K., Suzuki, H., Saito, T., Osugi, Y., Tokunaga, T., and Kishimoto, T. (1992). Human soluble IL-6 receptor: its detection and enhanced release by HIV infection. *J. Immunol.*, **148**, 2175–80.

Ip, N.Y., Nye, S.H., Boulton, T.G., Davis, S., Taga, T., Li, Y., Birren, S.J., Yasukawa, K., Kishimoto, T., Anderson, D.J., Stahl, N., and Yancopoulos, G.D. (1992). CNTF and LIF act on neuronal cells *via* shared signaling pathways that involve the IL-6 signal transducing receptor component gp130. *Cell*, **69**, 1121–32.

Ip, N.Y., McClain, J., Barrezueta, N.X., Aldrich, T.H., Pan, L., Li, Y., Wiegand, S.J., Friedman, B., Davis, S., and Yancopoulos, G.D. (1993). The alpha component of the CNTF receptor is required for signaling and defines potential CNTF targets in the adult and during development. *Neuron*, **10**, 89–102.

Kalberg, C., Yung, S.Y., and Kessler, J.A. (1993). The cholinergic stimulating effects of ciliary neurotrophic factor and leukaemia inhibitory factor are mediated by protein kinase C. *J. Neurochem.*, **60**, 145–52.

Kidd, V.J., Nesbitt, J.E., and Fuller, G.M. (1992). Chromosomal localization of the IL-6 receptor signal transducing subunit, gp130 (IL6ST). *Somat. Cell. Mol. Gen.*, **18**, 477–83.

Layton, M.J., Cross, B.A., Metcalf, D., Ward, L.D., Simpson, R.J., and Nicola, N.A. (1992). A major binding protein for leukaemia inhibitory factor in normal mouse serum: identification as a soluble form of the cellular receptor. *Proc. Natl. Acad. Sci. (USA)*, **89**, 8616–20.

Linsley, P.S., Bolton-Hanson, M., Horn, D., Malik, N., Kallestad, J.C., Ochs, V., Zarling, J.M., and Shoyab M. (1989). Identification and characterization of cellular receptors for the growth regulator, oncostatin M. *J. Biol. Chem.*, **264**, 4282–9.

Liu, J., Modrell, B., Aruffo, A., Marken, J.S., Taga, T., Yasukawa, K., Murakami, M., Kishimoto, T., and Shoyab, M. (1992). Interleukin-6 signal transducer gp130 mediates oncostatin M signaling. *J. Biol. Chem.*, **267**, 16763–6.

Lord, K.A., Abdollahi, A., Thomas, S.M., DeMarco, M., Brugge, J.S., Hoffmann-Liebermann, B., and Liebermann, D.A. (1991). Leukaemia inhibitory factor and interleukin-6 trigger the same immediate early response, including tyrosine phosphorylation, upon induction of myeloid leukaemia differentiation. *Mol. Cell. Biol.*, **11**, 4371–9.

Mereau, A., Grey, L., Piquet-Pellorce, C., and Heath, J.K. (1993). Characterization of a binding protein for leukaemia inhibitory factor localized in extracellular matrix. *J. Cell Biol.*, **122**, 713–9.

Michishita, M., Satoh, M., Yamaguchi, M., Hirayoshi, K., Okuma, M., and Nagata, K. (1991). Phosphorylation of the stress protein hsp27 is an early event in murine myelomonocytic leukemic cell differentiation induced by leukaemia inhibitory factor/D-factor. *Biochem. Biophys. Res. Commun.*, **176**, 979–84.

Müllberg, J., Schooltink, H., Stoyan, T., Günther, M., Graeve, L., Buse, G., Mackiewicz, A., Heinrich, P.C., and Rose-John, S. (1993). The soluble interleukin-6 receptor is generated by shedding. *Eur. J. Immunol.*, **23**, 473–80.

Murakami, M., Narazaki, M., Hibi, M., Yawata, H., Yasukawa, K., Hamaguchi, M., Taga, T., and Kishimoto, T. (1991). Critical cytoplasmic region of the interleukin 6 signal transducer gp130 is conserved in the cytokine receptor family. *Proc. Natl. Acad. Sci. (USA)*, **88**, 11349–53.

Murakami, M., Hibi, M., Nakagawa, N., Nakagawa, T., Yasukawa, K., Yamanishi, K., Taga, T., and Kishimoto, T. (1993). IL-6-induced homodimerization of gp130 and associated activation of a tyrosine kinase. *Science*, **260**, 1808–10.

Narazaki, M., Yasukawa, K., Saito, T., Ohsugi, Y., Fukui, H., Koishihara, Y., Yancopoulos, G.D., Taga, T., and Kishimoto, T. (1993). Soluble forms of the interleukin-6 signal-transducing receptor component gp130 in human serum possessing a potential to inhibit signals through membrane-anchored gp130. *Blood*, **82**, 1120–6.

Saito, M., Yoshida, K., Hibi, M., Taga, T., and Kishimoto, T. (1992). Molecular cloning of a murine IL-6 receptor-associated signal transducer, gp130, and its regulated expression *in vivo*. *J. Immunol.*, **148**, 4066–71.

Schieven, G.L., Kallestad, J.C., Brown, T.J., Ledbetter, J.A., and Linsley, P.S. (1992). Oncostatin M induces tyrosine phosphorylation in endothelial cells and activation of p62^{yes} tyrosine kinase. *J. Immunol.*, **149**, 1676–82.

Selvakumaran, M., Liebermann, D.A., and Hoffmann-Liebermann, B. (1992). Deregulated c-myb disrupts interleukin-6- or leukaemia inhibitory factor-induced myeloid differentiation prior to c-myc: role in leukemogenesis. *Mol. Cell. Biol.*, **12**, 2493–500.

Stahl, N., Davis, S., Wong, V., Taga, T., Kishimoto, T., Ip, N.Y., and Yancopoulos, G.D. (1993). Cross-linking identifies leukaemia inhibitory factor-binding protein as a ciliary neurotrophic factor receptor component. *J. Biol. Chem.*, **268**, 7628–31.

Taga, T., Hibi, M., Hirata, Y., Yamasaki, K., Yasukawa, K., Matsuda, T., Hirano, T., and Kishimoto, T. (1989). Interleukin-6 triggers the association of its receptor with a possible signal transducer, gp130. *Cell*, **58**, 573–81.

Thoma, B., Bird, T.A., Friend, D.J., Gearing, D.G., and Dower, S.K. (1994). Oncostatin M and leukaemia inhibitory factor trigger overlapping and different signals through partially shared receptor complexes. *J. Biol. Chem*. **269**, 6215–22.

Yamaguchi, M., Michishita, M., Hirayoshi, K., Yasukawa, K., Okuma, M., and Nagata, K. (1992). Down-regulation of interleukin 6 receptors of mouse myelomonocytic leukemic cells by leukaemia inhibitory factor. *J. Biol. Chem*., **267**, 22035–42.

Yin, T., Taga, T., Tsang, M.L., Yasukawa, K., Kishimoto, T., and Yang,

Y.-C. (1993). Involvement of IL-6 signal transducer gp130 in IL-11-mediated signal transduction. *J. Immunol*., **151**, 2555–61.

David P. Gearing:
Department of Molecular Biology,
Systemix,
3155 Porter Drive,
Palo Alto, CA 94304, USA

Ciliary Neurotrophic Factor (CNTF)

CNTF was first described as a neurotrophic activity that could promote the survival of embryonic chick ciliary neurones. Most of the currently defined actions of CNTF remain limited to cells of the nervous system, with much attention focused on the finding that CNTF can prevent motor neurone degeneration in vitro and in a variety of in vivo models of human motor neurone disease. Despite its specificity for cells of the nervous system, CNTF is a member of a subfamily of cytokines which includes leukaemia inhibitory factor (LIF), interleukin-6 (IL-6) and oncostatin M (OSM). CNTF shares signal transducing 'β' receptor components with these more generally acting cytokines. However, CNTF also requires a specificity-determining 'α' receptor component that is largely restricted to the nervous system in its distribution, thus limiting the cellular targets of CNTF action.

■ Mammalian CNTF genes and protein

Genes or cDNAs encoding CNTF have been cloned from rat (GenBank accession number X17457) (Stockli *et al*. 1989), rabbit (GenBank accession number M29828) (Lin *et al*. 1989), and human (GenBank accession numbers 60477–8, 60542, 55889–90) (McDonald *et al*. 1991; Masiakowski *et al*. 1991). The rat and human CNTF proteins are 200 amino acids in length, while the rabbit protein is 199 amino acids long; the three proteins share 152 (~76 per cent) identically placed residues which include only a single conserved cysteine (Fig. 1). The rat, rabbit and human CNTF proteins do not contain consensus secretory signal peptides at their amino termini, and CNTF is not secreted but rather found in the cytoplasm of cells, such as astrocytes (Rudge *et al*. 1992) and schwann cells (Rende *et al*. 1992; Sendtner *et al*. 1992a; Friedman *et al*. 1992), which express CNTF; the mechanism by which CNTF may be released from cells remains unknown. CNTF also lacks consensus glycosylation sites, and CNTF isolated from native sources is similar to active CNTF produced in recombinant form in *E. coli* in size (23–28 kDa) and biological activity (EC_{50}s of both native and recombinant rat CNTF are in the range 2–4 pM when assayed on chick ciliary neurones; human CNTF appears less potent in this assay) (Masiakowski *et al*. 1991). Analysis of the human and rat CNTF genes indicate that both contain a single intron in the same position, between the codons for the 38th and 39th amino acids. A rat CNTF cDNA has been used to map the mouse gene to chromosome 19 (Kaupman *et al*. 1991), while the human gene has been localized to a cyto-genetic band position of 11q12.2 by fluorescence *in situ* hybridization (Giovannini *et al*. 1993).

■ Potential chick homologue of CNTF (known as growth promoting activity [GPA])

A 195-amino-acid protein purified from chick sciatic nerve, using the same ciliary neurone survival assay used to purify mammalian CNTF, proved to have substantial homology to mammalian CNTF (about 50 per cent amino acid identity, see Fig. 1) (Leung *et al*. 1992); this protein was called growth promoting activity (GPA, GenBank accession number M80827). GPA may be the closest chick homologue of CNTF and thus might correspond to chick CNTF. Alternatively, GPA may represent a close relative of CNTF that shares some of its functions and perhaps its receptor components. GPA, like CNTF, also lacks a conventional signal sequence but instead shares considerable amino-terminal homology with the mammalian CNTFs (Leung *et al*. 1992). In contrast to the mammalian CNTFs, however, GPA can apparently be released from human 293 cells transiently over-expressing GPA expression plasmids (Leung *et al*. 1992); thus, mechanisms similar to those which allow for release of GPA from 293 cells may in other cell types allow for the release of the mammalian CNTFs.

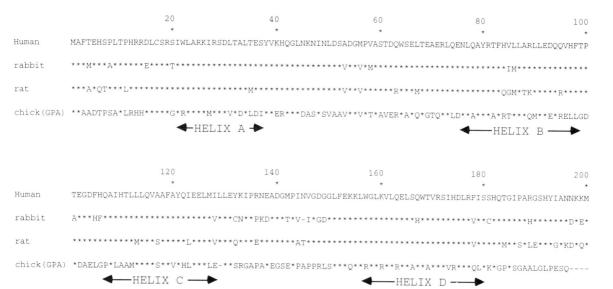

```
                    20              40              60              80              100
                    *               *               *               *               *
Human     MAFTEHSPLTPHRRDLCSRSIWLARKIRSDLTALTESYVKHQGLNKNINLDSADGMPVASTDQWSELTEAERLQENLQAYRTFHVLLARLLEDQQVHFTP

rabbit    ***M***A******E****T*************************************V**V*M*********************************IM************

rat       ***A*QT***L*******************************M****************V**V*****R***M****************QGM*TK*****R*****

chick(GPA)**AADTPSA*LRHH*****G*R****M***V*D*LDI**ER***DAS*SVAAV**V*T*AVER*A*Q*GTQ**LD**A***A*RT***QM**E*RELLGD
           ←——HELIX A—→                                              ←————HELIX B——→


                    120             140             160             180             200
                    *               *               *               *               *
Human     TEGDFHQAIHTLLLQVAAFAYQIEELMILLEYKIPRNEADGMPINVGDGGLFEKKLWGLKVLQELSQWTVRSIHDLRFISSHQTGIPARGSHYIANNKKM

rabbit    A***HF*********************V***CN**PKD***T*V-I*GD*****************H*********V**C******H******D*E*

rat       *************M***S*****L****V***Q***E******AT*************************V****M**S*LE***G*KD*Q*

chick(GPA)*DAELGP*LAAM****S**V*HL***LE-**SRGAPA*EGSE*PAPPRLS***Q**R**R**R**A**A***VR***QL*K*GP*SGAALGLPESQ----
           ←——HELIX C—→                           ←——HELIX D——→
```

Figure 1. Rabbit and rat CNTF amino acid sequences compared to human CNTF; potential chick homologue of CNTF, known as GPA, is also included in the comparison. Identities to the human sequence are indicated by an asterisk, missing residues compared to human indicated by a dash. Approximate location of four predicted alpha helices (Bazan 1991) are indicated.

CNTF as a member of a cytokine subfamily

A predictive structural analysis suggested that CNTF was a member of a cytokine subfamily that included LIF, IL-6, and OSM (Bazan 1991). Though these four factors exhibit minimal primary sequence homology, it was argued that they all shared secondary structural features which linked them and perhaps allowed them all to conform generally to the four-α helix bundle structure first described for growth hormone (Bazan 1991). Elucidation of the CNTF receptor structure supports the notion that CNTF is related to LIF, IL-6, and OSM by demonstrating that all four of these cytokines share the use of 'β' signal transducing receptor components (i.e. LIFRβ and gp130) (Davis et al. 1991; Davis et al. 1993a; Stahl et al. 1993; Baumann et al. 1993). Some of these factors, including CNTF, also require specificity-determining 'α' receptor components. The CNTF 'α' component (CNTFRα) is homologous to the 'α' component (IL-6Rα) used by IL-6 (Davis et al. 1991); however, in contrast to IL-6Rα, CNTFRα is generally restricted to the nervous system in its expression, thus largely limiting the actions of CNTF to cells of the nervous system (Davis et al. 1991; Ip et al. 1993a). LIF utilizes the same 'β' components as does CNTF, but does not require an 'α' component. Thus, while LIF mimics all of CNTF's actions in the nervous system, it also acts on a variety of other cell types throughout the body which do not normally respond to CNTF.

As mentioned above, if GPA does not correspond to chick CNTF, it represents a new member of this cytokine subfamily that is far more closely related to CNTF than to other members of the subfamily. The remarkable conservation of

CNTFRα has led to the suggestion that it might serve as a receptor for additional CNTF-like cytokines (Ip et al. 1993a), perhaps including close relatives such as GPA.

CNTF gene and protein expression

Schwann cells in peripheral nerves produce remarkably high levels of CNTF mRNA and protein (Stockli et al. 1989; Rende et al. 1992; Sendtner et al. 1992a; Friedman et al. 1992). Despite early results to the contrary, recent work has demonstrated that CNTF mRNA (as well as transcripts encoding its various receptor components) is also expressed throughout the adult brain, in skeletal muscle, and in the embryo (Ip et al. 1993a). Though these levels of CNTF mRNA are much lower than those observed in adult peripheral nerve, they could clearly result in physiologically relevant concentrations of this extremely potent factor. Recent findings also provide support for a role for CNTF as an injury factor both in the central and peripheral nervous systems. Thus, following injury to the central nervous system, reactive astrocytes near the lesion site markedly increase their expression of CNTF mRNA and protein (Ip et al. 1993b). Following peripheral nerve injury, Schwann cells distal to the lesion reduce their synthesis of CNTF mRNA and apparently release their intracellular stores of CNTF protein into the extracellular space (by an unknown mechanism), perhaps making it transiently available to regenerating axons or other cells (Rende et al. 1992; Sendtner et al. 1992a; Friedman et al. 1992). Skeletal muscle distal to the injured peripheral nerve simultaneously upregulates its expression

Table 1. Biological actions of CNTF

1. In the nervous system.
Neuronal cells:
- survival of ciliary neurones, sympathetic neurones, sensory neurones, motor neurones, preganglionic sympathetic neurones, hippocampal neurones and medial septal neurones;
- inhibits proliferation and promotes differentiation of sympathetic neurone precursors;
- promotes 'cholinergic switch' of mature sympathetic neurones.

Glial cells:
- promotes astrocytic differentiation of O2A precursor;
- survival and maturation of oligodendrocytes.

2. Outside the nervous system.
Embryonic stem cells:
- maintains undifferentiated, pluripotential state.

Skeletal muscle:
- express CNTF receptors, but role of CNTF not yet clarified.

All LIF responsive cells (not normally responsive to CNTF):
- will also respond to CNTF if soluble CNTFRα also provided.

---

of CNTFRα, and actually releases a soluble form of this receptor component (Davis et al. 1993b). Interestingly, this soluble form of CNTFRα is apparently capable of combining with CNTF to form a heterodimeric factor that can mimic LIF by acting on cells expressing the appropriate β components (i.e. LIFRβ and gp130) but which lack CNTFRα expression and, thus, do not normally respond to CNTF (Davis et al. 1993b).

■ Biological actions of CNTF

Table 1 summarizes the biological actions of CNTF, which was originally discovered based on its ability to promote the survival of chick parasympathetic motor neurones located in the ciliary ganglion (Adler et al. 1979). CNTF also promotes survival of sympathetic and sensory neurones in peripheral ganglia, as well as the survival of a variety of central nervous system neurones including motor neurones, preganglionic sympathetic spinal cord neurones, hippocampal neurones and medial septal neurones (Ip and Yancopoulos 1992). In addition to its neuronal survival actions, CNTF can also inhibit the proliferation and induce cholinergic properties of sympathetic neurone precursors, promote cholinergic differentiation of mature sympathetic neurones, induce the astrocytic differentiation of glial progenitors, and promote survival of oligodendrocytes (Ip and Yancopoulos 1992; Louis et al. 1993). CNTF has also been shown to act on a few cellular targets outside of the nervous system. Thus, CNTF mimics LIF's ability to maintain the undifferentiated state of embryonic stem cells, which have been shown to express functional CNTF receptors (Conover et al. 1993). CNTF receptors are also apparently expressed by skeletal muscle (Davis et al. 1991; Ip et al.

1993a; Davis et al. 1993b), although potential actions of CNTF on muscle have not yet been clarified. Furthermore, any cell that normally responds to LIF but not CNTF can be made responsive to CNTF if a soluble form of CNTFRα is also provided (Davis et al. 1993b).

The actions of CNTF on motor neurones in vitro and in vivo have generated a great deal of excitement. In particular, CNTF has dramatic salutory effects on mice homozygous for progressive motor neuroneopathy (pmn) (Sendtner et al. 1992b), a mouse model of human motor neurone disease, providing great promise for the therapeutic potential of CNTF in human diseases such as amyotrophic lateral sclerosis (also known as Lou Gehrig's disease).

■ References

Adler, R., Landa, K.B., Manthorpe, M., and Varon, S. (1979). Cholinergic neuroneotrophic factors: intraocular distribution of trophic activity for ciliary neurones. Science, 204, 1434–6.

Baumann, H., Ziegler, S.F., Mosley, B., Morella, K.K., Pajovic, S., and Gearing, D.P. (1993). Reconstitution of the response to leukaemia inhibitory factor, oncostatin M, and ciliary neurotrophic factor in hepatoma cells. J.Biol.Chem., 268, 8414–7.

Bazan, J.F. (1991). Neuropoietic cytokines in the hematopoietic fold. Neuron, 7, 197–208.

Conover, J.C., Ip, N.Y., Poueymirou, W.T., Bates, B., Goldfarb, M.P., DeChiara, T.M., and Yancopoulos, G.D. (1993). Ciliary neurotrophic factor maintains the pluripotentiality of embryonic stem cells. Development, 119, 559–65.

Davis, S., Aldrich, T.H., Valenzuela, D.M., Wong, V.V., Furth, M.E., Squinto, S.P., and Yancopoulos, G.D. (1991). The receptor for ciliary neurotrophic factor. Science, 253, 59–63.

Davis, S., Aldrich, T.H., Stahl, N., Taga, T., Pan, L., Kishimoto, T., Ip, N.Y., and Yancopoulos, G.D. (1993a). LIFR beta and gp130 as heterodimerizing signal transducers of the tripartite CNTF receptor. Science, 260, 1805–8.

Davis, S., Aldrich, T.H., Ip, N.Y., Stahl, N., Scherer, S., Farruggella, T., DiStefano, P.S., Curtis, R., Panayotatos, N., Gascan, H., Chevalier, S., and Yancopoulos, G.D. (1993b). Released form of CNTF receptor alpha component as a soluble mediator of CNTF responses. Science, 259, 1736–9.

Friedman, B., Scherer, S.S., Rudge, J.S., Helgren, M., Morrisey, D., McClain, J., Wang, D.Y., Wiegand, S.J., Furth, M.E., Lindsay, R.M., and Ip, N.Y. (1992). Regulation of ciliary neurotrophic factor expression in myelin-related Schwann cells in vivo. Neuron, 9, 295–305.

Giovannini, M., Romo, A.J., and Evans, G.A. (1993). Chromosomal localization of the human ciliary neurotrophic factor gene (CNTF) to 11q12 by fluorescence in situ hybridization. Cytogenet. Cell Genet., 63, 62–3.

Ip, N.Y., and Yancopoulos, G.D. (1992). Ciliary neurotrophic factor and its receptor complex. Prog. Growth Factor Res., 4, 139–55.

Ip, N.Y., McClain, J., Barrezueta, N.X., Aldrich, T.H., Pan, L., Li, Y., Wiegand, S.J., Friedman, B., Davis, S., and Yancopoulos, G.D. (1993a). The alpha component of the CNTF receptor is required for signaling and defines potential CNTF targets in the adult and during development. Neuron, 10, 89–102.

Ip, N.Y., Wiegand, S.J., Morse, J., and Rudge, J.S. (1993b). Injury-induced regulation of ciliary neurotrophic factor mRNA in the adult rat brain. Eur. J. Neurosci., 5, 25–33.

Kaupman, K., Sendtner, M., Stockli, K.A., and Jockusch, H. (1991). The gene for ciliary neurotrophic factor maps to murine chromosome 19 and its expression is not affected in the hereditary moto-neurone disease 'Wobbler' of the mouse. *Eur. J. Neurosci.*, **3**, 1182–6.

Leung, D.W., Parent, A.S., Cachianes, G., Esch, F., Coulombe, J.N., Nikolics, K., Eckenstein, F.P., and Nishi, R. (1992). Cloning, expression during development, and evidence for release of a trophic factor for ciliary ganglion neurones. *Neuron*, **8**, 1045–53.

Lin, L.F., Mismer, D., Lile, J.D., Armes, L.G., Butler, E.T. III, Vannice, J.L., and Collins, F. (1989). Purification, cloning, and expression of ciliary neurotrophic factor (CNTF). *Science*, **246**, 1023–5.

Louis, J.C., Magal, E., Takayama, S., and Varon, S. (1993). CNTF protection of oligodendrocytes against natural and tumour necrosis factor-induced death. *Science*, **259**, 689–92.

Masiakowski, P., Liu, H.X., Radziejewski, C., Lottspeich, F., Oberthuer, W., Wong, V., Lindsay, R.M., Furth, M.E., and Panayotatos, N. (1991). Recombinant human and rat ciliary neurotrophic factors. *J. Neurochem.*, **57**, 1003–12.

McDonald, J.R., Ko, C., Mismer, D., Smith, D.J., and Collins, F. (1991). Expression and characterization of recombinant human ciliary neurotrophic factor from *Escherichia coli. Biochim. Biophys. Acta*, **1090**, 70–80.

Rende, M., Muir, D., Ruoslahti, E., Hagg, T., Varon, S., and Manthorpe, M. (1992). Immunolocalization of ciliary neuroneotrophic factor in adult rat sciatic nerve. *Glia*, **5**, 25–32.

Rudge, J.S., Alderson, R.R., Pasnikowski, E., McClain, J., Ip, N.Y., and Lindsay, R.M. (1992). Expression of ciliary neurotrophic factor and the neurotrophins—nerve growth factor brain-derived neurotrophic factor and neurotrophin-3—in cultured rat hippocampal astrocytes. *Eur. J. Neurosci.*, **4**, 459–71.

Sendtner, M., Stockli, K.A., and Thoenen, H. (1992a). Synthesis and localization of ciliary neurotrophic factor in the sciatic nerve of the adult rat after lesion and during regeneration. *J. Cell Biol.*, **118**, 139–48.

Sendtner, M., Schmalbruch, H., Stockli, K.A., Carroll, P., Kreutzberg, G.W., and Thoenen, H. (1992b). Ciliary neurotrophic factor prevents degeneration of motor neurones in mouse mutant progressive motor neuroneopathy. *Nature*, **358**, 502–4.

Stahl, N., Davis, S., Wong, V., Taga, T., Kishimoto, T., Ip, N.Y., and Yancopoulos, G.D. (1993). Cross-linking identifies leukaemia inhibitory factor-binding protein as a ciliary neurotrophic factor receptor component. *J. Biol. Chem.*, **268**, 7628–31.

Stockli, K.A., Lottspeich, F., Sendtner, M., Masiakowski, P., Carroll, P., Gotz, R., Lindholm, D., and Thoenen, H. (1989). Molecular cloning, expression and regional distribution of rat ciliary neurotrophic factor. *Nature*, **342**, 920–3.

George D. Yancopoulos:
Regeneron Pharmaceuticals, Inc.,
777 Old Saw Mill River Road,
Tarrytown, New York 10591, USA

The Ciliary Neurotrophic Factor (CNTF) Receptor Complex

The CNTF receptor complex contains three proteins: a specificity-determining 'α' component (CNTFRα) that directly binds to CNTF, as well as two signal-transducing 'β' components (LIFRβ and gp130) that cannot bind CNTF on their own, but are required to initiate signalling in response to CNTF. CNTFRα expression is generally restricted to the nervous system, and thus largely limits CNTF actions to neurones and glia. The 'β' components are, on the other hand, widely distributed throughout the body. The three components of the CNTF receptor complex are normally unassociated on the cell surface; CNTF induces the stepwise assembly of a complete receptor complex by first binding to CNTFRα, then engaging gp130, and finally recruiting LIFRβ. It is this final step in receptor assembly—heterodimerization of the 'β' components—that initiates intracellular signalling by activating non-receptor tyrosine kinases (the JAK/TYK kinases) associated with the 'β' components. The widely expressed 'β' components used by CNTF are also utilized by cytokines distantly related to CNTF—including leukaemia inhibitory factor (LIF), interleukin-6 (IL-6) and oncostatin M (OSM)—which similarly initiate signalling by inducing either homodimerization or heterodimerization of these 'β' components. Remarkably, CNTFRα is not a transmembrane protein but is instead anchored to the cell surface via a glycosyl-phosphatidylinositol (GPI) linkage, and it can actually participate in forming activated receptor complexes when supplied in soluble form.

■ The CNTFRα component

cDNAs encoding CNTFRα have been cloned from both human (GenBank accession number M73238) (Davis *et al.* 1991) and rat (GenBank accession number S54212) (Ip *et al.* 1993). The cDNAs for human and rat CNTFRα both predict protein precursors of 372 amino acids with putative leader sequences of approximately 20 amino acids and four conserved glycosylation sites. Glycosylation at these sites partially accounts for the difference between the observed molecular weight of CNTFRα on SDS-PAGE gels (~70 kDa) and the molecular weight predicted from the amino acid

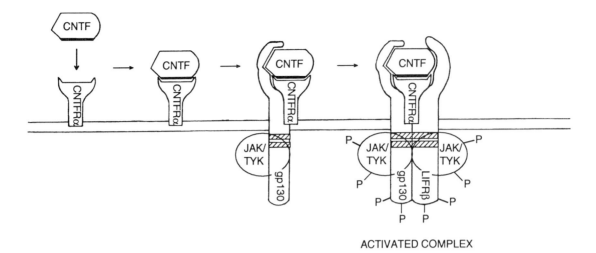

ACTIVATED COMPLEX

Figure 1. Stepwise formation of a CNTF receptor complex. Hatched boxes in β components represent conserved 'box 1' and 'box 2' sequences.

sequence (~40 kDa) (Davis *et al.* 1991). Unlike all other growth factor receptor components, CNTFRa lacks transmembrane and intracytoplasmic domains; instead, it is anchored to the cell membrane *via* a GPI linkage (Davis *et al.* 1991). The closest known relative to CNTFRa is IL-6Ra (~30 per cent amino acid identity) (Davis *et al.* 1991), which is a transmembrane protein. CNTFRa, like the extracellular portion of IL-6Ra, can be divided into two domains: the first is an immunoglobulin-like domain and the second is a 'cytokine receptor' domain (Davis *et al.* 1991); the 'cytokine receptor' domain, which contains features conserved in domains found in many cytokine receptors (including a conserved pattern of cysteine residues, as well as a 'WSXWS' box [Bazan 1990]), is presumed to be responsible for binding to CNTF. The remarkable degree of homology between the human and rat CNTFRas (approximately 94 per cent identity), which is unprecedented among cytokine receptors, has led to the suggestion that CNTFRa might serve as a receptor for additional as yet undiscovered cytokines (Ip *et al.* 1993). The absence of an intracytoplasmic domain for CNTFRa together with the homology of CNTFRa to IL-6Ra, which requires an additional signal-transducing receptor component (gp130) in order to mediate responses to IL-6, provided the first clues that CNTF receptor complex might require additional receptor components.

■ The signal-transducing 'β' components

The two 'β' receptor components in the CNTF receptor complex were originally discovered as receptor components for cytokines related to CNTF. gp130 was cloned as the signal-transducing partner to IL-6Ra (Hibi *et al.* 1990). Human gp130 is synthesized as a 918-amino-acid precursor containing a 22-amino-acid leader; it has an extracellular region of 597 amino acids, a transmembrane region of 22

amino acids, and a cytoplasmic region of 277 amino acids (Hibi *et al.* 1990). The extracellular region of gp130 contains a cytokine receptor domain followed by a 'contactin' domain (containing fibronectin type III modules) (Hibi *et al.* 1990); the intracellular region of gp130 lacks recognizable catalytic domains but instead contains two 'BOXES' in its membrane proximal region which are conserved with sequences found in corresponding portions of other cytokine receptors and are apparently required for signalling by gp130 (Murakami *et al.* 1991). LIFRβ was originally cloned as an LIF binding protein (Gearing *et al.* 1991). Human LIFRβ is synthesized as a 1097 amino acid precursor. Its extracellular region (~810 amino acids) has two tandem cytokine receptor domains, which are followed by a 'contactin' domain, transmembrane region (26 amino acids) and intracellular domain (238 amino acids) notably homologous to the corresponding regions of gp130 (Gearing *et al.* 1991).

■ Sequential assembly of the CNTF receptor complex results in activation of non-receptor tyrosine kinases

In the absence of CNTF, the receptor components comprising the CNTF receptor complex are unassociated on the cell surface (Davis *et al.* 1993a; Stahl and Yancopoulos 1993). CNTF induces step-wise formation of a receptor complex by first binding CNTFRa, then associating with gp130, and finally recruiting LIFRβ (Fig. 1) (Davis *et al.* 1993a; Stahl and Yancopoulos 1993). It is this last step in receptor assembly—which involves heterodimerization between related 'β' components—that is responsible for transducing a signal across the membrane (Davis *et al.* 1993a; Stahl and Yancopoulos 1993). Dimerization of the 'β' components apparently initiates signalling by activating non-receptor tyrosine kinases (the JAK/TYK kinases) associated with the 'β' components *via* the conserved 'BOX' sequences in their membrane-proximal regions (Stahl and Yancopoulos 1993; Stahl *et al*. 1994). Activation of receptor-associated tyrosine kinase activity appears to be required for all subsequent intracellular signalling responses to CNTF (Ip *et al*. 1992). The 'β' components used by CNTF are also utilized by the cytokines related to CNTF (Hibi *et al*. 1990; Gearing *et al*. 1991; Ip *et al*. 1992; Gearing *et al*. 1992; Baumann *et al*. 1993; Stahl *et al*. 1993) which similarly initiate signalling by either inducing homodimerization (of gp130 by IL-6 [Murakami *et al*. 1993]) or heterodimerization (of gp130 with LIFRβ by both LIF and OSM [Davis *et al*. 1993a]) of these 'β' components (Stahl and Yancopoulous 1993; Stahl *et al*. 1994). The coupling of ligand-induced β component dimerization to the activation of cytoplasmic tyrosine kinase activity by these cytokines is very reminiscent of the mechanism of activation of receptor tyrosine kinases, such as those for fibroblast growth factor or epidermal growth factor.

■ Expression of the CNTF receptor components

CNTFRα is the one component of the CNTF receptor complex that uniquely characterizes CNTF responding cells; the largely restricted expression of CNTFRα to cells of the nervous system underlies the discovery of CNTF as a neurotrophic factor as opposed to a more generally acting cytokine. Thus, while gp130 and LIFRβ are widely expressed in many cell types, CNTFRα expression specifically marks CNTF targets (Ip *et al*. 1993). During embryogenesis, CNTFRα is expressed by neuronal and glial precursors (Ip *et al*. 1993). In the adult, it is expressed throughout the central and peripheral nervous systems (Ip *et al*. 1993). While CNTFRα expression can be noted in all known targets of CNTF action (such as ciliary, sympathetic neurones, sensory and motor neurones) it is also found on a variety of neuronal populations that have not been examined for their CNTF responsivity (Ip *et al*. 1993), suggesting that CNTF may act on broader populations of neurones than previously appreciated. Outside of the nervous system, CNTFRα expression has only been noted in embryonic stem cells (CNTF, like LIF, can maintain the undifferentiated phenotype of these cells) (Conover *et al*. 1993) and in skeletal muscle (Davis *et al*. 1991, 1993b; Ip *et al*. 1993). Although the increased CNTFRα expression by skeletal muscle might be involved in directly mediating muscle responses to CNTF, it should be pointed out that the increased CNTFRα expression in response to nerve injury also results in release of a soluble form of CNTFRα from the muscle (Davis *et al*. 1993b); soluble CNTFRα is also constitutively found in the CNS (Davis *et al*. 1993b). Because the soluble form of CNTFRα can actually participate in forming active receptor complexes (it need not be membrane anchored) (Davis *et al*. 1993b), it is possible that it may under certain circumstances combine with CNTF *in vivo* to act on cells that normally respond to LIF (because they express gp130 and LIFRβ) but not to CNTF (because they do not express CNTFRα on their surface).

■ References

Baumann, H., Ziegler, S.F., Mosley, B., Morella, K.K., Pajovic, S., and Gearing, D.P. (1993). Reconstitution of the response to leukaemia inhibitory factor, oncostatin M, and ciliary neurotrophic factor in hepatoma cells. *J. Biol. Chem.*, **268**, 8414–7.

Bazan, J.F. (1990). Structural design and molecular evolution of a cytokine receptor superfamily. *Proc. Natl. Acad. Sci. (USA)*, **87**, 6934–8.

Conover, J.C., Ip, N.Y., Poueymirou, W.T., Bates, B., Goldfarb, M.P., DeChiara, T.M., and Yancopoulos, G.D. (1993). Ciliary Neurotrophic Factor maintains the pluripotentiality of embryonic stem cells. *Development*, **119**, 559–65.

Davis, S., Aldrich, T.H., Valenzuela, D.M., Wong, V.V., Furth, M.E., Squinto, S.P., and Yancopoulos, G.D. (1991). The receptor for ciliary neurotrophic factor. *Science*, **253**, 59–63.

Davis, S., Aldrich, T.H., Stahl, N., Pan, L., Taga, T., Kishimoto, T., Ip, N.Y., and Yancopoulos, G.D. (1993a). LIFR beta and gp130 as heterodimerizing signal transducers of the tripartite CNTF receptor. *Science*, **260**, 1805–8.

Davis, S., Aldrich, T.H., Ip, N.Y., Stahl, N., Scherer, S., Farruggella, T., DiStefano, P.S., Curtis, R., Panayotatos, N., Gascan, H., Chevalier, S., and Yancopoulos, G.D. (1993b). Released form of CNTF receptor alpha component as a soluble mediator of CNTF responses. *Science*, **259**, 1736–9.

Gearing, D.P., Thut, C.J., VandeBos, T., Gimpel, S.D., Delaney, P.B., King, J., Price, V., Cosman, D., and Beckmann, M.P. (1991). Leukemia inhibitory factor receptor is structurally related to the IL-6 signal transducer, gp130. *EMBO J.*, **10**, 2839–48.

Gearing, D.P., Comeau, M.R., Friend, D.J., Gimpel, S.D., Thut, C.J., McGourty, J., Brasher, K.K., King, J.A., Gillis, S., Mosley, B., Ziegler, S.F., and Cosman, D. (1992). The IL-6 signal transducer, gp130: an oncostatin M receptor and affinity converter for the LIF receptor. *Science*, **255**, 1434–7.

Hibi, M., Murakami, M., Saito, M., Hirano, T., Taga, T., and Kishimoto, T. (1990). Molecular cloning and expression of an IL-6 signal transducer, gp130. *Cell*, **63**, 1149–57.

Ip, N.Y., Nye, S.H., Boulton, T.G., Davis, S., Taga, T., Li, Y., Birren, S.J., Yasukawa, K., Kishimoto, T., Anderson, D.J., Stahl, N., and Yancopoulos, G.D. (1992). CNTF and LIF act on neuronal cells via shared signaling pathways that involve the IL-6 signal transducing receptor component gp130. *Cell*, **69**, 1121–32.

Ip, N.Y., McClain, J., Barrezueta, N.X., Aldrich, T.H., Pan, L., Li, Y., Wiegand, S.J., Friedman, B., Davis, S., and Yancopoulos, G.D. (1993). The alpha component of the CNTF receptor is required for signaling and defines potential CNTF targets in the adult and during development. *Neuron*, **10**, 89–102.

Murakami, M., Narazaki, M., Hibi, M., Yawata, H., Yasukawa, K., Hamaguchi, M., Taga, T., and Kishimoto, T. (1991). Critical cytoplasmic region of the interleukin 6 signal transducer gp130 is conserved in the cytokine receptor family. *Proc. Natl. Acad. Sci. (USA)*, **88**, 11349–53.

Murakami, M., Hibi, M., Nakagawa, N., Nakagawa, T., Yasukawa, K., Yamanishi, K., Taga, T., and Kishimoto, T. (1993). IL-6-induced homodimerization of gp130 and associated activation of a tyrosine kinase. *Science*, **260**, 1808–10.

Stahl, N., and Yancopoulos, G.D. (1993). The alphas, betas, and kinases of cytokine receptor complexes. *Cell*, **74**, 587–90.

Stahl, N., Davis, S., Wong, V., Taga, T., Kishimoto, T., Ip, N.Y., and Yancopoulos, G.D. (1993). Cross-linking identifies leukaemia inhibitory factor-binding protein as a ciliary neurotrophic factor receptor component. *J. Biol. Chem.*, **268**, 7628–31.

Stahl, N., Boulton, T.G., Farugello, T., Ip, N.Y., Davis, S., Witthuhn, B.A., Quelle, F.W., Silvennoinen, O., Barbieri, G., Pellegrini, S., Ihle, J.N., and Yancopoulos, G.D. (1984). Association and activation of Jak-Tyk kinases by CNTF-LIF-OSM-IL-6β receptor components. *Science*, **263**, 92–5.

George D. Yancopoulos:
Regeneron Pharmaceuticals, Inc.,
777 Old Saw Mill River Road,
Tarrytown, New York 10591, USA

The Nerve Growth Factor Family— Neurotrophins

Members of the nerve growth factor (NGF) family, also named neurotrophins, are small (about 120 amino acids), basic (pI 9–10) proteins which support the survival of embryonic neurones. These proteins are strongly related in their primary structures, and are probably all active as non-covalently linked dimers. Four proteins belonging to this family have been identified so far: NGF, brain-derived neurotrophic factor (BDNF), neurotrophin-3 (NT-3), and neurotrophin-4/5 (NT-4/5). Their genes are also expressed in the adult when the neurotrophins might maintain or enhance the functions of post-mitotic, differentiated neurones.

■ Nomenclature

The name 'Neurotrophin' was introduced subsequent to the finding that the sequence of BDNF was related to that of NGF. It is used to designate structurally related proteins which all act on cells of the nervous system. NT-4 was originally discovered in *Xenopus laevis*, and NT-5 subsequently identified in rat and man. The latter was thought to be different from NT-4 as its sequence diverges much more between these species than that of the other neurotrophins. However, NT-4 and NT-5 seem to serve the same function in different species, as can be deduced form binding studies: they both seem to bind to the same receptor (Ip *et al*. 1993), hence the designation NT-4/5.

■ Neurotrophin sequences

Nucleotide sequences are available for several species, including rat and human for all of them (GenBank accession numbers of the human neurotrophins are: NGF VO1511; BDNF:M61176; NT-3:M61180; and NT-4/5 M86528). About 50 per cent of the amino acids are common to all neurotrophins, including the six cysteine residues (Fig. 1). With the exception of NT-4/5, neurotrophins are highly conserved proteins between species. In all mammals analysed so far, no amino acid replacements have been found in the mature sequence of BDNF and NT-3.

■ Neurotrophin proteins

All neurotrophins are secretory proteins. Their cleavable leader sequence is followed by a pro-sequence of variable length (80 aa for the shortest, human NT-4/5) and cleavage of the pro-sequence occurs at a consensus sequence of the furin-type found in all neurotrophins (R–X–K/R–R) to yield the mature, biologically active neurotrophins (Barde 1990). Each mature monomer contains six cysteine residues involved in the formation of three disulphide bridges. The arrangement of these bridges (known for NGF and BDNF) is likely to be the same for all neurotrophins (Acklin *et al*. 1993). The crystal structure of the NGF dimer has been elucidated (at a resolution of 2.3Å) and has revealed that the disulphide bridges are all grouped at one end of the molecule in the homodimers (McDonald *et al*. 1991) (bottom end in Fig. 2). Their arrangement is reminiscent of that found in the TGFβ superfamily and the PDGFs (McDonald and Hendrickson 1993), both of which are also first translated as long biosynthetic precursors and cleaved at a furin-like consensus sequence. Three-quarters of the residues that differ between the neurotrophins are found in three variable β-hairpin loops and one reverse turn (Fig. 2), while most conserved residues seem to play a structural role. These are localized in three anti-parallel β-strands and some are involved as contact residues between the monomers (McDonald *et al*. 1991). The residues principally involved in the binding of NGF (and of BDNF) to the low-affinity neurotrophin receptor p75 have been localized (Ibáñez *et al*. 1992), as well as those binding to *trk*, the latter being grouped on one side of the NGF dimer, delineating a continuous surface extending approximately par-

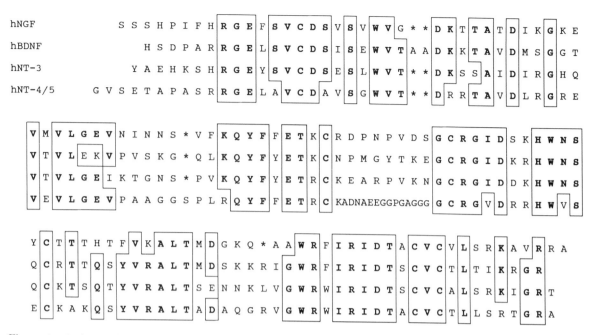

hNGF, hBDNF, hNT-3, hNT-4/5 amino acid sequences

Figure 1. Amino acid sequences corresponding to the processed, mature part of the human neurotrophins.

allel to the two-fold axis of the molecule (Ibáñez *et al.* 1993).

■ Neurotrophin genes and transcription

In human, NGF has been localized to chromosome 1q21–22.1, BDNF to 11q13, NT-3 to 12q13, and NT-4/5 to 19q13.3 (Ip *et al.* 1992). The NGF gene spans at least 40 kb and contains several small 5′ exons and a larger 3′ exon that contains most of the translated sequence (Selby *et al.* 1987). The major NGF transcript is about 1.3 kb. An intronic AP-1 binding site has been localized and shown to be involved in the regulation of NGF gene expression at least in some cells such as fibroblasts (Hengerer *et al.* 1990). In peripheral nerves after lesion, macrophage-derived interleukin-1β up-regulates NGF mRNA levels (but not those of BDNF) (Meyer *et al.* 1992). The size of the BDNF gene is not known, but the length of the longest transcription unit is larger than 40 kb (Timmusk *et al.* 1993). The gene consists of four short 5′ exons and one 3′ exon encoding the mature BDNF protein. Several transcription initiation sites have been described and have been mapped upstream of the four 5′ exons. Differential splicing controls tissue-specific and seizure-induced expression of the various BDNF mRNAs (Timmusk *et al.* 1993). The major BDNF transcripts are 1.6 and 4.2 kb in size, the use of two alternative polyadenylation sites being responsible for this size difference (Timmusk *et al.* 1993). In the NT-3 gene, one transcription start site has been identified, as well as a silencer element (Shintani *et al.* 1993). The major NT-3 transcript is about 1.4 kb in length. Four NT-4/5

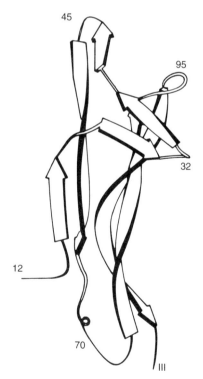

Figure 2. Schematic representation of the NGF monomer (McDonald *et al.* 1991). Three β-hairpin turns (located around residues 32, 45, and 95) connect three pairs of β-strands (indicated by arrows).

transcripts have been identified: 1.1, 2.1, 4.0, and 9 kb. In the brain, neurones are a major cellular site of neurotrophin gene expression and numerous *in situ* hybridization studies have examined the regional patterns of expression of these genes, specially in rodents. In some brain areas (cerebral cortex and hippocampal formation in particular), increased neuronal activity leads to an enhancement of the levels of NGF and BDNF transcripts, but not of NT-3 (see Merlio *et al*. 1993 and references therein).

◼ Biological functions

Neurotrophins typically support the survival of embryonic neurones that die in their absence. Thus, antibodies to NGF lead to the virtually complete destruction of the peripheral sympathetic nervous systems (Levi-Montalcini 1987), and the loss of many neural crest-derived sensory neurones. BDNF, NT-3, and NT-4/5 all support the survival of subpopulations of peripheral sensory neurones, including those derived from epidermal placodes, not supported by NGF (Barde 1989). Neurotrophins (except NGF) also support the survival of embryonic rat motoneurones *in vitro*, and BDNF can prevent their death *in vivo* (see Henderson *et al*. 1993 and references therein). In the brain, NGF and to a lesser degree BDNF prevent the loss of cholinergic function seen after axotomy in adult animals (Hefti 1986; Knüsel *et al*. 1992). BDNF also supports the survival of dopaminergic neurones dissociated from the rodent mesencephalon, as well as of retinal ganglion cells. In the mature nervous system, neurotrophins are involved in the maintenance of neuronal phenotypes: in particular, antibodies to NGF decrease the levels of enzymes synthesizing catecholamines, as well as neurotransmitters like substance P in the peripheral nervous system of adult animals (Barde 1989). *In vitro*, BDNF increases the number of neural crest cells differentiating along the sensory pathway (Sieber-Blum 1991). Finally, NT-3 acts on oligodendrocyte precursors and contributes to their division and survival (Barres *et al*. 1993).

◼ Pathology

In aged, learning-deficient rats, the intra-ventricular injection of NGF is able to revert the learning deficiencies in simple behavioural tests (Fischer *et al*. 1987). Decreased levels of BDNF mRNA have been measured in the hippocampus of patients with Alzheimer's disease (Phillips *et al*. 1991).

◼ References

Acklin, C., Stoney, K., Rosenfeld, R.A., Miller, J.A., Rohde, M.F., and Haniu, M. (1993). Recombinant human brain-derived neurotrophic factor (rHuBDNF). Disulphide structure and characterization of BDNF expressed in CHO cells. *Int. J. Peptide Res.*, **41**, 548–52.

Barde, Y.-A. (1989). Trophic factors and neuronal survival. *Neuron*, **2**, 1525–34.

Barde, Y.-A. (1990). The nerve growth factor family. *Prog. Growth Factor Res.*, **2**, 237–48.

Barres, B.A., Schmid, R., Sendtner, M., and Raff, M. C. (1993). Multiple extracellular signals are required for long-term oligodendrocyte survival. *Development*, 118, 283–95.

Fischer, W., Wictorin, K., Bjorklund, A., Williams, L.R., Varon, S., and Gage, F.H. (1987). Amelioration of cholinergic neurone atrophy and spatial memory impairment in aged rats by nerve growth factor. *Nature*, **329**, 65–8.

Hefti, F. (1986). Nerve growth factor promotes survival of septal cholinergic neurones after fimbrial transections. *J. Neurosci.*, **6**, 2155–62.

Henderson, C.E., Camu, W., Mettling, C., Gouin, A., Poulsen, K., Karihaloo, M., Rullamas, J., Evans, T., McMahon, S.B., Armanini, M.P., Berkemeier, L., Phillips, H.S., and Rosenthal, A. (1993). Neurotrophins promote motor neurone survival and are present in embryonic limb bud. *Nature*, **363**, 266–70.

Hengerer, B., Lindholm, D., Heumann, R., Rüther, U., Wagner, E.F., and Thoenen, H. (1990). Lesion-induced increase in nerve growth factor mRNA is mediated by c-*fos*. *Proc. Natl. Acad. Sci. (USA)*, **87**, 3899–903.

Ibáñez, C.F., Ebendal, T., Barbany, G., Murray-Rust, J., Blundell, T.L., and Persson, H. (1992). Disruption of the low affinity receptor-binding site in NGF allows neuronal survival and differentiation by binding to the *trk* gene product. *Cell*, **69**, 329–41.

Ibáñez, C.F., Ilag, L.L., Murray-Rust, J., and Persson, H. (1993). An extended surface of binding to trk tyrosine kinase receptors in NFG and BDNF allows the engineering of a multifunctional panneurotrophin. *EMBO J.*, **12**, 2281–93.

Ip, N.Y., Ibáñez,C.F., Nye,S.H., McClain,J., Jones, P.F., Gies, D.R., Belluscio, L., Le Beau, M.M., Espinosa III, R., Squinto, S.P., Persson, H., and Yancopoulos, G.D. (1992). Mammalian neurotrophin-4: structure, chromosomal localization, tissue distribution, and receptor specificity. *Proc. Natl. Acad. Sci. (USA)*, **89**, 3060–4.

Ip, N.Y., Stitt, T.N., Tapley, P., Klein, R., Glass, D.J., Fandl, J., Greene, L.A., Barbacid, M., and Yancopoulos, G. D. (1993). Similarities and differences in the way neurotrophins interact with the trk receptors in neuronal and nonneuronal cells. *Neuron*, **10**, 137–49.

Knüsel, B., Beck, K.D., Winslow, J.W., Rosenthal, A., Burton, L.E., Widmer, H.R., Nikolics, K., and Hefti, F. (1992). Brain-derived neurotrophic factor administration protects basal forebrain cholinergic but not nigral dopaminergic neurones from degenerative changes after axotomy in the adult rat brain. *J. Neurosci.*, **12**, 4391–402.

Levi-Montalcini, R. (1987). The nerve growth factor: thirty-five years later. *EMBO J.*, **6**, 1145–54. [Erratum (1987) *EMBO J.*, **6**, 2856.]

McDonald, N.Q., and Hendrickson, W.A. (1993). A structural superfamily of growth factors containing a cystine knot motif. *Cell*, **73**, 421–4.

McDonald, N.Q., Lapatto, R., Murray-Rust, J., Gunning, J., Wlodawer, A., and Blundell, T.L. (1991). New protein fold revealed by a 2.3-Å resolution crystal structure of nerve growth factor. *Nature*, **354**, 411–4.

Merlio, J.-P., Ernfors, P., Kokaia, Z. Middelmas, D.S., Bengzon, J., Kokaia, M., Smith, M.L., Siesjö, B.K., Hunter, T., Lindvall, O., and Persson, H. (1993). Increased production of the TrkB protein tyrosine kinase receptor after brain insults. *Neuron*, **10**, 151–64.

Meyer, M., Matsuoka, I., Wetmore, C., Olson, L., and Thoenen, H. (1992). Enhanced synthesis of brain-derived neurotrophic factor in the lesioned peripheral nerve: different mechanisms are responsible for the regulation of BDNF and NGF mRNA. *J. Cell Biol.*, **119**, 45–54.

Phillips, H.S., Hains, J.M., Armanini, M., Laramee, G.R., Johnson, S.A., and Winslow, J.W. (1991). BDNF mRNA is decreased in the hippocampus of individuals with Alzheimer's disease. *Neuron*, **7**, 695–702.

Selby, M.J., Edwards, R., Sharp, F., and Rutter, W.J. (1987). Mouse nerve growth factor gene: structure and expression. *Mol. Cell Biol.*, **7**, 3057–64.

Shintani, A., Ono, Y., Kaisho, Y., Sasada, R., and Igarashi, K. (1993). Identification of the functional regulatory region of the neurotrophin-3 gene promoter. *Mol. Brain Res.*, **17**, 129–34.

Sieber-Blum, M. (1991). Role of the neurotrophic factors BDNF and NGF in the commitment of pluripotent neural crest cells. *Neuron*, **6**, 949–55.

Timmusk, T., Palm, K., Metsis, M., Reintam, T., Paalme, V., Saarma, M., and Persson, H. (1993). Multiple promoters direct tissue-specific expression of the rat BDNF gene. *Neuron*, **10**, 475–89.

Y.-A. Barde:
Max-Planck Institute for Psychiatry,
Department of Neurobiochemistry,
82152 Planegg-Martinsried, FRG

The Neurotrophin Receptors

Two distinct forms of neurotrophin receptors have been identified. The p75 neurotrophin receptor binds all the neurotrophins with nanomolar affinity and is widely expressed. It is related to several other cell surface receptors by virtue of a tandemly repeated cysteine motif present extracellularly. Its contribution to the neurotrophin signal transduction cascade remains unclear. The trk receptors are transmembrane tyrosine kinases which bind and are activated by the neurotrophins by a mechanism involving homodimerization. Each of the trk receptors show clear ligand specificities, trkA binding NGF, trkB binding BDNF, and NT-4 and trkC binding NT-3 preferentially. Activation of the appropriate trk appears essential for biological activity of the neurotrophins. Multiple isoforms of each of the trks are generated by alternative splicing.

■ The trk receptors

Trk receptors are ligand-regulated transmembrane tyrosine kinase receptors that bind specific members of the neurotrophin family (Barbacid 1993). The first trk to be discovered, trkA, was originally observed as a transforming oncogene in colon carcinoma (Martin-Zanca *et al.* 1986). Its transforming potential was due to the replacement of its normal extracellular sequence with portions of a non-muscle tropomyosin gene. The trk family members are highly related, particularly within their tyrosine kinase domains. Numerous N-linked glycosylation sites are present extracellularly, which, in part, accounts for the fact that although the proteins have a predicted molecular weight of about 80 kDa, they run on SDS-PAGE at an apparent molecular weight of 140–145 kDa. The preferred ligand for trkA (GenBank accession number X03541, M85214) is nerve growth factor (NGF), for trkB (GenBank accession number X17647, M55291, M55292, M55293) is brain-derived neurotrophic factor (BDNF) and neurotrophin-4 (NT-4), and for trkC (GenBank accession number M80800, L03813) is neurotrophin-3 (NT-3) (Ip *et al.* 1993). Each of the mammalian trk family members has a signal peptide of about 32 amino acids, an extensively glycosylated extracellular domain of approximately 440 amino acids, a single transmembrane domain of 24 amino acids, and an intracellular domain, containing a juxtamembrane region, the tyrosine kinase domain and a short carboxy tail of approximately 360 amino acids. In addition to the mammalian trk receptors, a trk homologue from drosophilia, termed Dtrk, has been cloned (Pulido *et al.* 1992). The extracellular domain of this protein has a structural homology with adhesion molecules of the immunoglobulin superfamily and appears capable of mediating homotypic adhesion. Interestingly, adhesion activates this molecule's tyrosine kinase activity. The mammalian trk family members bear a limited homology to molecules of the immunoglobulin superfamily (Schneider and Schweiger 1991) but whether these are vestigial remnants of an ancestral trk or important functional domains remains to be determined.

■ Trk genes and alternative splicing

The human trkA gene is located on the long arm of chromosome 1 within bands 1q23–1q24. The chromosomal positions of the the other trks remain to be determined, with the exception of drosophilia Dtrk, which is located at position 48D of the right arm of the second chromosome. Each of the mammalian trk receptors undergoes alternative splicing (see Fig. 1). In both humans and rats, a six-amino-acid region is present within the extracellular domain of trkA expressed within neurones but not in trkA expressed outside the nervous system (Barker *et al.* 1993). Alternatively spliced forms of trk B and trkC which lack the tyrosine kinase domain have been identified and forms of trkC

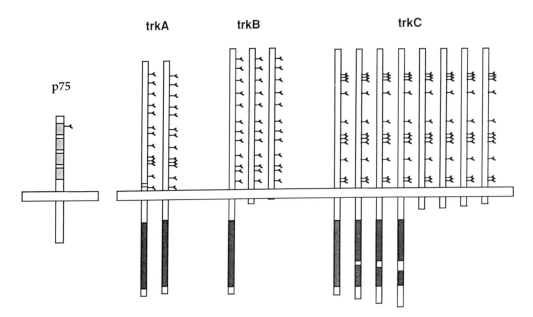

Figure 1. Each of the known mammalian neurotrophin receptors and their alternatively spliced forms are shown schematically. THe cysteine-rich domains of the p75 neurotrophin receptor are shown by light shading, the tyrosine kinase domain of the trk receptors is indicated by dark shading and alternatively spliced domains are shown by unshaded boxes. N-linked glycosylation sites are indicated by the forked lines to the right.

containing 14, 25, or 39 amino acid inserts within the tyrosine kinase domain have also been described (Barbacid 1993; Valenzuela *et al*. 1993; Tsoulfas *et al*. 1993). The isoforms of trkC which contain these inserts become phosphorylated in reponse to ligand binding yet show functional deficits. Only the kinase-lacking forms of trkB and trkC are expressed within astrocytes and non-neuronal tissues whereas both kinase-containing and -lacking forms are detected throughout the nervous system. Polyclonal anti-peptide antibodies directed against the COOH-terminal domain of various trks are available commercially (Santa Cruz Biotechnology).

■ The p75 neurotrophin receptor

The human p75 neurotrophin receptor is synthesized as a 427-amino-acid precursor containing a 28-amino-acid signal peptide (GenBank accession number M14764). In the processed protein, a 22-amino-acid transmembrane domain separates a 222-amino-acid extracellular domain from a 155-amino-acid intracellular domain. The protein contains both N- and O-linked carbohydrate extracellularly and the intracellular domain is phosphorylated on serine and threonine. The intracellular domain shows no homologies to known proteins. The extracellular domain contains four tandemly linked repeats, each containing six cysteines ($CX_{12-15}CX_{0-2}CX_2CX_9CX_7C$) in which the NGF-binding domain resides. The modular arrangement of these cysteine-rich repeats is shared by several other receptors, including the two tumour necrosis receptors, the fas antigen

receptor, CD27, CD30, CD40, and OX40. The crystal structure of the extracellular domain of the 55 kDa human TNF receptor has revealed that these repeats are linked to produce an elongated array of cysteine rich-domains, each containing three intradomain disulphide bridges (Banner *et al*. 1993). Compared to the human receptor, both the rat (GenBank accession number X05137) and chick forms of the p75 neurotrophin receptor are relatively well conserved, not only with respect to the cysteine-rich repeats but also across the transmembrane and intracellular juxtamembrane domains. Monoclonal antibodies directed against the extracellular domain of the human and rat p75 neurotrophin receptors are available commercially (Boehringer Mannheim).

The human p75 neurotrophin receptor gene is composed of six exons and is located on chromosome 17 at position q21. No alternatively spliced forms of p75 mRNA have been detected. However, a soluble form of the p75 neurotrophin receptor which arises from proteolysis of the intact receptor and is capable of binding neurotrophin has been described.

■ Binding characteristics of the neurotrophin receptors

The p75 neurotrophin receptor binds all of the neurotrophins with an approximately equal affinity (K_D about 1–5 nM for NGF), shows very rapid on/off kinetics at 4 °C and is internalized poorly at 37 °C. Some investigators have reported that trkA expressed on its own binds NGF with a relatively low affinity but that coexpression of p75 and trkA results in the production of a low number of high affinity binding sites (K_{D1} about 1 nM; K_{D2} about 60–100 pM) (Hempstead et al. 1991). However, others have found that the various trks display both high and low affinity interactions in the absence of the p75 neurotrophin receptor and that co-expression of the p75 neurotrophin receptor with trkA has no effect on dissociation constants nor the relative percentage of low and high affinity binding sites (Jing et al. 1992). The possibility that the trk receptors and the p75 neurotrophin receptor functionally interact remains open but cross-linking and co-immunoprecipitation analyses have failed to reveal a physical association between them (Meakin and Shooter 1991).

Signalling mechanisms

The trk receptors are transmembrane tyrosine kinases whose activity is regulated by neurotrophin binding. Ligand-dependent tyrosine kinase activation of the trk receptors does not require the coexpression of the p75 neurotrophin receptor. As for the PDGF and EGF tyrosine kinase receptors, trkA appears to become activated by forming transphosphorylating homodimers in response to ligand binding (Jing et al. 1992). Several SH2-domain-containing signal transduction elements have been shown to associate directly or indirectly with activated trkA, including phosphotidylinositol-3-kinase and phospholipase-C (Ohmichi et al. 1991, 1992). In addition, a complex that includes MAP kinase can be co-immunoprecipitated with trkA (Loeb et al. 1992). NGF treatment of PC12 cells results in activation of ras but GTPase activating protein does not associate directly with the receptor (Li et al. 1992).

The physiological function of the p75 neurotrophin receptor remains obscure. A direct role in neurotrophin signal transduction has been suggested by some studies. Expression of the p75 neurotrophin receptor confers an NGF response to some cell lines and EGF treatment of PC12 cells containing a chimeric receptor consisting of the EGF receptor extracellular domain and the p75 neurotrophin receptor transmembrane and intracellular domains results in neurite production (Yan et al. 1991). However, other work has shown that antibodies which block binding of NGF to the p75 neurotrophin receptor fail to inhibit the effect of NGF on PC12 cells or on trk-containing sensory or sympathetic neurones (Weskamp and Reichardt 1991) and NGF mutants which bind trk but not the p75 neurotrophin receptor still mediate NGF-specific effects on PC12 cells or on neurones at, or near, wild-type specific activities (Ibáñez et al. 1992). These data and the finding that the trkA receptor forms homodimers that appear to be crucial for NGF signalling suggest that the neurite- and survival-promoting activities of NGF may be mediated solely by the trkA receptor. This is supported by recent data showing that PC12 cells bearing a chimeric molecule consisting of the TNF receptor extracellular domain and the trkA receptor transmembrane and intracellular domains gain the ability to extend neurites and survive in response to TNF (Rovelli et al. 1993). A mouse line in which the p75 neurotrophin receptor has been deleted by homologous recombination has been created (Lee et al. 1992). The mice are viable but display increased sensitivity to heat and, with increasing age, appear to progressively lose sensory modalities in the limbs. One possible role for the receptor, consistent with its low affinity for neurotrophins and its fast on and off rates, may be to play some role in concentrating or presentation of neurotrophin for the appropriate trk receptor.

References

Banner, D.W., D'Arcy, A., Janes, W., Gentz, R. Schoenfeld, H.J., Broger, C., Loetscher, H., and Lesslauer, W. (1993). Crystal structure of the soluble human 55 kd TNF receptor–human TNF beta complex: implications for TNF receptor activation. *Cell*, **73**, 431–45.

Barbacid, M. (1993). Nerve growth factor: A tale of two receptors. *Oncogene*, **8**, 2033–42.

Barker, P.A., Lomen-Hoerth, C., Gensch, E.M., Meakin, S.O., Glass, D.J., and Shooter, E.M. (1993). Tissue-specific alternative splicing generates two isoforms of the trkA receptor. *J. Biol. Chem.*, **268**, 15150–7.

Hempstead, B.L., Martin-Zanca, D., Kaplan, D.R., Parada, L.F., and Chao M.V. (1991). High-affinity NGF binding requires coexpression of the trk proto-oncogene and the low-affinity NGF receptor. *Nature*, **350**, 678–83.

Ibáñez, C.F., Ebendal, T., Barbany, G., Murray-Rust, J., Blundell, T.L., and Persson H. (1992). Disruption of the low affinity receptor-binding site in NGF allows neuronal survival and differentiation by binding to the trk gene product. *Cell*, **69**, 329–41.

Ip, N.Y., Stitt, T.N., Taplet, P., Klein, R., Glass, D.J., Fandl, J., Greene, L.A., Barbacid, M., and Yancopoulos, G.D. (1993). Similarities and differences in the way neurotrophins interact with the trk receptors in neuronal and nonneuronal cells. *Neuron*, **10**, 137–49.

Jing, S., Taplet, P., and Barbacid, M. (1992). Nerve growth factor mediates signal transduction through trk homodimer receptors. *Neuron*, **9**, 1067–79.

Lee, K.F., Li, E., Huber, L.J., Landis, S.C., Sharpe, A.H., Chao, M.V., and Jaenisch R. (1992). Targeted mutation of the gene encoding the low affinity NGF receptor p75 leads to deficits in the peripheral sensory nervous system. *Cell*, **69**, 737–49.

Li, B.Q., Kaplan, D., Kung, H.F., and Kamata, T. (1992). Nerve growth factor stimulation of the ras–guanine nucleotide exchange factor and GAP activities. *Science*, **256**, 1456–9.

Loeb, D.M., Tsao, H., Cobb, M.H., and Greene, L.A. (1992). NGF and other growth factors induce an association between ERK1 and the NGF receptor gp 140prototrk. *Neuron*, **9**, 1953–65.

Martin-Zanca, D., Hughes, S.H., and Barbacid, M. (1986). A human oncogene formed by the fusion of truncated tropomyosin and protein tyrosine kinase sequences. *Nature*, **319**, 743–77.

Meakin, S.O., and Shooter, E.M. (1991). Molecular investigations on the high-affinity nerve growth factor receptor. *Neuron*, **6**, 153–63.

Ohmichi, M., Decker, S.J., Pang, L., and Saltiel, A.R. (1991). Phospholipase C-gamma 1 directly associates with the p70 trk oncogene product through its *src* homology domains. *J. Biol. Chem.*, **266**, 14858–61.

Ohmichi, M., Decker, S.J., and Saltiel, A.R. (1992). Activation of phosphatidylinositol-3 kinase by nerve growth factor involves direct coupling of the *trk* proto-oncogene with *src* homology 2 domains. *Neuron*, **9**, 769–77.

Pulido, D., Campuzano, S., Koda, T., Modolell, J., and Barbacid, M. (1992). D*trk*, a drosophilia gene related to the trk family of neurotrophin receptors, encodes a novel class of neural cell adhesion molecule. *Embo J.*, **11**, 391–404.

Rovelli, G., Heller, R.A., Canossa, M., and Shooter, E.M. (1993). Chimeric tumour necrosis factor-TrkA receptors reveal that ligand-dependent activation of the TrkA tyrosine kinase is sufficient for differentiation and survival of PC12 cells. *Proc. Natl. Acad. Sci. (USA)*, **90**, 8717–21.

Schneider, R., and Schweiger, M. (1991). A novel modular mosaic of cell adhesion motifs in the extracellular domains of the neurogenic trk and trkB tyrosine kinase receptors. *Oncogene*, **6**, 1807–11.

Tsoulfas, P., Soppet, D., Escandon, E., Tessarollo, T., Mendoza-Ramirez, J.-L., Rosenthal, A., Nikolics, K., and Parada, L.F. (1993). The rat *trk*C locus encodes multiple neurogenic receptors that exhibit differential response to neurotrophin-3 in PC12 cells. *Neuron*, **10**, 963–74.

Valenzuela, D.M., Maisonpierre, P.C., Glass, D.J., Rojas, E., Nunez, L.,

Kong, Y., Gies, G., Stitt, T.N., Ip, N.Y., and Yancopoulos, G.D. (1993). Alternative forms of rat TrkC with different functional capabilities. *Neuron*, **10**, 963–74.

Weskamp, G., and Reichardt L. (1991). Evidence that biological activity of NGF is mediated through a novel subclass of high affinity receptors. *Neuron*, **6**, 1–20.

Yan, H., Schlessinger, J., and Chao, M.V. (1991). Chimeric NGF-EGF receptors define domains responsible for neuronal differentiation. *Science*, **252**, 561–3.

Philip A. Barker:
Department of Neurobiology,
Stanford University,
Stanford, CA 94305–5401, USA

Neu Differentiation Factor (NDF) and the Neuregulin (NRG) Family

NDF is a 44 kDa glycoprotein that induces growth or differentiation of epithelial cells and affects certain neuronal functions. It belongs to the NRG family, which includes multiple secreted or membrane-bound proteins, all of which contain an EGF-like motif and arise by alternative splicing from a single gene. Various NRGs are normally expressed primarily by neuronal tissues but NDF production is induced in fibroblasts by a ras oncogene. Apparently, the receptor for NDF is comprised of the protein product of the neu/erbB-2 proto-oncogene and an additional receptor tyrosine kinase. NDF, as well as other NRGs, may have potential clinical utility in the treatment of nerve injuries and in certain adenocarcinomas.

■ Alternative names

Heregulin (HRG), glial growth factor (GGF), acetylcholine receptor inducing activity (ARIA), gp30.

■ NDF protein

The heat-stable rodent factor and its human homologue (named HRG) are secreted proteins that were purified by heparin-chromatography from the media conditioned by cancer cells (Peles *et al*. 1992; Holmes *et al*. 1992). NDF is sensitive to reduction but resists extremes of pH. N- and O-linked sugars account for approximately one-quarter of the mass of the mature 44 kDa glycoprotein, but they are not required for receptor recognition. Three glial growth factors, GGFI (34 kDa), GGFII (59 kDa), and GGFIII (45 kDa), which were isolated from bovine brain were found to be related to NDF/HRG (Marchionni *et al*. 1993). Similarly, a group of proteins in the range of 33 kDa to 44 kDa that were purified from chicken brain on the basis of their

acetylcholine receptor inducing activity (ARIA) is structurally related to NDF (Falls *et al*. 1993).

■ NRG sequences

On the basis of their amino acid and nucleotide sequences NDF (Wen *et al*. 1992) (GenBank accession number S35165), HRGs (Holmes *et al*. 1992) (M94165, M94168, M94166), GGFs (Marchionni *et al*. 1993) (L12259, L12260, L12261), and ARIA (Falls *et al*. 1993) (L11264) are members of the neuregulin (NRG) family (Fig. 1). This group contains at least 12 distinct molecules that are encoded by one gene, but its diversity is due to alternative splicing (Marchionni *et al*. 1993). The precursor forms of these proteins are mosaics of recognizable structural motifs. These include an N-terminal hydrophobic signal peptide followed by a kringle domain (248 amino acids) or an N-terminal non-hydrophobic stretch of 40 amino acids. Other blocks are an immunoglobulin-(Ig-) like domain (approximately 70 amino acids), a 'spacer' domain that contains multiple sites for N- and O-linked glycosylation and a site for glycosaminoglycan attachment, an epidermal growth factor-(EGF-) like domain (63–72 amino acids) that includes six conserved cysteine residues, a 23-amino-acid-long hydrophobic stretch that functions as a transmembrane domain, and a variable-length cytoplasmic tail (three forms: 157, 196 and 376 amino acids). Some of the transmembrane precursor forms undergo proteolytic cleavage at both the N-terminus and at the short stretch that connects the EGF-like-domain with the transmembrane sequence. Two eight-amino-acid-long sequences, which are confined to this juxtamembrane region, determine the identity of NRGs as subtype 1 or 2. Alternatively, NRG forms that terminate at this domain are designated subtype 3. Additional variations are added by two forms of the C-terminal loop of the EGF-like domain, corresponding to NRGα and NRGβ. The EGF-like domains of the various NRGs are responsible for receptor recognition and they act independently of other structural motifs (Holmes *et al.* 1992). Nevertheless, it appears that the α and the β forms differ only in their affinities while they share the same receptor specificity. The two distinct N-terminal sequences and the hydrophobic sequence control the transmembrane topology of NRGs. However, the functions of other domains and the role of post-translational modifications are unknown.

■ NRG receptors and signal transduction

Different NRG molecules elevate tyrosine phosphorylation of the 185 kDa neu/erbB-2 receptor tyrosine kinase, which is encoded by a proto-oncogene. In addition, NDF undergoes covalent cross-linking to a molecule that is immunologically related to neu. Nevertheless, the binding of NDF involves, in addition to neu, another receptor tyrosine kinase that functions as a co-receptor and whose cellular distribution is narrower than the expression of Neu (Peles *et al.* 1993). Apparently upon binding to these receptors NDF induces a cascade of signalling events which employs *src* homology-2 (SH2) containing proteins and serine/threonine kinases in analogy to other ligands of receptor tyrosine kinases.

■ NRG gene and expression

A single NRG gene is located on human chromosome 8p12–p21 (Orr-Urtreger *et al.* 1993) and it contains at least 13 exons whose precise organization has not been determined. Multiple transcripts of NRG exist and their relative abundance displays tissue specificity (Holmes *et al.* 1992; Wen *et al.* 1992). The largest mRNA is 6.8 kb long and its highest expression is exhibited by the spinal cord. Other organs that express NRGs are the brain, dorsal root ganglia, myenteric ganglia, adrenal cortex, embryonic liver, dermis, genital ridge, ovary, stomach, kidney, and heart (Holmes *et al.* 1992; Wen *et al.* 1992). *In situ* hybridization analysis indicated that NRG expression is predominantly confined to neuronal cells within some of these tissues. These were identified as motor neurones, sensory neurones and neuroepithelial cells (Marchionni *et al.* 1993; Orr-Urtreger *et al.*

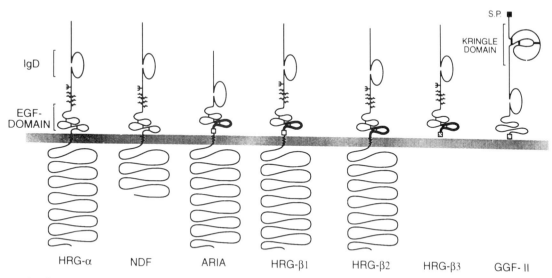

Figure 1. Schematic representation of the structures of various NRG molecules. The horizontal grey bar represents the plasma membrane. The extracellular portions of NRG molecules are shown above the plasma membrane and they contain the following motifs: an immunoglobulin-like domain (IgD) and EGF-like domain, a signal peptide (S.P.), and a kringle domain. The 'spacer' domain, which is rich in glycosylated groups (branched and horizontal lines), is missing in some forms. Note that the cytoplasmic portion displays both length and sequence heterogeneity. The variant form of the C-terminal loop of the EGF-like domain is shown by a solid line. The juxtamembrane squares distal to it represent variable sequences.

1993). The forms of NRG that are expressed by neuronal cells are characterized by the absence of the 'spacer' domain. In addition, some of their N-termini include a signal peptide that is linked to a kringle domain (Marchionni et al. 1993).

■ Biological actions of NRGs

In vitro, NDF and HRGs are weakly mitogenic for various epithelial cells including mammary, lung, and gastric cells (Holmes et al. 1992). However, certain mammary tumour cells undergo growth arrest at the G_2/M phase of the cell cycle, in response to NDF (Peles et al. 1992; Bacus et al. 1993), or to its human homologue, gp30 (Bacus et al. 1992). The factor-treated cells also exhibit a mature phenotype that includes flat morphology, synthesis of the intercellular cell adhesion molecule 1 (ICAM-1), and secretion of milk components. By contrast, recombinant GGFs are mitogenic for cultured Schwann cells, which otherwise divide very slowly even in the presence of mitogens (Marchionni et al. 1993). This may explain the effect of partially purified preparations of the factors, which were derived from brain and pituitary extracts, to stimulate limb regeneration in the newt (Brockes 1984). Another nerve-dependent function of NRGs involves the formation of the neuromuscular junction. Motor neurone-derived ARIA appears to induce the synthesis of acetylcholine receptors (AChRs), and possibly other molecules, by post-synaptic muscle cells (Falls et al. 1993). This, together with muscle activity, generates a remarkable gradient of AChR density in the synaptic cleft.

■ Pathology of NRGs

Several lines of evidence link the NRG family with neurodevelopment and cancer. A stage-specific and carcinogen-induced mutation activates the oncogenic potential of the neu/erbB-2 gene, which encodes a co-receptor for NDF (Bargmann et al. 1986). The oncogenic effect of the mutation is also tissue-specific as it affects only neu in Schwann cells of cranial and peripheral nerves and specifically gives rise to Schwannomas (Perantoni et al. 1987). Presumably, the mutation mimics a mid-gestation-specific mitogenic effect of a particular NRG molecule. This may be related to the relatively high concentrations of GGF activities in acoustic neuromas, a non-hereditary spontaneous Schwannoma (Brockes et al. 1986). The normal function of the corresponding activity may play a role in embryonic development of the nervous system to control the interaction of nerve cells with target tissues such as muscles. In addition, it may act postnatally to regulate Schwann-cell growth, migration, and myelination which accompanies post-injury reinnervation. Consistent with this paradigm, the neuronal expression of the neu gene falls dramatically after birth but it increases in nerves that undergo Wallerian degeneration (Cohen et al. 1992).

Overexpression of neu characterizes a portion of human adenocarcinomas arising at a number of sites such as colon, stomach, breast and ovary. Overexpression in primary breast tumours correlates with earlier relapse and with poor survival (Slamon et al. 1987). It is still unknown whether the poor prognosis of the neu-overexpressing subset of tumours is related to NRGs, but approximately one-third of breast cancers express NDF/HRG (Bacus et al. 1993). This may be due, at least in part, to oncogenic activation of an allele of the ras gene, which dramatically elevates transcription of several forms of NRGs in fibrosarcomas, melanomas and adenocarcinomas (Wen et al. 1992). Thus, neuregulins may have clinical value in diagnosis or treatments of certain types of human cancer.

■ References

Bacus, S.S., Huberman, E., Chin, D., Kiguchi, K., Simpson, S., Lippman, M., and Lupu, R. (1992). A ligand for the erbB-2 oncogene product (gp30) induces differentiation of human breast cancer cells. Cell Growth Diff., **3**, 401–11.

Bacus, S.S., Gudkov, A.V., Zelnick, C.R., Chin, D., Stern, R., Stancovski, I., Peles, E., Ben-Baruch, N., Farbstein, H., Lupu, R., Wen, D., Sela, M., and Yarden, Y. (1993). Neu differentiation factor (heregulin) induces expression of intercellular adhesion molecule 1: Implications for mammary tumours. Cancer Res., **53**, 5251–61.

Bargmann, C.I., Hung, M.C., and Weinberg, R.A. (1986). Multiple independent activations of the neu oncogene by a point mutation altering the transmembrane domain of p185. Cell, **45**, 649–57.

Brockes, J.P. (1984). Mitogenic growth factors and nerve dependence of limb regeneration. Science, **225**, 1280–7.

Brockes, J.P., Breakefield, X.O., and Martuza, R.L. (1986). Glial growth factor-like activity in Schwann cell tumours. Ann. Neurol., **20**, 317–22.

Cohen, J.A., Yachnis, A.T., Arai, M., Davis, J.G., and Scherer, S.S. (1992). Expression of the neu proto-oncogene by Schwann cells during peripheral nerve development and Wallerian degeneration. J. Neurosci. Res., **31**, 622–34.

Falls, D.L., Rosen, K.M., Corfas, G., Lane, W.S., and Fischbach, G.D. (1993). ARIA, a protein that stimulates acetylcholine receptor synthesis, is a member of the neu ligand family. Cell, **72**, 801–15.

Holmes, W.E., Sliwkowski, M.X., Akita, R.W., Henzel, W.J., Lee, J., Park, J.W., Yansura, D., Abadi, N., Raab, H., Lewis, G.D., Shepard, H.M., Kuang, W.J., Wood, W.I., Goeddel, D.V., and Vandlen, R.L. (1992). Identification of heregulin, a specific activator of p185erbB2. Science, **256**, 1205–10.

Marchionni, M.A., Goodearl, A.D.J., Chen, M.S., Bermingham-McDonogh, O., Kirk, C., Hendricks, M., Danehy, F., Misumi, D., Sudhalter, J., Kobayashi, K., Wroblewski, D., Lynch, C., Baldassre, M., Hiles, I., Davis, J.B., Hsuan, J.J., Totty, N.F., Otsu, M., McBurney, R.N., Waterfield, M.D., Stroobant, P., and Gwynne, D. (1993). Glial growth factors are alternatively spliced erbB2 ligands expressed in the nervous system. Nature, **362**, 312–8.

Orr-Urtreger, A., Trakhtenbrot, L., Ben-Levi, R., Wen, D., Rechavi, G., Lonai, P., and Yarden, Y. (1993). Neural expression and chromosomal mapping of neu differentiation factor to 8p12–p21. Proc. Natl. Acad. Sci. (USA), **90**, 1867–71.

Peles, E., Bacus, S.S., Koski, R.A., Lu, H.S., Wen, D., Ogden, S.G., Levy, R.B., and Yarden, Y. (1992). Isolation of the neu/HER-2 stimulatory ligand: a 44 kd glycoprotein that induces differentiation of mammary tumour cells. Cell, **69**, 205–16.

Peles, E., Ben-Levy, R., Tzahar, E., Liu, N., Wen, D., and Yarden, Y. (1993). Cell-type specific interaction of Neu differentiation factor (NDF/heregulin) with neu/HER-2 suggests complex ligand–receptor relationships. EMBO J., **12**, 961–71.

Perantoni, A.O., Rice, J.M., Reed, C.D., Watatani, M., and Wenk, M.L. (1987). Activated *neu* oncogene sequences in primary tumours of the peripheral nervous system induced in rats by transplacental exposure to ethylnitrosourea. *Proc. Natl. Acad. Sci. (USA)*, **84**, 6317–21.

Slamon, D.J., Clark, G.M., Wong, S.G., Levin, W.J., Ullrich, A., and McGuire, W.L. (1987). Human breast cancer: correlation of relapse and survival with amplification of the HER-2/*neu* oncogene. *Science*, **235**, 177–82.

Wen, D., Peles, E., Cupples, R., Suggs, S.V., Bacus, S.S., Luo, Y., Trail, G., Hu, S., Silbiger, S.M., Levy, R.B., Koski, R.A., Lu, H.S., and Yarden, Y. (1992). *Neu* differentiation factor: a transmembrane glycoprotein containing an EGF domain and an immunoglobulin homology unit. *Cell*, **69**, 559–72.

Yosef Yarden:
Department of Chemical Immunology,
The Weizmann Institute of Science,
Rehovot 76100, Israel

Dunzhi Wen:
Amgen Center,
Thousand Oaks, California 91320, USA

Neu/ErbB-2 Receptor

The neu/erbB-2 tyrosine kinase is a single-chain receptor that belongs to the epidermal growth factor (EGF) receptor subfamily. The 1255 amino-acid long glycoprotein spans the plasma membrane once. Its extracellular domain consists of two cysteine-rich domains whereas the intracytoplasmic portion carries tyrosine-specific protein kinase activity. Neu functions as a co-receptor for the neu differentiation factor and other neuregulins. However, ligand(s) that recognize neu with no involvement of other surface molecules may exist. Signal transduction by neu is initiated by autophosphorylation of several tyrosine residues located at the C-terminus of the receptor, and it employs src homology 2-containing proteins. In addition, neu interacts with the EGF receptor, and possibly with other related proteins, to generate distinct functions. The receptor is expressed in most tissues and organs, excluding the haemopoietic system. Its putative functions include the regulation of growth and differentiation of secretory epithelia, muscle, and Schwann cells. Pathologically, neu is involved in malignancies of cells of epithelial and neuronal origins: whereas overexpression of the protein characterizes a subset of human adenocarcinomas with apparently poor prognosis, a point mutation activates the oncogenic potential of neu in the embryonic rodent nervous system.

■ Alternative names

ErbB-2, HER-2.

■ The neu protein

The rodent 185 kD glycoprotein consists of 1260 amino acids (Bargmann *et al.* 1986) (GenBank accession number X03662), whereas the human protein is 1255 amino acids long (Coussens *et al.* 1985; Yamamoto *et al.* 1986) (GenBank accession number M11730). An insect homologue of neu has been identified in *Drosophila melanogaster* (Livneh *et al.* 1985). The protein is bisected by a single hydrophobic stretch that functions as a transmembrane domain and separates the extracellular ligand-binding portion from the cytoplasm-facing tyrosine kinase domain (Fig. 1). Neu is related in its molecular architecture and amino acid sequence to three other receptor tyrosine kinases. These are the receptor for EGF, erbB-3/HER3 (Kraus *et al.* 1989; Plowman *et al.* 1990) and erbB-4/HER-4 (Plowman *et al.* 1993). They constitute a subfamily of receptor tyrosine kinases which is characterized by an ectodomain, which contains two cysteine-rich domains and an intervening spacer that most likely functions as the ligand-binding cleft. In addition, as in other members of its group, the tyrosine kinase sequence of neu is flanked at its C-terminus by an approximately 250-amino-acid long tail that contains at least five tyrosine autophosphorylation sites.

■ Ligands of neu

Because neu was first identified as an oncogenic protein, its endogenous ligand is still unknown. A few candidate ligands have been isolated. The only one that was molecularly cloned is the neu differentiation factor (NDF)/heregulin (Peles *et al.* 1992; Wen *et al.* 1992; Holmes *et al.* 1992). This EGF-related glycoprotein is a member of a large family of factors, collectively termed neuregulins, that arise by alternative splicing from a single gene (see corresponding entry in this volume). Several other activities that putatively interact with neu were isolated and partially

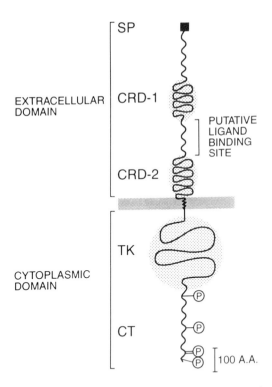

EXTRACELLULAR
DOMAIN

SP

CRD-1

PUTATIVE
LIGAND
BINDING
SITE

CRD-2

CYTOPLASMIC
DOMAIN

TK

CT

100 A.A.

Figure 1. Schematic representation of the neu protein. The horizontal grey bar represents the plasma membrane. The protein is drawn to scale and the following structural motifs are indicated: signal peptide (SP), cysteine-rich domains (CRDs), the tyrosine kinase (TK) portion, and the C-terminus (CT). The locations of the major autophosphorylation sites are shown by the letter P.

characterized. They include T-cell-derived 8–24 kDa factors (Dobashi et al. 1991), a 25 kDa kidney protein (Huang and Huang 1992), a 75 kDa glycoprotein from mammary tumour cells (Lupu et al. 1992), and a macrophage-derived protein (Tartakhovsky et al. 1991). Although NDF becomes covalently cross-linked to neu and it binds specifically and with high affinity ($K_D = 0.5 \times 10^{-9}$M) to the receptor (Holmes et al. 1992; Peles et al. 1993), this interaction is cell-type specific. For example, fibroblasts and ovarian cells that express neu do not interact with NDF, implying that a still unknown tissue-specific co-receptor is essential for the interaction of neu with NDF (Peles et al. 1993).

◼ Neu signal transduction

The signalling mechanism that is employed by neu is analogous to the pathways utilized by other growth factor receptors (Ullrich and Schlessinger 1990) and it includes the following steps: ligand binding to monomeric receptors is followed by rapid formation of receptor dimers in which the tyrosine kinase undergoes catalytic activation, and

subsequently allows autophosphorylation of the receptor on five C-terminally located tyrosine residues (Hazan et al. 1990). These behave as docking sites for various src homology-2 (SH-2) containing proteins, which activate multiple biochemical pathways in the cytoplasm and culminate in the regulation of gene expression. A unique feature of neu is its relatively high basal level of autophosphorylation in living cells and a concomitant high mitogenic activity in the absence of a ligand. This activity was attributed to a single amino acid, within the intracellular juxtamembrane region, which apparently enables recruitment of substrates that are involved in mitogenesis (Di Fiore et al. 1992). On the basis of the analysis of signal transduction by chimeric EGF-receptor-neu proteins, in which the cytoplasmic domain was derived from neu, it was inferred that this receptor tyrosine kinase mediates changes in intracellular Ca^{2+} concentrations, activates turnover of phosphatidylinositol and transcriptionally induces fos and jun expression (Sistonen et al. 1989). The underlying molecular mechanism includes physical associations of the receptor with several signalling proteins (e.g., phospholipase Cγ, the GTPase activating protein of ras, and phosphatidylinositol 3'-kinase), and activation of the serine and threonine-specific MAP-kinase (Peles and Yarden 1993).

In addition to this signalling pathway, which is initiated by neu homodimer formation, an alternative mechanism involves heterodimerization with the ligand-occupied EGF-receptor (Wada et al. 1990; Goldman et al. 1990). Although it is still unknown whether distinct substrates are coupled to the heterodimers, it appears that these types of dimers are characterized by relatively high affinity for EGF, very high kinase activity, and remarkably potent coupling to mitogenesis (Kokai et al. 1989).

◼ Neu gene and transcription

The neu gene was localized to the long arm of human chromosome 17 (Schechter et al. 1985). During embryonic development the gene is transcribed in a variety of tissues, including the nervous system, connective tissues and secretory epithelium (Kokai et al. 1987). In adults the protein was identified on cell membranes in the gastrointestinal, respiratory, reproductive, and urinary tracts, as well as in the skin, breast, liver and pancreas. This tissue-specific expression is driven by a promoter that is G–C rich, includes no TATA box and contains several binding sites for cis-acting elements. One of the latter mediates transcriptional repression of neu by a ligand-stimulated oestrogen receptor (Russell and Hung 1992).

◼ Biological functions

The physiological role of neu is unknown. However, in vitro studies suggest that it regulates cell growth in response to binding of NDF and possibly other ligands. Under certain conditions neu may mediate inhibitory signals for cell proliferation. These include NDF- or gp30-induced growth arrest and differentiation of certain human mammary tumour cells that overexpress the receptor (Peles et al. 1992; Bacus et al. 1992). Alternatively, some monoclonal antibodies to neu can inhibit cellular proliferation, probably by downregulation of the relatively high basal activity of the receptor. In addition to its role in secretory epithelia, neu transmits a mitogenic signal to Schwann cells subsequent to its activation by the glial growth factors (GGFs) (Marchionni et al. 1993). Another putative ligand of neu, the acetylcholine-receptor inducing activity (ARIA), is secreted by motor neurones and elevates the synthesis of acetylcholine receptors by muscle cells (Falls et al. 1993). Thus, neu fulfills a role in the formation of the neuromuscular junction, and probably also in neurite outgrowth and ensheathment by Schwann cells. Consistent with this possibility, neu undergoes transcriptional activation following nerve degeneration.

■ Pathological aspects

Neu has been linked to cancer development and metastasis in a variety of tissues (Stancovski et al. 1994). A point mutation which affects a valine residue of the transmembrane domain of neu and replaces it with glutamic acid causes Schwannomas in rats (Perantoni et al. 1987). The defective receptor undergoes constitutive catalytic activation as a result of dimer formation (Weiner et al. 1989). The mutation can be induced by a carcinogenic agent that acts across the placenta in mid-gestation embryos. Presumably, this reflects a specific developmental stage in which neu and its ligands reach a peak of activity. Although similar mutations were not identified in human cancers, the neu gene was found to be amplified, or otherwise overexpressed, in adenocarcinomas of the breast, ovary, colon, lung, and stomach (Stancovski et al. 1994). Similar levels of overexpression lead to oncogenic transformation of rodent fibroblasts both in vitro and in vivo (Di Fiore et al. 1987). Consistent with this observation, the subset of mammary and ovarian tumours that overexpress neu (20–25 per cent of all tumours) is more virulent than non-overexpressors (Slamon et al. 1987, 1989). In breast tumours, positivity is higher in in situ ductal carcinomas and correlates with a higher growth rate. In addition, neu overexpression correlates with several prognostic factors that include nodal status and an undifferentiated phenotype. Importantly, overexpression can be used as a predictor of poor prognosis (disease-free survival and overall survival). This identifies neu as a potential target for anti-tumour therapies that utilize monoclonal antibodies and ligands in their native or drug-conjugated forms.

■ References

Bacus, S.S., Huberman, E., Chin, D., Kiguchi, K., Simpson, S., Lippman, M., and Lupu, R. (1992). A ligand for the erbB-2 oncogene product (gp30) induces differentiation of human breast cancer cells. Cell Growth Differ., 3, 401–11.

Bargmann, C.I., Hung, M.-C., and Weinberg, R.A. (1986). Multiple independent activations of the neu oncogene by a point mutation altering the transmembrane domain of p185. Cell, 45, 649–57.

Coussens, L., Yang-Feng, T.L., Liao, Y.C., Chen, E., Gray, A., McGrath, J., Seeburg, P.H., Libermann, T.A., Schlessinger, J., Francke, U., Levinson, A., and Ullrich, A. (1985). Tyrosine kinase receptor with extensive homology to EGF receptor shares chromosomal location with neu oncogene. Science, 230, 1132–9.

Di Fiore, P.P., Pierce, J.H., Kraus, M.H., Segatto, O., King, C.R., and Aaronson, S.A. (1987). erbB-2 is a potent oncogene when overexpressed in NIH/3T3 cells. Science, 237, 178–82.

Di Fiore, P.P., Helin, K., Kraus, M.H., Pierce, J.H., Artrip, J., Segatto, O., and Bottaro, DP. (1992). A single amino acid substitution is sufficient to modify the mitogenic properties of the epidermal growth factor receptor to resemble that of gp185erbB-2. EMBO J., 11, 3927–33.

Dobashi, K., Davis, J.G., Mikami, Y., Freeman, J.K., Hamuro, J., and Greene, M.I. (1991). Characterization of a neu/c-erbB-2 protein-specific activating factor. Proc. Natl. Acad. Sci. (USA), 88, 8582–6.

Falls, D.L., Rosen, K.M., Corfas, G., Lane, W.S., and Fischbach, G.D. (1993). ARIA, a protein that stimulates acetylcholine receptor synthesis, is a member of the neu ligand family. Cell, 72, 801–15.

Goldman, R., Levy, R.B., Peles, E., and Yarden, Y. (1990). Heterodimerization of the erbB-1 and erbB-2 receptors in human breast carcinoma cells: a mechanism for receptor transregulation. Biochemistry, 29, 11024–8.

Hazan, R., Margolis, B., Dombalagian, M., Ullrich, A., Zilberstein, A., and Schlessinger, J. (1990). Identification of autophosphorylation sites of HER2/neu. Cell Growth Differ., 1, 3–7.

Holmes, W.E., Sliwkowski, M.X., Akita, R.W., Henzel, W.J., Lee, J., Park, J.W., Yansura, D., Abadi, N., Raab, H., Lewis, G.D., Shepard, H.M., Kuang, W.-J., Wood, W.I., Goeddel, D.V., and Vandlen, R.L. (1992). Identification of heregulin, a specific activator of p185erbB2. Science, 256, 1205–10.

Huang, S.S., and Huang, J.S. (1992). Purification and characterization of the neu/erb B2 ligand-growth factor from bovine kidney. J. Biol. Chem., 267, 11508–12.

Kokai, Y., Cohen, J.A., Drebin, J.A., and Greene, M.I. (1987). Stage- and tissue-specific expression of the neu oncogene in rat development. Proc. Natl. Acad. Sci. (USA), 84, 8498–501.

Kokai, Y., Myers, J.N., Wada, T., Brown, V.I., LeVea, C.M., Davis, J.G., Dobashi, K., and Greene, M.I. (1989). Synergistic interaction of p185c-neu and the EGF receptor leads to transformation of rodent fibroblasts. Cell, 58, 287–92.

Kraus, M., Issing, W., Miki, T., Popescu, N.C., and Aaronson, S.A. (1989). Isolation and characterization of ERBB3, a third member of the ERBB/epidermal growth factor receptor family: evidence for overexpression in a subset of human mammary tumours. Proc. Natl. Acad. Sci. (USA), 86, 9193–7.

Livneh, E., Glazer, L., Segal, D., Schlessinger, J., and Shilo, B.-Z. (1985). The Drosophila EGF receptor gene homolog: conservation of both hormone binding and kinase domains. Cell, 40, 599–607.

Lupu, R., Colomer, R., Kannan, B., and Lippman, M.E. (1992). Characterization of a growth factor that binds exclusively to the erbB-2 receptor and induces cellular responses. Proc. Natl. Acad. Sci. (USA), 89, 2287–91.

Marchionni, M.A., Goodearl, A.D., Chen, M.S., Bermingham-McDonogh, O., Kirk, C., Hendricks, M., Danehy, F., Misumi, D., Sudhalter, J., Kobayashi, K., Wroblewski, D., Lynch, C., Baldassare, M., Hiles, I., Davis, J.B., Hsuan, J.J., Totty, N.F., Otsu, M., McBurney, R.N., Waterfield, M.D., Stroobant, P., and Gwynne, D. (1993). Glial growth factors are alternatively spliced erbB2 ligands expressed in the nervous system. Nature, 362, 312–8.

Peles, E., and Yarden, Y. (1993). Neu and its ligands: From an onco-gene to neural factors. *Bioessays*, **15**, 815–24.

Peles, E., Bacus, S.S., Koski, R.A., Lu, H.S., Wen, D., Ogden, S.G., Levy, R.B., and Yarden, Y. (1992). Isolation of the neu/HER-2 stimulatory ligand: a 44 kd glycoprotein that induces differentiation of mam-mary tumour cells. *Cell*, **69**, 205–16.

Peles, E., Ben-Levy, R., Tzahar, E., Liu, N., Wen, D., and Yarden, Y. (1993). Cell-type specific interaction of Neu differentiation factor (NDF/heregulin) with Neu/HER-2 suggests complex ligand-recep-tor relationships. *EMBO J.*, **12**, 961–71.

Perantoni, A.O., Rice, J.M., Reed, C.D., Watatani, M., and Wenk, M.L. (1987). Activated *neu* oncogene sequences in primary tumours of the peripheral nervous system induced in rats by transplacental exposure to ethylnitrosourea. *Proc. Natl. Acad. Sci. (USA)*, **84**, 6317–21.

Plowman, G.D., Whitney, G.S., Neubaur, M.G., Green, J.M., Mc-Donald, V.L., Todaro, G.J., and Shoyab, M. (1990). Molecular clon-ing and expression of an additional epidermal growth factor receptor-related gene. *Proc. Natl. Acad. Sci. (USA)*, **87**, 4905–9.

Plowman, G.D., Culouscou, J.-M., Whitney, G.S., Green, J.M., Carl-ton, G.W., Foy, L., Neubauer, M.G., and Shoyab, M. (1993). Ligand-specific activation of HER4/p180erbB4, a fourth member of the epidermal growth factor receptor family. *Proc. Natl. Acad. Sci. (USA)*, **90**, 1746–50.

Russell, K.S., and Hung, M.C. (1992). Transcriptional repression of the neu protooncogene by estrogen stimulated estrogen recep-tor. *Cancer Res.*, **62**, 6624–9.

Schechter, A.L., Hung, M.C., Vaidyanathan, L., Weinberg, R.A., Yang Feng, T.I., Francke, U., Ullrich, A., and Coussens, L. (1985). The *neu* gene: an *erb*B-homologous gene distinct from and unlinked to the gene encoding the EGF receptor. *Science*, **229**, 976–8.

Sistonen, L., Holtta, E., Lehvaslaiho, H., Lehtola, L., and Alitalo, K. (1989). Activation of the neu tyrosine kinase induces the fos/jun transcription factor complex, the glucose transporter and ornith-ine decarboxylase. *J. Cell. Biol.*, **109**, 1911–9.

Slamon, D.J., Clark, G.M., Wong, S.G., Levin, W.J., Ullrich, A., and

McGuire, W.L. (1987). Human breast cancer: correlation of relapse and survival with amplification of the HER-2/*neu* oncogene. *Science*, **235**, 177–82.

Slamon, D.J., Godolphin, W., Jones, L.A., Holt, J.A., Wong, S.G., Keith, D.E., Levin, W.J., Stuart, S.G., Udove, J., Ullrich, A., and Press, M.F. (1989). Studies of the HER-2/*neu* proto-oncogene in human breast and ovarian cancer. *Science*, **244**, 707–12.

Stancovski, I., Sela, M., and Yarden, Y. (1994). Molecular and clinical aspects of the Neu/ErbB-2 receptor tyrosine kinase. In *Regulatory mechanisms of breast cancer* (ed. M. Lippman and R. Dikson). Kluwer, Boston.

Tartakhovsky, A., Zaichuk, T., Prassolov, V., and Butenko, Z.A. (1991). A 25 kDa polypeptide is the ligand for p185neu and is secreted by activated macrophages. *Oncogene*, **6**, 2187–96.

Ullrich, A., and Schlessinger, J. (1990). Signal transduction by recep-tors with tyrosine kinase activity. *Cell*, **61**, 203–12.

Wada, T., Qian, X.L., and Greene, M.I. (1990). Intermolecular associ-ation of the p185neu protein and EGF receptor modulates EGF receptor function. *Cell*, **61**, 1339–47.

Weiner, D.B., Liu, J., Cohen, J.A., Williams, W.V., and Greene, M.I. (1989). A point mutation in the neu oncogene mimics ligand induction of receptor aggregation. *Nature*, **339**, 230–1.

Wen, D., Peles, E., Cupples, R., Suggs, S.V., Bacus, S.S., Luo, Y., Trail, G., Hu, S., Silbiger, S.M., Levy, R.B., Koski, R.A., Lu, H.S., and Yarden, Y. (1992). Neu differentiation factor: a transmembrane glycoprotein containing an EGF domain and an immunoglobulin homology unit. *Cell*, **69**, 559–72.

Yamamoto, T., Ikawa, S., Akiyama, T., Semba, K., Nomura, N., Miya-jima, N., Saito, T., and Toyoshima, K. (1986). Similarity of protein encoded by the human c-*erb*-B-2 gene to epidermal growth factor receptor. *Nature*, **319**, 230–4.

Yosef Yarden and Rachel Ben Levy:
Department of Chemical Immunology,
The Weizmann Institute of Science,
Rehovot 76100, Israel

Erythropoietin

Erythropoietin (epo) is a glycoprotein (M_r = 30400 Da) that stimulates proliferation and differen-tiation of erythroid progenitor cells in bone marrow. It is produced by kidney cells at a low level normally and in increased amount in response to anaemia. It is used clinically to treat anaemia of chronic renal failure and has the potential to correct the anaemias of chronic inflammatory and malignant diseases.

■ Epo protein

Epo was purified from the urine of anaemic patients (Miyake *et al.* 1977), and the amino acid sequence derived from the purified protein (Lai *et al.* 1986) led to the cloning of the human epo gene (Lin *et al.* 1985; Jacobs *et al.* 1985). Epo is a secreted glycoprotein containing 193 amino acids (human) or 192 amino acids (monkey and mouse) including a hydrophobic leader sequence of 27 amino acids. Approxi-mately 40 per cent of the human recombinant mature

protein is composed of carbohydrate. It has one site (Ser 126) with O-linked glycosylation and three sites (Asn 24, 38, 83) with N-linked glycosylation. The N-linked glycosylation is necessary for biological activity, solubility, secretion and structural stability of the molecule. The O-linked glycosyla-tion appears to be unnecessary for correct processing and secretion. Internal disulphide bridges exist in human epo between Cys 7 and Cys l61 and between Cys 29 and Cys 33. At least one of these disulphide bridges is important for the secondary structure, since the biological activity of human epo can be inactivated reversibly by reduction or irreversibly by alkylation (Wang *et al.* 1985). The higher order structure of epo is not yet known.

■ Epo gene and transcription

Epo genes or cDNAs have been isolated from human, monkey (Lin *et al.* 1986), rat (Nagao *et al.* 1992), mouse (McDonald *et al.* 1986; Shoemaker and Mitsock 1986; GenBank MUSERPA), sheep, dog, cat, and pig (Wen *et al.* 1993). The epo gene is located on chromosome 7 q11–q22 in human and chromosome 5 in mouse. It exists as a single copy and comprises five exons and four introns. One of the intervening sequences occurs within the sequence encoding the signal peptide. The basic organization of the epo gene is similar between species and the proximal promoter region shows 90 per cent sequence homology in mouse and human genes, but intronic regions show a high degree of divergence. Within the coding region, the sequence homology between different species is more than 80 per cent. It generates a single mRNA of 1.39 kb. Epo gene expression shows tissue specificity as well as developmental stage specificity. In mice, it is expressed in liver during the fetal and postnatal period after which kidney plays a progressively greater role in producing epo (Koury and Koury 1993). The basal level of epo mRNA is undetectable in adult liver and at the limit of detection by Northern analysis in adult kidney. Epo mRNA is also detected by PCR in brain, liver and spleen; the physiological significance of these findings is uncertain (Fandrey and Bunn 1993). Anti-sense experiments and PCR demonstrate that its presence in bone marrow haemopoietic precursor cells may have an internal autocrine role in erythropoiesis (Hermine *et al.* 1991). Its expression in kidney is induced many fold following hypoxia, anaemia, or cobalt chloride administration (Caro *et al.* 1982). In addition, hepatoma cells HepG2 or Hep3B are also capable of sensing hypoxia and producing epo (Goldberg *et al.* 1987).

The regulation of epo production is primarily under transcriptional control, but post transcriptional stabilization of epo mRNA is also suggested (Schuster *et al.* 1989; Goldberg *et al.* 1991). Both the 5′ and 3′ regions of epo gene have been shown to contain regulatory elements that confer oxygen responsiveness and tissue specific expression of the epo gene (Blanchard *et al.* 1992). A hypoxia-responsive enhancer element has been defined in the 3′ untranslated region that interacts with Hep3B nuclear factors for the inducible expression, whereas in kidney additional sequences are required for oxygen responsiveness of the epo gene (Semenza *et al.* 1991a,b). Murine kidneys were shown to contain two nuclear factors, a 47 kDa protein and a ribonucleoprotein that bind to the same sequence in the 5′ region of the epo gene. The 47 kDa protein was identified as a positive transacting factor whose action in the uninduced state appeared to be blocked by the ribonucleoprotein. This negative transacting factor was reduced in amount following *in vivo* activation of the epo gene by cobalt or hypoxia treatment of the mice (Beru *et al.* 1990). Transgenic mice studies reveal that the expression of the human epo transgene is not restricted to liver or kidney tissues and separate elements confer hypoxia inducibility of the injected gene in liver and kidney (Semenza *et al.* 1991a,b). Thus, complex control mechanisms seem to operate in regulating the epo gene.

■ Biological activity

Erythropoietin, as the name implies stimulates erythropoiesis by stimulating proliferation and differentiation of progenitor cells present in fetal liver, spleen and in bone marrow. The target cells for epo action are committed erythroid progenitor cells and early erythroblasts, which have specific receptors for the hormone. The most primitive erythrocytic progenitor used for *in vitro* studies is the burst forming unit–erythroid (BFU-E), which gives rise to large multi-clustered bursts of haemoglobin-synthesizing progeny when cultured in the presence of epo. More mature cells in the erythroid pathway are those called colony forming units–erythroid (CFU-E). CFU-E are very sensitive to epo and are totally dependent on epo for colony formation and terminal differentiation, as well as for their survival (Koury and Bondurant 1992).

■ Pathophysiology

Secondary erythrocytosis can result from excessive production of epo. This may be associated with tissue hypoxia as is the case at high altitude, in patients with severe cyanotic congenital heart disease or as an inherited disorder in association with mutant haemoglobin (Jelkmann 1992).

Epo deficiency is the primary cause of the anaemia in chronic renal failure. Mild anaemia associated with chronic infections, autoimmune disease or malignancy develops because of inhibition of proliferation of erythrocytic progenitors. In addition, plasma epo levels are lower than expected from the red cell mass in patients with rheumatoid arthritis, AIDS and cancer. The mediators (IL-1 and TNFα) of anaemia of chronic disorder may inhibit the action of epo on its target cells (Jelkmann 1992).

Recombinant epo is used clinically to treat the anaemia of chronic renal failure. It raises hematocrit and blood haemoglobin concentration in a dose dependent and predictable way, resulting in increased well being and exercise tolerance of the patients with chronic renal failure. Epo treatment replaces blood transfusions and their attendant risks.

■ References

Beru, N., Smith, D., and Goldwasser, E. (1990). Evidence suggesting negative regulation of the erythropoietin gene by ribonucleoprotein. *J. Biol. Chem.*, **265**, 14100–4.

Blanchard, K.L., Acquaviva, A.M., Galson, D.L., and Bunn, H.F. (1992). Hypoxic induction of the human erythropoietin gene: cooperation between the promoter and enhancer, each of which contains steroid receptor response elements. *Mol. Cell. Biol.*, **12**, 5373–85.

Caro, J., Erslev, A.J., Silver, R., Miller, O., and Birgegard, G. (1982). Erythropoietin production in response to anemia or hypoxia in the newborn rat. *Blood*, **60**, 984–8.

Fandrey, J., and Bunn, H.F. (1993). *In vivo* and *in vitro* regulation of erythropoietin mRNA: measurement by competitive polymerase chain reaction. *Blood*, **81**, 617–23.

Goldberg, M.A., Glass, G.A., Cunningham, J.M., and Bunn, H.F. (1987). The regulated expression of erythropoietin by two human hepatoma cell lines. *Proc. Natl. Acad. Sci. (USA)*, **84**, 7972–6.

Goldberg, M.A., Gaut, C.C., and Bunn, H.F. (1991). Erythropoietin mRNA levels are governed by both the rate of gene transcription and posttranscriptional events. *Blood*, **77**, 271–7.

Hermine, O., Beru, N., Pech, N., and Goldwasser, E. (1991). An autocrine role for erythropoietin in mouse hematopoietic cell differentiation [published erratum appears in *Blood* 1992 79, 3397]. *Blood*, **78**, 2253–60.

Jacobs, K., Shoemaker, C., Rudersdorf, R., Neill, S.D., Kaufman, R.J., Mufson, A., Seehra, J., Jones, S.S., Hewick, R., Fritsch, E.F., Kawakita, M., Shimizu, T., and Miyake, T. (1985). Isolation and characterization of genomic and cDNA clones of human erythropoietin. *Nature*, **313**, 806–10.

Jelkmann, W. (1992). Erythropoietin: structure, control of production, and function. *Physiol. Rev.*, **72**, 449–89.

Koury, M.J., and Bondurant, M.C. (1992). The molecular mechanism of erythropoietin action. *Eur. J. Biochem.*, **210**, 649–63.

Koury, S.T., and Koury, M.J. (1993). Erythropoietin production by the kidney. *Semin. Nephrol.*, **13**, 78–86.

Lai, P.-H., Everett, R., Wang, F.F., Arakawa, T., and Goldwasser, E. (1986). Structural characterization of human erythropoietin. *J. Biol. Chem.*, **261**, 3116–21.

Lin, F.K., Suggs, S., Lin, C.H., Browne, J.K., Smalling, R., Egrie, J.C., Chen, K.K., Fox, G.M., Martin, F., Stabinsky, Z., Badrawi, S.M., Lai, P.-H., and Goldwasser, E. (1985). Cloning and expression of the human erythropoietin gene. *Proc. Natl. Acad. Sci. (USA)*, **82**, 7580–4.

Lin, F.K., Lin, C.H., Lai, P.H., Browne, J.K., Egrie, J.C., Smalling, R., Fox, G.M., Chen, K.K., Castro, M., and Suggs, S. (1986). Monkey erythropoietin gene: cloning, expression and comparison with the human erythropoietin gene. *Gene*, **44**, 201–9.

McDonald, J.D., Lin, F.K., and Goldwasser, E. (1986). Cloning, sequencing, and evolutionary analysis of the mouse erythropoietin gene. *Mol. Cell. Biol.*, **6**, 842–8.

Miyake, T., Kung, C.K., and Goldwasser, E. (1977). Purification of human erythropoietin. *J. Biol. Chem.*, **252**, 5558–64.

Nagao, M., Suga, H., Okano, M., Masuda, S., Narita, H., Ikura, K., and Sasaki, R. (1992). Nucleotide sequence of rat erythropoietin. *Biochem. Biophys. Acta*, **1171**, 99–101.

Schuster, S.J., Badiavas, E.V., Costa-Giomi, P., Weinmann, R., Erslev, A.J., and Caro, J. (1989). Stimulation of erythropoietin gene transcription during hypoxia and cobalt exposure. *Blood*, **73**, 13–16.

Semenza, G.L, Koury, S.T., Nejfelt, M.K., Gearhart, J.D., and Antonarakis, S.E. (1991a) Cell-type-specific and hypoxia-inducible expression of the human erythropoietin gene in transgenic mice. *Proc. Natl. Acad. Sci. (USA)*, **88**, 8725–9.

Semenza, G.L, Nejfelt, M.K., Chi, S.M., and Antonarakis, S.E. (1991b) Hypoxia-inducible nuclear factors bind to an enhancer element located 3' to the human erythropoietin gene. *Proc. Natl. Acad. Sci. (USA)*, **88**, 5680–4.

Shoemaker, C.B., and Mitsock, D. (1986). Murine erythropoietin gene: cloning, expression and human gene homology. *Mol. Cell. Biol.*, **6**, 849–58.

Wang, F.F., Kung, C.K-H., and Goldwasser, E. (1985). Some chemical properties of human erythropoietin. *Endocrinol.*, **116**, 2286–92.

Wen, D., Boissel, J.P.R., Tracy, T.E., Gruninger, R.H., Mulcahey, L.S., Czelusniak, J., Goodman, M., and Bunn, H.F. (1993). Erythropoietin structure-function relationship: high degree of sequence homology among mammals. *Blood*, **82**, 1507–16.

Madhu Gupta and Eugene Goldwasser:
Department of Biochemistry and Molecular Biology,
The University of Chicago,
Chicago, IL, USA

Erythropoietin Receptor

Receptors for erythropoietin (epo) are expressed on a wide range of epo-responsive cells, including primary erythroid progenitors from the yolk sac, fetal liver and bone marrow and a variety of cell lines. A receptor which binds epo has been cloned. Stable expression of this receptor also allows unresponsive cell lines to respond to epo. The cloned epo receptor is a single transmembrane-spanning glycoprotein and is a member of the cytokine receptor family. The defining features of this family are a series of conserved amino acids in the extracellular domain including four cysteines and the five-amino acid motif WSXWS. This primary sequence similarity is thought to underlie a common tertiary structure comprising two domains of seven beta sheets. The mechanism by which the epo receptor transduces a biological signal is unclear—however, homodimerization and activation of the 130kDa cytoplasmic kinase Jak-2 appear to be important early events.

■ Expression patterns and binding properties

Epo receptors are detectable in binding experiments using ^{125}I-epo. Primary erythroid progenitors in the yolk sac, fetal liver, bone marrow and spleen express small numbers of receptors, as do megakaryocytes (300–1000 per cell) (Youssoufian et al. 1993). Similar numbers of receptors are expressed on a variety of epo-responsive and epo-unresponsive cell lines. The equilibrium dissociation constant (K_D) for the interaction between epo and its receptor is usually reported to be between 400 and 1000 pM (Youssoufian et al. 1993). In some instances, however, cells appear to express receptors with two affinities—a high affinity component with a K_D of 100 pM and a second component with a K_D of 600 pM. Epo receptors have also been described in some non-haemopoietic tissues, notably the placenta (Youssoufian et al. 1993).

Using cross-linking reagents, two proteins have been identified which are capable of binding epo; these have an apparent molecular weight of 85 000 and 100 000 Da (Youssoufian et al. 1993). The relationship between these species is unclear, although they are suggested to be related (Youssoufian et al. 1993). The epo receptor appears to be glycosylated rapidly after synthesis. In many cell types the majority of the receptor remains sensitive to endoglycosidase-H digestion, suggesting that it is retained within the endoplasmic reticulum (Youssoufian et al. 1993). This proposition is supported by the demonstration that in some haemopoietic cells and in COS cells the epo receptor is stably associated with the heavy-chain binding protein BiP (Murray et al., unpublished observations).

■ Epo receptor cDNA and gene

In 1989 a cDNA library from the mouse erythroleukaemia cell line (MEL) was screened by expression in COS cells using ^{125}I-epo as a probe. A cDNA encoding a protein of 507 amino acids, with a single transmembrane region and with the capacity to bind epo specifically and with high affinity was isolated (D'Andrea et al. 1989). Stable expression of the epo receptor cDNA in a variety of IL-3-dependent cell lines allows these cells to proliferate in response to epo (Youssoufian et al. 1993). cDNA for the human and rat epo receptors have also been cloned, and exhibit a high degree of sequence similarity with the mouse epo receptor (Fig. 1) (Jones et al. 1990). The epo receptor is a member of the cytokine receptor superfamily, which includes the receptors for growth hormone, prolactin, interleukins-2,3,4,5,6,7,9,11, and 13, GM-CSF, G-CSF, LIF, CNTF, NKSFp40 and oncostatin-M, as well as the orphan receptor mpl (Bazan 1990). The defining features of this family are the presence, in the extracellular domain of four cysteines, a series of aromatic residues, and the motif Trp-Ser-Xaa-Trp-Ser (Bazan 1990). The epo receptor, as with other members of this family, lacks a cytoplasmic kinase domain.

The genes encoding the murine and human epo receptors have been characterized. Both have eight exons and share a common genomic structure with the genes for interleukin-2 receptor β chain, AIC2A, AIC2B and the human and mouse interleukin-7 receptors. The genes for murine and human epo are on chromosomes 9 and 19, respectively (Youssoufian et al. 1993).

The GenBank accession numbers for epo receptor cDNA sequences are M60459, S04843, and D13566. The GenBank accession numbers for the epo receptor genomic sequences are S45332 and X53081.

■ Structure, function, and signal transduction

The structure of the growth hormone/growth hormone receptor complex has been solved recently (DeVos et al. 1992). By analogy, it is likely that the extracellular domain of the epo receptor is also composed of two domains containing seven β-strands, in a sandwich arrangement similar to CD-4 and pap-D. Little is known about the identity of residues responsible for binding epo. Despite suggestions that the WSXWS motif of cytokine receptors lies in the ligand binding groove (Bazan 1990), mutation of these

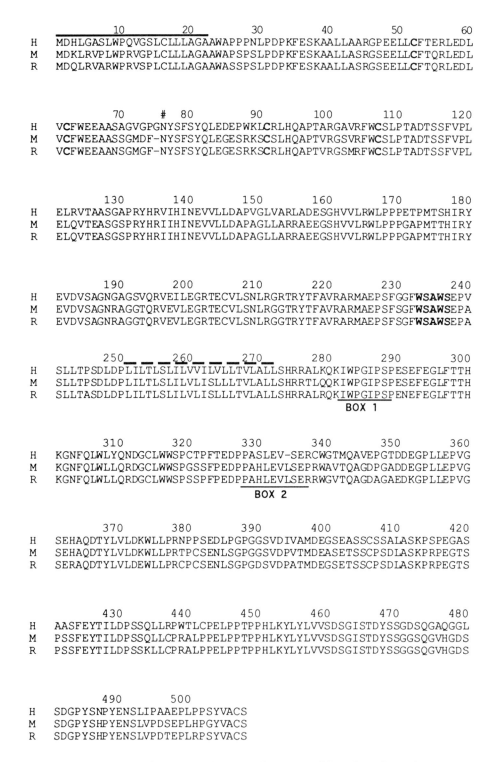

Figure 1. Alignment of the amino acid sequence of human (H), murine (M), and rat (R) erythropoietin receptors. The following predicted features are highlighted: signal sequence (solid line), transmembrane domain (broken bar), N-linked glycosylation site ('#'). The cysteine residues and WSXWS motif conserved among members of the cytokine receptor family cysteine are shown in bold. The cytoplasmic elements termed box 1 and box 2 are also labelled.

residues in the epo receptor does not appear to affect cytokine binding—rather, protein folding and exit from the endoplasmic reticulum are affected (Yoshimura et al. 1992).

In an attempt to isolate cells expressing elevated numbers of murine epo receptors, two receptor mutants were discovered (Yoshimura et al. 1990). The first allowed cells to grow in low doses of epo and the second conferred epo-independent growth. The first mutation was a truncation of the C-terminal 42 amino acids. Interestingly, a similar human epo receptor mutation has been recently described in a case of familial erythrocytosis (de la Chappelle et al. 1993a,b). The second mutation results in change of an arginine at position 129 to cysteine (Yoshimura et al. 1990). Receptors containing this mutation, but not wild-type receptors, form disulphide linked homodimers (Watowich et al. 1992). It is attractive to speculate that constitutive dimerization recapitulates an event normally driven by epo and that dimerization represents the first step in the signal transduction process. Dimerization also has been demonstrated for the growth hormone and G-CSF receptors (DeVos et al. 1992; Fukunaga et al. 1990). Epo independent growth can also be achieved by another route: co-expression with the envelope protein (gp[55]) of the polycythemia strain of the spleen focus-forming virus (Li et al. 1990). While gp[55] is thought to associate with the epo receptor the mechanism of receptor activation is not clear.

Although the epo receptor does not itself exhibit intrinsic tyrosine kinase activity, stimulation of cells with epo results in phosphorylation of a variety of proteins (Youssoufian et al. 1993). One of these proteins has an apparent molecular weight of 130 000 Da and can be shown by immunoprecipitation studies to associate with the epo receptor. This molecule has been recently identified as the tyrosine kinase, Jak-2 (Argetsinger et al. 1993). It has been suggested that Jak-2 interacts with a membrane proximal cytoplasmic element of the epo receptor. This region of the epo receptor has been defined by mutagenic studies to be necessary for epo to elicit a proliferative response.

■ References

Argetsinger, L.S., Campbell, G.S., Yang, X., Witthuhn, B.A., Silvennoinen, O., Ihle, J.N., and Carter-Su, C. (1993). Identification of JAK2 as a growth hormone receptor-associated tyrosine kinase. Cell, 74, 237–44.

Bazan, J.F. (1990). Structural design and molecular evolution of a cytokine receptor superfamily. Proc. Natl. Acad. Sci. (USA), 87, 6934–8.

D'Andrea, A.D., Lodish, H.F., and Wong, G.G. (1989). Expression cloning of the murine erythropoietin receptor. Cell, 57, 277–85.

de la Chappelle, A., Sistonen, P., Lehvaslaiho, H., Ikkala, E., and Juvonen, E. (1993a). Familial erythrocytosis genetically linked to erythropoietin receptor gene. Lancet, 341, 82–4.

de la Chappelle, A., Traskelin, A.-L., and Juvonen, E. (1993b). Truncated erythropoietin receptor causes dominantly inherited benign human erythrocytosis. Proc. Natl. Acad. Sci. (USA), 90, 4495–9.

DeVos, A.M., Ultsch, M., and Kossiakoff, A.A. (1992). Human growth hormone and extracellular domain of its receptor: crystal structure of the complex. Science, 255, 306–12.

Fukunaga, R., Ishizaka-Ikeda, E., and Nagata, S. (1990). Purification and characterization of the receptor for murine granulocyte colony-stimulating factor. J. Biol. Chem., 265, 14008–15.

Jones, S.S., D'Andrea, A.D., Haines, L.L., and Wong, G.G. (1990). Human erythropoietin receptor: cloning, expression, and biologic characterization. Blood, 76, 31–5.

Li, J.-P., D'Andrea, A.D., Lodish, H.F., and Baltimore, D. (1990). Activation of cell growth by binding of Friend spleen focus-forming virus gp[55] glycoprotein to the erythropoietin receptor. Nature, 348, 762–4.

Murray, P.J., Hilton, D.J., and Watowich, S.S. Unpublished observations.

Watowich, S.S., Yoshimura, A., Longmore, G.D., Hilton, D.J., Yoshimura, Y., and Lodish, H.F. (1992). Homodimerization and constitutive activation of the erythropoietin receptor. Proc. Natl. Acad. Sci. (USA), 89, 2140–4.

Yoshimura, A., Longmore, G., and Lodish, H.F. (1990). Point mutation in the exoplasmic domain of the erythropoietin receptor resulting in hormone-independent activation and tumourigenicity. Nature, 348, 647–9.

Yoshimura, A., Zimmers, T., Neumann, D., Longmore, G., Yoshimura, Y., and Lodish, H.F. (1992). Mutations in the Trp-Ser-X-Trp-Ser motif of the erythropoietin receptor abolish processing, ligand binding, and activation of the receptor. J. Biol. Chem., 267, 11619–25.

Youssoufian, H., Longmore, G., Neumann, D., Yoshimura, A., and Lodish, H.F. (1993). Structure, function, and activation of the erythropoietin receptor. Blood, 81, 2223–36.

Douglas J. Hilton:
The Walter and Eliza Hall Institute of Medical Research and The Cooperative Research Centre for Cellular Growth Factors,
Melbourne, Australia

Stephanie S. Watowich:
The Whitehead Institute for Biomedical Research, Cambridge, MA, USA.

Granulocyte Colony-Stimulating Factor (G-CSF)

G-CSF is a 20–25 kDa glycoprotein that specifically regulates the production of neutrophilic granulo-cytes as well as enhancing the functional activities of mature neutrophils. It is produced by activated macrophages, endothelial cells, and fibroblasts. G-CSF is widely used clinically in the treatment of patients with neutropenia after cancer chemotherapy.

■ Alternative names

Colony-stimulating factor 3 (CSF-3), macrophage and granulocyte inducer type 1, granulocyte (MGI-1G), granulocyte macrophage colony-stimulating factor β (GM-CSFβ), pluripotent colony-stimulating factor (pluripoetin).

■ G-CSF sequence

The nucleotide sequences of human G-CSF cDNAs (Nagata *et al.* 1986*a*; Souza *et al.* 1986) (GenBank accession numbers X03655 and M13008) predict a protein of 204 amino acids containing a 30-amino-acid leader sequence (Fig. 1). A minor alternatively spliced form of the human G-CSF cDNA coding for 207 amino acids (GenBank accession number X03438) is also known (Nagata *et al.* 1986*b*). Mouse G-CSF cDNA (GenBank accession number M13926) codes for a protein of 208 amino acids with a 30-amino-acid leader sequence (Tsuchiya *et al.* 1986). Human and mouse G-CSFs are 73.6 per cent identical at the amino acid sequence level and fully cross-react both biologically and at receptor-binding level (Nicola *et al.* 1985; Fukunaga *et al.* 1990). There is a significant sequence homology between G-CSF and interleukin-6 (Fig. 2).

■ G-CSF protein

G-CSF is a secreted glycoprotein monomer containing 174 (human) or 178 (mouse) amino acids in the mature protein. The molecular weight of the core protein is 18671 Da (human) or 19061 Da (mouse). The human G-CSF synthesized from the alternatively spliced mRNA contains 177 amino acids and is at least 10 times less active than the one consisting of 174 amino acids. Human G-CSF contains two intramolecular disulphide bonds (Cys 36–Cys 42 and Cys 64–Cys 74) both of which are essential for proper folding and biological activity (Lu *et al.* 1989). The positions of these cysteine residues are conserved between human and mouse G-CSFs. Human G-CSF is O-glycosylated at Thr-133 (Oheda *et al.* 1988) giving rise to a protein of molecular weight 19000 Da. The structure of the sugar moiety attached to the human G-CSF in N-acetylneuraminic acid α(2–6)[galactose β (1–3)] N-acetylgalactosamine (Souza *et al.* 1986; Oheda *et al.* 1988). Mouse G-CSF may also be O-glycosylated and has a molecular weight of 25000 Da

(Nicola *et al.* 1983). The sugar moiety is not necessary for biological activity but it contributes to the stability of the molecule by preventing its aggregation (Oheda *et al.* 1990). G-CSF is an acidic glycoprotein with isoelectric points between 5.5 and 6.1 (human) (Nomura *et al.* 1986) or between 4.5 and 5.7 (mouse) (Simpson *et al.* 1987) depending on the degree of sialylation. G-CSF is relatively stable to extreme pH levels (pH 2–10), temperature (50 per cent loss of activity at 70 °C for 30 min), and strong denaturing agents (6M guanidine hydrochloride, 8M urea, 0.1 per cent SDS) as long as the disulphide bonds are intact (Nicola *et al.* 1983). The structure of G-CSF has been determined by X-ray diffraction analysis, and it shows a four α-helical bundle structure. A mutational analysis of the human G-CSF indicated that the N-terminal 11 amino acids are dispensable for biological activity (Kuga *et al.* 1989). An epitope mapping by neutralizing antibodies indicated that the amino acid residues from 20 to 46, as well as the carboxyl terminus, are important for binding to the receptor (Layton *et al.* 1991).

G-CSF can be purified using conventional chromatography steps and HPLC (Nicola *et al.* 1983; Nomura *et al.* 1986; Welte *et al.* 1985).

■ G-CSF gene and transcription

There is only one G-CSF chromosomal gene per haploid genome in the human and murine systems. The G-CSF gene is localized on human chromosome 17q21–22 and mouse chromosome 11 (Kanda *et al.* 1987; Buchberg *et al.* 1988). Human and mouse G-CSF genes consist of about 2.5 kb and contain five exons. (Nagata *et al.* 1986*a*; Tsuchiya *et al.* 1987) (GenBank accession numbers X003656 and X05402). Human but not mouse G-CSF gene can produce two alternatively spliced mRNAs of about 1.5 kb. The minor form codes for G-CSF of 177 amino acids. Like other cytokine mRNAs, the G-CSF mRNA contains several AUUUA sequences in the 3′ non-coding sequence which confers instability on the mRNA.

The G-CSF gene is transcriptionally activated in monocytes and macrophages by bacterial endotoxins (Nishizawa and Nagata 1990), and in endothelial cells and fibroblasts by interleukin-1 or tumour necrosis factor (Seelentag *et al.* 1987). The production of G-CSF is also regulated post-transcriptionally by these inducers which stabalize the G-CSF mRNA (Koeffler *et al.* 1988; Ernst *et al.* 1989). The G-CSF gene is constitutively activated in some carcinoma cells such

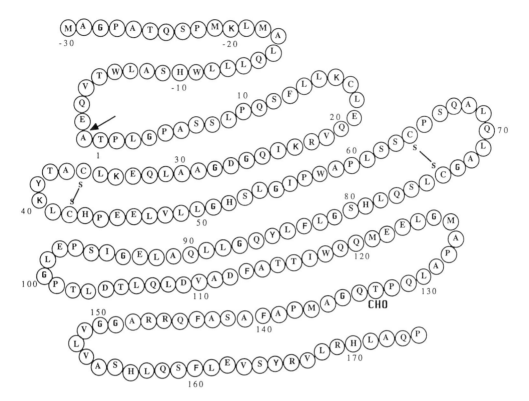

Figure 1. A primary structure of human G-CSF.

as human squamous carcinoma CHU-2, bladder carcinoma 5637, glioblastoma U87MG, and hepatoma SKHEP-1 cell lines (Tweardy *et al.* 1987; Nishizawa *et al.* 1990; Nagata 1990).

The 300 nucleotides upstream of the transcription initiation site are well conserved between human and mouse G-CSF genes, and this region is responsible for expression. In this region, at least three regulatory elements (GPE1–3) were identified (Nishizawa and Nagata 1990; Nishizawa *et al.* 1990). The GPE1 comprises an NF-κB-like element (CSF

box) and an NF-IL-6 binding site, and the GPE2 is an OTF (octamer transcription factor) binding site while the GPE3 has no apparent homology with known promoter elements.

■ Biological functions

G-CSF has an ability to stimulate the colony formation of neutrophilic granulocytes in semi-solid cultures of bone

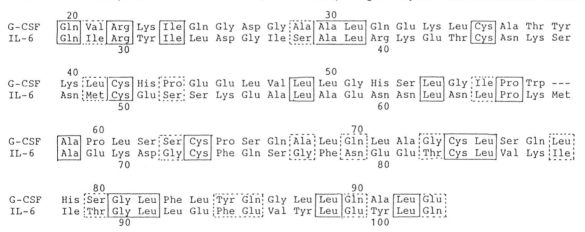

Figure 2. Similarity of G-CSF with interleukin-6.

marrow cells. In this assay, G-CSF has a specific activity of about 2×10^8 units per mg protein (where 50 units/ml is the concentration required for half-maximal stimulation). Unlike other CSFs such as GM-CSF and IL-3, G-CSF is rather specific to progenitor cells of neutrophilic granulocytes. G-CSF stimulates not only proliferation and differentiation of the progenitors but also prolongs the survival of the mature neutrophils and enhances the functional capacity of the mature neutrophils (Nagata 1990; Demetri and Griffin 1991). Several myeloid leukaemia cell lines proliferate in response to G-CSF while some other myeloid cell lines can be induced to differentiate into neutrophilic granulocytes by GCSF. *In vivo* administration of G-CSF specifically induces an increase of neutrophilic granulocytes without apparent pathological effects. It is now widely used clinically for patients with neutropenia (Demetri and Griffin 1991; Davis and Morstyn 1991). G-CSF is administered to cancer patients receiving chemotherapy with or without bone marrow transplantation, patients receiving immunosuppressive agents after organ transplanation, and patients with agranulocytosis.

■ References

Buchberg, A.M., Bedigian, H.G., Taylor, B.A., Brownell, E., Ihle, J.N., Nagata, S., Jenkins, N.A., and Copeland, N.G. (1988). Localization of Evi-2 to chromosome 11: linkage to other proto-oncogene and growth factor loci using interspecific backcross mice. *Oncogene Res.*, **2**, 149–65.

Davis, I., and Morstyn, G. (1991). The role of granulocyte colony-stimulating factor in cancer chemotherapy. *Semi. Hematol.*, **28**, 25–33.

Demetri, G.D., and Griffin, J.D. (1991). Granulocyte colony-stimulating factor and its receptor. *Blood*, **78**, 2791–808.

Ernst, T.J., Ritchie, A.R., Demetri, G.D., and Griffin, J.D. (1989). Regulation of granulocyte- and monocyte-colony stimulating factor mRNA levels in human blood monocytes is mediated primarily at a post-transcriptional level. *J. Biol. Chem.*, **264**, 5700–3.

Fukunaga, R., Ishizaka-Ikeda, E., Seto, Y., and Nagata, S. (1990). Expression cloning of a receptor for murine granulocyte colony-stimulating factor. *Cell*, **61**, 341–50.

Kanda, N., Fukushige, S., Murotsu, T., Yoshida, M.C., Tsuchiya, M., Asano, S., Kaziro, Y., and Nagata, S. (1987). Human gene coding for granulocyte-colony stimulating factor is assigned to the q21–q22 region of chromosome 17. *Somat. Cell Mol. Genet.*, **13**, 679–84.

Koeffler, H.P., Gasson, J., and Tobler, A. (1988). Transcriptional and posttranscriptional modulation of myeloid colony-stimulating factor expression by tumour necrosis factor and other agents. *Mol. Cell. Biol.*, **8**, 3432–8.

Kuga, T., Komatsu, Y., Yamasaki, M., Sekine, S., Miyaji, H., Nishi, T., Sato, M., Yokoo, Y., Asano, M., Okabe, M., Morimoto, M., and Itoh, S. (1989). Mutagenesis of human granulocyte colony stimulating factor. *Biochem. Biophys. Res. Commun.*, **159**, 103–11.

Layton, J.E., Morstyn, G., Fabri, L.J., Reid, G.E., Burgess, A.W., Simpson, R.J., and Nice, E.C. (1991). Identification of a functional domain of human granulocyte colony stimulating factor using neutralizing monoclonal antibodies. *J. Biol. Chem.*, **266**, 23815–23.

Lu, H.S., Boone, T.C., Souza, L.M., and Lai, P.H. (1989). Disulfide and secondary structures of recombinant human granulocyte colony stimulating factor. *Arch. Biochem. Biophys.*, **268**, 81–92.

Nagata, S. (1990). In *Handbook of experimental pharmacology* 95/I, (ed. M.B. Sporn and A.B. Roberts), pp. 699–722. Springer, Heidelberg.

Nagata, S., Tsuchiya, M., Asano, S., Yamamoto, O., Hirata, Y., Kubota, N., Oheda, M., Nomura, H., and Yamazaki, T. (1986a). The chromosomal gene structure and two mRNAs for human granulocyte colony-stimulating factor. *EMBO J.*, **5**, 575–81.

Nagata, S., Tsuchiya, M., Asano, S., Kaziro, Y., Yamazaki, T., Yamamoto, O., Hirata, Y., Kubota, N., Oheda, M., Nomura, H., and Ono, M. (1986b). Molecular cloning and expression of cDNA for human granulocyte colony-stimulating factor. *Nature*, **319**, 415–8.

Nicola, N.A., Metcalf, D., Matsumoto, M., and Johnson, G.R. (1983). Purification of a factor inducing differentiation in murine myelomonocytic leukaemia cells. Identification as granulocyte colony-stimulating factor. *J. Biol. Chem.*, **258**, 9017–23.

Nicola, N.A., Begley, C.G., and Metcalf, D. (1985). Identification of the human analogue of a regulator that induces differentiation in murine leukaemic cells. *Nature*, **314**, 625–8.

Nishizawa, M., and Nagata, S. (1990). Regulatory elements responsible for inducible expression of the granulocyte colony-stimulating factor gene in macrophages. *Mol. Cell. Biol.*, **10**, 2002–11.

Nishizawa, M., Tsuchiya, M., Watanabe-Fukunaga, R., and Nagata, S. (1990). Multiple elements in the promoter of granulocyte colony-stimulating factor gene regulate its constitutive expression in human carcinoma cells. *J. Biol. Chem.*, **265**, 5897–902.

Nomura, H., Imazeki, I., Oheda, M., Kubota, N., Tamura, M., Ono, M., Ueyama, Y., and Asano, S. (1986). Purification and characterization of human granulocyte colony-stimulating factor (G-CSF). *EMBO J.*, **5**, 871–6.

Oheda, M., Hase, S., Ono, M., and Ikenaka, T. (1988). Structures of the sugar chains of recombinant human granulocyte-colony stimulating factor produced by Chinese hamster ovary cells. *J. Biochem.*, **103**, 544–6.

Oheda, M., Hasegawa, M., Hattori, K., Kuboniwa, H., Kojima, T., Orita, T., Tomonou, K., Yamazaki, T., and Ochi, N. (1990). O-linked sugar chain of human granulocyte colony-stimulating factor protects it against polymerization and denaturation allowing it to retain its biological activity. *J. Biol. Chem.*, **265**, 11432–5.

Seelentag, W.K., Mermod, J.J., Montesano, R., and Vassalli, P. (1987). Additive effects of interleukin 1 and tumour necrosis factor-alpha on the accumulation of the three granulocyte and macrophage colony-stimulating factor mRNAs in human endothelial cells. *EMBO J.*, **6**, 2261–5.

Simpson, R.J., Nice, E.C., and Nicola, N.A. (1987). Structural studies on the murine granulocyte colony-stimulating factor. *Biol. Chem. Hoppe-Seyler*, **368**, 1327–31.

Souza, L.M., Boone, T.C., Gabrilove, J., Lai, P.H., Zsebo, K.M., Murdock, D.C., Chazin, V.R., Bruszewski, J., Lu, H., Chen, K.K., Barendt, J., Platzer, E., Moore, M.A., Mertelsmann, R., and Welte, K. (1986). Recombinant human granulocyte colony-stimulating factor: effects on normal and leukemic myeloid cells. *Science*, **232**, 61–5.

Tsuchiya, M., Asano, S., Kaziro, Y., and Nagata, S. (1986). Isolation and characterization of the cDNA for murine granulocyte colony-stimulating factor. *Proc. Natl. Acad. Sci. (USA)*, **83**, 7663–37.

Tsuchiya, M., Kaziro, Y., and Nagata, S. (1987). The chromosomal gene structure for murine granulocyte colony-stimulating factor. *Eur. J. Biochem.*, **165**, 7–12.

Tweardy, D.J., Caracciolo, D., Valtieri, M., and Rovera, G. (1987). Tumor-derived growth factors that support proliferation and differentiation of normal and leukemic haemopoietic cells. *Ann. NY Acad. Sci.*, **511**, 30–8.

Welte, K., Platzer, E., Lu, L., Gabrilove, J.L., Levi, E., Mertelsmann, R., and Moore, M.A. (1985). Purification and biochemical characterization of human pluripotent hematopoietic colony-stimulating factor. *Proc. Natl. Acad. Sci. (USA)*, **82**, 1526–30.

Shigekazu Nagata:
Osaka Bioscience Institute, Japan

Receptor for Granulocyte Colony-Stimulating Factor (G-CSF)

The G-CSF receptor contains a single transmembrane domain which divides the molecule into the extracellular and cytoplasmic regions. The extracellular region contains an about 200 amino acid domain which is conserved among the haemopoietic growth factor receptors. The G-CSF receptor in homodimer form binds G-CSF with high affinity, and transduces the signal into cells. The detailed mechanism of the signal transduction is unknown. Since the G-CSF receptor does not contain apparent signalling elements, it is presumably coupled to other signalling components such as tyrosine kinases. The G-CSF receptor is expressed on bone marrow cells, mature neutrophils and placental trophoblasts.

■ G-CSF receptor sequence

The nucleotide sequences of mouse and human G-CSF receptor cDNAs (GenBank accession numbers M58288 and M59820) predict a protein of 837 (mouse) or 836 (human) amino acids each containing a 25- or 23-amino-acid leader sequence (Fukunaga et al. 1990a, 1990b; Larsen et al. 1990). Human and mouse G-CSF receptors have marked homology (62.5 per cent) at the amino acid sequence level, and the two G-CSF receptors cross-react in their G-CSF binding and signal transducing activities. The G-CSF receptor has a significant overall similarity with the gp130 (the interleukin 6 receptor signal transducer) and leukaemia inhibitory factor receptor (Fig. 1) (Nagata and Fukunaga 1991, 1993).

■ The G-CSF receptor protein

The core protein of the mouse G-CSF receptor is composed of 812 amino acids with a predicted molecular weight of 90 814 Da (Fig. 2) (Fukunaga et al. 1990a). The mouse G-CSF receptor contains a single transmembrane domain of 24 amino acids which divides the molecule into an extracellular region of 601 amino acids and a cytoplasmic region of 187 amino acids. There are eleven predicted N-glycosylation sites in the extracellular region. Modification of the protein by glycosylation produces the native G-CSF receptor of molecular weight of 100 000–130 000 Da (Fukunaga et al. 1990c; Nicola and Peterson 1986). The extracellular region of the G-CSF receptor has a composite structure. The N-terminal domain of 100 amino acids has an immunoglobulin-like sequence. The following domain of 200 amino acids (CRH domain) is related to the domain of the extracellular region of various cytokine receptors (Bazan 1990) and contains four conserved cysteine residues at the N-terminal half and the consensus Trp-Ser-X-Trp-Ser element at the C-terminus. The 300-amino-acid domain proximal to the transmembrane region is composed of three fibronectin type III-like sequences. The cytoplasmic region of the G-CSF receptor does not carry an apparent signalling domain such as tyrosine kinase, but it contains a proline-rich region

proximal to the transmembrane domain which is conserved among the members of the cytokine receptor family (Fukunaga et al. 1991).

The mature human G-CSF receptor is composed of 813 amino acids with a calculated molecular weight of 89 743 (Fukunaga et al. 1990b; Larsen et al. 1990). This molecular weight differs from that of native human G-CSF receptor by 30 000–60 000 (Uzumaki et al. 1989), which is explained by the occurrence of N-glycosylation on some of the nine N-glycosylation sites in the extracellular region. The overall structure of the human G-CSF receptor is similar to that of the mouse G-CSF receptor.

The G-CSF receptor can be purified from the solubilized membrane fraction by G-CSF affinity chromatography and gel filtration (Fukunaga et al. 1990c).

■ The G-CSF receptor gene and expression

The human G-CSF receptor gene is located on chromosome 1p35–34.3 (Inazawa et al. 1991), and is composed of 17 exons spanning about 17 kb (Seto et al. 1992), whereas the mouse G-CSF receptor gene is on chromosome 4 (Ito et al. 1994). In human, four alternative transcripts of the G-CSF receptor gene have been observed (Fukunaga et al. 1990b; Larsen et al. 1990). The main one codes for the authentic G-CSF receptor described above. A small percentage of G-CSF receptor mRNAs expressed in some myeloid leukaemia cells, such as the U937 cell line, codes for a soluble receptor in which a 88 bp internal deletion removes the transmembrane domain and changes the reading frame resulting in 149 different amino acids at the C-terminus (Fukunaga et al. 1990b). The third type, which comprises about 20 per cent of the G-CSF receptor mRNAs expressed in placenta, contains an insertion of 27 amino acids in the cytoplasmic region (Fukunaga et al. 1990b). The fourth type of G-CSF receptor mRNA was also isolated from a placental cDNA library. This mRNA carries a sequence coding for 34 amino acids at the C-terminus which are different from those of the authentic G-CSF receptor (Larsen et al. 1990).

```
hG-CSFR   -CLNWGNSLQILDQVELRAGYPPAIPHNLSCLMNLTTSSLICQWEPGPETHLPTSFTLKSFKSRGNCQTQGDSILDCVPK
           # +#+  # +  + + +# ## #+####+#+#  +  + #+#+#+ #### # # #####      +  +   ## +#
hIL-6β    NILTFGQLEQNVYGITIISGLPPEKPKNLSCIVN-EGKKMRCEWDGGRETHLETNFTLKS-------EWATHKFADCKAK


hG-CSFR   -DGQSHCCIPRKHLLLYQNMGIWVQAENALGTSMSPQLCLDPMDVVKLEPPMLRTMDPSPEAAPPQAGCLQLCWEPWQPG
           #+ + # +    + + #+ +##+####### #   ##+  ## +## + #    +    ++ # # #
hIL-6β    RDTPTSCTVDYSTV-YFVNIEVWVEAENALGKVYSDHINFDPVYKVKPNPP----HNLSVINSEELSSILKLTWTNPSIK


hG-CSFR   LHINQKCELRHKPQRGEASWALVGPLPLEALQ--YELCGLLPATAYTLQIRCIRWPLPGHWSDWSPSLELRTTERAPTVR
           #  # ++  ++ + ++#+ ++#   +   + + # # # ###++  # #####     # # #+
hIL-6β    SVIILKYNIQYRT-KDASTWSQIPPEDTASTRSSFTVQDLKPFTEYVFRIRCMKEDGKGYWSDWSEEASGITYEDRPSKA


hG-CSFR   LDTWWR---QRQLDPRTVQLFWKPVPLEEDSGRIQGY---VVSWRPSGQAGAILPLCNTTELSCTFHLPSEAQEVALVAY
           #++    +    ##### ##++# #  #+# #   + #++  # ++ #  # +#+  #++ +   ##+
hIL-6β    PSFWYKIDPSHTQGYRTVQLVWKTLPPFEANGKILDYEVTLTRWKSHLQNYTVNATKLTVNLTNDRYLATLTVR-NLVGK


hG-CSFR   NSAGTSRPTPVVFSESRGPALTRLHAMARDPHSLWVCWEPPNPWPQGYVIEWGLGPPSASNSNKTWRMEQNGRATGFLLK
           #+    ++ # + + + #+# + # ### # +#  #++## + + #+  # #+ #  ++ #+
hIL-6β    SDAAVLTIPACDFQATHP--VMDLKAFPKD-NMLWVEWTTPRESVKKYILEWCVLSDKAP-CITDWQQEDGTVHRTY-LR


hG-CSFR   ENIRPFQLYEIIVTPLYQDTMGPSQHVYAYSQEMAPSHAPELHLKHIGKTWAQLEWVPEPPELGKSPLTHYTIFWTNAQN
           #+    # #####+# #+ #+++ + ## + +##++# ++ #++## # ###   # ++ + + ####+
hIL-6β    GNLAESKCYLITVTPVYADGPGSPESIKAYLKQAPPSKGPTVRTKKVGKNEAVLEWDQLPVDVQNGFIRNYTIFYRTIIG


hG-CSFR   QSFSAILNASSRGFVLHGLEPASLYHIHLMAASQAGATNSTVLTLMTLTPEGSELHIILGLFGLLLLLTCLCGTAWLCCS
           +  +  +++#   + # +# + +## +++ # ++ #+ +++  #  #    #+ #+   # ### # #      #
hIL-6β    NETAVNVDSSHTEYTLSSLTSDTLYMVRMAAYTDEGGKDGPEFTFTTPKFAQGEIEAIVVPVCLAFLLTTLLG-VLFCFN


hG-CSFR   --PNRKNPLWPSVPDPAHSSLGSWWPTIMEEDAFQLPG-----GTPPITKLTVLEEDEKKPVP--------WESHNSSET
            #  +## ####++# ++ # #      #+     # +   + +# ++### #         +
hIL-6β    KRDLIKKHIWPNVPDPSKSHIAQWSPHTPPRHNFNSKDQMYSDGNFTDVSVVEIEANDKKPFPEDLKLLDLFKKEKINTE


hG-CSFR   CGLPTLVQTYVLQGDPRAL--STQPQSQSGTSDQVLYGQLLGS---PTSPGPGHYLRCDSTQPLLAGLTPSPKSYENLWF
           +++ +   +   ++ # + +# ## # # ++ #      #+  + # +###### # +  +
hIL-6β    GHSSGIGGSSCMSSSRPSISSSDENESSQNTSSTVQYSTVVHSGYRHQVPSVQVFSRSESTQPLL-DSEERPEDLQLV-D


hG-CSFR   QASPLGTLVTPAPSQEDDCVFGPLL-NFPLLQ-GIRVHGMEALGSF
           +  ++++   ++# + +  + # +++
hIL-6β    HVDGGDGILPRQQYFKQNCSQHESSPDISHFERSKQVSSVNEEDFVRLKQQISDHISQSCGSGQMKMFQEVSAADAFGPG


hIL-6β    TEGQVERFETVGMEAATDEGMPKSYLPQTVRQGGYMPQ
```

Figure 1. Alignment of amino acid sequences of the mouse G-CSF receptor and the IL-6 signal transducer, gp130.

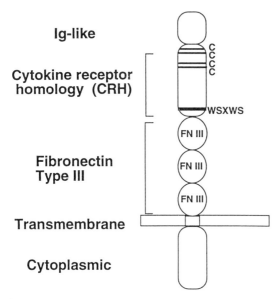

Ig-like

Cytokine receptor homology (CRH)

C
C
C
C

WSXWS

FN III

FN III

Fibronectin Type III

FN III

Transmembrane

Cytoplasmic

Figure 2. Domain structure of the G-CSF receptor.

The functional significance of these aberrant forms of the G-CSF receptor is unknown.

The G-CSF receptor gene is expressed in neutrophilic progenitor cells in the bone marrow, mature neutrophils and various myeloid leukaemia cells (Fukunaga *et al.* 1990a,b; Larsen *et al.* 1990; Nicola and Peterson 1986; Park *et al.* 1989) as well as non-haemopoietic tissues such as placenta and endothelial cells (Uzumaki *et al.* 1989; Bussolino *et al.* 1989). The regulatory elements in the promoter of the G-CSF receptor gene is not yet well-characterized. One 18 bp element in the promoters of neutrophil-specific genes such as myeloperoxidase and neutrophil elastase genes can be found at a similar position of the G-CSF receptor gene (Seto *et al.* 1992; Ito *et al.* 1994).

■ Binding characteristics of the G-CSF receptor

The purified G-CSF receptor exists as a monomer or a homodimer (Fukunaga *et al.* 1990c). G-CSF specifically binds to the monomer form of the receptor with low affinity (K_D = 2.6–4.2 nM), whereas it binds to the dimer form with high affinity (K_D = about 100 pM). The G-CSF receptor on the cell surface binds G-CSF with high affinity suggesting that the G-CSF receptor on the cell surface exists as a dimer. The immunoglobulin-like domain and fibronectin type III domains in the extracellular region, as well as the entire cytoplasmic region of the G-CSF receptor, are not necessary for G-CSF binding (Fukunaga *et al.* 1991). The mutation or deletion of the CRH domain in the G-CSF receptor reduces or abolishes G-CSF binding activity (Fukunaga *et al.* 1991), and replacement of this domain with the corresponding domain of the growth hormone receptor renders the chimeric receptor able to bind growth hormone (Fuh *et al.*

1992; Ishizaka-Ikeda *et al.* 1993). These results indicate that the CRH domain of the G-CSF receptor is responsible for binding of the ligand (G-CSF).

■ Signalling mechanisms

Unlike other cytokine receptors, a single polypetide of the G-CSF receptor which binds G-CSF is sufficient to transduce the signal (Fukunaga *et al.* 1991). G-CSF seems to induce dimerization of the receptor, which is an active form in transducing the signal. The G-CSF receptor can transduce proliferation as well as neutrophilic differentiation signals into cells (Fukunaga *et al.* 1993). For the proliferation signal, a 76-amino-acid region proximal to the transmembrane domain is sufficient (Fukunaga *et al.* 1991, 1993), whereas the C-terminal region of the receptor is indispensable for transducing the differentiation signal (Fukunaga *et al.* 1993). Activation of the G-CSF receptor by G-CSF induces tyrosine phosphorylation of several proteins including the G-CSF receptor itself (Pan *et al.* 1993), suggesting that some unidentified tyrosine kinases are involved in the signalling pathway mediated by the G-CSF receptor. The detailed signal transduction mechanism through the G-CSF receptor is unknown.

■ References

Bazan, J.F. (1990). Structural design and molecular evolution of a cytokine receptor superfamily. *Proc. Natl. Acad. Sci. (USA)*, **87**, 6934–8.

Bussolino, F., Wang, J.M., Defilippi, P., Turrini, F., Sanavio, F., Edgell, C.J., Aglietta, M., Arese, P., and Mantovani, A. (1989). Granulo-cyte- and granulocyte-macrophage-colony stimulating factors induce human endothelial cells to migrate and proliferate. *Nature*, **337**, 471–3.

Fuh, G., Cunningham, B.C., Fukunaga, R., Nagata, S., Goeddel, D.V., and Wells, J.A. (1992). Rational design of potent antagonists to the human growth hormone receptor. *Science*, **256**, 1677–80.

Fukunaga, R., Ishizaka-Ikeda, E., Seto, Y., and Nagata, S. (1990a). Expression cloning of a receptor for murine granulocyte colony-stimulating factor. *Cell*, **61**, 341–50.

Fukunaga, R., Seto, Y., Mizushima, S., and Nagata, S. (1990b). Three different mRNAs encoding human granulocyte colony-stimulating factor receptor. *Proc Natl Acad Sci (USA)*, **87**, 8702–6.

Fukunaga, R., Ishizaka-Ikeda, E., and Nagata, S. (1990c). Purification and characterization of the receptor for murine granulocyte colony-stimulating factor. *J. Biol. Chem.*, **265**, 14008–15.

Fukunaga, R., Ishizaka-Ikeda, E., Pan, C., Seto, Y., and Nagata, S. (1991). Functional domains of the granulocyte colony-stimulating factor receptor. *EMBO J.*, **10**, 2855–65.

Fukunaga, R., Ishizaka-Ikeda, E., and Nagata, S. (1993). Growth and differentiation signals mediated by two different regions in the cytoplasmic domain of G-CSF receptor. *Cell*, **74**, 1079–87.

Inazawa, J., Fukunaga, R., Seto, Y., Nakagawa, H., Misawa, S., Abe, T., and Nagata, S. (1991). Assignment of the human granulocyte colony-stimulating factor receptor gene (CSF3R) to chromosome 1 at region p35-p34.3. *Genomics*, **10**, 1075–8.

Ishizaka-Ikeda, E., Fukunaga, R., Wood, W.I., Goeddel, D.V., and Nagata, S. (1993). Signal transduction mediated by growth hormone receptor and its chimeric molecules with the granulocyte colony-stimulating factor receptor. *Proc. Natl. Acad. Sci. (USA)*, **90**, 123–7.

Ito, Y., Seto, Y., Brennan, C.I., Copeland, N.G., Jenkins, N.A., Fukunaga, R., and Nagata, S. (1994). Structural analysis of the functional gene and pseudogene for the murine granulocyte colony-stimulating factor receptor. *Eur. J. Biochem.* **220**, 881–91.

Larsen, A., Davis, T., Curtis, B.M., Gimpel, S., Sims, J.E., Cosman, D., Park, L., Sorensen, E., March, C.J., and Smith, C.A. (1990). Expression cloning of a human granulocyte colony-stimulating factor receptor: a structural mosaic of hematopoietin receptor, immunoglobulin, and fibronectin domains. *J. Exp. Med.*, **172**, 1559–70.

Nagata, S., and Fukunaga, R. (1991). Granulocyte colony-stimulating factor and its receptor. *Prog. Growth Factor Res.*, **3**, 131–41.

Nagata, S., and Fukunaga, R. (1993). Granulocyte colony-stimulating factor receptor and its related receptors. *Growth Factors*, **8**, 99–107.

Nicola, N.A., and Peterson, L. (1986). Identification of distinct receptors for two haemopoietic growth factors (granulocyte colony-stimulating factor and multipotential colony-stimulating factor) by chemical cross-linking. *J. Biol. Chem.*, **261**, 12384–9.

Pan, C.-X., Fukunaga, R., Yonehara, S., and Nagata, S. (1993). Unidirectional cross-phosphorylation between the granulocyte colony-stimulating factor and interleukin 3 receptors. *J. Biol. Chem.*, **268**, 25818–23.

Park, L.S., Waldron, P.E., Friend, D., Sassenfeld, H.M., Price, V., Anderson, D., Cosman, D., Andrews, R.G., Bernstein, I.D., and Urdal, D.L. (1989). Interleukin-3, GM-CSF, and G-CSF receptor expression on cell lines and primary leukaemia cells: receptor heterogeneity and relationship to growth factor responsiveness. *Blood*, **74**, 56–65.

Seto, Y., Fukunaga, R., and Nagata, S. (1992). Chromosomal gene organization of the human granulocyte colony-stimulating factor receptor. *J. Immunol.*, **148**, 259–66.

Uzumaki, H., Okabe, T., Sasaki, N., Hagiwara, K., Takaku, F., Tobita, M., Yasukawa, K., Ito, S., and Umezawa, Y. (1989). Identification and characterization of receptors for granulocyte colony-stimulating factor on human placenta and trophoblastic cells. *Proc. Natl. Acad. Sci. (USA)*, **86**, 9323–6.

Shigekazu Nagata:
Osaka Bioscience Institute, Japan

Colony-Stimulating Factor-1 (CSF-1)

CSF-1 regulates the production of mononuclear phagocytes and osteoclasts and the function of other cell types in the female reproductive tract. It is secreted as a 80–100 kDa glycoprotein or 130–160 kDa chondroitin–SO4 proteoglycan or expressed on the cell surface as a biologically active membrane spanning protein of 68–86 kDa. The glycoprotein and proteoglycan forms are found in the circulation. The cell surface form is involved in local regulation and the proteoglycan form may be selectively sequestered from the circulation. CSF-1 is synthesized by a variety of different cell types, including fibroblasts, endothelial cells, bone marrow stromal cells, osteoblasts, keratinocytes, astrocytes, myoblasts and, during pregnancy, by uterine epithelial cells. The target cells it regulates may have trophic as well as scavenger functions and are involved in organogenesis and tissue turnover. CSF-1 has potential clinical utility in enhancing the rate of haemopoietic recovery after cancer chemotherapy and as an anti-fungal agent. Pathologically, it may be involved in the progression of myelo-proliferative diseases and neoplasms of the ovary, endometrium and breast for which elevated levels of CSF-1 are a tumour marker.

■ Alternative names

Colony-stimulating factor (CSF), macrophage growth factor (MGF), macrophage and granulocyte inducer IM (MGI-IM), macrophage colony-stimulating factor (M-CSF).

■ CSF-1 sequence

The nucleotide sequences of CSF-1 c-DNAs cloned from mouse and man (GenBank accession numbers M21149, M21952, J03862 and M64592, M37435, respectively) predict a full length CSF-1 precursor protein of 522 (man) and 520 (mouse) amino acids, including a 32-amino-acid leader sequence, and, towards the carboxy terminus, a stretch of 23 hydrophobic amino acids, followed by a charged 'stop transfer' sequence (reviewed by Sherr and Stanley 1990). The human and mouse proteins are 69.5 per cent identical at the amino acid level. Alternative splicing within the coding region generates human precursor proteins from which amino acids 150–447 or 332–447 have been deleted. Precursors bearing the 150–447 deletion lack the intracellular proteolytic cleavage sites that generate the secreted forms (see below). Amino acids 150–522 are not required for *in vitro* biological activity and the amino acids 1–149 show a greater degree of sequence identity (80.5 per cent) between mouse and human CSF-1. Human CSF-1 binds to the mouse CSF-1 receptor but mouse CSF-1 exhibits no or low binding to the human CSF-1 receptor. In the region responsible for *in vitro* biological activity there is 16 per cent identity (32 per cent similarity) between CSF-1 and stem cell factor.

■ CSF-1 protein

CSF-1 encoded by an mRNA possessing a full-length coding region is co-translationally N-glycosylated in the endoplasmic reticulum to yield a membrane-spanning monomeric precursor that rapidly forms an interchain disulphide-linked homodimer. In the Golgi, the N-linked oligosaccharides are converted to the complex type and O-linked oligosaccharides are added. The O-linked oligosaccharides include one ~18 000 kDa chondroitin sulphate chain per monomer which is added at serine 276 (mouse) or 277

(human) in the single consensus sequence for glycosaminoglycan addition. In secretory vesicles, the mature forms of secreted CSF-1 are cleaved from the dimeric precursor. They are either glycoprotein (80–100 kDa) or proteoglycan (130–160 kDa), depending on whether the proteolytic cleavage takes place on the amino terminal side or the carboxy terminal side of the glycosaminoglycan addition site. Both forms rapidly accumulate extracellularly ($t_{1/2}$ = 40 minutes) (Fig. 1). A smaller dimeric precursor (monomeric unit: 224 amino acids) encoded by an mRNA in which a large segment of the coding region, including the glycosaminoglycan addition and proteolytic cleavage sites, has been spliced out, fails to be processed in the same way and is instead stably expressed ($t_{1/2}$ = 11 h) as a membrane-spanning, biologically-active species on the cell surface (Fig. 1) (reviewed by Price *et al.* 1992). The extracellular proteolytic cleavage of soluble CSF-1 from this form is possibly regulated *via* activation of protein kinase C (Stein and Rettenmier 1991).

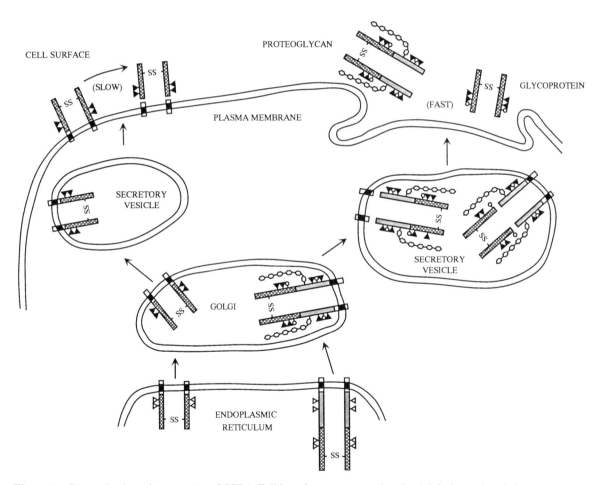

Figure 1. Biosynthesis and expression of CSF-1. Full length precursor molecules (*right bottom*) and short precursors (*left bottom*) are processed and expressed as described in detail in the text. Potential N-linked (*triangles*) and O-linked (*circles*) glycosylation sites and the glycosaminoglycan chain (*linked hexagons*) are also shown. Open and filled triangles designate high mannose and complex structures, respectively.

While these various forms of CSF-1 have different *in vivo* roles, their shared amino terminal 149 amino acids are sufficient for *in vitro* biological activity (Sherr and Stanley 1990). Despite its lack of sequence similarity with members of the cytokine family, this portion of CSF-1 possesses the four α-helical bundle/anti-parallel β-ribbon structure shared by granulocyte-macrophage colony stimulating factor, growth hormone and other cytokine family members (Pandit *et al*. 1992). Interestingly, only the intrachain disulphide bonds (CYS 7–CYS 90, CYS 48–CYS 139, and CYS 102–CYS 146) and not the interchain disulphide bonds (CYS 31–CYS 31, CYS 157–CYS 157, and CYS 159–CYS 159) (Glocker *et al*. 1993) are essential for biological and antibody binding activity of this region (Krautwald and Baccarini 1993). All forms of CSF-1 can be conveniently purified by conventional chromatographic procedures coupled with affinity chromatography using a monoclonal anti-mouse CSF-1 antibody (Stanley 1985).

■ CSF-1 gene and transcription

The CSF-1 gene is localized to human chromosome 1p13–p21 (Morris *et al*. 1991). In the mouse it is localized to chromosome 3 and is the *op* locus (Yoshida *et al* 1990; Wiktor-Jedrzejczak *et al*. 1990). The gene is approximately 21 kb in length, comprising 10 exons. Alternative use of different 3′ untranslated regions encoded by exons 9 and 10 results in 4 kb and 2.5 kb mRNAs that may possess different half-lives, since exon 10 contains an AU rich sequence that confers mRNA instability. These mRNAs have identical full-length coding regions and give rise to a protein precursor that is proteolytically processed to yield the mature secreted forms. Alternative splicing within exon 6 can yield mRNAs in which the regions encoding these intracellular proteolytic cleavage sites are lost including one which encodes the cell-surface membrane-spanning glycoprotein form. Regulation of CSF-1 transcription is not well understood (Sherr and Stanley 1990).

■ Biological functions

CSF-1 alone stimulates the survival, proliferation and differentiation of mononuclear phagocytic cells from the undifferentiated but committed bone marrow precursor cell to the mature, non-dividing macrophage (Stanley *et al*. 1983). Combinations of CSF-1 with interleukin-1 (IL-1), interleukin-3 (IL-3) and interleukin-6 (IL-6) synergistically stimulate the proliferation and differentiation of even more primitive cells than the mononuclear phagocyte precursor and this may represent a mechanism by which the growth factor can increase the number of mononuclear phagocyte precursor cells. CSF-1 is also involved in the regulation of cells of the female reproductive tract during pregnancy, osteoclast progenitor cell differentiation and microglial cell proliferation (reviewed by Roth and Stanley 1992).

A variety of different cell types synthesize CSF-1, including fibroblasts, endothelial cells, bone marrow stromal cells, osteoblasts, keratinocytes, astrocytes, myoblasts and, during pregnancy, uterine epithelial cells (Roth and Stanley 1992). It is found at biologically active concentrations in the circulations of both mouse ($\sim$12 ng/ml) and man ($\sim$4 ng/ml). Circulating CSF-1 (half-life $\sim$10 minutes) is synthesized by endothelial cells and cleared by sinusoidally located macrophages of the liver and spleen (Bartocci *et al*. 1987). Thus, the number of these macrophages, whose development and maintenance is predominantly controlled by CSF-1, actually determines its circulating concentration in the normal steady state. Circulating CSF-1 is elevated in response to bacterial endotoxin, bacterial, viral and parasitic infections, and injection of CSF-1 can cause a 10-fold elevation in the blood monocyte concentration and increase macrophage numbers in certain areas of the periphery (Roth and Stanley 1992). The proteoglycan form of CSF-1 is found in the circulation and may be recruited to specific tissue sites by virtue of its association with particular types of extracellular matrix (Price *et al*. 1992).

The role of CSF-1 in the development and regulation of mononuclear phagocytes has been studied in the CSF-1-less, osteopetrotic *(op/op)* mutant mouse (Yoshida *et al*. 1990; Wiktor-Jedrzejczak *et al*. 1990). Compared with control mice, *op/op* mice exhibit impaired bone resorption associated with a paucity of osteoclasts, have no incisors, poor fertility, a lower body weight, a shorter average life span, and deficiencies in blood monocytes and certain tissue macrophages (reviewed by Wiktor-Jedrzejczak *et al*. 1990). Postnatal restoration of circulating CSF-1 in *op/op* mice cures their osteopetrosis and monocytopenia and restores some, but not all of the tissue macrophage populations, consistent with local as well as humoral regulation by this growth factor (reviewed in Wiktor-Jedrzejczak *et al*. 1991). Other studies suggest that CSF-1 generates macrophages that have trophic as well as scavenger (i.e. physiological) functions and are involved in organogenesis and tissue turnover. The development of macrophages that are more involved in inflammatory and immunological (i.e. pathological) functions is apparently dependent on other growth factors, such as GM-CSF (Cecchini *et al*. 1994).

CSF-1 also plays an important role in fertility, probably via its action on other cell types as well as the macrophage (Pollard *et al*. 1991; Daiter and Pollard 1992). In pregnancy, it is synthesized by the oviduct and uterine epithelium under the control of the female endocrine system and can influence several CSF-1 receptor-expressing cell types, including oocytes, pre-implantation embryonic cells, macrophages, decidual cells and trophoblastic cells. Local production of CSF-1, including the cell-surface form, is important for post-implantation development (Wiktor-Jedrzejczak *et al*. 1991). CSF-1 is also critically involved in bone metabolism through its regulation of osteoclast progenitor cell differentiation.

Administration of CSF-1 to animals has been reported to lower plasma cholesterol, accelerate recovery of the haemopoietic system following cytotoxic drug treatment, increase host defence to yeast infections and to reduce the number of metastases and prolong survival of mice bearing a melanoma (reviewed by Roth and Stanley 1992).

Pathology

Circulating levels of CSF-1 are elevated in patients with myeloid and lymphoid malignancies and carcinomas of the ovary, endometrium and breast (Janowska-Wieczorek *et al.* 1991; Kacinski *et al.* 1991). In ovarian cancer, elevated levels of CSF-1 in the circulation or ascitic fluid predict a poor prognosis (Price *et al.* 1993). At least in some cases, the malignant cells themselves co-express CSF-1 and its receptor, raising the possibility of autocrine growth control by CSF-1 and an aetiological role in tumour development and progression.

Acknowledgements

This work was supported by NIH grants CA 26504, CA 32551, an award from the Lucille P. Markey Charitable Trust, and the Albert Einstein Core Cancer Grant P30-CA 1330.

References

Bartocci, A., Mastrogiannis, D.S., Migliorati, G., Stockert, R.J., Wolkoff, A.W., and Stanley, E.R. (1987). Macrophages specifically regulate the concentration of their own growth factor in the circulation. *Proc. Natl. Acad. Sci. (USA)*, **84**, 6179–83.

Cecchini, M.G., Dominguez, M.G., Mocci, S., Wetterweld, A., Felix, R., Fleisch H., Chisholm, O., Hofstetter, W., Pollard, J.W., and Stanley, E.R. (1994). Role of colony stimulating factor-1 in the establishment and regulation of tissue macrophages during postnatal development of the mouse. *Development*, **120**, 1357–72.

Daiter, E., and Pollard, J.W. (1992). Colony stimulating factor-1 (CSF-1) in pregnancy. *Reprod. Med. Rev.*, **1**, 83–97.

Glocker, M.O., Arbogast, B., Schreurs, J., and Deinzer, M.L. (1993). Assignment of the inter- and intramolecular disulphide linkages in recombinant human macrophage colony stimulating factor using fast atom bombardment mass spectrometry. *Biochemistry*, **32**, 482–8.

Janowska-Wieczorek, A., Belch, A.R., Jacobs, A., Bowen, D., Padua, R.A., Paietta, E., and Stanley, E.R. (1991). Increased circulating colony-stimulating factor-1 in patients with preleukaemia, leukaemia, and lymphoid malignancies. *Blood*, **77**, 1796–803.

Kacinski, B.M., Scata, K.A., Carter, D., Yee, L.D., Sapi, E., King, B.L., Chambers, S.K., Jones, M.A., Pirro, M.H., Stanley, E.R., and Rohrschneider, L.R. (1991). FMS (CSF-1 receptor) and CSF-1 transcripts and protein are expressed by human breast carcinomas *in vivo* and *in vitro*. *Oncogene*, **6**, 941–52.

Krautwald, S., and Baccarini, M. (1993). Bacterially expressed murine CSF-1 possesses agonistic activity in its monomeric form. *Biochem. Biophys. Res. Commun.*, **192**, 720–7.

Morris, S.W., Valentine, M.B., Shapiro, D.N., Sublett, J.E., Deaven, L.L., Foust, J.T., Roberts, W.M., Cerretti, D.P., and Look, A.T. (1991). Reassignment of the human CSF1 gene to chromosome 1p13-p21. *Blood*, **78**, 2013–20.

Pandit, J., Bohm, A., Jancarik, J., Halenbeck, R., Koths, K., and Kim, S.H. (1992). Three-dimensional structure of dimeric human recombinant macrophage colony-stimulating factor. *Science*, **258**, 1358–62.

Pollard, J.W., Hunt, J.S., Wiktor-Jedrzejczak, W., and Stanley, E.R. (1991). A pregnancy defect in the osteopetrotic (*op/op*) mouse demonstrates the requirement for CSF-1 in female fertility. *Dev. Biol.*, **148**, 273–83.

Price, F.V., Chambers, S.K., Chambers, J.T., Carcangiu, M.L., Schwartz, P.E., Kohorn, E.I., Stanley, E.R., and Kacinski, B.M. (1993). Colony-stimulating factor-1 in primary ascites of ovarian cancer is a significant predictor of survival. *Am. J. Obstet. Gynecol.*, **168**, 520–7.

Price, L.K., Choi, H.U., Rosenberg, L., and Stanley, E.R. (1992). The predominant form of secreted colony stimulating factor-1 is a proteoglycan. *J. Biol. Chem.*, **267**, 2190–9.

Roth, P., and Stanley, E.R. (1992). The biology of CSF-1 and its receptor. *Curr. Topics Microbiol. Immunol.*, **181**, 141–67.

Sherr, C.J., and Stanley, E.R. (1990). In *Handbook of experimental pharmacology*, Vol. 95/I, Peptide growth factors and their receptors 1 (ed. M.B. Sporn and A.B. Roberts), pp. 667–98. Springer, Berlin.

Stanley, E.R. (1985). The macrophage colony-stimulating factor, CSF-1. *Methods Enzymol.*, **116**, 564–87.

Stanley, E.R., Guilbert, L.J., Tushinski, R.J., and Bartelmez, S.H. (1983). CSF-1—a mononuclear phagocyte lineage-specific haemopoietic growth factor. *J. Cell. Biochem.*, **21**, 151–9.

Stein, J., and Rettenmier, C.W. (1991). Proteolytic processing of a plasma membrane-bound precursor to human macrophage colony-stimulating factor (CSF-1) is accelerated by phorbol ester. *Oncogene*, **6**, 601–5.

Wiktor-Jedrzejczak, W., Bartocci, A., Ferrante, A.W. Jr., Ahmed-Ansari, A., Sell, K.W., Pollard, J.W., and Stanley, E.R. (1990). Total absence of colony-stimulating factor 1 in the macrophage-deficient osteopetrotic (*op/op*) mouse [published erratum appears in *Proc. Natl. Acad. Sci. (USA)*, **88**, 5937]. *Proc. Natl. Acad. Sci. (USA)*, **87**, 4828–32.

Wiktor-Jedrzejczak, W., Urbanowska, E., Aukerman, S.L., Pollard, J.W., Stanley, E.R., Ralph, P., Ansari, A.A., Sell, K.W., and Szperl, M. (1991). Correction by CSF-1 of defects in the osteopetrotic *op/op* mouse suggests local, developmental, and humoral requirements for this growth factor. *Exp. Hematol.*, **19**, 1049–54.

Yoshida, H., Hayashi, S., Kunisada, T., Ogawa, M., Nishikawa, S., Okamura, H., Sudo, T., Shultz, L.D., and Nishikawa, S. (1990). The murine mutation osteopetrosis is in the coding region of the macrophage colony stimulating factor gene. *Nature*, **345**, 442–4.

E. Richard Stanley:
Department of Developmental and Molecular Biology,
Albert Einstein College of Medicine,
Bronx, NY 10461–1975, USA

The Macrophage Colony-Stimulating Factor (M-CSF) Receptor

The M-CSF receptor is a member of the tyrosine kinase class of growth factor receptors and is identical to the product of the normal c-fms proto-oncogene. The c-fms proto-oncogene is, in turn, related to the v-fms oncogene contained in the genome of both Susan McDonough and Hardy-Zuckerman 5 strains of feline sarcoma virus. Thus, both normal and neoplastic versions of the M-CSF receptor are known. The extracellular ligand-binding domain of the M-CSF receptor contains five tandom immunoglobulin-like loops, and the cytoplasmic portion contains a tyrosine kinase domain with an insert of about 72 hydrophilic amino acids. The M-CSF receptor is expressed on mature and progenitor cells of the monocyte and macrophage lineage where it functions in promoting cell survival, growth stimulation, differentiation induction, and in the activation of mature cell functions. It is also expressed on maternal decidual cells and trophoblast cells of the developing placenta.

■ Structure and function

The M-CSF receptor (Fms) is a member of the class III family of tyrosine kinase growth factor receptors and is most highly related to the PDGF receptors, c-*kit*, kdr/flk1 (vascular endothelial cell growth factor receptor), the *torso* gene product, *flt1*, *flt3/flk2*, and *flt4*. The cDNAs for feline, murine, and human c-*fms* genes have been cloned (Rohrschneider and Woolford 1991) (GenBank accession numbers J03149, X06368, and X03663, respectively). The predicted amino acid sequences exhibit the general structural features shown in Fig. 1. For the murine receptor, the extracellular, M-CSF-binding domain is composed of 510 amino acids (512 amino acids for human Fms), with a 26-amino-acid transmembrane sequence, and a 440 amino acid cytoplasmic domain (434 amino acids for human Fms). The primary translation product contains a hydrophobic leader sequence of which the N-terminal 19 amino acids are cleaved. N-linked glycosylation is added at up to nine potential sites in murine Fms, 10 sites in feline Fms, and 11 in the human Fms. Seven of these sites are common among all three Fms proteins and no O-linked glycosylation is detectable on Fms. The size of the mature cell surface form of Fms is 165 kDa for murine and 150 kDa for the human protein. The immature high mannose forms are about 15–20 kDa smaller in each case.

The extracellular domain of Fms contains five tandem arrays of immunoglobulin(Ig)-like loops designated D1-D5 from the N-terminal end (Fig. 1). Ig-like domains D1–D4 exhibit C2 type Ig loops, whereas Ig-like domain D5 is larger, containing an extra fold typical of variable-type Ig-like domains. All Ig-like loops in Fms, with the exception of the D4 loop, contain an intradomain disulphide bond. The M-CSF binding site is contained in the three N-terminal Ig-like loops D1–D3 (Wang *et al.* 1993). The D4 Ig-like loop is more flexible than the others and may undergo a conformational change upon M-CSF binding thus permitting receptor dimerization and subsequent steps in signal transduction (Carlberg and Rohrschneider 1994). The func-

tion of the D5 Ig-like loop is not understood, but it may interact with other cellular proteins.

The cytoplasmic domain of Fms can be subdivided into five distinct subdomains: (1) the juxtamembrane region, (2) the ATP-binding lobe, (3) the kinase insert (KI) region, (4) the main tyrosine kinase domain, and (5) the C-terminal tail. The juxtamembrane region encompasses the first approximately 50 amino acids of the Fms cytoplasmic domain. This region exhibits two tyrosine-based activation motifs (Alber *et al.* 1993) common to murine feline and human Fms. Tyrosine 559 within this motif of the murine Fms is autophosphorylated and is thought to associate with other tyrosine kinases of the Src-family. It is not yet clear whether the src-family kinase associates with Y559 before or after its phosphorylation. In addition, tyrosine 569 is implicated as part of the internalization signal for endocytosis. The ATP-binding lobe begins before the GxGxxG motif just C-terminal to the juxtamembrane region and extends for approximately 100 amino acids. The KI region interrupts homology of the catalytic domain with other Src family members beginning around amino acid 682 (684 in human Fms) and extending for approximately 72 amino acids. The KI region contains three tyrosine autophosphorylation sites at amino acids 697, 706 and 721 in the murine receptor (Rohrschneider and Woolford 1991). Mutation of the tyrosine at amino acid 721 to phenylalanine drastically reduces association of Fms with phosphatidylinositol 3′-kinase (PI 3′-K) activity by preventing the binding *via* an SH2 domain of the noncatalytic p85 'adaptor' subunit of PI 3′-K (Reedijk *et al.* 1992). Another adaptor molecule, Grb2, binds to the tyrosine phosphorylated Y697 site of activated Fms (van der Geer and Hunter 1993; Lioubin and Rohrschneider 1994). These interactions and the KI region may function in augmenting growth stimulation and/or directing the internalized receptor from endosomal to lysosomal compartments. In addition to the ATP-binding lobe, the Fms catalytic domain is composed of a larger kinase lobe containing a major tyrosine autophosphorylation site at amino acid 807 in murine Fms (amino acid 809 in the human receptor). This

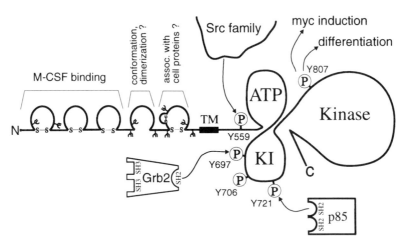

Figure 1. Structural features of Fms.

site has a homologue in all tyrosine kinases (amino acid 416 in chicken Src), and in Fms, autophosphorylation of this site is linked to the M-CSF-dependent induction of endogenous c-myc expression (Roussel et al. 1991). Phosphorylation of tyrosine 807 also controls the ability of Fms to initiate a differentiation signal (Bourette et al. 1994). No proteins have yet been found to bind to this phosphorylated site although associated Src activity is affected when this site is phosphorylated (Courtneidge et al. 1993). Finally, the C-terminal tail of all Fms proteins contains sequences within the C-terminal 24 amino acids that negatively regulate the tyrosine kinase activity of the catalytic domain. The structure of the catalytic domain of Fms shown in Fig. 1 is based on the crystal structure of the catalytic subunit of the cAMP-dependent protein kinase (Knighton et al. 1991).

■ Expression

Multiple alleles for fms exist in both feline and human genomes and are located on human chromosome 5q33.3–34 (Sherr 1990), and murine chromosome 18 (Hoggan et al. 1988). The locus consists of 21 exons spread over more than 30 kb of the genome and exhibits a tissue specific regulation of transcription. Fms is expressed primarily in monocytes/macrophages and their progenitors, and in various tissues (primarily trophoblast cells) of the developing placenta, but lesser amounts are detectable in B cells (Baker et al. 1993), osteoclast cells, and in a few uncharacterized cells scattered throughout the developing embryo. Fms is also expressed in the more aggressive forms of breast cancer (Kacinski et al. 1991), sometimes in conjunction with M-CSF, and in smooth muscle cells from arteriosclerotic lesions of the rabbit (Inaba et al. 1992).

■ Signalling mechanism

Fms exhibits a single high affinity binding site for M-CSF. Growth signalling is initiated by the M-CSF-induced dimer-

ization of cell surface receptors resulting in transphosphorylation of receptor subunits on the multiple tyrosine residues listed above. The tyrosine phosphorylated residues act as docking sites for SH2-containing proteins that couple to other signal transduction molecules which amplify and propagate the signal. In this regard, Fms associates with Src-family members, with Grb2, and with the p85 subunit of PI 3′K, but unlike other growth factor receptors of the same class, it does not bind phospholipase Cγ1 and exhibits only weak associations with the GTPase activating protein (GAP) (Rohrschneider and Woolford 1991; Sherr 1990). Several cellular proteins are phosphorylated transiently on tyrosine after M-CSF stimulation of Fms-expressing cells. The Shc protein and a unique haemopoietic cell-specific 150 kDa protein (not the mSOS1 protein) are prominent examples. Shc, Grb2, and the 150 kDa protein form complexes after M-CSF stimulation of myeloid cells (Lioubin et al. 1994), and may couple to various signalling mechanisms as occurs with the EGF (Egan et al. 1993) or insulin (Skolnik et al. 1993) receptors. Growth signalling requires multiple tyrosine autophosphorylation sites including Y559, Y697, Y706, and Y721, and probably proceeds, in part, through the Ras-MAP kinase pathway. The single Y807 site, however, regulates a separate differentiation signal that antagonizes cell growth in response to M-CSF or other growth factors such as IL-3 and GM-CSF (Bourette et al. 1994). Subsequent to signalling at the cell surface membrane, receptors undergo endocytosis via clathrin-coated vesicles with eventual transfer to secondary lysosomes for degradation of both receptor and ligand. This step is believed to downmodulate the signal but a potential role in the signalling process has not been eliminated.

■ Transforming potential

Fms was originally isolated as the oncogene carried by the SM- and HZ5 strains of feline sarcoma virus (GenBank accession number K01643). The modifications that created the fully active oncogene from the normal c-*fms* proto-oncogene have been recently reviewed (Rohrschneider and Woolford 1991). The feline v-Fms differs from the feline c-Fms by twelve single amino acid changes across the body of the protein and a large deletion near the C-terminal tail resulting in the replacement of the normal 50 terminal amino acids with 14 unrelated amino acids. Of these modifications, only the C-terminal tail alteration plus amino acid changes at both positions 301 and 374 are needed to fully activate the oncogenic potential of either feline or murine c-Fms proteins. The deletion of the C-terminal tail eliminates a negative regulatory element for the kinase activity, and the point mutations in the extracellular D4 Ig-like domain renders Fms M-CSF independent. For human c-Fms a number of single point mutations around the D4 Ig-like domain plus a C-terminal tail modification is sufficient (Rohrschneider and Woolford 1991; Sherr 1990). So far, none of these mutations is consistently associated with Fms expressed in various human haemopoietic malignancies.

■ References

Alber, G., Kim, K.-M., Weiser, P., Riesterer, C., Carsetti, R., and Reth, M. (1993). Molecular mimicry of the antigen receptor signalling motif by transmembrane proteins of the Epstein–Barr virus and the bovine leukaemia virus. *Curr. Biol.*, **3**, 333–9.

Baker, A.H., Ridge, S.A., Hoy, T., Cachia, P.G., Culligan, D., Baines, P., Whittaker, J.A., Jacobs, A., and Padua, R.A. (1993). Expression of the colony-stimulating factor 1 receptor in B lymphocytes. *Oncogene*, **8**, 371–8.

Bourette, R.P., Myles, G.M., Carlberg, K., and Rohrschneider, L.R. (1994). Uncoupling proliferation and differentiation signals mediated by the murine macrophage-colony-stimulating factor receptor expressed in myeloid FDC-PI cells. *Cell*, submitted.

Carlberg, K., and Rohrschneider, L.R. (1994). The effect of activating mutations on dimerization, tyrosine phosphorylation and internalization of the M-CSF receptor. *Mol. Biol. Cell*, **5**, 81–95.

Courtneidge, S.A., Dhand, R., Pilat, D., Twamley, G.M., Waterfield, M.D., and Roussel, M.F. (1993). Activation of Src family kinases by colony stiumulating factor-1, and their association with its receptor. *EMBO J.*, **12**, 943–50.

Egan, S.E., Giddings, B.W., Brooks, M.W., Buday, L., Sizeland, A.M., and Weinberg, R.A. (1993). Association of Sos Ras exchange protein with Grb2 is implicated in tyrosine kinase signal transduction and transformation. *Nature*, **363**, 45–51.

Hoggan, M.D., Halden, N.F., Buckler, C.E., and Kozak, C.A. (1988). Genetic mapping of the mouse c-fms proto-oncogene to chromosome 18. *J. Virol.*, **62**, 1055–6.

Inaba, T., Yamada, N., Gotoda, T., Shimano, H., Shimada, M., Momomura, K., et al. (1992). Expression of M-CSF receptor encoded by c-*fms* on smooth muscle cells derived from arteriosclerotic lesion. *J. Biol. Chem.*, **267**, 5293–699.

Kacinki, B.M., Scata, K.A., Carter, D., Yee, L.D., Sapi, E., King, B.L., et al. (1991). FMS (CSF-1 receptor) and CSF-1 transcripts and protein are expressed by human breast carconmias in vivo and in vitro. *Oncogene*, **6**, 941–52.

Knighton, D.R., Zheng, J.H., Teneyck, L.F., Ashford, V.A., Xuong, N-H., Taylor, S.S., and Sowadski, J.M. (1991). Crystal structure of the catalytic subunit of cyclic adenosine monophosphate-dependent protein kinase. *Science*, **253**, 407–14.

Lioubin, M.N., Myles, G.M., Carlberg, K., Bowtell, D., and Rohrschneider, L.R. (1994). Shc, Grb2, Sos1, and a 150-kilodalton tyrosine-phosphorylated protein form complexes with Fms in hematopoietic cells. *Mol. Cell. Biol.* **14**, in press.

Reedijk, M., Liu, X., van der Geer, P., Letwin, K., Waterfield, M.D., Hunter, T., and Pawson, T. (1992). Tyr721 regulates specific binding of the CSF-1 receptor kinase insert to PI 3'-kinase SH2 domains: a model for SH2-mediated receptor-target interactions. *EMBO J.*, **11**, 1365–72.

Rohrschneider, L.R., and Woolford, J. (1991). Structural and functional comparison of viral and cellular fms. *Virology*, **2**, 385–95.

Roussel, M.F., Cleveland, J.L., Shurtleff, S.A., and Sherr, C.J. (1991). Myc rescue of a mutant CSF-1 receptor impaired in mitogenic signalling. *Nature*, **353**, 361–3.

Sherr, C.J. (1990). Colony-stimulating factor-1 receptor. *Blood*, **75**, 1–12.

Skolnik, E.Y., Batzer, A., Li, N., Lee, C.-H., Lowenstein, E., Mohammadi, M., Margolis, B., and Schlessinger, J. (1993). The function of GRB2 in linking the insulin receptor to Ras signaling pathways. *Science*, **260**, 1953–5.

van der Geer, P., and Hunter, T. (1993). Mutation of tyrosine 697, a Grb2 binding site, and tyrosine 721, a PI-3' kinase binding site, abrogates signal transduction by the murine CSF-1 receptor expressed in Rat2 fibroblasts. *EMBO J.*, **12** (In press).

Wang, Z., Myles, G.M., Brandt, C.S., Lioubin, M.N., and Rohrschneider, L. (1993). Identification of the ligand-binding regions in the macrophage colony-stimulating factor receptor extracellular domain. *Mol. Cell. Biol.*, **13**, 5348–59.

Larry R. Rohrschneider:
Fred Hutchinson Cancer Research Center,
1124 Columbia Street,
Seattle, Washington, USA

Granulocyte-Macrophage Colony-Stimulating Factor (GM-CSF)

GM-CSF is a 20–30 kDa glycoprotein that primarily regulates the production of neutrophilic and eosinophilic granulocytes and macrophages as well as enhancing the functional activities of the mature cells involved in host defence. Its production and secretion is induced in endothelial cells, fibroblasts and macrophages by bacterial cell wall products and in T-lymphocytes by immune activation. It has potential clinical utility in enhancing the rate of haemopoietic recovery after cancer chemotherapy and in enhancing immune responses. Pathologically it might be involved in inflammatory reactions and autoimmunity.

■ Alternative names

Colony-stimulating factor 2 (CSF-2), macrophage-granulocyte inducer-1GM (MGI-1GM), colony-stimulating factor-α (CSFα), pluripoietin α.

■ GM-CSF sequence

The nucleotide sequences of GM-CSF cDNAs cloned from mouse and man (GenBank accession numbers X03020 and X03021) predict a protein of 144 (human) or 141 (mouse) amino acids each containing a 17-amino-acid leader sequence (Gough and Nicola 1990; Walter *et al.* 1992). These proteins are 54 per cent identical at the amino acid level (Fig. 1) and do not cross-react either biologically or at the receptor binding level between the two species. There is some very limited sequence homology between GM-CSF and interleukin-4.

■ GM-CSF protein

GM-CSF is a secreted glycoprotein monomer containing 127 (human) or 124 (mouse) amino acids in the mature protein (core protein molecular weight of about 15 000). It contains two intramolecular disulphide bonds (Cys54–Cys96 and Cys88–Cys21 in human GM-CSF; Cys51–Cys93 and Cys85–Cys118 in mouse GM-CSF) the first of which is essential for correct folding and biological activity. GM-CSF contains both N- and O-linked glycosylation giving rise to several molecular weight species ranging from 18–30 kDa (positions 27 and 37 in human GM-CSF and 66 and 75 in mouse GM-CSF are N-glycosylated). N-glycosylation appears to reduce both the specific biological activity and receptor

```
                               -------A-------                 -S
                10       +20        30        40        50      60
HUMAN    MWLQSLLLLGTVACSISAPARSPSPSTQPWEHVNAIQEARRLLNLSRDTAAEMNETVEVI
MOUSE    MWLQNLLFLGIVVYSLSAPTRSPITVTRPWKHVEAIKEALNLLD---DMPVTLNEEVEVV
BOVINE   MWLQNLLLLGTVVCSFSAPTRPPNTATRPWQHVDAIKEALSLLNHSSDTDAVMNDT-EVV


                 -          -----B-----      ------C------      -S- -
                70        80        90       100       110      120
HUMAN    SEMFDLQEPTCLQTRLELYKQGLRGSLTKLKGPLTMMASHYKQHCPPTPETSCATQIITF
MOUSE    SNEFSFKKLTCVQTRLKIFEQGLRGNFTKLKGALNMTASYYQTYCPPTPETDCETQVTTY
BOVINE   SEKFDSQEPTCLQTRLKLYKNGLQGSLTSLMGSLTMMATHYEKHCPPTPETSCGTQFISF


                 -----D-------
                130       140
HUMAN    ESFKENLKDFLLVIPFDCWEPVQE
MOUSE    ADFIDSLKTFLTDIPFECKKPSQK
BOVINE   KNFKEDLKEFLFIIPFDCWEPAQK
```

Figure 1. Alignment of amino acid sequences for human, mouse, and bovine GM-CSF. The first amino acid of the mature protein is indicated by a +. Location of the four α-helices A–D and the two β-strands (S) are shown.

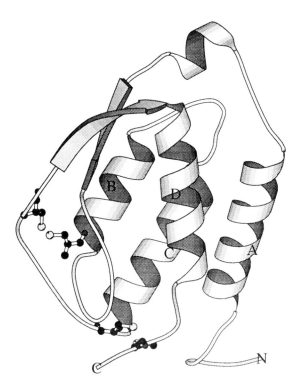

Figure 2. Ribbon diagram of the X-ray-determined structure for human GM-CSF (Diederichs *et al.*, 1991). The two disulphides are shown as ball and stick models. The four major anti-parallel α-helices are labelled A,B,C,D starting from the N-terminus. The two short β-strands are indicated as twisted arrows. The diagram was drawn using the MOLSCRIPT software of PJ Kraulis (J. Appl. Cryst., 1991, **24**, 946–50).

binding affinity of human GM-CSF relative to the non-glycosylated molecule (Cebon *et al.* 1990). GM-CSF is an acidic glycoprotein with isoelectric points between 3.5 and 4.5 due to variable sialylation. It is a highly stable protein not easily denatured irreversibly by heat, extremes of pH, proteases or chaotropic salts so long as the disulphide bonds are intact. The structure of GM-CSF has been determined by X-ray crystallography and like several other cytokines it takes up the conformation of an antiparallel four α-helical bundle with two antiparallel β-strands in the loops between helices A and B and helices C and D (see Fig. 2) (Walter *et al.* 1992; Diederichs *et al.* 1991). A variety of structure–function studies have suggested that the C and D helices are important for binding to the α-chain of the GM-CSF receptor while residues in the A-helix (particularly Glu21) are important for binding to the β-chain of the GM-CSF receptor (Walter *et al.* 1992; Diederichs *et al.* 1991; Kaushansky *et al.* 1989; Shanafelt *et al.* 1991; Lopez *et al.* 1992).

GM-CSF can be purified using conventional chromatography steps and HPLC (Burgess and Nice 1985) or by affinity chromatography using monoclonal antibodies and acid elution (Cebon *et al.* 1990).

■ GM-CSF gene and transcription

The human GM-CSF gene is localized to chromosome 5q21–q32 in the human and chromosome 11B1 in the mouse. In both cases it is very close to the interleukin-3 gene (within 10 kbp) and appears to be coregulated with it. The gene encompasses about 2.5 kbp and contains four exons (Miyatake *et al.* 1985). It generates a single mRNA of 0.8 kbp with no alternate transcripts described. The mRNA contains an AU rich sequence in the 3' untranslated region which confers instability on the message due to an RNA binding activity (Shaw and Kamen 1986).

The GM-CSF gene is constitutively active in monocytes, endothelial cells, and fibroblasts but GM-CSF production can be induced both by an increase in transcription and by message stabilization. A large range of inducers has been described including bacterial endotoxin, phorbol ester, tumour necrosis factor, interleukin-1, and macrophage-activating agents. T cell activation by antigen, concanavalin A, anti-CD3, antibody, or calcium ionophore and phorbol esters results in a large increase in GM-CSF production (Gough and Nicola 1990).

Both human and mouse GM-CSF genes contain a variety of consensus sequences upstream of the TATA box which bind nuclear transcription factors. These include cytokine consensus sequences CK-1 and CK-2, an NFKB site, GC rich regions, and CATTA boxes (Gasson 1991). In addition, the region between the GM-CSF and IL-3 genes appears to contain a DNAse1 hypersensitive site that is important for inducible expression of the GM-CSF and interleukin-3 genes in T cells. This site (3 kbp upstream of the GM-CSF gene) contains an enhancer region with four AP-1 and three NFAT binding sites and is inhibitable by the immunosuppressant cyclosporin A (Cockerill *et al.* 1993). Cyclosporin A inhibits the translocation of NFATc from the cytoplasm to the nucleus where it normally associates with an AP-1 like transcription factor to initiate transcription.

■ Biological functions

GM-CSF is named for its capacity to stimulate the formation of granulocyte and macrophage colonies in semi-solid cultures of bone marrow cells. In this assay it has a specific biological activity of approximately 10^8 units mg^{-1} (where 50 units ml^{-1} is the concentration required for half-maximal stimulation). GM-CSF, in fact, has a rather broad haemopoietic specificity with actions on neutrophils, eosinophils, macrophages, erythroid progenitors, megakaryocyte progenitors, and antigen-presenting dendritic cells. These actions include prolongation of the survival (inhibition of apoptosis) of progenitor cells and, in the case of neutrophils, eosinophils and macrophages, of the mature cells; stimulation of proliferation of the progenitor cells; and enhancement of the functional capacity of the mature cells (Table 1). *In vivo*, GM-CSF administration results in an increase in circulating levels of neutrophils, eosinophils, monocytes, and all progenitor cells and increased numbers and activation status of tissue macrophages. It has been

Table 1. Biological actions of GM-CSF

1. Enhances survival of neutrophils, eosinophils, macrophages, and their progenitor cells.
2. Stimulates proliferation of neutrophil, eosinophil, macrophage, megakaryocyte, and early erythroid progenitor cells, as well as some leukaemic and non-haemopoietic tumour cells.
3. Induces differentiation coupled to proliferation of neutrophil, eosinophil, macrophage and megakaryocyte progenitor cells, as well as some myeloid leukaemic cell lines.
4. Stimulates secretion of cytokines and soluble inflammatory mediators from neutrophils (IL-1, G-CSF, M-CSF, PAF, LTB_4) and macrophages (G-CSF, M-CSF, IL-1, IL-6, PGE).
5. Increases phagocytosis and antibody dependent cell-mediated cytotoxicity (ADCC) of neutrophils, eosinophils and macrophages and primes these cells for enhanced oxidative burst formation to primary stimuli.
6. Enhances bactericidal activity in neutrophils and macrophages, and the killing of parasites in eosinophils and macrophages.
7. Enhances antigen-presenting activity of macrophages and Langerhans cells and stimulates immune reactions to foreign antigens.

Gough and Nicola 1990; Gasson 1991; Tao and Levy 1993; Dranoff *et al*. 1993.

used clinically to enhance haemopoietic recovery in cancer patients receiving chemotherapy with or without bone marrow transplantation (Grant and Heel 1992).

■ Pathology

Excess levels of GM-CSF in transgenic mice (Lang *et al*. 1987) or mice reconstituted with bone marrow cells constitutively expressing high levels of GM-CSF (Johnson *et al*. 1989) result in macrophage infiltration in various organs including the eye, muscle and lungs. In extreme cases, this can lead to blindness and muscle wasting. In patients, high doses of administered GM-CSF can lead to mild flu-like symptoms, rashes, and in a few cases, capillary-leak syndrome and respiratory distress (Tao and Levy 1993). Animal models suggest that GM-CSF expression can lead to a myeloproliferative syndrome but does not in itself lead to leukaemia unless an immortalization event has occurred in the pre-leukaemic clones (Lang *et al*. 1987; Johnson *et al*. 1989).

■ References

Burgess, A.W., and Nice, E.C. (1985). Murine granulocyte-macrophage colony-stimulating factor. *Methods Enzymol*., **116**, 588–600.

Cebon, J., Nicola, N., Ward, M., Gardner, I., Dempsey, P., Layton, J., Duhrsen, U., Burgess, A.W., Nice, E., and Morstyn, G. (1990). Granulocyte-macrophage colony stimulating factor from human lymphocytes. The effect of glycosylation on receptor binding and biological activity. *J. Biol. Chem*., **265**, 4483–91.

Cockerill, P.N., Shannon, M.F., Bert, A.G., Ryan, G.R., and Vadas, M.A. (1993). The granulocyte-macrophage colony-stimulating factor/interleukin 3 locus is regulated by an inducible cyclosporin A-sensitive enhancer. *Proc. Natl. Acad. Sci. (USA)*, **90**, 2466–70.

Diederichs, K., Boone, T., and Karplus, P.A. (1991). Novel fold and putative receptor binding site of granulocyte-macrophage colony-stimulating factor. *Science*, **254**, 1779–82.

Dranoff, G., Jaffee, E., Lazenby, A., Columbek, P., Levitsky, H., Brose, K., Jackson, V., Hamada, H., Pardoll, D., and Muligan, R.C. (1993). Vaccination with irradiated tumour cells engineered to secrete murine granulocyte-macrophage colony-stimulating factor stimulates potent, specific, and long-lasting anti-tumour immunity. *Proc. Natl. Acad. Sci. (USA)*, **90**, 3539–43.

Gasson, J.C. (1991). Molecular physiology of granulocyte-macrophage colony-stimulating factor. *Blood*, **77**, 1131–45.

Gough, N.M., and Nicola, N.A. (1990). Granulocyte-macrophage colony-stimulating factor. In *Colony-stimulating factors* (eds T.M. Dexter, J.M. Garland, and N.G. Testa), pp. 111–53. Marcel Dekker, New York.

Grant, S.M., and Heel, R.C. (1992). Recombinant granulocyte-macrophage colony-stimulating factor (rGM-CSF). A review of its pharmacological properties and prospective role in the management of myelosuppression. *Drugs*, **43**, 516–60.

Johnson, G.R., Gonda, T.J., Metcalf, D., Hariharan, I.K., and Cory, S. (1989). A lethal myeloproliferative syndrome in mice transplanted with bone marrow cells infected with a retrovirus expressing granulocyte-macrophage colony stimulating factor. *EMBO J.*, **8**, 441–8.

Kaushansky, K., Shoemaker, S.G., Alfaro, S., and Brown, C. (1989). Hematopoietic activity of granulocyte/macrophage colony-stimulating factor is dependent upon two distinct regions of the molecule: functional analysis based upon the activities of interspecies hybrid growth factors. *Proc. Natl. Acad. Sci. (USA)*, **86**, 1213–7.

Lang, R.A., Metcalf, D., Cuthbertson, R.A., Lyons, I., Stanley, E., Kelso, A., Kannourakis, G., Williamson, D.J., Klintworth, G.K., Gonda, T.J. et al. (1987). Transgenic mice expressing a haemopoietic growth factor gene (GM-CSF) develop accumulations of macrophages, blindness, and a fatal syndrome of tissue damage. *Cell*, **51**, 675–86.

Lopez, A.F., Shannon, M.F., Hercus, T., Nicola, N.A., Cambareri, B., Dottore, M., Layton, M.J., Eglinton, L., and Vadas, M.A. (1992). Residue 21 of human granulocyte-macrophage colony-stimulating factor is critical for biological activity and for high but not low affinity binding. *EMBO J.*, **11**, 909–16.

Miyatake, S., Otsuka, T., Yokota, T., Lee, F., and Arai, K. (1985). Structure of the chromosomal gene for granulocyte-macrophage colony stimulating factor: comparison of the mouse and human genes. *EMBO J.*, **4**, 2561–8.

Shanafelt, A.B., Johnson, K.E., and Kastelein, R.A. (1991). Identification of critical amino acid residues in human and mouse granulocyte-macrophage colony-stimulating factor and their involvement in species specificity. *J. Biol. Chem.*, **266**, 13804–10.

Shaw, G., and Kamen, R. (1986). A conserved AU sequence from the 3′ untranslated region of GM-CSF mRNA mediates selective mRNA degradation. *Cell*, **46**, 659–67.

Tao, M.-H., and Levy, R. (1993). Idiotype/granulocyte-macrophage colony-stimulating factor fusion protein as a vaccine for B-cell lymphoma. *Nature*, **362**, 755–8.

Walter, M.R., Cook, W.J., Ealick, S.E., Nagabhushan, T.L., Trotta, P.P., and Bugg, C.E. (1992). Three-dimensional structure of recombinant human granulocyte-macrophage colony-stimulating factor. *J. Mol. Biol.*, **224**, 1075–85.

Nicos A. Nicola:

The Walter and Eliza Hall Institute for Medical Research and, The Cooperative Research Centre for Cellular Growth Factors, Melbourne, Victoria, Australia

Receptor for Granulocyte-Macrophage Colony-Stimulating Factor (GM-CSF)

The GM-CSF receptor consists of a ligand-specific α-subunit that confers low-affinity binding and a β-subunit that by itself does not bind GM-CSF but endows the receptor complex with the capacity to bind GM-CSF with high affinity. The β-subunit is shared with the receptor complexes for interleukin-3 and interleukin-5 and is essential for biological signalling. Both the α- and β-subunits of the GM-CSF receptor contain a 200-amino-acid extracellular domain whose structure is conserved among the members of the haemopoietin receptor family and is typified by the growth hormone receptor. The details of biological signalling through the GM-CSF receptor are unknown but since neither the α- nor β-subunits contain recognizable signalling elements the receptor presumably couples to tyrosine kinases and other signalling components present in the cell. The GM-CSF receptor is expressed on haemopoietic progenitor cells, neutrophils, eosinophils, and macrophages as well as some non-haemopoietic cells including placental trophoblasts and endothelial cells.

■ The GM-CSF receptor α-subunit protein

The α-chain of the human GM-CSF receptor is produced as a 400-amino-acid precursor with a predicted 22-amino-acid leader sequence (GenBank accession numbers M73832 and M64445) (Fig. 1). There is a single predicted transmembrane segment of 26 amino acids, an extracellular domain of 298 amino acids and a short cytoplasmic domain of 54 amino acids. The predicted molecular weight of the protein is 44 000 but the observed molecular weight on SDS-PAGE gels is 75 000–85 000 due to N-glycosylation at several of the predicted eleven N-glycosylation sites (Gearing *et al.* 1989). The 200 amino acids proximal to the transmembrane region (on the extracellular side) form a haemopoietin receptor domain that is presumed to bind GM-CSF and which is homologous in many cytokine receptors (see Introductory chapters). This domain consists of several conserved short sequence elements including four conserved cysteine residues and the element Trp-Ser-X-Trp-Ser (Gearing *et al.* 1989; Bazan 1990). The structure of this homologous domain has been solved for the growth hormone receptor by X-ray crystallography and shown to form two β-barrels, each consisting of seven antiparallel β-sheets (de Vos *et al.* 1992).

The murine GM-CSF receptor α-chain (GenBank accession number M85078) is 387 amino acids long and shows only 35 per cent sequence identity with the human protein. The conserved WSSWS motif in the human α-chain is replaced with WGEWS (Park *et al.* 1992). The human α-chain contains one cysteine residue in its transmembrane domain while the murine protein contains three. The human β-chain also contains a cysteine residue in the transmembrane region although there is no evidence yet for disulphide-linked dimers.

■ Associated receptor subunits

The β-subunit of the human GM-CSF receptor (also known as KH97) (GenBank accession numbers M59941 and M38275) shows no detectable binding affinity for GM-CSF but, in association with the α-subunit, generates a high-affinity GM-CSF binding site ($K_D \sim 30$pM) (Hayashida *et al.* 1990). The β-subunit is an 897-amino-acid polypeptide that contains a duplicated haemopoietin domain in the extracellular segment, a leader sequence of 16 amino acids, a single 27-amino-acid transmembrane region and a 350-amino-acid cytoplasmic tail. The predicted molecular weight of the protein is 110 000 and on SDS-PAGE gels it has an apparent molecular weight of 120 000 suggesting that it is only lightly glycosylated. Like the GM-CSF receptor α-subunit, it has no classical cytoplasmic sequence characteristic of tyrosine kinases but it does contain two sequence elements proximal to the transmembrane region which are conserved in several cytokine receptors and may be required for proliferative signalling. The same β-subunit is able to confer high-affinity binding to IL-5 and IL-3 receptor α-chains and is required for proliferative signalling in each of these receptors (Nicola and Metcalf 1991).

In the mouse the common β-chain for the GM-CSF, IL-3, and IL-5 receptors is known as AIC2B (GenBank accession number M34397) and is closely related to another β-chain (AIC2A) (GenBank accession number M29855) that is specific for IL-3 receptors (91 per cent amino acid identity). The mouse and human common β-chains have 55 per cent identity at the amino acid level (Hayashida *et al.* 1990).

■ GM-CSF receptor gene and expression

The GM-CSF receptor α-chain gene is located on the pseudoautosomal region of the human X and Y chromosomes

```
         10        20        30        40
MLLLVTSLLL CELPHPAFLL IPEKSDLRTV APASSLNVRF

         50        60        70        80
DSRTMNLSWD CQENTTFSKC FLTDKKNRVV EPRLSNNECS

         90       100       110       120
CTFREICLHE GVTFEVHVNT SQRGFQQKLL YPNSGREGTA

        130       140       150       160
AQNFSCFIYN ADLMNCTWAR GPTAPRDVQY FLYIRNSKRR

        170       180       190       200
REIRCPYYIQ DSGTHVGCHL DNLSGLTSRN YFLVNGTSRE

        210       220       230       240
IGIQFFDSLL DTKKIERFNP PSNVTVRCNT THCLVRWKQP

        250       260       270       280
RTYQKLSYLD FQYQLDVHRK NTQPGTENLL INVSGDLENR

        290       300       310       320
YNFPSSEPRA KHSVKIRAAD VRILNWSSWS EAIEFGSDDG

        330       340       350       360
NLGSVYIYVL LIVGTLVCGI VLGFLF KRFL RIQRLFPPVP

        370       380       390       400
QIKDKLNDNH EVEDEMGPQR HHRCGWNLYP TPGPSPGSGS

        410
SPRLGSESSL
```

Figure 1. Amino acid sequence of the human GM-CSF Receptor α-chain. The N-terminal leader sequence is underlined and the transmembrane region is boxed (Gearing *et al.*, 1989). The amino acid sequence of the β-chain of the receptor can be found in the chapter on interleukin-3 receptors.

exons spanning about 28 kb and, although two potential alternate transcripts have been described, their significance is unclear (Gorman *et al.* 1992).

The GM-CSF receptor α-chain gene is expressed at low levels in neutrophils, eosinophils, macrophages, placenta, many myeloid leukaemic cells and some non-haemopoietic tumour cells and in several cell lines (Gough and Nicola 1990). The β-chain gene is expressed more broadly and, in the mouse, AIC2A and AIC2B are co-expressed. The regulatory elements of these genes have not been well defined but it is clear that the α- and β-chains of the receptor can be expressed and induced independently.

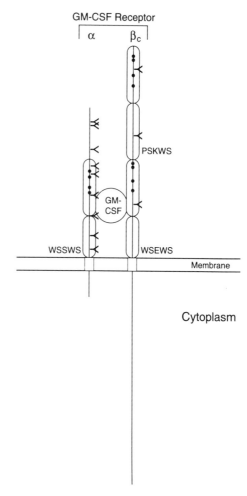

Figure 2. Structure of the heterodimeric human GM-CSF Receptor. The α-chain is shown to the left and the β-chain is shown to the right, drawn to scale. Glycosylation sites are shown as –<, conserved cysteines as dots and the sequences of the WSXSW motifs are shown. The proposed binding site for GM-CSF is shown.

about 1100–1300 kbp from the telomeres (Gough *et al.* 1990) but, in the mouse, is located on an autosomal chromosome 19 (Disteche *et al.* 1992). The β-chain is at human chromosome 22q12.3–13.1 (Shen *et al.* 1992) and on mouse chromosome 15 within 28 kbp of the AIC2A gene (Gorman *et al.* 1992). In humans three alternate transcripts of the α-chain gene have been described; the main one codes for the protein described above; about 20 per cent of the mRNA in placenta and choriocarcinoma cell lines which codes for α-chain codes for a soluble receptor in which a 97 bp internal deletion removes the transmembrane and cytoplasmic domains and changes the reading frame so that 16 different amino acids are at the C-terminus (Raines *et al.* 1991). About 10 per cent of transcripts in receptor-positive cells code for a 410-mino-acid protein in which the C-terminal 25 amino acids have been replaced by 35 new amino acids (Crosier *et al.* 1991). The soluble receptor binds GM-CSF with the same affinity as the two longer transcripts and both longer transcripts code for functional protein in delivering proliferative signals. All three transcripts probably arise from alternate splicing of a single gene spread out over 40–50 kbp. The murine β-chain gene consists of 14

■ Binding characteristics of the GM-CSF receptors

GM-CSF binds to isolated α-chains with high specificity but low affinity ($K_D \sim 2$–5 nM) characterized by rapid 'on' ($k_a \sim 10^8 M^{-1}min^{-1}$) and rapid 'off' ($k_d \sim 0.3$ min^{-1}) kinetics at 4 °C (Gearing et al. 1989). The isolated α-chain on cells is only poorly internalized at 37 °C. The isolated β-chain does not bind GM-CSF (nor IL-3 or IL-5) with detectable affinity but the α-β complex expressed in cells binds GM-CSF with high affinity (Hayashida et al. 1990) ($K_D \sim 30$ pM) with fast on kinetics ($k_a \sim 10^8 M^{-1}min^{-1}$) and slow off kinetics ($k_d \sim 0.003$ min^{-1}) measured at 4 °C (Fig. 2). The slower off rate results in a higher probability of internalization of the ligand complex at 37 °C ($k_e \sim 0.01$–0.06 min^{-1}) and a rapid rate of intracellular degradation ($k_h \sim 0.02$–0.1 min^{-1}) (Nicola et al. 1988).

As a result of the shared β-chain of the receptors, interleukin-3 and interleukin-5 compete for the binding of GM-CSF to high-affinity receptors on human cells where the appropriate α-chains are expressed. Conversely, GM-CSF competes for the high-affinity binding of IL-3 and IL-5 (Nicola and Metcalf 1991). In the mouse, the situation is complicated by the existence of two types of β-chain for the IL-3 receptor so that IL-3 can compete for high affinity GM-CSF receptors at 37 °C but not vice versa (Walker et al. 1985). GM-CSF receptors can also be downregulated by other compounds such as phorbol esters, bacterial lipopolysaccharide, and tumour necrosis factor although the mechanisms for these processes have not been defined (Gough and Nicola 1990). Alternatively, GM-CSF receptor α- and β-chains can be independently upregulated by synthesis by the action of other growth factors and differentiation inducers (Watanabe et al. 1992).

■ Signalling mechanisms

The α-chain, or the β-chain, in isolation is unable to signal in response to GM-CSF (Kitamura et al. 1991). The α-β complex is required for signalling although the signalling capacity of the complex can be dissociated from the capacity to form a high-affinity complex (Kitamura et al. 1991). Signalling requires both the cytoplasmic domains of the α- and β-chains, and proliferative signalling through the β-chain cytoplasmic domain has been localized to two short sequences (Arg456 to Phe487 and to a lesser extent Val518 to Asp544) proximal to the transmembrane region which are conserved in other haemopoietin receptor β-chains (Sakamaki et al. 1992). Receptor activation is associated with rapid tyrosine phosphorylation of several cellular substrates (including the β-chain itself, although this does not appear to be essential for proliferation). A C-terminal domain of the β-chain (residues 626–763) is responsible for tyrosine phosphorylation and activation of the MAP kinase pathway (Sakamaki et al. 1992). Common substrates suggest common signalling mechanisms by the cytokines that share the same β-subunit. GM-CSF activation of the α–β receptor complex induces physical association of the cytoplasmic

tyrosine kinase c-fes with the β-chain and activation of fes tyrosine kinase activity (Hanazono et al. 1993). Similarly, another cytoplasmic tyrosine kinase (JAK-2) associates with the membrane proximal cytoplasmic domain of the β-chain and is activated by GM-CSF (Quelle et al. 1994) By analogy with the interferon system, this kinase may directly activate cytoplasmic transcription factors of the ISGF-3 or STAT family. Although numerous other potential signalling events have been associated with GM-CSF action none have been definitively associated with biological function.

■ References

Bazan, J.F. (1990). Structural design and molecular evolution of a cytokine receptor superfamily. Proc. Natl. Acad. Sci. (USA), **87**, 6934–8.

Crosier, K.E., Wong, G.G., Mathey-Prevot, B., Nathan, D.G., and Sieff, C.A. (1991). A functional isoform of the human granulocyte/macrophage colony-stimulating factor receptor has an unusual cytoplasmic domain. Proc. Natl. Acad. Sci. (USA), **88**, 7744–8.

de Vos, A.M., Ultsch, M., and Kossiakoff, A.A. (1992). Human growth hormone and extracellular domain of its receptor: crystal structure of the complex. Science, **255**, 306–12.

Disteche, C.M., Brannan, C.I., Larsen, A., Adler, D.A., Schorderet, D.F., Gearing, D., Copeland, N.G., Jenkins, N.A., and Park, L. (1992). The human pseudoautosomal GM-CSF receptor alpha subunit gene is autosomal in mouse. Nature Genetics, **1**, 333–6.

Gearing, D.P., King, J.A., Gough, N.M., and Nicola, N.A. (1989). Expression cloning of a receptor for human granulocyte-macrophage colony-stimulating factor. EMBO J., **8**, 3667–76.

Gorman, D.M., Itoh, N., Jenkins, N.A., Gilbert, D.J., Copeland, N.G., and Miyajima, A. (1992). Chromosomal localization and organization of the murine genes encoding the beta subunits (AIC2A and AIC2B) of the interleukin 3, granulocyte/macrophage colony-stimulating factor, and interleukin 5 receptors. J. Biol. Chem., **267**, 15842–8.

Gough, N.M., and Nicola, N.A. (1990). Granulocyte-macrophage colony-stimulating factor. In Colony-stimulating factors (eds T.M. Dexter, J.M. Garland, and N.G. Testa), pp. 111–53. Marcel Dekker, New York.

Gough, N.M., Gearing, D.P., Nicola, N.A., Baker, E., Pritchard, M., Callen, D.F., and Sutherland, G.R. (1990). Localization of the human GM-CSF receptor gene to the X–Y pseudoautosomal region. Nature, **345**, 734–6.

Hanazono, Y., Chiba, S., Sasaki, K., Mano, H., Miyajima, A., Arai, K., Yazaki, Y., and Hirai, H. (1993). c-fps/fes protein-tyrosine kinase is implicated in a signaling pathway triggered by granulocyte-macrophage colony-stimulating factor and interleukin-3. EMBO J., **12**, 1641–6.

Hayashida, K., Kitamura, T., Gorman, D.M., Arai, K., Yokota, T., and Miyajima, A. (1990). Molecular cloning of a second subunit of the receptor for human granulocyte-macrophage colony-stimulating factor (GM-CSF): reconstitution of a high-affinity GM-CSF receptor. Proc. Natl. Acad. Sci. (USA), **87**, 9655–9.

Kitamura, T., Hayashida, K., Sakamaki, K., Yokota, T., Arai, K., and Miyajima, A. (1991). Reconstitution of functional receptors for human granulocyte/macrophage colony-stimulating factor (GM-CSF): evidence that the protein encoded by the AIC2B cDNA is a subunit of the murine GM-CSF receptor. Proc. Natl. Acad. Sci. (USA), **88**, 5082–6.

Nicola, N.A., and Metcalf, D. (1991). Subunit promiscuity among haemopoietic growth factor receptors. Cell, **67**, 1–4.

Nicola, N.A., Peterson, L., Hilton, D.J., and Metcalf, D. (1988). Cellular processing of murine colony-stimulating factor (Multi-CSF,

GM-CSF, G-CSF) receptors by normal haemopoietic cells and cell lines. *Growth Factors*, **1**, 41–9.

Park, L.S., Martin, U., Sorensen, R., Luhr, S., Morrissey, P.J., Cosman, D., and Larsen, A. (1992). Cloning of the low-affinity murine granulocyte-macrophage colony-stimulating factor receptor and reconstitution of a high-affinity receptor complex. *Proc. Natl. Acad. Sci. (USA)*, **89**, 4295–9.

Quelle, F.W., Sato, N., Witthun, B.A., Inhorn, R.C., Eder, M., Miyajima, A., Griffin, J.D., and Ihle, J.N. (1994). JAK2 associates with the β_c chain of the receptor for granulocyte-macrophage colony-stimulating factor, and its activation requires the membrane-proximal region. *Mol. Cell. Biol.* **14**, 4335–41.

Raines, M.A., Liu, L., Quan, S.G., Joe, V., DiPersio, J.F., and Golde, D.W. (1991). Identification and molecular cloning of a soluble human granulocyte-macrophage colony-stimulating factor receptor. *Proc. Natl. Acad. Sci. (USA)*, **88**, 8203–7.

Sakamaki, K., Miyajima, I., Kitamura, T., and Miyajima, A. (1992). Critical cytoplasmic domains of the common beta subunit of the human GM-CSF, IL-3 and IL-5 receptors for growth signal transduction and tyrosine phosphorylation. *EMBO J.*, **11**, 3541–9.

Shen, Y., Baker, E., Callen, D.F., Sutherland, G.R., Willson, T.A., Rakar, S., and Gough, N.M. (1992). Localization of the human GM-CSF receptor beta chain gene (CSF2RB) to chromosome 22q12.2–γq13.1. *Cytogenet. Cell Genet.*, **61**, 175–7.

Walker, F., Nicola, N.A., Metcalf, D., and Burgess, A.W. (1985). Hierarchical down-modulation of haemopoietic growth factor receptors. *Cell*, **43**, 269–76.

Watanabe, Y., Kitamura, T., Hayashida, K., and Miyajima, A. (1992). Monoclonal antibody against the common beta subunit (beta c) of the human interleukin-3 (IL-3), IL-5, and granulocyte-macrophage colony-stimulating factor receptors shows upregulation of beta c by IL-1 and tumour necrosis factor-alpha. *Blood*, **80**, 2215–20.

Nicos A. Nicola:
The Walter and Eliza Hall Institute of Medical Research, and The Cooperative Research Centre for Cellular Growth Factors, Melbourne, Victoria, Australia

Stem Cell Factor (SCF)

Native stem cell factor (SCF) is a 30 kDa glycoprotein which has both N-linked and O-linked carbohydrate residues with two intramolecular disulphide bonds. This factor exhibits a spectrum of biological activities on various tissues including haemopoietic cells, melanocytes, and developing primordial germ cells. It is secreted by bone marrow stromal cells, fibroblasts, and endothelial cells. The expression of SCF mRNA is downregulated by TNF, IL-1 and $TGF_{\beta1}$. Human SCF has potential clinical utility in enhancing the ability of haemopoietic growth factors (CSFs) to stimulate the production of haemopoietic cells, and to mobilize haemopoietic progenitor cells to peripheral blood. Pathologically, mice bearing mutations at either of two loci, the dominant white spotting (W) locus (allelic to c-kit protooncogene; receptor for SCF) or the steel (Sl) locus (allelic to the ligand of c-kit; SCF), exhibit profound developmental defects in germ cells, melanocytes, and haemopoietic lineages. In addition, the expression of SCF mRNA has been detected in brain, lung, heart, and kidney of developing mouse embryos.

■ Alternative names

c-*kit* ligand (KL), steel factor (SLF), and mast cell growth factor (MGF).

■ SCF sequence

The cDNA sequence for human SCF predicts a protein with a 25-amino-acid signal peptide, a 185-amino-acid extracellular domain, a 27-amino-acid transmembrane segment, and a 36-amino-acid intracellular domain (Martin *et al.* 1990). Secreted and membrane-bound forms of SCF have been identified and are due to alternative splicing of exon 6 resulting in a 248-amino-acid protein and a 220-amino-acid protein (Flanagan *et al.* 1991). The 248 residue form of SCF is thought to contain a proteolytic cleavage site which results in the secreted 164-amino-acid protein. Due to the lack of this proteolytic site in the 220-amino-acid form, the protein remains membrane associated. Soluble SCF is highly conserved across species at the amino acid level (Fig. 1). Human and murine SCF are 83 per cent identical; human and rat, 79 per cent identical; and mouse and rat, 95 per cent identical (Martin *et al.* 1990; Zsebo *et al.* 1990a). SCFs from various species also cross-react at both the biological and receptor-binding levels. Rat and murine SCF retain biological activity on human bone marrow cells and receptor binding activities to human c-*kit*, however, human ^{125}I-SCF binding to human **c**-*kit* is six-fold higher than that to murine c-*kit* (Zsebo *et al.* 1990a).

■ SCF protein

Native SCF is secreted as a soluble glycoprotein monomer with an isoelectric point of 3.8 (Williams *et al.* 1990). The core protein has a M_r of 18.4 kDa, but the addition of N-linked and O-linked carbohydrate residues increases the M_r to about 30–35 kDa (Lu *et al.* 1991). However, glycosylation is not necessary for biological activity because

Human Glu Gly Ile Cys Arg Asn Arg Val Thr Asn Asn Val Lys Asp Val[15]
Mouse Lys Glu Ile Cys Gly Asn Pro Val Thr Asp Asn Val Lys Asp Ile
Rat   Gln Gly Ile Cys Arg Asn Pro Val Thr Asp Asn Val Lys Asp Ile

Human Thr Lys Leu Val Ala Asn Leu Pro Lys Asp Tyr Met Ile Thr Leu[30]
Mouse Thr Lys Leu Val Ala Asn Leu Pro Asn Asp Tyr Met Ile Thr Leu
Rat   Thr Lys Leu Val Ala Asn Leu Pro Asn Asp Tyr Met Ile Thr Leu

Human Lys Tyr Val Pro Gly Met Asp Val Leu Pro Ser His Cys Trp Ile[45]
Mouse Asn Tyr Val Ala Gly Met Asp Val Leu Pro Ser His Cys Trp Leu
Rat   Asn Tyr Val Ala Gly Met Asp Val Leu Pro Ser His Cys Trp Leu

Human Ser Glu Met Val Val Gln Leu Ser Asp Ser Leu Thr Asp Leu Leu[60]
Mouse Arg Asp Met Val Ile Gln Leu Ser Leu Ser Leu Thr Thr Leu Leu
Rat   Arg Asp Met Val Thr His Leu Ser Val Ser Leu Thr Thr Leu Leu

Human Asp Lys Phe Ser Asn Ile Ser Glu Gly Leu Ser Asn Tyr Ser Ile[75]
Mouse Asp Lys Phe Ser Asn Ile Ser Glu Gly Leu Ser Asn Tyr Ser Ile
Rat   Asp Lys Phe Ser Asn Ile Ser Glu Gly Leu Ser Asn Tyr Ser Ile

Human Ile Asp Lys Leu Val Asn Ile Val Asp Asp Leu Val Glu Cys Val[90]
Mouse Ile Asp Lys Leu Gly Lys Ile Val Asp Asp Leu Val Leu Cys Met
Rat   Ile Asp Lys Leu Gly Lys Ile Val Asp Asp Leu Val Ala Cys Met

Human Lys Glu Asn Ser Ser Lys Asp Leu Lys Lys Ser Phe Lys Ser Pro[105]
Mouse Glu Glu Asn Ala Pro Lys Asn Ile Lys Glu Ser Pro Lys Arg Pro
Rat   Glu Glu Asn Ala Pro Lys Asn Ile Lys Glu Ser Pro Lys Arg Pro

Human Glu Pro Arg Leu Phe Thr Pro Glu Glu Phe Phe Arg Ile Phe Asn[120]
Mouse Glu Thr Arg Ser Phe Thr Pro Glu Glu Phe Phe Ser Ile Phe Asn
Rat   Glu Thr Arg Asn Phe Thr Pro Glu Glu Phe Phe Ser Ile Phe Asn

Human Arg Ser Ile Asp Ala Phe Lys Asp Phe Val Val Ala Ser Glu Thr[135]
Mouse Arg Ser Ile Asp Ala Phe Lys Asp Phe Met Val Ala Ser Asp Thr
Rat   Arg Ser Ile Asp Ala Phe Lys Asp Phe Met Val Ala Ser Asp Thr

Human Ser Asp Cys Val Val Ser Ser Thr Leu Ser Pro Glu Lys Asp Ser[150]
Mouse Ser Asp Cys Val Leu Ser Ser Thr Leu Gly Pro Glu Lys Asp Ser
Rat   Ser Asp Cys Val Leu Ser Ser Thr Leu Gly Pro Glu Lys Asp Ser

Human Arg Val Ser Val Thr Lys Pro Phe Met Leu Pro Pro Val Ala Ala[165]
Mouse Arg Val Ser Val Thr Lys Pro Phe Met Leu Pro Pro Val Ala Ala
Rat   Arg Val Ser Val Thr Lys Pro Phe Met Leu Pro Pro Val Ala Ala

Figure 1. Amino acid sequences derived from cDNA of soluble human, mouse, and rat SCF.

Escherichia coli-derived material is highly active (Zsebo *et al.* 1990b). This soluble factor probably exists as a non-convalently linked dimer in solution (Zsebo *et al.* 1990a) and contains two intramolecular disulphide bonds (Cys4–Cys89 and Cys43–Cys138) (Lu *et al.* 1991), which may be involved in the folding of the protein. Both soluble SCF and membrane-bound SCF are active in promoting the formation of colonies from murine bone marrow cells (Anderson *et al.* 1990), and the number of human progenitor cells in the liquid culture on stromal cells (Toksoz *et al.* 1992). However, the membrane form is more effective in supporting haemopoiesis than the soluble form (Toksoz *et al.* 1992), implying that the membrane-bound SCF may play an important role in the cellular interactions occurring between haemopoietic cells and stromal cells both *in vivo* and *in vitro*.

■ SCF gene, transcription, and translation

The genes for murine and human SCF map to chromosome 10 (Zsebo *et al.* 1990a; Copeland *et al.* 1990) and chromosome 12 in the region of 12q22–12q24 (Anderson *et al.* 1991), respectively. It generates a single mRNA of 6.5 kb,

but smaller forms were also observed (Zsebo *et al.* 1990a; Anderson *et al.* 1990; Huang *et al.* 1990). Multiple cDNA clones corresponding to alternative mRNA splicing events have been reported for both humans and mice. In humans, a cDNA encodes a protein with a deletion of 28 amino acids in the extracellular domain of the protein (Zsebo *et al.* 1990a). In mice, cDNAs encoding proteins with a deletion of 16 and 28 amino acids have been cloned (Lyman and Williams 1992).

SCF is produced by bone marrow stromal cells (Andrews *et al.* 1992; Heinrich *et al.* 1992), fibroblasts (Heinrich *et al.* 1992; Buzby *et al.* 1992), and endothelial cells (Buzby *et al.* 1992; Broudy *et al.* 1992). *In vitro*, TNF (Andrews *et al.* 1992; Buzby *et al.* 1992), TGFαβ1 (Heinrich *et al.* 1992), and IL-1 (Buzby *et al.* 1992) suppress the expression of SCF mRNA in the above producing cells. Further studies showed that 12-myristate-13-acetate (PMA; protein kinase C activator) or calcium ionophore A23187 facilitates proteolytic cleavage of cell membrane associated SCF (Huang *et al.* 1992).

■ Biological actions

Stem cell factor was originally identified as an activity produced by buffalo rat liver cells (BRL-3A) (Martin *et al.* 1990; Zsebo *et al.* 1990b), which synergized with IL-6 or CSF-I to stimulate the formation of colonies from primitive progenitor cells, termed high proliferative potential colony-forming cells (HPP-CFC). It was subsequently purified from medium conditioned by BRL-3A cells (Martin *et al.* 1990; Zsebo *et al.* 1990b), and named for its capacity, in combination with other haemopoietic growth factors, to enhance the formation of colonies in semi-solid medium and the generation of colony-forming cells (CFC) in liquid culture by cells with a stem cell phenotype, such as murine lin⁻/Sca-I⁺/Rh dull cells (Williams *et al.* 1992) and human lin⁻/CD34⁺/CD33⁻ cells (Bernstein *et al.* 1991). Several factors including kit ligand (KL) (Huang *et al.* 1990) and mast cell growth factor (MGF) (Williams *et al.* 1990) were purified and cloned based upon various activities, and have subsequently shown to be identical to SCF. *In vitro*, SCF alone or combined with IL-3 can induce the proliferation and differentiation of cells of the mast cell lineage (Tsai *et al.* 1991). In combination with other growth factors, such as CSFs, IL- I, IL-3, IL-6, IL-7, IL-11, or Epo, SCF can synergize to stimulate the production of myeloid (Martin *et al.* 1990), lymphoid (McNiece *et al.* 1991), megakaryocytic (Briddell *et al.* 1991), and erythroid cells (Martin *et al.* 1990). Additionally, SCF enhances IL-2 to stimulate the proliferation and cytotoxic activity of human NK cell subset constitutively expressing the high-affinity IL-2R and the c-kit (Matos *et al.* 1993). *In vivo*, SCF has been shown to reduce markedly the severity of the macrocytic anaemia of *Sl/Sl*ᵈ mice (Hunt *et al.* 1992), inducing a leucocytosis and thrombocytosis with accumulation of large numbers of mast cells at injection sites. From animal studies, SCF alone or in combination with other haemopoietic growth factors stimulates various haemopoietic cells (Urich *et al.* 1991), protects against lethal irradiation (Zsebo *et al.* 1992), and mobilizes progenitor cells to peripheral blood (Briddell *et al.* 1993). These biological properties suggest that SCF may be useful clinically for enhanced haemopoietic recovery following chemotherapy and radiotherapy.

■ Pathology

Mice bearing mutations in either of two genetic loci, the dominant white spotting locus(W) (which encodes c-kit; SCF receptor) (Geissler et al. 1988) or the steel locus (Sl) (which encodes c-kit ligand; SCF) (Martin et al. 1990; Williams et al. 1990; Copeland et al. 1990; Huang et al. 1990), exhibit alternation of coat colour, varying degrees of anaemia, and are defective in gonadal development (Silver 1979). These mutations have been shown to be intrinsic (W) or microenvironmental (Sl) defects. It has also been suggested that patients with cutaneous mastocytosis have increased levels of soluble SCF compared to the membrane-bound SCF (Longley et al. 1993). Consistent with the above pathological facts, high doses of SCF resulted in a mastcytosis and activation of melanocytes in animal studies (Hunt et al. 1991). In addition, expression of SCF mRNA in developing mouse embryos indicated that SCF is expressed in the tissues of haemopoiesis, gonads, brain, heart, lung, and kidney (Matsui et al. 1990). However, Sl mice only exhibit defects in haemopoietic and gonadal tissues. The further pathological role(s) for SCF remain to be defined.

■ References

Anderson, D.M., Lyman, S.D., Baird, A., Wignall, J.M., Eisenman, J., Rauch, C., March, C.J., Boswell, H.S., Gimpel, S.D., Cosman, D., and Williams, D.E. (1990). Molecular cloning of mast cell growth factor, a hematopoietin that is active in both membrane bound and soluble forms. Cell, 63, 235–43.

Anderson, D.M., Williams, D.E., Tushinski, R., Gimpel, S., Eisenman, J., Cannizzaro, L.A., Aronson, M., Croce, C.M., Huebner, K., Cosman, D., et al. (1991). Alternate splicing of mRNAs encoding human mast cell growth factor and localization of the gene to chromosome 12q22–q24. Cell Growth Differ., 2, 373–8.

Andrews, D.F., Montgomery, R.B., Moran, D.J., Leung, D., Harris, W.E., Bursten, S.L., Bianco, J.A., and Singer, J.W. (1992). Tumor necrosis factor-α (TNFα) suppression of c-kit ligand(KL) is mediated by phospholipid (PL) intermediates in marrow stromal cells (MSC). Blood, 80, 365A.

Bernstein, I.D., Andrews, R.G., and Zsebo, K.M. (1991). Recombinant human stem cell factor enhances the formation of colonies by CD34+ and CD34+lin⁻ cells, and the generation of colony-forming cell progeny from CD34+lin⁻ cells cultured with interleukin-3, granulocyte colony-stimulating factor, or granulocyte-macrophage colony-stimulating factor. Blood, 77, 2316–21.

Briddell, R.A., Bruno, E., Cooper, R.J., Brandt, J.E., and Hoffman, R. (1991). Effect of c-kit ligand on in vitro human megakaryocytopoiesis. Blood, 78, 2854–9.

Briddell, R.A., Hartley, C.A., Smith, K.A., and McNiece, I.K. (1993). Recombinant rat stem cell factor synergizes with recombinant human granulocyte colony-stimulating factor in vivo in mice to mobilize peripheral blood progenitor cells that have enhanced repopulating potential. Blood, 82, 1720–3.

Broudy, V.C., Kovach, N., Lin, N., Jacobsen, F.W., and Bennett, L.G. (1992). Human umbilical vein endothelial cells (HUVES) display high affinity c-kit receptors and produce stem cell factor (SCF). Blood, 80, 362A.

Buzby, J.S., Knoppel, E., Yancik, S., Bhullar, A., and Cairo, M.S. (1992). Differential expression of steel factor, kit receptor, ICAM-1 and ELAM-1 in mesenchymal cells from newborns compared to adults. Blood, 80, 1571A.

Copeland, N.G., Gilbert, D.J., Cho, B.C., Donovan, P.J., Jenkins, N.A., Cosman, D., Anderson, D.M., Lyman, S.D., and Williams, D.E. (1990). Mast cell growth factor maps near the steel locus on mouse chromosome 10 and is deleted in a number of steel alleles. Cell, 63, 175–83.

Flanagan, J.G., Chan, D.C., and Leder, P. (1991). Transmembrane form of the kit ligand growth factor is determined by alternative splicing and is missing in the Sld mutant. Cell, 64, 1025–35.

Geissler, E.N., Ryan, M.A., and Housman, D.E. (1988). The dominant-white spotting (W) locus of the mouse encodes the c-kit proto-oncogene. Cell, 55, 185–92.

Heinrich, M.C., Dooley, D.C., Freed, A.C., Band, L., Keeble, W.K., Oppernlander, B., Spurgin, P., and Bagby, G.C. (1992). TGF-β 1 represses steel factor (SF) gene expression in long term human bone marrow culture (LTBMC) adherent cells. Blood, 80(10), 369A.

Huang, E., Nocka, K., Beier, D.R., Chu, T.Y., Buck, J., Lahm, H.W., Wellner, D., Leder, P., and Besmer, P. (1990). The hematopoietic growth factor KL is encoded by the Sl locus and is the ligand of the c-kit receptor, the gene product of the W locus. Cell, 63, 225–33.

Huang, E.J., Nocka, K.H., Buck, J., and Besmer, P. (1992). Differential expression and processing of two cell associated forms of the kit-ligand: KL-1 and KL-2. Mol. Biol. Cell, 3, 349–62.

Hunt, P., Zsebo, K.M., Hokom, M.M., Hornkohl, A., Birkett, N.C., del Castillo, J.C., and Martin, F. (1992). Evidence that stem cell factor is involved in the rebound thrombocytosis that follows 5-fluoroura-cil treatment. Blood, 80, 904–11.

Longley, J., Tyrrell, L., Anderson, D., Williams, D., and Halaban, R. (1993). Alternation of mast cell growth factor (MGF) metabolism in human mastocytes. J. Cell. Biochem., supplement 17B, E126A.

Lu, H.S., Clogston, C.L, Wypych, J., Fausset, P.R., Lauren, S., Mendiaz, E.A., Zsebo, K.M., and Langley, K.E. (1991). Amino acid sequence and post-translational modification of stem cell factor isolated from buffalo rat liver cell-conditioned medium. J. Biol. Chem., 266, 8102–7.

Lyman, S.D., and Williams, D.E. (1992). Biological activities and potential therapeutic uses of steel factor. A new growth factor active in multiple hematopoietic lineages. Amer. J. Ped. Hematol. Oncol., 14, 1–7.

Martin, F.H., Suggs, S.V., Langley, K.E., Lu, H.S., Ting, J., Okino, K.H., Morris, C.F., McNiece, I.K., Jacobsen, F.W., Mendiaz, E.A., Birkett, N.C., Smith, K.A., Johnson, M.J., Parker, V.P., Flores, J.C., Patel, A.C., Fisher, E.F., Erjavec, H.O., Herrera, C.J., Wypych, J., Sachdev, R.K., Pope, J.A., Leslie, I., Wen, D., Lin, C.-H., Cupples, R.L., and Zsebo, K.M. (1990). Primary structure and functional expression of rat and human stem cell factor DNAs. Cell, 63, 203–11.

Matos, M.E., Schnier, G.S., Beecher, M.S., Ashman, L.K., William, D.E., and Caligiuri, M.A. (1993). Expression of a functional c-kit receptor on a subset of natural killer cells. J. Exp. Med., 178, 1079–84.

Matsui, Y., Zsebo, K.M., and Hogan, B.L. (1990). Embryonic expression of a haemopoietic growth factor encoded by the Sl locus and the ligand for c-kit. Nature, 347, 667–9.

McNiece, I.K., Langley, K.E., and Zsebo, K.M. (1991). The role of recombinant stem cell factor in early B cell development. Synergistic interaction with IL-7. J. Immunol., 146, 3785–90.

Silver, W.K. (1979). In The coat colors of mice: A model for gene action and interaction, pp. 206–41. Springer, New York.

Toksoz, D., Zsebo, K.M., Smith, K., Hu, S., Brankow, D., Suggs, S.V., Martin, F.H., and Williams, D.A. (1992). Support of human hematopoiesis in long-term bone marrow cultures by murine stromal cells selectively expressing the membrane-bound and secreted forms of the human homolog of the steel gene product, stem cell factor. Proc. Natl. Acad. Sci. (USA), 89, 7350–4.

Tsai, M., Takeishi, T., Thompson, H., Langley, K.E., Zsebo, K.M., Metcalf, D.D., Geissler, E.N., and Galli, S.J. (1991). Induction of mast cell proliferation, maturation, and heparin synthesis by the rat c-kit ligand, stem cell factor. *Proc. Natl. Acad. Sci. (USA)*, **88**, 6382–6.

Ulich, T.R., del Castillo, J., McNiece, I.K., Yi, E.S., Alzona, C.P., Yin, S.M., and Zsebo, K.M. (1991). Stem cell factor in combination with granulocyte colony-stimulating factor (CSF) or granulocyte-macrophage CSF synergistically increases granulopoiesis *in vivo. Blood*, **78**, 1954–62.

Williams, D.E., Eisenman, J., Baird, A., Rauch, C., Van Ness, K., March, C.J., Park, L.S., Martin, U., Mochizuki, D.Y., Boswell, H.S., Burgess, G.S., Cosman, D., and Lyman, S.D. (1990). Identification of a ligand for the *c-kit* proto-oncogene. *Cell*, **63**, 167–74.

Williams, N., Bertoncello, I., Kavnoudias, H., Zsebo, K.M., and McNiece, I.K. (1992). Recombinant rat stem cell factor stimulates the amplification and differentiation of fractionated mouse stem cell populations. *Blood*, **79**, 58–64.

Zsebo, K.M., Williams, D.A., Geissler, E.N., Broudy, V.C., Martin, F.H., Atkins, H.L., Hsu, R.Y., Birkett, N.C., Okino, K.H., Murdock, D.C.,

Jacobsen, F.W., Langley, K.E., Smith, K.A., Takeishi, T., Cattanach, B.M., Galli, S.J., and Suggs, S.V. (1990*a*). Stem cell factor is encoded at the Sl locus of the mouse and is the ligand for the c-kit tyrosine kinase receptor. *Cell*, **63**, 213–24.

Zsebo, K.M., Wypych, J., McNiece, I.K., Lu, H.S., Smith, K.A., Kakare, S.B., Sachdev, R.K., Yuschenkoff, V.N., Birkett, N.C., Williams, L.R., Satyagal, V.S., Tung, W., Bosselman, R.A., Mediaz, E.A., and Langley, K.E. (1990*b*). Identification, purification, and biological characterization of hematopoietic stem cell factor from buffalo rat liver-conditioned medium. *Cell*, **63**, 195–201.

Zsebo, K.M., Smith, K.A., Hartley, C.A., Greenblatt, M., Cooke, K., Rich, W., and McNiece, I.K. (1992). Radioprotection of mice by recombinant rat stem cell factor. *Proc. Natl. Acad. Sci. (USA)*, **89**, 9464–8.

Ian McNiece and Jae-Hung Shieh:
Developmental Hematology Group,
Amgen Inc.,
1840 DeHavilland Drive,
Thousand Oaks, California 91320–1789, USA

Kit (Stem Cell or Steel Factor) Receptor

The kit (stem cell factor) receptor is a member of the type III tyrosine kinase growth factor receptor family, which also includes the CSF-1 and PDGF receptors, flk-1, flk-2, flt-1, and flt-4. The family is characterized by the presence of five immunoglobulin-like domains in the extracellular portion and by an intracellular kinase domain that is split by a 70–100-amino-acid hydrophilic insert. The binding of the cognate kit ligand, variously called stem cell factor, mast cell growth factor, kit ligand, and steel factor, to the kit receptor results in receptor dimerization and autophosphorylation and subsequent association with and activation/relocalization of a specific set of signalling molecules including phosphatidylinositol 3' kinase and phospholipase Cγ. The c-kit gene is allelic with the murine W locus, mutations in which lead to intrinsic defects in the development of the melanocyte, germ cell, and haemopoietic lineages during embryonic development and in the adult. Analysis of the various W mutations has defined the amino acid residues which are essential for kinase activity. Mutations in the human c-kit gene are associated with piebaldism, a rare dominantly inherited hypopigmentation spotting trait.

■ The kit protein

The human c-*kit* receptor gene encodes a 972-amino-acid precursor protein with a predicted 24-amino-acid leader sequence (GenBank accession number X06182) (Yarden *et al*. 1987). A single predicted transmembrane domain of 23 amino acids divides the protein into two major domains. The amino terminal extracellular 487 amino acids form the ligand-binding domain and are composed of five immunoglobulin-like loops. The predicted molecular weight of the receptor is 109 000 but the observed molecular weight on SDS-PAGE gels is 145–160 000 due to N-glycosylation at several of the nine potential glycosylation sites. The C-terminal intracellular 439-amino-acids domain contains the phosphotyrosine kinase domain. The kinase domain is split by a 68-amino-acid kinase insert domain between the ATP binding site and the catalytic phosphotransferase domain.

The murine kit receptor (GenBank accession number Y00864) (Qui *et al*. 1988) contains 975 amino acids and shows 89 per cent identity over the length of the entire protein, and 98 per cent identity in the kinase domain, with the human protein.

■ c-*kit* gene and expression

The human c-*kit* gene is located on chromosome 4q11–q12 (Yarden *et al*. 1987) while the murine gene is located on chromosome 5 (*W* locus) (Chabot *et al*. 1988; Geissler *et al*. 1988). The c-*kit* gene is in a cluster with the genes for two related receptor tyrosine kinases: PDGFRA and *flk*-1. The human c-*kit* transcript is 5.2 kb while the murine transcript is 5.5 kb. Both the human and murine genes are composed of 21 exons spanning at least 70 kb and have very similar

exon/intron boundaries to each other and to the human CSF-1R gene (Vandenbark *et al.* 1992; Andre *et al.* 1992; Gokkel *et al.* 1992). Alternative splicing at the 3' end of exon 9 creates an isoform of the receptor (kitA) differing by an additional four amino acids (Gly-Asp-Asp-Lys) within the extracellular domain (Reith *et al.* 1991). Both kit and kitA bind stem cell factor and initiate activation of downstream substrates, although only the kit isoform shows some low constitutive (ligand independent) activity when expressed at high levels (Reith *et al.* 1991). In addition, 2.3 and 3.2 kb spermatid-specific transcripts have been described encoding for truncated, probably kinase deficient, proteins.

The murine *c-kit* gene is expressed in a spatially and temporally regulated fashion in a variety of tissues during embryogenesis (Orr-Urtreger *et al.* 1990; Keshet *et al.* 1991; Motro *et al.* 1991). These include the tissues in which the *c-kit* receptor is known to have a vital function: melanoblasts, fetal liver haemopoietic cells, and germ cells (see discussion below on the *W* phenotype). In addition, *c-kit* is highly expressed in the placenta, restricted compartments of the embryonic peripheral and central nervous system, kidney, lung, and gut. These embryonic expression patterns are generally maintained in the adult where *c-kit* is expressed in melanocytes, pluripotent haemopoietic stem cells and progenitor cells, mast cells, germ cells, specific neurones and glial cells, lung, and gut (Motro *et al.* 1991). The contiguous nature of the sites expressing *c-kit* and those expressing its ligand (stem cell factor) suggests a possible function for the kit signal transduction pathway in these tissues, as well as those three lineages known to be affected by *W* mutations (Keshet *et al.* 1991; Motro *et al.* 1991).

High levels of *c-kit* expression have been reported in several malignancies, including acute myeloblastic leukaemia, mastocytoma, testicular germ cell tumours and small-cell lung cancer. Frequently, stem cell factor is also co-expressed in these tumours, suggesting that a possible autocrine growth control loop has been established in these tumours.

■ Downstream substrates and signalling mechanisms

Binding of stem cell factor to the kit receptor induces dimerization of the receptor and activation of the catalytic tyrosine kinase domain. Subsequently, the autophosphorylated receptor associates with and phosphorylates a number of SH2-containing proteins, including the 85 kDa subunit of phosphatidylinositol 3' kinase (PI3K), phospholipase Cγ and GTPase-activating protein (GAP) (Reith and Bernstein 1991; Rottapel *et al.* 1991; Herbst *et al.* 1991; Lev *et al.* 1991). Like other receptors with intrinsic tyrosine kinase activity, the binding is mediated through interaction of the SH2-domains of these proteins with specific phosphotyrosine residues in the kit receptor located, at least for PI3K, within the kinase insert domain of kit. In addition, activation of the kit receptor is associated with the rapid, but probably not direct, phosphorylation of Vav,

MAP2 kinase, and raf-1. The activated kit receptor also interacts with PTP-1C, a tyrosine-specific protein phosphatase (Yi and Ihle 1993).

■ The *W/c-kit* locus

Mice bearing severe mutations in the *c-kit* gene (dominant *white-spotting* (*W*) mutants) are white, sterile, and anaemic as the result of cell-autonomous dominant defects in the development of the melanocyte, germ gell, and haemopoietic lineages (Russell 1979; Reith and Bernstein 1991). An identical phenotype is caused by mutations in the *Sl* locus that encodes the kit ligand, stem cell, or Steel factor (see Stem Cell Factor entry p. 177). The severity of the *W* defect is variable and correlates with both the residual kinase activity of the mutated receptor and its ability to interact with downstream substrates (Reith and Bernstein 1991). In addition, alleles that act in a strongly dominant-negative manner in heterozygous animals (e.g. W^{37}, W^{42}) contain point mutations in the cytoplasmic domain that completely abolish detectable kinase activity, whereas alleles that are mild in the heterozygous state (e.g. W, W^{57}) are generally the result of regulatory or deletion mutations that result in the synthesis of lower levels of the wild-type receptor. These observations are consistent with a model for signalling that involves transphosphorylation interactions between dimerized receptor molecules (Reith and Bernstein 1991). Administration of inhibiting monoclonal antibodies against kit has also demonstrated a requirement for the kit receptor in intestinal pacemaker activity (Maeda *et al.* 1992).

Inherited mutations in the human *c-kit* gene cause piebaldism, a genetic disorder characterized by hypopigmented spots resembling the murine *W* locus (Giebel and Spritz 1991; Fleischman *et al.* 1991). However, no haemopoietic deficiency has been implicated in this disorder.

■ References

Andre, C., Martin, E., Cornu, F., Hu, W.X., Wang, X. P., and Galibert, F. (1992). Genomic organization of the human *c-kit* gene: evolution of the receptor tyrosine kinase subclass III. *Oncogene*, **7**, 685–91.

Chabot, B., Stephenson, D.A., Chapman V.M., Besmer, P., and Bernstein, A. (1988). The proto-oncogene *c-kit* encoding a transmembrane tyrosine kinase receptor maps to the mouse W locus. *Nature*, **335**, 88–9.

Fleischman, R.A., Saltman, D.L., Stastny, V., and Zneimer, S. (1991). Deletion of the *c-kit* protooncogene in the human developmental defect piebald trait. *Proc. Natl. Acad. Sci. (USA)*, **88**, 10885–9.

Geissler, E.N., Ryan, M.A., and Housman, D.E. (1988). THe dominant-white spotting (W) locus of the mouse encodes the *c-kit* protooncogene. *Cell*, **55**, 185–92.

Giebel, L.B., and Spritz, R.A. (1991). Mutation of the KIT (mast/stem cell growth factor receptor) protooncogene in human piebaldism. *Proc. Natl. Acad. Sci. (USA)*. **88**, 8696–9.

Gokkel, E., Grossman, Z., Ramot, B., Yarden, Y., Rechavi, G., and Givol, D. (1992). Structural organization of the murine *c-kit* protooncogene. *Oncogene*, **7**, 1423–9.

Herbst, R., Lammers, R., Schlessinger, J., and Ullrich, A. (1991). Substrate phosphorylation specificity of the human c-kit receptor tyrosine kinase. *J. Biol. Chem.*, **266**, 19908–16.

Keshet, E., Hyman, S.D., Williams, D.E., Anderson, D.M., Jenkins, N.A., Copeland, N.G., and Parada, L.F. (1991). Embryonic RNA expression patterns of the c-kit receptor and its cognate ligand suggest multiple functional roles in mouse development. *EMBO J.*, **10**, 2425–35.

Lev, S., Givol, D., and Yarden, T. (1991). A specific combination of substrates is involved in signal transduction by the *kit*-encoded receptor. *EMBO J.*, **10**, 647–54.

Maeda, H., Yamagata, A., Nishikawa, S., Yoshinaga K., Kobayashi, S., Nishi, K., and Nishikawa S. (1992). Requirement of c-kit for development of intestinal pacemaker system. *Development*, **116**, 369–75.

Motro, B., van der Kooy, D., Rossant, J., Reith, A., and Bernstein, A. (1991). Contiguous patterns of c-kit and *steel* expression: analysis of mutations at the W and Sl loci. *Development*, **113**, 1207–21.

Orr-Urtreger, A., Avivi, A., Zimmer, Y., Givol, D., Yarden, Y., and Lonai, P. (1990). Developmental expression of c-kit, a proto-oncogene encoded by the W locus. *Development*, **109**, 911–23.

Qiu, F.H., Ray, P., Brown, K., Barker, P.E., Jhanwar, S., Ruddle, F.H., and Besmer, P. (1988). Primary structure of c-kit: relationship with the CSF-1/PDGF receptor kinase family—oncogenic activation of v-kit involves deletion of extracellular domain and C terminus. *EMBO J.*, **7**, 1003–11.

Reith, A.D., and Bernstein, A. (1991). In *Genome analysis, Vol. 3, Genes and phenotypes*, pp. 105–33. Cold Spring Harbor Laboratory Press, New York.

Reith, A.D., Ellis, C., Lyman, S.D., Anderson, D.M., Williams, D.E., Bernstein, A., and Pawson, T. (1991). Signal transduction by normal isoforms and W mutant variants of the kit receptor tyrosine kinase. *EMBO J.*, **10**, 2451–9.

Rottapel, R., Reedijk, M., Williams, D.E., Lyman, S.D., Anderson, D.M., Pawson, T., and Bernstein, A. (1991). The steel/W transduction pathway: kit autophosphorylation and its association with a unique subset of cytoplasmic signalling proteins is induced by the steel factor. *Mol. Cell. Biol.*, **11**, 3043–51.

Russell, E.S. (1979). Hereditary anemias of the mouse: a review for geneticists. *Adv. Genet.*, **20**, 357–459.

Vandenbark, G.R., deCastro, C.M., Taylor, H., Dew-Knight, S., and Kaufman, R.E. (1992). Cloning and structural analysis of the human c-kit gene. *Oncogene*, **7**, 1259–66.

Yarden, Y., Kuang, W.J., Yang-Feng, T., Coussens, L., Munemitsu, S., Dull, T.J., Chen, E., Schlessinger, J., Francke, E., and Ullrich, A. (1987). Human proto-oncogene c-kit: a new cell surface receptor tyrosine kinase for an unidentified ligand. *EMBO J.*, **6**, 3341–51.

Yi, T., and Ihle, J.N. (1993). Association of hematopoietic cell phosphatase with c-kit after stimulation with c-kit ligand. *Mol. Cell. Biol.*, **13**, 3350–8.

Alan Bernstein:
Department of Molecular and Developmental Biology
Samuel Lunenfeld Research Institute,
Mount Sinai Hospital, Toronto

Benny Motro:
Division of Molecular and Developmental Biology,
Samuel Lunenfeld Research Institute,
Mount Sinai Hospital,
Toronto, Canada

Hepatocyte Growth Factor

Hepatocyte growth factor (HGF) and scatter factor (SF) are identical αβ heterodimeric glycoproteins secreted by cells of mesodermal origin. HGF stimulates mitogenesis, cell motility and dissociation of epithelial sheets, it promotes matrix invasion, is a morphogen for kidney tubule cells, is a potent angiogenic factor, and is considered a major mediator in liver regeneration in vivo. HGF is secreted as a single-chain biologically inactive precursor (pro-HGF), most of which is sequestered on the cell surface or bound to extracellular matrix. Maturation into the active αβ heterodimer is necessary to induce biological responses in target cells. The primary determinant for the receptor binding activity is located within the α chain. The same domain leads to activation of the downstream signal cascade involved in the motility response. However, the complete HGF protein is required for mitogenic activity. Pro-HGF is correctly processed to the mature heterodimer by pure urokinase-type plasminogen activator, which also acts as a pro-HGF convertase in vivo at the cell surface.

■ Alternative names

Scatter factor (SF), human lung fibroblast-derived mitogen, hepatopoietin A, hepatotrophin, tumour cytotoxic factor.

■ HGF structure and structure–function relationship

HGF is a disulphide-linked heterodimer of a heavy (α) subunit of 60 kDa and a light (β) subunit of 32 or 36 kDa, originating from proteolytic cleavage of a single 92 kDa precursor (Fig. 1) (Miyazawa *et al.* 1989; Nakamura *et al.* 1989; Weidner *et al.* 1991; Rubin *et al.* 1991). HGF is structurally homologous to plasminogen, showing 38 per cent

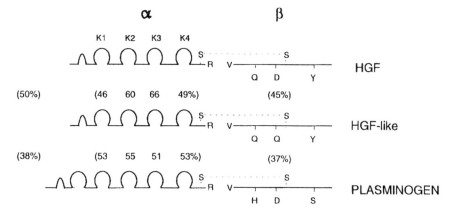

Figure 1. The hepatocyte growth factor (HGF) and related molecules. Schematic representation of the heterodimeric structure of HGF and of a related molecule (HGF-like). The α chain contains four 'kringle' domains (K1–K4) and it is covalently linked to the β chain which shares homology with serine proteases. The bottom line contains a scheme of plasminogen: numbers in parentheses indicate the percentage of homology.

primary sequence homology. The α subunit consists of an N-terminal hydrophobic leader sequence, a putative hairpin loop and four kringle modules. The β subunit has homology to the catalytic domain of serine proteases but lacks enzymatic activity due to the substitution of two critical amino acids in the active site. HGF originates from a single 6 kb transcript, whose open reading frame encodes a pre-pro polypeptide of 728 amino acids. The α subunit is preceded at the N-terminal by a leader sequence of 29 amino acids (the first is a Met) mostly hydrophobic, and by a 25-amino-acid pro-sequence. The predicted sequence indicates the presence of a putative proteolytic cleavage site following Arg_{494}. This site, including a basic residue followed by two hydrophobic amino acids (Arg-Val-Arg), fits the consensus sequence for cleavage by serine proteases. The N-terminal amino acid sequence of the isolated β subunit corresponds to the residues following Val_{495}. Site directed mutagenesis of Arg_{494} generates an uncleaved HGF, which still retains binding to the HGF receptor, but is biologically inactive (Hartmann *et al.* 1992; Lokker *et al.* 1992). Some variant clones lack a stretch of 15 nucleotides at the 5' end of the fifth exon (481–495), causing an in frame deletion of five amino acids in the first kringle domain (GenBank accession number M55379) (Rubin *et al.* 1991; Seki *et al.* 1990). The resulting HGF variant is as active as the wild-type in binding to the receptor and eliciting biological responses. From embryonic human fibroblasts was isolated also a mRNA transcript of 1.2kb, whose open reading frame encodes a molecule of 290 amino acids consisting of the signal sequence, an N-terminal hairpin loop domain, and the first two kringle domains of the α subunit (GenBank accession number L02931) (Hartmann *et al.* 1992; Chan *et al.* 1991; Miyazawa *et al.* 1991). The resulting HGF molecule, a monomer of 28 kDa, as well as the separate α subunit, stimulates tyrosine phosphorylation, and induces scattering of epithelial cells, but not mitogenesis (Hartmann *et al.* 1992). Furthermore, the two kringle domain variant was reported to inhibit the mitogenicity of the full-size HGF (Chan *et al.*

1991). The separately expressed β subunit is unable to bind to the receptor and does not elicit any response (Hartmann *et al.* 1992). HGF can be purified to homogeneity using conventional chromatography steps including heparin affinity chromatography and HPLC as final steps (see references in Gherardi and Stoker 1991).

■ Distribution of HGF

HGF is produced by non-parenchymal liver, kidney and lung cells. Parenchymal cells in these tissues express high level of the HGF receptor. Several fibroblast cell lines, which do not express the HGF receptor, also produce HGF, suggesting that HGF acts as a paracrine effector in mesenchymal–epithelial interactions (Gherardi and Stoker 1991).

■ Activation of HGF

HGF is secreted as a single-chain biologically inactive precursor, mostly found in a matrix associated form. Maturation of the precursor into the active $\alpha\beta$ heterodimer takes place in the extracellular environment and results from proteolytic cleavage (Naka *et al.* 1992; Naldini *et al.* 1992). Urokinase acts as a pro-HGF convertase and some of the growth and invasive cellular responses mediated by this enzyme may involve activation of HGF (Naldini *et al.* 1992).

■ HGF binding properties

HGF binds to two classes of sites, a low affinity and a high affinity form. The low affinity site (K_D: 10^{-9}M) most likely refers to matrix- or cell-associated heparin sulphate proteoglycans, and is thought to be instrumental for the recruitment of HGF to the matrix or to the membrane (Naldini *et al.* 1991). The high affinity binding site is the HGF receptor, identified as the product of the proto-oncogene c-*Met* (see HGF receptor entry).

■ Structure and chromosomal localization of the HGF coding gene

The gene of human HGF has been cloned (Seki *et al*. 1991). It consists of 18 exons and 17 introns, and is 70 kb long. The first exon comprises the 5′ untranslated region and the signal for the start of transcription. The following 10 exons code for the heavy chain containing the four kringle domains, each of them encoded by two exons. The proteolytic site is located in the 13th exon. The remaining six exons encode the β chain. The 18th exon contains the C-terminal region and the long untranslated tail. The locus of the HGF gene was mapped by *in situ* hybridization on chromosome 7q21.1 of the human genome (Fukuyama *et al*. 1991; Saccone *et al*. 1992).

■ Biological functions

HGF is the most potent mitogen for hepatocytes in primary cultures (Nakamura *et al*. 1984) and is considered the major mediator of liver regeneration *in vivo*. HGF is a multi-functional factor for several epithelial cells. It stimulates growth of kidney tubular epithelium and keratinocytes, endothelial cells, and melanocytes (Rubin *et al*. 1991; Bussolino *et al*. 1992). It dissociates layers of epithelial cells (scattering effect) increasing their motility and invasiveness (Stoker *et al*. 1987; Weidner *et al*. 1990). It promotes progression of carcinoma cells toward malignant invasive phenotypes (Weidner *et al*. 1990). It is also a morphogen that stimulates the three-dimensional organization of kidney tubular cells (Montesano *et al*. 1991) and is a potent angiogenic factor (Bussolino *et al*. 1992).

■ References

Bussolino, F., Di Renzo, M.F., Ziche, M., Bocchietto, E., Olivero, M., Naldini, L., Gaudino, G., Tamagnone, L., Coffer, A., and Comoglio, P.M. (1992). Hepatocyte growth factor is a potent angiogenic factor which stimulates endothelial cell motility and growth. *J. Cell Biol.*, **119**, 629–41.

Chan, A.M.L., Rubin, J.S., Bottaro, D.P., Hirschfield, D.W., Chedid, M., and Aaronson, S.A. (1991). Identification of a competitive HGF antagonist encoded by an alternative transcript. *Science*, **254**, 1382–5.

Fukuyama, R., Ichijoh, Y., Minoshima, S., Kitamura, N., and Shimizu, N. (1991). Regional localization of the hepatocyte growth factor (HGF) gene to human chromosome 7 band q21.1. *Genomics*, **11**, 410–5.

Gherardi, E., and Stoker, M. (1991). Hepatocyte growth factor—scatter factor: mitogen, motogen, and met. *Cancer Cells*, **3**, 227–32.

Hartmann, G., Naldini, L., Weidner, K.M., Sachs, M., Vigna, E., Comoglio, P.M., and Birchmeier, W. (1992). A functional domain in the heavy chain of scatter factor/hepatocyte growth factor binds the c-Met receptor and induces cell dissociation but not mitogenesis. *Proc. Natl. Acad. Sci. (USA)*, **89**, 11574–8.

Lokker, N.A., Mark, M.R., Luis, E.A., Bennet, G.L., Robbins, K.A., Baker, J.B., and Godowski, P.J. (1992). Structure–function analysis of hepatocyte growth factor: identification of variants that lack mitogenic activity yet retain high affinity receptor binding. *EMBO J.*, **11**, 2503–10.

Miyazawa, K., Tsubouchi, H., Naka, D., Takahashi, K., Okigaki, M., Arakaki, N., Nakayama, H., Hirono, S., Sakiyama, O., Takahashi, K., Gohda, E., Daikuhara, Y., and Kitamura, N. (1989). Molecular cloning andsequence analysis of cDNA for human hepatocyte growth factor. *Biochem. Biophys. Res. Commun.*, **163**, 967–73.

Miyazawa, K., Kitamura, A., Naka, D., and Kitamura, N. (1991). An alternatively processed mRNA generated from human hepatocyte growth factor gene. *Eur. J. Biochem.*, **197**, 15–22.

Montesano, R., Matsumoto, K., Nakamura, T., and Orci, L. (1991). Identification of a fibroblast-derived epithelial morphogen as hepatocyte growth factor. *Cell*, **67**, 901–8.

Naka, D., Ishii, T., Yoshiyama, Y., Miyazawa, K., Hara, H., Hishida, T., and Kitamura, N. (1992). Activation of hepatocyte growth factor by proteolytic conversion of a single chain form to a heterodimer. *J. Biol. Chem.*, **267**, 20114–9.

Nakamura, T., Nawa, K., and Ichihara, A. (1984). Partial purification and characterization of hepatocyte growth factor from serum of hepatectomized rats. *Biochem. Biophys. Res. Commun.*, **122**, 1450–9.

Nakamura, T., Nishizawa, T., Hagiya, M., Seki, T., Shimonishi, M., Sugimura, A., Tashiro, K., and Shimizu, S. (1989). Molecular cloning and expression of human hepatocyte growth factor. *Nature*, **342**, 440–3.

Naldini, L., Weidner, K.M., Vigna, E., Gaudino, G., Bardelli, A., Ponzetto, C., Narsimhan, R.P., Hartmann, G., Zarnegar, R., Michalopoulos, G.K., Birchmeier, W., and Comoglio, P.M. (1991). Scatter factor and hepatocyte growth factor are indistinguishable ligands for the MET receptor. *EMBO J.*, **10**, 2867–78.

Naldini, L., Tamagnone, L., Vigna, E., Sachs, M., Hartmann, G., Birchmeier, W., Daikuhara, Y., Tsubouchi, H., Blasi, F., and Comoglio, P.M. (1992). Extracellular proteolytic cleavage by urokinase is required for activation of hepatocyte growth factor/scatter factor. *EMBO J.*, **11**, 4825–33.

Rubin, J.S., Chan, A.M-L., Bottaro, D.P., Burgess, W.H., Taylor, W.G., Cech, A.C., Hirschfield, D.W., Wong, J., Miki, T., Finch, P.W., and Aaronson, S.A. (1991). A broad-spectrum human lung fibroblast-derived mitogen is a variant of hepatocyte growth factor. *Proc. Natl. Acad. Sci. (USA)*, **88**, 415–9.

Saccone, S., Narsimhan, R.P., Gaudino, G., Dalpra, L., Comoglio, P.M., and Della Valle, G. (1992). Regional mapping of the human hepatocyte growth factor (HGF)-scatter factor gene to chromosome 7q21.1. *Genomics*, **13**, 912–4.

Seki, T., Ihara, I., Sugimura, A., Shimonshi, M., Nishizawa, T., Asami, O., Hagiya, M., Nakamura, T., and Shimizu, S. (1990). Isolation and expression of cDNA for different forms of hepatocyte growth factor from human leukocyte. *Biochem. Biophys. Res. Commun.*, **172**, 321–7.

Seki, T., Hagiya, M., Shimonishi, M., Nakamura, T., and Shimizu, S. (1991). Organization of the human hepatocyte growth factor-encoding gene. *Gene*, **102**, 213–9.

Stoker, M., Gherardi, E., Perryman, M., and Gray, J. (1987). Scatter factor is a fibroblast-derived modulator of epithelial cell mobility. *Nature*, **327**, 239–42.

Weidner, K.M., Behrens, J., Vandekerckhove, J., and Birchmeier, W. (1990). Scatter factor: molecular characteristics and effect on the invasiveness of epithelial cells. *J. Cell Biol.*, **111**, 2097–108.

Weidner, K.M., Arakaki, N., Hartmann, G., Vandekerckhove, J., Weingart, S., Rieder, H., Fonatsch, C., Tsubouchi, H., Hishida, T., Daikuhara, Y., and Birchmeier, W. (1991). Evidence for the identity of human scatter factor and human hepatocyte growth factor. *Proc. Natl. Acad. Sci. (USA)*, **88**, 7001–5.

Paolo M. Comoglio and Andrea Graziani:
Department of Biomedical Sciences and Oncology,
University of Torino, Italy

Hepatocyte Growth Factor (HGF) Receptor

The HGF receptor is a 190 kDa heterodimer of two ($\alpha\beta$) disulphide-linked protein subunits. The α subunit is extracellular, while the β subunit bears an extracellular portion involved in ligand binding, a membrane-spanning segment and a cytoplasmic tyrosine kinase domain with phosphorylation sites regulating its activity. Both subunits originate from glycosylation and proteolytic cleavage of a common precursor, encoded by a unique gene, the proto-oncogene c-Met. Negative regulation of the receptor kinase activity occurs independently through protein kinase C activation and an increase of intracellular Ca^{2+} concentration. The signal transduction pathways following ligand-induced receptor autophosphorylation involve the generation of the D-3 phosphorylated inositol lipids and the stimulation of the ras pathway through the activation of a ras–guanine nucleotide release activity. Tyrosine phosphorylated HGF receptor associates in vitro and in vivo upon HGF/SF binding with several SH2 containing proteins. The HGF receptor is highly expressed in liver, in epithelial cells from several tissues and in endothelial cells. The expression of the HGF receptor is increased in several epithelial tumours.

■ Alternative names

p190^{Met}, scatter factor (SF) receptor.

■ Structure and biosynthesis of the HGF receptor

The HGF receptor was identified as the product of the proto-oncogene c-Met, the p190^{Met} protein (Bottaro et al. 1991; Naldini et al. 1991a,b). p190^{Met} is composed of a 50 kDa (α) chain disulphide linked to a 145 kDa (β) chain in an $\alpha\beta$ complex of 190 kDa (GenBank accession number X54559) (Giordano et al. 1989a; Ponzetto et al. 1991). Both the α chain and the β chain are exposed at the cell surface; the β chain contains a 23-amino-acid membrane-spanning segment and an intracellular region containing the tyrosine kinase domain. The extracellular region of the β chain features a small cysteine-rich domain (11 cysteines) between residues 520 and 625; the remaining 22 cysteine residues are dispersed through the extracellular domain of both the α and β chains. In the extracellular domain there are 4 sites for N-linked glycosylation in the α chain and 9 sites in the β chain. The translation product of the c-Met gene is a single chain precursor of 170 kDa, which before reaching the plasma membrane undergoes glycosylation, disulphide bond formation and proteolytic cleavage (Giordano et al. 1989b). The c-Met sequence features a consensus site for proteic cleavage Lys-Arg-Lys-Lys-Arg-Ser (amino acids 303–308) which is highly homologous to the cleavage site of the precursor of the insulin receptor. N-linked glycosylation is essential for correct cleavage of the pro-receptor and appearance of the $\alpha\beta$ heterodimer: treatment of cells with tunicamycin or HGF receptor digestion with endoglycosydases, yield a protein of apparent molecular weight 150 kDa (Giordano et al. 1989b). The unglycosylated precursor is unable to reach the plasma membrane. In a colon carcinoma cell (LoVo) the pro-receptor is not cleaved, due to defective posttranslational processing, and the Met protein is exposed at the cell surface as a single chain polypeptide of 190 kDa (Mondino et al. 1991). The intracellular region of the β chain features an unusually long juxtamembrane region (128 amino acids), between the transmembrane and the tyrosine kinase domain. The HGF receptor is the prototype of a distinct subfamily of heterodimeric protein–tyrosine kinases, which includes the putative receptor products of the Ron (Ronsin et al. 1993) and Sea (Huff et al. 1993) genes (Fig. 1). These receptors share significant sequence similarity including the consensus for proteolytic cleavage in the extracellular domain, the protein-tyrosine kinase domain and a short C-terminal tail harbouring only two conserved tyrosine phosphorylation sites. Two C-terminal truncated forms of the HGF receptor have been detected in several cell lines and identified: a 140 kDa transmembrane protein (p140^{Met}) and a 130 kDa soluble protein (p130^{Met}) which is released from the cell (Prat et al. 1991a; Giordano et al. 1993). Both forms have the same heterodimeric structure of the intact receptor and completely lack the intracellular domain. The α chain of the truncated forms are indistinguishable from the α chain of p190^{Met}. p140^{Met} originates from proteolytic cleavage of the p170 precursor occurring within the endoplasmic lumen. The soluble p130 is generated by proteolytic cleavage at the cell surface.

■ HGF receptor gene and expression

The c-Met proto-oncogene is located on human chromosome 7 band 7q21–q31, very close to the cystic fibrosis locus. The Met gene was originally isolated as a transforming gene activated by a DNA rearrangement in an osteosarcoma cell line treated with a chemical carcinogen (for a review see Cooper 1992). The 3' region of the Met gene is rearranged with the 5' region of Tpr (translocation

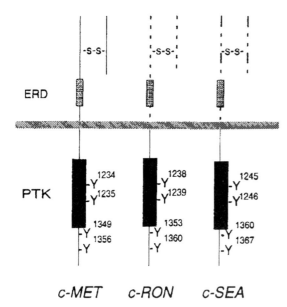

Figure 1. The HGF receptor tyrosine kinase family. Schematic representation of *MET* receptor tyrosine kinase family. Conserved tyrosine residues are indicated. S–S: disulphide bridges; PTK: protein tyrosine kinase domain; ERD: extracellular related domain.

promoter region, 41). The fused Tpr/Met protein features a cytoplasmic localization. The *Met* gene was found amplified, over-expressed, and the protein activated in a human gastric carcinoma cell line. The full size protein is entailed in a 9 kb transcript. Three additional transcripts, of 7, 5.2, and 3.4 kb, all containing the cytoplasmic domain, have been observed in several cell lines (Giordano *et al.* 1989a; Park *et al.* 1986). In humans the HGF receptor is expressed at high levels in liver, epithelial cells of the gastrointestinal tract, kidney, ovary, and endometrium (Di Renzo *et al.* 1991; Prat *et al.* 1991b). p190^{Met} is also expressed in endothelium (Bussolino *et al.* 1992), and in some areas of the brain (Di Renzo *et al.* 1993). The expression of the *Met* gene is increased in the majority of carcinomas of the gastrointestinal tract (Di Renzo *et al.* 1991; Prat *et al.* 1991b). A hundred-fold increase was found in thyroid tumours of the papillary histotype (Di Renzo *et al.* 1992).

■ Binding characteristics of the HGF receptor

HGF binds to p190^{Met} with high specificity and affinity (K_D around 0.2 nM) (Naldini *et al.* 1991b). The binding region involves the extracellular portion of the β chain, which is chemically cross-linked to the ligand. The involvement of the α chain cannot be ruled out. The uncleaved receptor features the same HGF binding properties as the mature

heterodimeric receptor (Naldini *et al.* 1991b), as well as the soluble C-terminal truncated form (p130^{Met}) (Naldini L., Prat M., and Comoglio P., unpublished observation).

■ Regulation of the HGF receptor tyrosine kinase activity

HGF binding to the extracellular domain of p190^{Met} triggers tyrosine autophosphorylation of the receptor β subunit in intact cells (Bottaro *et al.* 1991; Naldini *et al.* 1991a; Graziani *et al.* 1991). Autophosphorylation upregulates the kinase activity of the receptor, increasing several-fold the V_{max} of the phosphotransfer reaction (Naldini *et al.* 1991c). The major phosphorylation site has been mapped to Tyr$_{1235}$, both *in vitro* and *in vivo* (Ferracini *et al.* 1991). Tyr$_{1235}$ is part of a 'three tyrosine' motif, including also Tyr$_{1230}$ and Tyr$_{1234}$ conserved in other tyrosine kinase receptors, such as the insulin receptor. Substitution of either Tyr$_{1234}$ or Tyr$_{1235}$ with Phe residues severely reduces the *in vitro* kinase activity, without preventing autophosphorylation. Only the replacement of both tyrosines yielded a protein lacking any autokinase activity (Longati *et al.* 1994). Negative regulation of the receptor kinase activity occurs through two independent pathways involving protein kinase C activation (Gandino *et al.* 1990) or increase in the intracellular Ca^{2+} concentration up to 0.2–0.76 μM (Gandino *et al.* 1991). Both pathways lead to the phosphorylation of a unique phosphopeptide of the receptor containing Ser$_{985}$ and to a decrease in its kinase activity (Gandino *et al.* 1994). The serine kinase directly phosphorylating the receptor has not been formally identified. The negative control by protein kinase C seems to operate at a steady level also in physiological conditions, because down-regulation of protein kinase C results in increased tyrosine phosphorylation of p190^{Met}.

■ Signalling mechanisms

The HGF receptor is fully responsible for the biological events elicited by its ligand, which include mitogenesis, morphogenesis, cell dissociation of epithelial cell sheets, and promotion of matrix invasion (see HGF entry). The signal transduction mechanisms following ligand-induced receptor autophosphorylation involve the generation of the D-3 phosphorylated inositol lipids through the activation of PI 3-kinase activity (Graziani *et al.* 1991), the stimulation of the Ras pathway through the activation of a Ras-guanine nucleotide releasing factor (Graziani *et al.* 1993), and the activation of p60^{src} tyrosine kinase activity (Ponzetto *et al.* 1994). Tyrosine phosphorylated p190^{Met} associates *in vitro* and *in vivo* upon HGF/SF binding with a number of SH2 containing proteins, namely PI-3 kinase, RasGAP, phospholipase C-γ, and Src-related tyrosine kinases (Graziani *et al.* 1991; Bardelli *et al.* 1992). The binding site for the SH2 domains of PI 3-kinase have been mapped to Tyr$_{1349}$ and Tyr$_{1356}$, the two tyrosines which are conserved in the C-terminal tail of the HGF receptor family (Ponzetto *et al.* 1993). The phosphorylated receptor associates with Phospholipase C-γ, p60[src], the Grb2/Sos complex and Shc (Ponzetto et al. 1994). Moreover, a protein tyrosine phosphatase activity co-precipitates with the HGF receptor after stimulation with HGF/SF (Villa Moruzzi et al. 1993).

■ References

Bardelli, A., Maina, F., Gout, I., Fry, M.J., Waterfield, M.D., Comoglio, P.M., and Ponzetto, C. (1992). Autophosphorylation promotes complex formation of recombinant hepatocyte growth factor receptor with cytoplasmic effectors containing SH2 domains. Oncogene, 7, 1973–8.

Bottaro, D.P., Rubin, J.S., Faletto, D.L., Chan, A.M., Kmiecik, T.E., Vande Woude, G.F., and Aaronson, S.A. (1991). Identification of the hepatocyte growth factor receptor as the c-met proto-oncogene product. Science, 251, 802–4.

Bussolino, F., Di Renzo, M.F., Ziche, M., Bocchietto, E., Olivero, M., Naldini, L., Gaudino, G., Tamagnone, L., Coffer, A., and Comoglio, P.M. (1992). Hepatocyte growth factor is a potent angiogenic factor which stimulates endothelial cell motility and growth. J. Cell Biol., 119, 629–41.

Cooper, C. S. (1992). The met oncogene: from detection by transfection to transmembrane receptor for hepatocyte growth factor. Oncogene, 7, 3–7.

Di Renzo, M.F., Narsimhan, R.P., Olivero, M., Bretti, S., Giordano, S., Medico, E., Gaglia, P., Zara, P., and Comoglio, P.M. (1991). Expression of the Met/HGF receptor in normal and neoplastic human tissues. Oncogene, 6, 1997–2003.

Di Renzo, M.F., Olivero, M., Ferro, S., Prat, M., Bongarzone, I., Pilotti, S., Belfiore, A., Costantino, A., Vigneri, R., Pierotti, M.A., and Comoglio, P.M. (1992). Overexpression of the c-MET/HGF receptor gene in human thyroid carcinomas. Oncogene, 7, 2549–53.

Di Renzo, M.F., Bertolotto, A., Olivero, M., Putzolu, P., Crepaldi, T., Schiffer, D., Pagni, C.A., and Comoglio, P.M. (1993). Selective expression of the Met/HGF receptor in human central nervous system microglia. Oncogene, 8, 219–22.

Ferracini, R., Longati, P., Naldini, L., Vigna, E., and Comoglio, P.M. (1991). Identification of the major autophosphorylation site of the Met/hepatocyte growth factor receptor tyrosine kinase. J. Biol. Chem., 266, 19558–64.

Gandino, L., Di Renzo, M.F., Giordano, S., Bussolino, F., and Comoglio, P.M. (1990). Protein kinase-c activation inhibits tyrosine phosphorylation of the c-met protein. Oncogene, 5, 721–5.

Gandino, L., Munaron, L., Naldini, L., Ferracini, R., Magni, M., and Comoglio, P.M. (1991). Intracellular calcium regulates the tyrosine kinase receptor encoded by the MET oncogene. J. Biol. Chem., 266, 16098–104.

Gandino, L., Longati, P., Medico, E., Prat, M., and Comoglio, P. M. (1994). Phosphorylation of Ser[985] negatively regulates the hepatocyte growth factor receptor kinase. J. Biol. Chem., 269, 1815–20.

Giordano, S., Ponzetto, C., Di Renzo, M.F., Cooper, C.S., and Comoglio, P.M. (1989a). Tyrosine kinase receptor indistinguishable from the c-met protein. Nature, 339, 155–6.

Giordano, S., Di Renzo, M.F., Narsimhan, R.P., Cooper, C.S., Rosa, C., and Comoglio, P.M. (1989b). Biosynthesis of the protein encoded by the c-met proto-oncogene. Oncogene, 4, 1383–8.

Giordano, S., Zhen, Z., Medico, E., Gaudino, G., Galimi, F., and Comoglio, P.M. (1993). Transfer of motogenic and invasive response to scatter factor/hepatocyte growth factor by transfection of human MET protooncogene. Proc. Natl. Acad. Sci. (USA), 90, 649–53.

Graziani, A., Gramaglia, D., Cantley, L.C., and Comoglio, P.M. (1991). The tyrosine-phosphorylated hepatocyte growth factor/scatter factor receptor associates with phosphatidylinositol 3-kinase. J. Biol. Chem., 266, 22087–90.

Graziani, A., Gramaglia, D., dalla Zonca, P., and Comoglio, P.M. (1993). Hepatocyte growth factor/scatter factor stimulates the Ras-guanine nucleotide exchanger. J. Biol. Chem., 268, 9165–8.

Huff, J.L., Jelinek, M.A., Borgman, C.A., Lansing, T.J., and Parsons, J.T. (1993). The protooncogene c-sea encodes a transmembrane protein-tyrosine kinase related to the Met/hepatocyte growth factor/scatter factor receptor. Proc. Natl. Acad. Sci. (USA), 90, 6140–4.

Longati, P., Bardelli, A., Ponzetto, C., Naldini, L., and Comoglio, P. (1994). Tyrosines[1234–1235] are critical for activation of the tyrosine kinase encoded by the met proto-oncogene (HGF receptor). Oncogene, 9, 49–57.

Mondino, A., Giordano, S., and Comoglio, P.M. (1991). Defective posttranslational processing activates the tyrosine kinase encoded by the MET proto-oncogene (hepatocyte growth factor receptor). Mol. Cell. Biol., 11, 6084–92.

Naldini, L., Vigna, E., Narsimhan, R.P., Gaudino, G., Zarnegar, R., Michalopoulos, G.K., and Comoglio, P.M. (1991a). Hepatocyte growth factor (HGF) stimulates the tyrosine kinase activity of the receptor encoded by the proto-oncogene c-MET. Oncogene, 6, 501–4.

Naldini, L., Weidner, K.M., Vigna, E., Gaudino, G., Bardelli, A., Ponzetto, C., Narsimhan, R.P., Hartmann, G., Zarnegar, R., Michalopoulos, G.K., Birchmeier, W., and Comoglio, P.M. (1991b). Scatter factor and hepatocyte growth factor are indistinguishable ligands for the MET receptor. EMBO J., 10, 2867–78.

Naldini, L., Vigna, E., Ferracini, R., Longati, P., Gandino, L., Prat, M., and Comoglio, P.M. (1991c). The tyrosine kinase encoded by the MET proto-oncogene is activated by autophosphorylation. Mol. Cell. Biol., 11, 1793–803.

Park, M., Dean, M., Cooper, C.S., Schmidt, M., O'Brien, S.J., Blair, D.G., and Vande Woude, G.F. (1986). Mechanism of met oncogene activation. Cell, 45, 895–904.

Ponzetto, C., Giordano, S., Peverali, F., Della Valle, G., Abate, M.L., Vaula, G., and Comoglio, P.M. (1991). c-met is amplified but not mutated in a cell line with an activated met tyrosine kinase. Oncogene, 6, 553–9.

Ponzetto, C., Bardelli, A., Maina, F., Longati, P., Panayotou, G., Dhand, R., Waterfield, M.D., and Comoglio, P.M. (1993). A novel recognition motif for phosphatidylinositol 3-kinase binding mediates its association with the hepatocyte growth factor/scatter factor receptor. Mol. Cell. Biol., 13, 4600–8.

Ponzetto, C., Bardelli, A., Zhen, Z., Maina, F., dalla Zonca, P., Giordano, S., Graziani, A., Panayotou, G., and Comoglio, P. (1994). A multifunctional docking site mediates signalling and transformation by the hepatocyte growth factor/scatter factor receptor family. Cell, 77, 261–71.

Prat, M., Crepaldi, T., Gandino, L., Giordano, S., Longati, P., and Comoglio, P.M. (1991a) C-terminal truncated forms of Met, the hepatocyte growth factor receptor. Mol. Cell. Biol., 11, 5954–62.

Prat, M., Narsimhan, R.P., Crepaldi, T., Nicotra, M.R., Natali, P.G., and Comoglio, P.M. (1991b). The receptor encoded by the human c-MET oncogene is expressed in hepatocytes, epithelial cells and solid tumours. Int. J. Cancer, 49, 323–8.

Ronsin, C., Muscatelli, F., Mattei, M.G., and Breathnach, R. (1993). A novel putative receptor protein tyrosine kinase of the met family. Oncogene, 8, 1195–202.

Villa Moruzzi, E., Lapi, S., Prat, M., Gaudino, G., and Comoglio, P.M. (1993). A protein tyrosine phosphatase activity associated with the hepatocyte growth factor/scatter factor receptor. J. Biol. Chem., 268, 18176–80.

Paolo M. Comoglio and Andrea Graziani:
Department of Biomedical Sciences and Oncology,
University of Torino, Italy

The Growth Hormone/ Prolactin Family

The pituitary hormones growth hormone (GH) and prolactin (PRL) define an expanding family of ~22 kDa growth factors that also encompasses the placental lactogens (PLs), proliferins, and somatolactin. Crystallography reveals human GH as a four helix bundle with two distinct receptor-binding sites. In addition to its somatogenic activity GH has metabolic effects, especially in adipocytes; similarly PRL is lactogenic but has many other functions. Recombinant GH is administered to GH-deficient children to promote statural growth.

■ Alternative names

GH is alternatively referred to as somatotropin, and PL as choriomammosomatotropin or chorionic somatomammotropin (CS).

■ Sequences and evolutionary relationships

The cDNA sequence of human pituitary GH (hGH; GenBank accession number V00519), human PL, and GHs from most other species code for a 191-amino-acid (aa) protein preceded by a signal sequence of 26 aa (Wallis 1992). Human PL (hPL) is encoded identically from two non-allelic genes, hPL-3/hCS-B (V00573) and hPL-4/hCS-A (J00118). The human pituitary PRL (hPRL) cDNA sequence (J00299) predicts a leader of 28 aa and a mature protein of 199 aa. In primates GH is highly homologous to PL (85 per cent identity in humans), whereas GH and PRL are more distantly related (23 per cent in humans: see Fig. 1). By comparison human and porcine GH are 68 per cent identical.

Recently cDNA cloning has identified numerous placental factors in non-primates, apparently derived from separate genes (Wallis 1992). In addition to GH-related PLs, the placentae of rodents and particularly ruminants express several prolactin-related proteins (PRPs) which have high homology to PRLs. It appears that a diversity of placentally-expressed growth factors evolved rapidly after separation of the major mammalian orders. The identification of the related proteins proliferin (from mice) and somatostatin (from teleost fish) hints at further diversity (Wallis 1992).

■ The proteins

GH/PRL family members are generally secreted 22–25 kDa monomers and, with rare exceptions (e.g. bovine PL), appear not to be glycosylated. Two disulphide bonds, Cys53–Cys165 and Cys182–Cys189 in hGH, are conserved throughout GH and PRL sequences and most PLs and PRPs. However, several variant protein forms occur. Circulating hGH includes two-chain isomers resulting from proteolytic cleavage, and oligomers ('big' GH), both of which are cleared more slowly than the intact molecule (Baumann *et*

```
                              Helix 1                                    Mini-helix 1
        1                    10          20              30                40
hGH   F P T I P - - - - - - - - - L S R L F D N A M L R A H R L H Q L A F D T Y Q E F E E A Y I P K E Q K Y
hPL   V Q T V P - - - - - - - - - L S R L F D H A M L Q A H R A H Q L A I D T Y Q E F E E T Y I P K D Q K Y
hPRL  L P I C P G G A A R C Q V T L R D L F D R A V V L S H Y I H N L S S E M F S E F D K R Y T - H G R G -

                                      Mini-helix 2                 Helix 2
              50           60              70              80              90
hGH   S F L Q N P Q T S L C F S E S I P T P S N R E E T Q Q K S N L E L L R I S L L L I Q S W L E P V Q F L
hPL   S F L H D S Q T S F C F S D S I P T P S N M E E T Q Q K S N L E L L R I S L L L I E S W L E P V R F L
hPRL  - F I T K A - I N S C H T S S L A T P E D K E Q A Q Q M N Q K D F L S L I V S I L R S W N E P L Y H L

      Mini-helix 3               Helix 3
              100          110             120             130             140
hGH   - R S V F A N S L V Y G A S D S N V Y D L L K D L E E G I Q T L M G R L E D G S P R T G Q I F K Q T Y
hPL   - R S M F A N N L V Y D T S D S D D Y H L L K D L E E G I Q T L M G R L E D G S R R T G Q I L K Q T Y
hPRL  V T E V R G M Q E A P E A I L S K A V E I E E Q T K R L L E G M E L I V S Q V H P E T K E N E I Y P V

                                      Helix 4
              150          160             170             180             190
hGH   S K F D T N S H N D D A L L K N - - - Y G L L Y C F R K D M D K V E T F L R I V Q C R S - V E G S C G F
hPL   S K F D T N S H N H D A L L K N - - - Y G L L Y C F R K D M D K V E T F L R M V Q C R S - V E G S C G F
hPRL  W S G L P S L Q M A D E E S R L S A Y Y N L L H C L R R D S H K I D N Y L K L L K C R I I H N N N C
```

Figure 1. Sequence alignment of hGH, hPRL, and hPL.

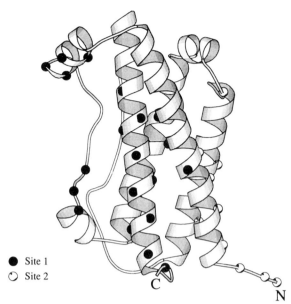

● Site 1
○ Site 2

C

N

Figure 2. hGH structure showing sites 1 and 2.

al. 1986). Around 10 per cent of pituitary hGH is in a '20k' form, generated by alternative splicing, that lacks residues 32–46 and may retain only a subset of GH activities. Proteolysis of PRL in the pituitary yields a '22k' form (missing the C-terminal 23 amino acids) apparently secreted specifically in females (Anthony *et al.* 1993), and separately a '16k' form (residues 1–123) that has potent anti-angiogenic activity (Clapp *et al.* 1993).

Human GH, PRL, and PL can be efficiently expressed in *E.coli*, and purified recombinant extracellular binding domains of receptors are available for affinity chromatography and direct binding assays (Wells *et al.* 1993). Alternatively, the proliferation of myeloid cell lines transfected with full-length receptors affords a cell-based assay for bioactivity (Wells *et al.* 1993; Fuh *et al.* 1993). The *in vivo* somatogenic effects of GH have been assessed using the weight gain or cartilage growth of hypophysectomized rats (Friesen 1990).

Unlike most other mammalian GHs, primate GHs have broad specificity and bind tightly to both the GH (somatogenic) and PRL (lactogenic) receptors. Human PRL and PL only bind tightly to the PRL receptor. High affinity binding of hPL and hGH (but not hPRL) to the PRL receptor requires zinc (Wells *et al.* 1993).

■ Structure–function analysis

hGH has been crystallized in complex with two molecules of the human GH receptor extracellular domain (de Vos *et al.* 1992). The basic structure is a four α-helix bundle of an 'up-up-down-down' topology also seen in several other cytokines (Wells and de Vos 1993), with three additional 'mini-helices' in the connecting loops (see Fig. 2). Detailed mutagenesis had previously identified the two receptor-binding sites (Wells *et al.* 1993; Wells and de Vos 1993). Site 1 binds first and comprises about 30 residues mainly in helix 4 but also helix 1 and two mini-helices. Only a small subset of site 1 residues appears functionally important (Cunningham and Wells 1993). An overlapping but different site 1 binds the human PRL receptor. Site 2 is smaller, including about 15 residues distributed over helices 1, 3, and the N-terminus. Binding is strictly sequential and dimerization is essential for signal transduction: a single mutation within site 2 (Gly120 to Arg in hGH) prevents dimerization and creates an antagonist to both receptors (Fuh *et al.* 1993; Wells and de Vos 1993). hPRL also acts through sequential receptor dimerization (Fuh *et al.* 1993).

Mutagenesis has been used to recruit the GH receptor-binding activity of hGH site 1 into hPL and hPRL, and to engineer receptor-specific analogues of hGH (Wells *et al.* 1993). Variants of hGH with extremely high site 1 affinities for the hGH receptor (~1 pM) have been generated by random mutagenesis and *in vitro* selection using phage display (Wells *et al.* 1993; Lowman and Wells 1993).

■ GH/PRL family genes and transcription

The gene for human pituitary GH, hGH-N, is on chromosome 17q22–24 and comprises five exons spread over around 2 kb (Chen *et al.* 1989). Alternative splicing to a cryptic splice site in exon 3 yields mRNA encoding the 20k variant. hGH-N is in a cluster of five related genes including the two for hPL, a probable hPL pseudogene (hPL-1 or hCS-L), and hGH-V, which is expressed only in placenta and encodes a protein 93 per cent identical to the pituitary hormone. This clustering is absent in non-primates; however the genes for the bovine PRPs and PL are clustered (on chromosome 23). The genes for human PRL, on chromosome 6p23–21.1 (Taggart *et al.* 1987), and rat PRL also have five exons but are much larger (~10 kb) due to longer introns. No alternative splices have been reported.

Upstream of the hGH-N gene TATA box are binding sites for the ubiquitous transcription factors NF-1, AP-2, USF and Sp1 (Rousseau 1992). Somatotroph-specific expression results from occupation of binding sites for POU-box factor GHF-1/Pit-1. GH expression can be induced by thyroid hormone (T3), glucocorticoids, and GH releasing hormone (GHRH). Glucocorticoids act through response elements in the upstream region and also through increased mRNA stability. GHRH acts indirectly via cyclic AMP which stimulates GHF-1/Pit-1 transcription. The rat PRL gene responds similarly to steroid hormones: a distal enhancer contains an oestrogen response element (Cullen *et al.* 1993).

■ Biological functions

GH is synthesized and stored in discrete somatotrophs in the pituitary, possibly as the zinc-mediated dimers observed for hGH *in vitro* (Wells *et al.* 1993). Its release is primarily controlled by the opposing actions of the hypothalamic hormones GHRH and somatostatin, and is pulsatile and mainly nocturnal. Significant amounts of plasma GH are complexed with binding proteins related to the GH receptor extracellular domain and derived by proteolysis (man) and/or alternative splicing (rodents) (Kelly *et al.* 1993). The placental hGH-V gene product, which lacks residues required for zinc-dependent binding to the PRL receptor (Wells *et al.* 1993), replaces the pituitary hormone in plasma during pregnancy (Frankenne *et al.* 1988).

Several GH functions, including skeletal growth, are at least partly mediated by the 'somatomedin' IGF-1: binding to liver GH receptors stimulates IGF-1 gene transcription. However, the relative contributions of direct and indirect effects to different GH activities are not fully defined (Behringer *et al.* 1990). Among the activities that are probably direct is the initiation of preadipocyte differentiation, in which c-*fos* and c-*jun* transcription is rapidly induced (Kelly *et al.* 1993). GH has both acute insulin-like and long term lipolytic effects on adipocytes *in vitro* and *in vivo*. Recombinant hGH has clinical utility in treating short stature in children resulting from GH deficiency, chronic renal insufficiency, or Turner's syndrome, and potentially in maintaining nitrogen balance in normal aging adults (Lippe and Nakomoto 1993). Bovine GH is used to stimulate milk production in dairy cows.

Release of PRL from pituitary lactotrophs is stimulated in humans by factors such as thyrotrophin-releasing hormone (TRH) and vasoactive intestinal peptide (VIP) and inhibited neurologically by dopamine. The lactogenic effects of PRL in mammary epithelia include induction of milk protein gene expression, and in humans of the secreted glycoprotein PIP (PRL-inducible protein) (Kelly *et al.* 1993). PRL also has a multitude of effects on metabolism, development and behaviour. GH and PRL are also potential immunostimulatory factors: normal lymphocytes secrete the hormones and bear receptors for them (Kelly *et al.* 1993).

The biological roles of the various PLs and PRPs remain obscure although they are presumably involved in regulating development and metabolism during pregnancy. To date no specific 'PL receptor' has been cloned.

■ Pathology

GH overproduction results in acromegaly and gigantism, and its deficiency results in dwarfism. GH overdosing in humans can result in insulin resistance, joint pain, oedema and acromegalic features (Lippe and Nakomoto 1993). Overexpression of GH and particularly PRL is associated with pituitary and possibly some breast tumours (Tornell *et al.* 1991).

■ References

Anthony, P.K., Stoltz, R.A., Pucci, M.L., and Powers, C.A. (1993). The 22K variant of rat prolactin: evidence for identity to prolactin-(1–173), storage in secretory granules, and regulated release. *Endocrinology*, **132**, 806–14.

Baumann, G., Stolar, M.W., and Buchanan, T.A. (1986). The metabolic clearance, distribution, and degradation of dimeric and monomeric growth hormone (GH): implications for the pattern of circulating GH forms. *Endocrinology*, **119**, 1497–501.

Behringer, R.R., Lewin, T.M., Quaife, C.J., Palmiter, R.D., Brinster, R.L., and D'Ercole, A.J. (1990). Expression of insulin-like growth factor I stimulates normal somatic growth in growth hormone-deficient transgenic mice. *Endocrinology*, **127**, 1033–40.

Chen, E.Y., Liao, Y.-C., Smith, D.H., Barrera-Saldaña, H.A., Gelinas, R.E., and Seeburg, P.H. (1989). The human growth hormone locus: nucleotide sequence, biology, and evolution. *Genomics*, **4**, 479–97.

Clapp, C., Martial, J.A., Guzman, R.C., Rentier-Delure, F., and Weiner, R.I. (1993). The 16-kilodalton N-terminal fragment of human prolactin is a potent inhibitor of angiogenesis. *Endocrinology*, **133**, 1292–9.

Cullen, K.E., Kladde, M.P., and Seyfred, M.A. (1993). Interaction between transcription regulatory regions of prolactin chromatin. *Science*, **261**, 203–6.

Cunningham, B.C., and Wells, J.A. (1993). Comparison of a structural and a functional epitope. *J. Mol. Biol.*, **233**, 554–63.

de Vos, A.M., Ultsch, M., and Kossiakoff, A.A. (1992). Human growth hormone and the extracellular domain of its receptor: crystal structure of the complex. *Science*, **255**, 306–12.

Frankenne, F., Closset, J., Gomez, F., Scippo, M.L., Smal, J., and Hennen, G. (1988). The physiology of growth hormones (GHs) in pregnant women and partial characterization of the placental GH variant. *J. Clin. Endocrinol. Metab.*, **66**, 1171–80.

Friesen, H.G. (1990). Receptor assays for growth hormone. *Acta Paediatr. Scand.*, [Suppl] **370**, 87–91.

Fuh, G., Colosi, P., Wood, W.I., and Wells, J.A. (1993). Mechanism-based design of prolactin receptor antagonists. *J. Biol. Chem.*, **268**, 5376–81.

Kelly, P.A., Ali, S., Rozakis, M., Goujon, L., Nagano, M., Pellegrini, I., Gould, D., Djiane, J., Edery, M., Finidori, J., and Postel-Vinay, M.C. (1993). The growth hormone/prolactin receptor family. *Recent Prog. Hormone Res.*, **48**, 123–64.

Lippe, B.M., and Nakomoto, J.M. (1993). Conventional and nonconventional uses of growth hormone. *Recent Prog. Hormone Res.*, **48**, 179–235.

Lowman, H.B., and Wells, J.A. (1993). Affinity maturation of human growth hormone by monovalent phage display. *J. Mol. Biol.*, **233**, 564–78.

Rousseau, G.G. (1992). Growth hormone gene regulation by trans-acting factors. *Hormone Res.*, **37 (suppl 3)**, 88–92.

Taggart, R.T., Mohandas, T.K., and Bell, G.I. (1987). Assignment of human preprogastricsin (PGC) to chromosome 6 and regional localization of PGC (6pter-p21.1), prolactin PRL (6pter-p21.1). *Cytogenet. Cell Genet.*, **46**, 701.

Tornell, J., Rymo, L., and Isaksson, O.G. (1991). Induction of mammary adenocarcinomas in metallothionein promoter–human growth hormone transgenic mice. *Int. J. Cancer*, **49**, 114–7.

Wallis, M. (1992). The expanding growth hormone/prolactin family. *J. Mol. Endocrinol.*, **9**, 185–8.

Wells, J.A., and de Vos, A.M. (1993). Structure and function of human growth hormone: implications for the hematopoietins. *Ann. Rev. Biophys. Biomol. Struct.*, **22**, 329–51.

Wells, J.A., Cunningham, B.C., Fuh, G., Lowman, H.B., Bass, S.H., Mulkerrin, M.G., Ultsch, M., and de Vos, A.M. (1993). The molecular basis for growth hormone–receptor interactions. *Recent Prog. Hormone Res.*, **48**, 253–75.

Tim Clackson:
Department of Protein Engineering,
Genentech, Inc., South San Francisco, CA, USA

Growth Hormone/ Prolactin Receptors

The growth hormone (GH) receptor and the prolactin (PRL) receptor are single transmembrane domain glycoproteins of around 600 amino acids which transduce respectively the somatogenic and lactogenic effects of the GH/PRL family of hormones in a wide range of tissues. The receptors are activated by homodimerization through sequential binding of two receptors to distinct sites on a single hormone molecule. In the crystal structure of the ternary complex of human GH and two GH receptor extracellular domains, the receptors adopt a tandem β-barrel structure and there is extensive inter-receptor contact near the C-termini. Both GH and PRL receptors can activate the cytoplasmic tyrosine kinase JAK2, and GH receptor also signals through protein kinase C. Some types of human dwarfism are caused by GH receptor defects.

■ Alternative names

As the hormones of some species bind both receptors, GH receptor is often termed the somatogenic receptor, and PRL receptor the lactogenic receptor, to distinguish them by downstream metabolic effects rather than their ligands.

■ Receptor proteins

The mature human GH receptor cloned from liver comprises 620 amino acids (aa) and is expressed with a signal sequence of 18 aa (GenBank accession number X06562). A 246-aa extracellular domain containing seven cysteine residues is followed by a transmembrane domain of 24 residues and a 350-aa intracellular domain. The extracellular domain (residues 1–238) is found in serum as a GH-binding protein (hGHbp) (Leung *et al*. 1987). In rodents the soluble binding protein bears a short (17 or 27 aa) C-terminal extension not present in the membrane receptor (Mathews 1991; Kelly *et al*. 1993).

The 538- aa human liver PRL receptor (M31661 M60727) is about 30 per cent identical to the GH receptor but lacks a 25-aa region at the N-terminus. However several variant proteins have been identified in rats (Kelly *et al*. 1993). The 'short' (291 aa) and 'long' (591 aa) forms differ in the length of their intracellular domains (57 aa and 357 aa, respectively), and an 'intermediate' form (393 aa, intracellular domain 159 aa) has been cloned from the lymphoma cell line Nb2. All PRL receptors have a single conserved cysteine in the transmembrane domain. A soluble PRL receptor has been identified in milk (Kelly *et al*. 1993).

In all cases examined, GH and PRL receptors are glycosylated and some GH receptors are mono-ubiquitinated: the human GH receptor for example runs at ~130 kDa in SDS gels compared to a predicted ~70 kDa (Leung *et al*. 1987). Ligand binding is unaffected by deglycosylation, and functional extracellular domains can be expressed in high yield by secretion from *E.coli* as well as from insect and mammalian cells (Wells *et al*. 1993; Wells and de Vos 1993).

The GH and PRL receptors are members of a large family of cytokine receptors with extracellular domains that share moderate homology (14–25 per cent overall), and characteristic motifs including two cysteine pairings and the WSXWS box. There is also limited homology in the intracellular domain, at the proline-rich 'box 1' and less conserved 'box 2' sequences near the transmembrane domain (Colosi *et al*. 1993). Between species each receptor type shows high homology (e.g. 84 per cent identity between the human and rabbit GH receptors), and both antibodies and hormones frequently cross-react.

■ Structure–function analysis

In solution a 2:1 complex is formed between bacterially expressed hGHbp and human GH (hGH) (Wells *et al*. 1993; Wells and de Vos 1993). Binding is strictly sequential (Cunningham *et al*. 1991): first a receptor binds at hGH site 1 and then a second at site 2. The crystal structure of the complex (de Vos *et al*. 1992) reveals that each receptor monomer contains two domains of seven β-strands forming a sandwich of two antiparallel sheets (Fig. 1). Residues from both domains and also the intervening hinge fashion the hGH binding site. Interestingly, in each receptor the same residues are involved in binding the very different hGH sites 1 and 2. There is also an extensive interface between the two C-terminal domains which would presumably drive intracellular domain dimerization in the full length receptor. The YGEFS sequence (analogous to WSXWS) is not near any binding interface. Residues 1–30 are not visible in either receptor.

Mutagenesis reveals that two tryptophan side chains (one from each domain) highly buried at the hormone–receptor interface contribute most of the binding energy for binding hGH site 1. The N-terminal region not apparent in the crystal structure, which is absent in PRL receptors, can be deleted without affecting binding (Bass *et al*. 1991; Clackson and Wells 1994).

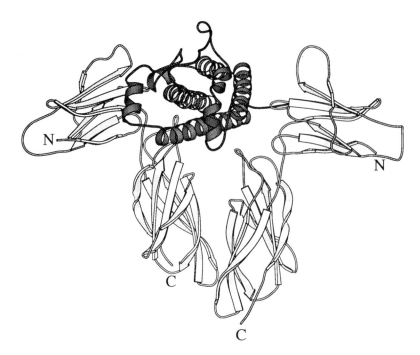

Figure 1. Structure of hGH-(hGHbp)₂ complex.

■ GH and PRL receptor genes and transcription

The gene for human GH receptor has been assigned to chromosome 5p12–14 and colocalizes with the gene for the PRL receptor (Kelly *et al.* 1993). The genes also colocalize in the mouse (chromosome 15) and rat (chromosome 2), interestingly near an integration locus for Moloney murine leukaemia virus (Barker *et al.* 1992). The human GH receptor gene consists of nine coding and several non-coding exons spanning at least 87 kb (Godowski *et al.* 1989). In human liver (and rat tissues) the major mRNA is around 4.5 kb with a small amount of a 1.2 kb message being detectable. Splices to alternative 5′ non-coding exons have been identified and also a variant lacking exon 3 which deletes residues 6–29 (known not to be involved in binding). In rodents the smaller transcript is more abundant and codes for the binding protein, with the unique C-terminal extension derived from a separate exon.

More heterogeneity has been observed for the PRL receptor: in humans transcripts of 2.5, 3 and 7.3 kb are derived from a gene of at least 11 exons spread over >70 kb. In humans and rabbits alternative splices involve non-coding exons but in rodents they account for the short and long receptor forms (Kelly *et al.* 1993).

The GH receptor is expressed at highest levels in liver and adipose tissue but in rats its mRNA can also be detected in other tissues including intestine, brain, testis, heart, and skeletal muscle (Mathews 1991; Kelly *et al.* 1993). Expression rises during postnatal growth and during preg-

nancy, and is regulated by a variety of hormonal and nutritional factors including oestrogens, insulin, and GH itself. PRL receptor expression is also widespread, and regulated by factors including GH and PRL as well as steroids. Expression rises during pregnancy although in rats levels in mammary gland do not rise until lactation, probably due to suppression by progesterone, and the short receptor form predominates (Mathews 1991; Kelly *et al.* 1993). Regulation of both genes is mainly at the level of transcription although regulatory elements and factors have not been well defined.

■ Binding characteristics

The binding specificity of the receptors varies across species (Wells *et al.* 1993). In primates the GH receptor binds only GH, whereas the PRL receptor binds GH and placental lactogen (PL) as well as PRL. In other mammals the PRL receptor does not bind GH but probably does bind the various prolactin-related proteins (PRPs) expressed in placenta. The bovine GH receptor in addition binds bovine PL.

Human GH receptor binds hGH site 1 with K_D ~0.4 nM; binding of the second receptor is difficult to monitor directly but the EC$_{50}$ for hGH-induced dimerization of the hGHbp *in vitro* is ~0.5 nM (Cunningham *et al.* 1991). Zinc enhances the binding of hGH (site 1) to the human PRL receptor from K_D 270 nM to 0.03 nM. A single zinc ion at the interface is coordinated by side chains from both molecules. Binding of human PL to the PRL receptor is also enhanced by zinc (K_D 0.05 nM compared to >10 nM) but PRL binding (K_D 2.8 nM) is not. In the absence of hormone, hGHbp does not dimerize at concentrations up to 0.1 mM (Wells *et al.* 1993).

The half-life of GH and PRL receptors is very short and more receptors are present in endosomal and other intracellular compartments than on the plasma membrane, reflecting high receptor turnover (Kelly *et al.* 1993). Recycling occurs independently of occupancy. Although the rodent serum GH binding proteins derive from alternative splicing, the human GHbp is thought to arise from full-length receptor by proteolysis (Mathews 1991; Kelly *et al.* 1993).

■ Signalling mechanisms

As for GH, 2:1 complexes of soluble PRL receptors with PRL have been observed in several cases (Hooper *et al.* 1993). Dimerization is required for signalling of both human GH and PRL receptors in cell-based (Fuh *et al.* 1992, 1993) and (for GH) transgenic animal (Chen *et al.* 1991) systems: some monoclonal antibodies elicit receptor signalling, and hormones with site 2 mutations are inactive. Characteristically, high concentrations of hormones are therefore antagonistic (Fig. 2). However, GH and PRL intracellular domains contain no known signalling motifs.

In mouse fibroblasts, GH binding to the GH receptor leads to binding of the 130 kDa cytoplasmic tyrosine kinase JAK2 to the receptor intracellular domain, and phosphorylation of both JAK2 and the receptor on tyrosine (Argetsinger *et al.* 1993). By analogy with the EPO receptor, JAK2 may bind to the membrane-proximal region of the intracellular domain that encompasses the box 1 and box 2 sequences. Only the first 54 residues of the intracellular domain are required for GH-dependent cell growth in a transfected myeloid cell line (Colosi *et al.* 1993). The PRL receptor also activates JAK2 as do a number of related cytokine receptors including those for IL-3, GM-CSF, G-CSF, and IFN-γ.

In preadipocytes the rapid GH-mediated induction of c-*fos* is mediated by diacylglyerol production and protein kinase C activation independently of inositol lipid turnover (Doglio *et al.* 1989). IGF-1 gene expression is not induced by this pathway. The receptor region(s) responsible for protein kinase C activation have not been defined. Several other potential signalling events have been reported (Kelly *et al.* 1993).

■ Pathology

Several (but not all) examples of Laron-type dwarfism, a rare autosomal recessive syndrome characterized by resistance to GH, have been attributed to GH receptor mutations (Goossens *et al.* 1993). These include a deletion of exons 3, 5, and 6 (which encode large regions of the extracellular domain) (Godowski *et al.* 1989), and nonsense mutations near the N-terminus. PRL receptor overexpression may contribute to some breast cancers (Bonneterre *et al.* 1990).

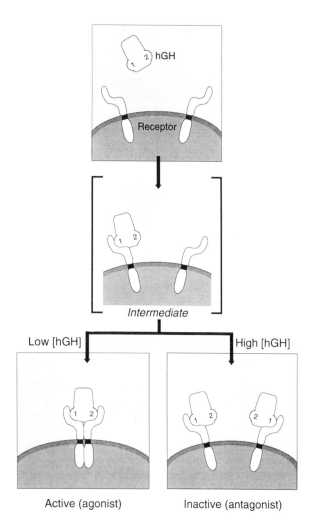

Figure 2. Receptor activation by sequential dimerization.

■ References

Argetsinger, L.S., Campbell, G.S., Yang, X., Witthuhn, B.A., Silvennoinen, O., Ihle, J.N., and Carter-Su, C. (1993). Identification of JAK2 as a growth hormone receptor-associated tyrosine kinase. *Cell*, **74**, 237–44.

Barker, C.S., Bear, S.E., Keler, T., Copeland, N.G., Gilbert, D.J., Jenkins, N.A., Yeung, R.S., and Tsichlis, P.N. (1992). Activation of the prolactin receptor gene by promoter insertion in a Moloney murine leukaemia virus-induced rat thymoma. *J. Virol.*, **66**, 6763–8.

Bass, S.H., Mulkerrin, M.G., and Wells, J.A. (1991). A systematic mutational analysis of hormone-binding determinants in the human growth hormone receptor. *Proc. Natl. Acad. Sci. (USA)*, **88**, 4498–502.

Bazan, J.F. (1990). Structural design and molecular evolution of a cytokine receptor superfamily. *Proc. Natl. Acad. Sci. (USA)*, **87**, 6934–8.

Bonneterre, J., Peyrat, J.P., Beuscart, R., and Demaille, A. (1990).

Biological and clinical aspects of prolactin receptors (PRL-R) in human breast cancer. *J. Steroid Biochem. Molec. Biol.*, **37**, 977–81.

Chen, W.Y., White, M.E., Wagner, T.E., and Kopchick, J.J. (1991). Functional antagonism between endogenous mouse growth hormone (GH) and a GH analog results in dwarf transgenic mice. *Endocrinology*, **129**, 1402–8.

Clackson, T., and Wells, J.A. (1994). A small part of a hormone-receptor interface dominates binding affinity. (In preparation).

Colosi, P., Wong, K., Leong, S.R., and Wood, W.I. (1993). Mutational analysis of the intracellular domain of the human growth hormone receptor. *J. Biol. Chem.*, **268**, 12617–23.

Cunningham, B.C., Ultsch, M., de Vos, A.M., Mulkerrin, M.G., Clausner, K.R., and Wells, J.A. (1991). Dimerization of the extracellular domain of the human growth hormone receptor by a single hormone molecule. *Science*, **254**, 821–5.

de Vos, A.M., Ultsch, M., and Kossiakoff, A.A. (1992). Human growth hormone and the extracellular domain of its receptor: crystal structure of the complex. *Science*, **255**, 306–12.

Doglio, A., Dani, C., Grimaldi, P., and Ailhaud, G. (1989). Growth hormone stimulates c-fos gene expression by means of protein kinase C without increasing inositol lipid turnover. *Proc. Natl. Acad. Sci. (USA)*, **86**, 1148–52.

Fuh, G., Cunningham, B.C., Fukunaga, R., Nagata, S., Goeddel, D.V., and Wells, J.A. (1992). Rational design of potent antagonists to the human growth hormone receptor. *Science*, **256** 1677–80.

Fuh, G., Colosi, P., Wood, W.I., and Wells, J.A. (1993). Mechanism-based design of prolactin receptor antagonists. *J. Biol. Chem.*, **268**, 5376–81.

Godowski, P.J., Leung, D.W., Meacham, L.R., Galgani, J.P., Hellmiss, R., Keret, R., Rotwein, P.S., Parks, J.S., Laron, Z., and Wood, W.I. (1989). Characterization of the human growth hormone receptor gene and demonstration of a partial gene deletion in two patients with Laron-type dwarfism. *Proc. Natl. Acad. Sci. (USA)*, **86**, 8083–7.

Goossens, M., Amselem, S., Duquesnoy, P., and Sobrier, M.-L. (1993). Molecular genetics of Laron-type GH insensitivity syndrome. *Recent Prog. Hormone Res.*, **48**, 165–78.

Hooper, K.P., Padmanabhan, R., and Ebner, K.E. (1993). Expression of the extracellular domain of the rat liver prolactin receptor and its interaction with ovine prolactin. *J. Biol. Chem.*, **268**, 22347–52.

Kelly, P.A., Ali, S., Rozakis, M., Goujon, L., Nagano, M., Pellegrini, I., Gould, D., Djiane, J., Edery, M., Finidori, J., and Postel-Vinay, M.C. (1993). The growth hormone/prolactin receptor family. *Recent Prog. Hormone Res.*, **48**, 123–64.

Leung, D.W., Spencer, S.A., Cachianes, G., Hammonds, R.G., Collins, C., Henzel, W.J., Barnard, R., Waters, M.J., and Wood, W.I. (1987). Growth hormone receptor and serum binding protein: purification, cloning and expression. *Nature*, **330**, 537–43.

Mathews, L.S. (1991). Molecular biology of growth hormone receptors. *Trends Endocrinol. Metab.*, **2**, 176–80.

Wells, J.A., and de Vos, A.M. (1993). Structure and function of human growth hormone: implications for the hematopoietins. *Ann. Rev. Biophys. Biomol. Struct.*, **22**, 329–51.

Wells, J.A., Cunningham, B.C., Fuh, G., Lowman, H.B., Bass, S.H., Mulkerrin, M.G., Ultsch, M., and de Vos, A.M. (1993). The molecular basis for growth hormone-receptor interactions. *Recent Prog. Hormone Res.*, **48**, 253–75.

Tim Clackson:
Department of Protein Engineering,
Genentech Inc., South San Francisco, CA, USA

The Epidermal Growth Factor (EGF) Family

EGF is low molecular weight (M$_r$=6045) polypeptide that when administered to animals produces marked proliferation of several epithelial tissues. In cell culture systems, EGF stimulates the proliferation of a wide range of cell types with the notable exclusion of haemopoietic cells. EGF is produced as a much larger transmembrane precursor molecule and is related in sequence and function to other growth factors, such as transforming growth factor-α, amphiregulin, heparin-binding EGF, and betacellulin, as well as viral gene products.

■ Family members

Based on sequence similarity, mitogenic activity, and capacity to interact with the EGF receptor, there are five distinct mammalian gene products that can be identified as EGF-like molecules (EGF, transforming growth factor-α (TGFα), amphiregulin, heparin-binding EGF (HB-EGF), betacellulin) (Carpenter and Wahl 1990; Massagué and Pandiella 1993; Shing et al. 1993). A viral gene frequently encoded by members of the pox virus (vaccinia, myxoma, Shope fibroma) family produces an EGF-like protein (PVGF) (Carpenter and Wahl 1990; Massagué and Pandiella 1993), and genetic analyses (Rutledge et al. 1992; Hill and Sternberg 1992; Neuman-Silberberg and Schüpbach 1993) have identified genes encoding EGF-like molecules in *Drosophila* (spitz and gurken) and *Caenorhabditis elegans* (*lin*-3). GenBank accession numbers for available sequences are as follows: EGF (human, X04571; mouse, J00380; rat, M63585), TGFα (human, X70340; mouse, M9240; rat, M31076), amphiregulin (human, M30704), HB-EGF (human, M60278; monkey, M93012; mouse, L07264; rat, L05489; pig, X67295), betacellulin (mouse, L08394), *lin*-3 (L11148), *spitz* (M95199), *gurken* (L22531), and pox viruses (vaccinia, J02421; Shope, M15921; myxoma, M15806 and M35234).

■ Precursor structures

All primary translation products of EGF-like genes exist as transmembrane glycoproteins (Massagué and Pandiella 1993; Derynck 1992) in which the EGF-like sequence is external to the plasma membrane (Fig. 1). Relatively little is known about the exact mechanisms by which mature EGF-like motifs are released from the various precursor molecules (Massagué and Pandiella 1993), nor are there conserved endoprotease sites at each end of the EGF-like sequences. However, there is evidence that juxtacrine interaction can occur between cells bearing EGF-like precursors and neighbouring cells that express EGF receptors (Massagué and Pandiella 1993; Derynck 1992). In one case, it appears that the precursor has a function distinct from the generation of soluble growth factor: the HB-EGF precursor is a receptor component for diphtheria toxin (Massagué and Pandiella 1993). The precursor molecule for EGF is, by far, the largest precursor and contains eight repeats of EGF-like sequences, but there is no evidence that any of these repeats are released from the precursor to function as soluble growth factors.

■ Growth factor structure

The mature form of EGF and TGFα are polypeptides of 53 and 50 residues, respectively. Other members of this growth factor family are somewhat larger due to N-terminal extensions. Numerous studies have shown that the N-terminus of EGF can be modified in several ways without affecting receptor binding capacity or biological activity (Carpenter and Wahl 1990). NMR studies (Campbell et al. 1990; Hommel et al. 1992; Kohda and Inagaki 1992) have resolved the high resolution structure of the mature forms of EGF and TGFα (Fig. 1 insert). The central features are two anti-parallel β-sheets comprising residues 18–23 and 28–34, plus the three intramolecular disulphide bonds. Together, these secondary structures restrict possible tertiary conformations. Residues that interact with the EGF receptor have not been defined. However, receptor-binding is particularly sensitive to mutations at Tyr 13, Tyr 22, Ile 23, Leu 26, Tyr 29, Tyr 37, Leu 47, and all six cysteinyl residues (Carpenter and Wahl 1990; Campbell et al. 1990). The mature growth factor is quite stable to heat and acidic pH, and denaturation requires strenuous conditions (Carpenter and

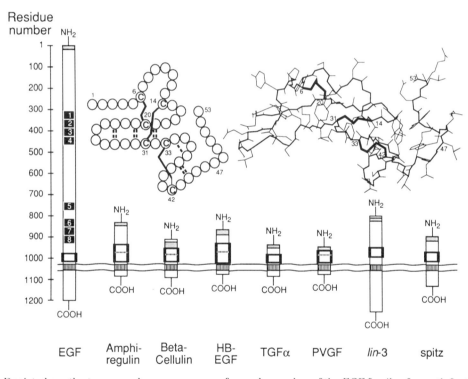

Figure 1. Depicted are the transmembrane precursors for each member of the EGF family of growth factors. The transmembrane sequences are denoted by vertical stripes, amino terminal signal sequences (which are probably cleaved) by stippled boxes, and mature EGF-like molecules are indicated by bold boxes just above the membrane. Within some of the mature EGF-like sequences is a dotted line to divide N-terminal extensions to the 'core' EGF-like sequence. The inserts show representations of the secondary and tertiary structures of EGF, based on Campbell et al. (1990), Hommel et al. (1992), and Kohda and Inagaki (1992). In these drawings, disulphide bonds are indicated by heavy bold lines, the peptide backbone by a dark line, and side chain groups by light lines. Numbering is according to the sequence of EGF.

Wahl 1990). While precursor molecules are glycosylated, the mature growth factors are usually not or, if so, are glycosylated at N-terminal extensions. There is no evidence for physiological oligomerization of the mature growth factors.

Mature EGF can be isolated rapidly in milligram quantities from the mouse submaxillary gland (Savage and Cohen 1972), but is available also as a recombinant protein (including human EGF and TGFα) from numerous commercial sources. The mature growth factors can be radiolabelled with ^{125}I using chloramine T (Carpenter and Cohen 1976) or other oxidizing procedures.

■ Assays

Molecular cloning has uncovered a large number of proteins that contain an EGF-like sequence motif (Carpenter and Wahl 1990; Bork 1991). However, many of these do not exhibit EGF-like activity in biological or biochemical assays. Critical assays include the capacity to form a high affinity complex with the EGF receptor, often detected by radio-competition assay with ^{125}I-EGF. In receptor binding assays, there is little species dependency, though some family members such as amphiregulin bind less tightly than EGF (Shoyab et al. 1989). Others are approximately equivalent to EGF in ligand-binding assays. A second biological criterion is the capacity of EGF family members to stimulate DNA synthesis and cell division. The most specific of these mitogenesis assays may be the induction of precocious eye-lid opening in the newborn mouse (Cohen 1962).

Immunological reagents are usually not very cross-reactive between different members of the EGF family and may even exhibit species preference (low cross-reactivity) within one group. For example, anti-mouse EGF does not usually recognize human EGF well, though this may depend on the particular antibody. Antibodies to EGF and TGFα are commercially available, but antibodies to other family members are not yet reported.

■ EGF genes and transcripts

In humans the genes for EGF, TGFα, amphiregulin, and HB-E6F are localized to chromosomes 4q25→q29, 2p11→p13, 4q13→q21, and 5, respectively (Carpenter and Wahl 1990; Plowman et al. 1990; Fen et al. 1993). The gene for EGF is approximately 120 kb and includes 24 exons, while the genes for TGFα, amphiregulin, and HB-E6F are 18 kb, 10.2 kb, and 14 kb, respectively, and contain six exons each. For each of these three growth factors, two exons encode the mature form of the molecule, such that approximately residues 1–32 and 33–53 are encoded by separate exons. The TGFα gene is near the breakpoint of Burkitt's lymphoma t(2;8) variant translocation (Carpenter and Wahl 1990), and the amphiregulin gene is close to a common breakpoint for acute lymphoblastic leukaemia (Plowman et al. 1990).

Mature mRNAs for these growth factors are 4.7–4.9 kb for EGF, 4.5–4.8 kb for TGFα, 1.4 kb for amphiregulin, 2.5 kb for HB-EGF, and 3.0 kb for betacellulin. While expression is often higher in a few tissues, most of these growth factors are expressed in most tissues. Interestingly, TGFα and HB-EGF expression are induced by members of this growth factor family, including the homologous growth factor (Massagué and Pandiella 1993; Dluz et al. 1993). The tumour promoting phorbol ester, TPA, is a potent inducer of amphiregulin (Shoyab et al. 1989), HB-EGF (Temizer et al. 1992), and TGFα expression (Massagué and Pandiella 1993).

TGFα and amphiregulin expression are high in tumour cells, suggesting that they may perform an autocrine function in the growth of some transformed cells (Massagué and Pandiella 1993). TGFα expression is also high in the embryo, while EGF expression commences in neonatal life.

The 5′ regulatory regions of these genes suggest multiple transcription start sites for TGFα and amphiregulin. The EGF gene has a recognizable TATA sequence and CAAT box (Carpenter and Wahl 1990), the amphiregulin promoter has a TATA sequence and no distinguishable CAAT box (Plowman et al. 1990), while the TGFα and HB-E6F promoters contain neither of these features (Carpenter and Wahl 1990). Other features of transcriptional regulation are unknown.

■ Biological responses

Administration of EGF to animals results in the hyperproliferation of various epithelial tissues (Carpenter and Wahl 1990). This has led to the idea that EGF may be a 'maintenance' factor for the continuous renewal of epithelial cell populations. In addition to mitogenic responses, the administration of EGF produces several non-mitogenic responses such as the inhibition of gastric acid secretion by parietal cells (Carpenter and Wahl 1990). In cell culture systems, EGF stimulates mitogenesis in a large number of cell types (Carpenter and Wahl 1990). However, no haemopoietic cells are responsive to EGF or EGF-like factors. Homologous recombination has been employed to eliminate the expression of TGFα in mice (Mann et al. 1993; Luettke et al. 1993). The only abnormality detected was a 'wavy' hair coat and abnormal hair follicles. TGFα deficiency due to disruption of TGFα alleles is the basis of the previously described wave-1 mutant mouse. Disruption of the spitz gene) in Drosophila (Rutledge et al. 1992) and the lin-3 gene in C. elegans (Hill and Sternberg 1992) does produce serious developmental defects.

■ References

Bork, P. (1991). Shuffled domains in extracellular proteins. FEBS Lett., **286**, 47–54.

Campbell, I.D., Baron, M., Cooke, R.M., Dudgeon, T.J., Fallon, A., Harvey, T.S., and Tappin, M.J. (1990). Structure–function relationships in epidermal growth factor (EGF) and transforming growth factor-alpha (TGF-alpha). Biochem. Pharmacol., **40**, 35–40.

Carpenter, G., and Cohen, S. (1976). ^{125}I-labeled human epidermal growth factor. Binding, internalization, and degradation in human fibroblasts. *J. Cell Biol.*, **71**, 159–71.

Carpenter, G., and Wahl, M.I. (1990). The epidermal growth factor family. In *Handbook Exptl. Pharmacol.*, **95** part I, 69–171.

Cohen, S. (1962). Isolation of a mouse submaxillary gland protein accelerating incisor eruption and eyelid opening in the newborn animal. *J. Biol. Chem.*, **237**, 1555–62.

Cohen, S., and Carpenter, G. (1975). Human epidermal growth factor: isolation and chemical and biological properties. *Proc. Natl. Acad. Sci. (USA)*, **72**, 1317–21.

Derynck, R. (1992). The physiology of transforming growth factor-alpha. *Adv. Cancer Res.*, **58**, 27–52.

Dluz, S.M., Higashiyama, S., Damm, D., Abraham, J.A., and Klagsbrun, M. (1993). Heparin-binding epidermal growth factor-like growth factor expression in cultured fetal human vascular smooth muscle cells. Induction of mRNA levels and secretion of active mitogen. *J. Biol. Chem.*, **268**, 18330–4.

Fen, Z., Dhadly, M., Yoshizumi, M., Hilkert, R.J., Quertermos, T., Eddy, R.L., Shows, T.B., and Lee, M.-E. (1993). Structural organization and chromosomal assignment of the gene encoding the human heparin-binding epidermal growth factor-like growth factor diphtheria toxin receptor. *Biochemistry*, **32**, 7932–8.

Hill, R.J., and Sternberg, P.W. (1992). The gene *lin*-3 encodes an inductive signal for vulval development in *C. elegans*. *Nature*, **358**, 470–6.

Hommel, U., Harvey, T.S., Driscoll, P.C., and Campbell, I.D. (1992). Human epidermal growth factor. High resolution solution structure and comparison with human transforming growth factor alpha. *J. Mol. Biol.*, **227**, 271–82.

Kohda, D., and Inagaki, F. (1992). Three-dimensional nuclear magnetic resonance structures of mouse epidermal growth factor in acidic and physiological pH solutions. *Biochemistry*, **31**, 11928–39.

Luettke, N.C., Qiu, T.H., Pfeiffer, R.L., Oliver, P., Smithies, O., and Lee, D.C. (1993). TGF alpha deficiency results in hair follicle and eye abnormalities in targeted and waved-1 mice. *Cell*, **73**, 263–78.

Mann, G.B., Fowler, K.J., Gabriel, A., Nice, E.C., Williams, R.L., and Dunn, A.R. (1993). Mice with a null mutation of the TGF alpha gene have abnormal skin architecture, wavy hair, and curly whiskers and often develop corneal inflammation. *Cell*, **73**, 249–61.

Massagué, J., and Pandiella, A. (1993). Membrane-anchored growth factors. *Annu. Rev. Biochem.*, **62**, 515–41.

Neuman-Silberberg, F.S., and Schüpbach, T. (1993). The *Drosophila* dorsoventral patterning gene *gurken* produces a dorsally localized RNA and encodes a TGFa-like protein. *Cell*, **75**, 165–74.

Plowman, G.D., Green, J.M., McDonald, V.L., Neubauer, M.G., Disteche, C.M., Todaro, G.J., and Shoyab, M. (1990). The amphiregulin gene encodes a novel epidermal growth factor-related protein with tumour-inhibitory activity. *Mol. Cell. Biol.*, **10**, 1969–81.

Rutledge, B.J., Zhang, K., Bier, E., Jan, Y.N., and Perrimon, N. (1992). The *Drosophila* spitz gene encodes a putative EGF-like growth factor involved in dorsal-ventral axis formation and neurogenesis. *Genes & Dev.*, **6**, 1503–17.

Savage, C.R. Jr., and Cohen, S. (1972). Epidermal growth factor and a new derivative. Rapid isolation procedures and biological and chemical characterization. *J. Biol. Chem.*, **247**, 7609–11.

Shing, Y., Christofori, G., Hanahan, D., Ono, Y., Sasada, R., Igarashi, K., and Folkman, J. (1993). Betacellulin: a mitogen from pancreatic beta cell tumours. *Science*, **259**, 1604–7.

Shoyab, M., Plowman, G.D., McDonald, V.L., Bradley, J.G., and Todaro, G.J. (1989). Structure and function of human amphiregulin: a member of the epidermal growth factor family. *Science*, **243**, 1074–6.

Temizer, D.H., Yoshizumi, M., Perrela, M.A., Susanni, E.E., Quertermous, T., and Lee, M.E. (1992). Induction of heparin-binding epidermal growth factor-like growth factor mRNA by phorbol ester and angiotensin II in rat aortic smooth muscle cells. *J. Biol. Chem.*, **267**, 24892–6.

■ Acknowledgements

The authors appreciate the assistance of Dr Cheryl Guyer in the preparation of the detailed structure of EGF in Fig. 1 and Edna Kunkel for the graphics in Fig. 1. Also, the assistance of Laura Hein in the preparation of this manuscript is acknowledged. The authors are supported by funding from the National Cancer Institute (G.C.) and the Ministerio de Educaciùn y Ciencia, Spain (C.S.).

Concepció Soler and Graham Carpenter:
Department of Biochemistry,
Vanderbilt University School of Medicine,
Nashville, TN 37232–0146, USA

The EGF receptor is displayed as a 170 kDa, glycoprotein on the surface of many cell types in both high and low affinity forms, and is characterized by its ligand-dependent tyrosine kinase activity (Carpenter and Wahl 1990). Mitogenic stimulation involves the ligand-induced dimerization of the high affinity receptors and the activation of the intracellular receptor kinase activity (Schlessinger 1993). Although the activated EGF receptor can deliver a mitogenic signal to cells in culture, the physiological role of this receptor is yet to be identified. Several ligands are capable of binding to the EGF receptor: EGF, transforming growth factor-α (TGFα), amphiregulin, heparin-binding EGF, vaccinia growth factor, cripto and β-cellulin (see the chapter on the EGF family, p. 194). Furthermore, the cell surface display of the EGFR can also be modulated indirectly by other growth factors (Zachary and Rozengurt 1985).

Three other members of the EGF receptor family have been identified: c-neu, erbB3 and erbB4 (see Fig. 1). Except for c-neu (see the chapter on neu/erbB-2 receptor, p. 149), there are no clues to the ligand-binding specificity for these other family members. Mutation, deletion, or overexpression of both the EGFR and c-neu have been detected in many tumour types, and there has been recent progress in using the level of the EGFR and c-neu proteins as prognostic markers for helping with the optimization of treatment for a number of cancers.

■ Nucleotide sequences

The complete nucleotide sequences have been determined for the human EGFR (GenBank accession No X00588), human c-neu (GenBank accession No X03363), human EGFR B3 (HER3, Genbank accession No M29366) and human EGFR B4 (GenBank accession No L07868). Partial sequence information is available for the mouse and chicken EGFRs (Avivi et al. 1991).

■ Alternative names and abbreviations

Human EGF receptor is also called HER or erbB1; human c-neu is also called HER2 or erbB2; HER3 is also called erbB3 and HER4 is also called erbB4.

■ Purification

Purification of the EGF receptor has been achieved by immunoaffinity chromatography with immobilized anti-receptor antibody (Yarden et al. 1985) and by ligand chromatography on Affigel-EGF (Cohen et al. 1982). In both purifications the initial step is followed by absorption to and specific elution from a wheat germ agglutinin (WGA) column. Since the EGF receptor is cleaved easily to a molecular form of approximately 150 000 Da by a calcium-dependent protease during the purification procedures, it is important to include EDTA or EGTA in all of the buffers.

■ Antibodies

Most of the commercially available antibodies which recognize the EGF receptor have been raised against the human receptor from A431 cells, and react poorly with the murine EGF receptor. Table 1 lists some of the properties for EGF receptor antibodies.

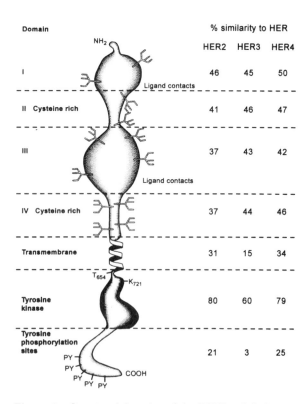

Figure 1. Structural domains of the EGFR and their similarity to HER2 (c-neu), HER3, and HER4.

■ Signal transduction

Tyrosine kinase activity appears to be necessary for mitogenic signal transduction by the EGF receptor (Chen et al. 1987). Whether a kinase-defective receptor is mitogenically impaired because of a failure to autophosphorylate, or a failure to phosphorylate other cellular proteins, is still not clear. Receptor autophosphorylation occurs very rapidly but transiently following EGF binding, and it may represent the first step in EGF signal transduction. Many intracellular signalling proteins contain SH2 (src homology 2) domains, through which they interact selectively with the autophosphorylation sites on activated EGF receptors (Koch et al. 1991). At least five such proteins bind directly to the autophosphorylated EGF receptor through separate and distinct phosphotyrosines: the regulatory subunit of PI3 kinase, phospholypase C-γ, nck, shc, and GBR2 (see Fig. 2). Two of these proteins, shc and GRB2, link the EGF receptor to SOS, one of the GDP exchange factors for ras-GDP. Thus, activation of the EGF receptor converts inactive ras-GDP to active ras-GTP (Gale et al. 1993). The activation of ras leads to the activation of a cascade of cytosolic serine/threonine kinases, including raf-1 (Rapp 1991) and MAP kinase (Blenis 1993). The activation of this signalling network by the EGF receptor appears to be dependent upon autophosphorylation. However, C-terminally truncated EGF receptors or mutated receptors which lack all of the major autophosphorylation sites can still transmit a mitogenic signal. It is therefore unclear whether signal transduction by the EGF receptor occurs solely via the ras pathway, or whether there are other pathways activated by the EGF receptor which can lead to mitogenesis.

■ Ligand binding

In most cell types, analysis of EGF binding to its receptor yields two classes of binding sites for EGF. The high affinity binding site (K_D 20–50 pM) represents between 1 and 10 per cent of the total receptor population, but appears to be responsible for the mitogenic actions of EGF (Defize et al. 1989). Transfection of the cloned EGF receptor in mammalian cells is sufficient to generate both high and low affinity sites. High affinity binding can be abolished by treating the cells with EGF, by activators of protein kinase C or by treatment of the cells with PDGF in a process called 'transmodulation' (Zachary and Rozengurt 1985). The molecular nature of the high affinity EGF receptor has not yet been elucidated. EGF receptors dimerize in the presence of EGF and receptor dimers bind EGF with higher affinity than the monomers, leading to the suggestion that high affinity EGF binding is associated with the receptor dimers (Schlessinger 1988). However, high affinity EGF binding is abolished during transmodulation without a corresponding decrease in dimer formation (Northwood and Davis 1989), and can be restored by dephosphorylation of non-membrane associated components (Walker and Burgess 1991). Presumably EGF receptor dimerization is not sufficient to form the high affinity binding site. Recently it has been shown that co-expression of pl85^{neu} and EGF receptor causes the appearance of a very high affinity EGF binding site (Wada et al. 1990).

■ Receptor trafficking

Stimulation of cells by EGF causes an increase in EGF receptor mRNA levels, but ligand binding to the EGF receptor causes a rapid and profound change in the surface display of EGF receptors: ligand–receptor complexes cluster in coated pits and are internalized (Carpenter and Wahl 1990). Tyrosine kinase activity is necessary for the appropriate cellular routing of the EGF receptor, since kinase-deficient receptors are recycled to the cell surface rather than degraded. Deletion of the amino acids between 993 and 1022 at the C-terminus produces a receptor which is still capable of signalling but is no longer internalized (Chang et al. 1991).

■ Mutational analysis

Many mutant EGF receptors have been constructed in an attempt to elucidate the role of key residues (Carpenter and Wahl 1990). Briefly, the mutations can be classified by the EGF receptor domain in which they have been introduced.

■ Extracellular domain

Removal of the entire ligand-binding domain is sufficient to create an oncogenic form of the protein. Exchange of

Table 1.

Company	Antibody	Epitope	Specificity	Uses
Sigma	29.1	Carbohydrate residues	Human>mouse	IP,WB,ELISA
Sigma	E3138	Residues 985–996	human, mouse	IP,WB,ELISA,IHS
Zymed	035600	hEGFR	human	IP,WB,ELISA,IHS
Zymed	035700	Activated hEGFR	human	IP,WB,ELISA,IHS
ATCC	528	Blocks EGF binding to hEGFR	human	IP,WB

IP = immunoprecipitation; WB = Western blotting; ELISA = enzyme-linked imunoassay; IHS = immunohistological staining.

domains within the extracellular region between the human and avian EGF receptors has helped to identify domain III as the major site for ligand binding (see Fig. 1).

■ Kinase domain

Lysine 721 in the EGF receptor has been implicated in ATP binding and is conserved throughout all of the tyrosine and Ser/Thr kinases. Mutations at this position in the EGF receptor abolish the tyrosine kinase activity. Kinase deficient EGF receptor is incapable of mediating mitogenesis, regulation of Ca^{2+} flux, or activation of gene transcription. However, a tyrosine kinase negative receptor can still activate MAP kinase (Selva *et al.* 1993).

■ C-terminal (regulatory) domain

Many studies have addressed the functional role of autophosphorylation sites in the C-terminal domain, sometimes with conflicting results. Generally, the mutation of individual autophosphorylation sites does not markedly alter kinase activity or signal transduction by the EGF receptor. Mutation of all five major autophosphorylation sites or limited deletion of the C-terminal domain only slightly reduces biological activity of the receptor, while completely abolishing phosphorylation of PLCγ and activation of PI3 kinase (Decker 1993). Further truncation of the C-terminus, however, alters the mitogenic signalling capacity of the receptor (Chen *et al.* 1989). Other mutations have been engineered to define the role of Ser/Thr phosphorylation. Mutation of Thr 654 to Ala or Tyr blocks PKC-induced

desensitization of the EGF receptor tyrosine kinase, but does not abolish PKC-dependent transmodulation of the high affinity EGF binding (Davis 1988; Livneh *et al.* 1988). Replacement of Ser 1046/1047 (substrates for the calmodulin-dependent protein kinase II) with Ala inhibits the EGF-dependent downmodulation of the EGF receptor and blocks the desensitization of its kinase activity (Countaway *et al.* 1992).

■ Pathology

Modification of the EGFR or its over-expression has been associated with several tumour types (Fleming *et al.* 1992). The EGFR gene is amplified in 20 per cent of human gliomas and appears to be involved in the maintenance of the tumour phenotype. Breast and colorectal tumours have been reported to overexpress EGF receptors. A point mutation in the transmembrane domain of c-neu leads to uncontrolled activation of the receptor kinase and brain tumours (Bargmann *et al.* 1986). Overexpression of c-neu in breast and ovarian cancers appears to correlate with a poor prognosis (Horak *et al.* 1991). HER3 expression has been detected in both breast and pancreatic tumours. HER4 expression has been detected in a variety of human carcinoma cell lines; the first cDNA clones for HER4 were actually isolated from a mammary carcinoma. Tyrosine kinase defective mutants of the EGFR and anti-EGF receptor monoclonal antibodies have both been shown to inhibit cell proliferation. Reagents which inhibit specifically the action of an oncogenically activated EGF receptor kinase are potential anti-cancer agents.

Mitogenic Signalling from the EGF Receptor

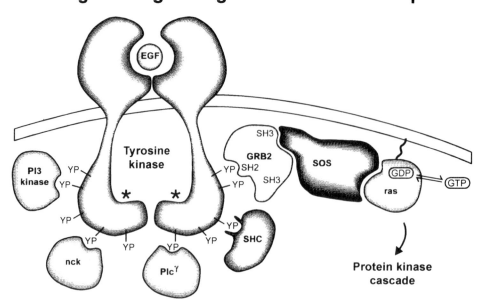

Figure 2. Model for the EGFR-associated proteins and the link between receptor-dependent activation of ras.

References

Avivi, A., Lax, I., Ullrich, A., Schlessinger, J., Givol, D., and Morse, B. (1991). Comparison of EGF receptor sequences as a guide to study the ligand binding site. *Oncogene*, **6**, 673–6.

Bargmann, C.I., Hung, M.C., and Weinberg, R.A. (1986). Multiple independent activations of the neu oncogene by a point mutation altering the transmembrane domain of p185. *Cell*, **45**, 649–57.

Blenis, J. (1993). Signal transduction via the MAP kinases: proceed at your own RSK. *Proc. Natl. Acad. Sci. (USA)*, **90**, 5889–92.

Carpenter, G. and, Wahl, M.I. (1990). In *Peptide growth factors and their receptors* (ed. M.B. Sporn and A.B. Roberts), pp. 69–171. Springer, Berlin.

Chang, C.P., Kao, J.P., Lazar, C.S., Walsh, B.J., Wells, A., Wiley, H.S., Gill, G.N., and Rosenfeld, M.G. (1991). Ligand-induced internalization and increased cell calcium are mediated via distinct structural elements in the carboxyl terminus of the epidermal growth factor receptor. *J. Biol. Chem.*, **266**, 23467–70.

Chen, W.S., Lazar, C.S., Poenie, M., Tsien, R.Y., Gill, G.N., and Rosenfeld, M.G. (1987). Requirement for intrinsic protein tyrosine kinase in the immediate and late actions of the EGF receptor. *Nature*, **328**, 820–3.

Chen, W.S., Lazar, C.S., Lund, K.A., Welsh, J.B., Chang, C.P., Walton, G.M., Der, C.J., Wiley, H.S., Gill, G.N., and Rosenfeld, M.G. (1989). Functional independence of the epidermal growth factor receptor from a domain required for ligand-induced internalization and calcium regulation. *Cell*, **59**, 33–43.

Cohen, S., Fava, R.A., and Sawyer, S.T. (1982). Purification and characterization of epidermal growth factor receptor/protein kinase from normal mouse liver. *Proc. Natl. Acad. Sci. (USA)*, **79**, 6237–41.

Countaway, J.L., Nairn, A.C., and Davis, R.J. (1992). Mechanism of desensitization of the epidermal growth factor receptor protein–tyrosine kinase. *J. Biol. Chem.*, **267**, 1129–40.

Davis, R.J. (1988). Independent mechanisms account for the regulation by protein kinase C of the epidermal growth factor receptor affinity and tyrosine–protein kinase activity. *J. Biol. Chem.*, **263**, 9462–9.

Decker, S.J. (1993). Transmembrane signaling by epidermal growth factor receptors lacking autophosphorylation sites. *J. Biol. Chem.*, **268**, 9176–9.

Defize, L.H., Boonstra, J., Meisenhelder, J., Kruijer, W., Tertoolen, L.G., Tilly, B.C., Hunter, T., van Bergen en Henegouwen, P.M., Moolenaar, W.H., and de Laat, S.W. (1989). Signal transduction by epidermal growth factor occurs through the subclass of high affinity receptors. *J. Cell Biol.*, **109**, 2495–507.

Fleming, T.P., Saxena, A., Clark, W.C., Robertson, J.T., Oldfield, E H., Aaronson, S.A., and Ali, I.U. (1992). Amplification and/or overexpression of platelet-derived growth factor receptors and epidermal growth factor receptor in human glial tumours. *Cancer Res.*, **52**, 4550–3.

Gale, N.W., Kaplan, S., Lowenstein, E.J., Schlessinger, J., and Bar-Sagi, D. (1993). Grb2 mediates the EGF-dependent activation of guanine nucleotide exchange on Ras. *Nature*, **363**, 88–92.

Horak, E., Smith, K., Bromley, L., LeJeune, S., Greenall, M., Lane, D., and Harris, A.L. (1991). Mutant p53, EGF receptor and c-erbB-2 expression in human breast cancer. *Oncogene*, **6**, 2277–84.

Koch, C.A., Anderson, D., Moran, M.F., Ellis, C., and Pawson, T. (1991). SH2 and SH3 domains: elements that control interactions of cytoplasmic signaling proteins. *Science*, **252**, 668–74.

Livneh, E., Dull, T.J., Berent, E., Prywes, R., Ullrich, A., and Schlessinger, J. (1988). Release of a phorbol ester-induced mitogenic block by mutation at Thr-654 of the epidermal growth factor receptor. *Mol. Cell. Biol.*, **8**, 2302–8.

Northwood, I.C., and Davis, R.J. (1989). Protein kinase C inhibition of the epidermal growth factor receptor tyrosine protein kinase activity is independent of the oligomeric state of the receptor. *J. Biol. Chem.*, **264**, 5746–50.

Rapp, U.R. (1991). Role of Raf-1 serine/threonine protein kinase in growth factor signal transduction. *Oncogene*, **6**, 495–500.

Schlessinger, J. (1988). The epidermal growth factor receptor as a multifunctional allosteric protein. *Biochemistry*, **27**, 3119–23.

Schlessinger, J. (1993). How receptor tyrosine kinases activate Ras. *TIBS* **18**, 273–5.

Selva, E., Raden, D.L., and Davis, R.J. (1993). Mitogen-activated protein kinase stimulation by a tyrosine kinase-negative epidermal growth factor receptor. *J. Biol. Chem.*, **268**, 2250–4.

Wada, T., Qian, X.L., and Greene, M.I. (1990). Intermolecular association of the p185neu protein and EGF receptor modulates EGF receptor function. *Cell*, **61**, 1339–47.

Walker, F., and Burgess, A.W. (1991). Reconstitution of the high affinity epidermal growth factor receptor on cell-free membranes after transmodulation by platelet-derived growth factor. *J. Biol. Chem.*, **266**, 2746–52.

Yarden, U., Harari, I., and Schlessinger, J. (1985). Purification of an active EGF receptor kinase with monoclonal antireceptor antibodies. *J. Biol. Chem.*, **260**, 315–9.

Zachary, I., and Rozengurt, E. (1985). Modulation of the epidermal growth factor receptor by mitogenic ligands: effects of bombesin and role of protein kinase C. *Cancer Surveys*, **4**, 729–65.

F. Walker and A.W. Burgess:
Ludwig Institute for Cancer Research,
Melbourne, Australia 3050

Platelet-derived Growth Factor (PDGF)

PDGF is a 30 kDa homo- or heterodimeric protein consisting of disulphide-bonded A- and B-polypeptide chains. It stimulates growth and chemotaxis of fibroblasts, smooth muscle cells and certain other cell types. PDGF is likely to have a function during embryonal development and has also been shown to stimulate wound healing in the adult. Overproduction of PDGF has been associated with the development of several disorders involving excessive cell growth, such as atherosclerosis, rheumatoid arthritis, fibrotic conditions as well as malignant diseases (for reviews, see Heldin and Westermark 1990; Raines et al. 1990).

■ PDGF sequence

cDNAs for human PDGF A–chain (Betsholtz *et al.* 1986) and B–chain (Josephs *et al.* 1984) have been cloned (GenBank accession numbers X03795 and M12782, respectively). The PDGF A-chains have also been cloned from mouse (Rorsman *et al.* 1992) and *Xenopus* (Mercola *et al.* 1988) and the B-chain from mouse (Bonthron *et al.* 1991) and cat (Van den Ouweland *et al.* 1987); the conservation of the sequences between different species is very high. The human B–chain sequence predicts a 241-amino-acid polypeptide including a 20-amino-acid long leader sequence (Fig. 1). The A-chain occurs in two forms as a result of differential splicing. The short form has 191 amino acid residues including a 20-amino-acid leader sequence; the C-terminal three amino acids in the short form have in the long form been replaced with 18 different amino acids (Fig. 1).

Both the A- and the B-chains undergo proteolytic processing after synthesis; the mature proteins of about 100 amino acids each are about 60 per cent similar with a perfect conservation of the eight cysteine residues. The cysteine residues are also conserved in vascular endothelial growth factor/vascular permeability factor and in placenta growth factor.

■ PDGF protein

PDGF is a dimeric protein of about 30 kDa. Homodimers (PDGF-AA and -BB) as well as the heterodimer (PDGF–AB) have been identified from natural sources (Stroobant and Waterfield 1984; Heldin *et al.* 1986; Hart *et al.* 1990). The short form of PDGF-AA is secreted normally; in contrast, the long form of PDGF-AA and of the PDGF-BB proform contain

```
                            ↓
B-chain   MNRCWALFLSLCCYLRLVSAEGDPIPEELYEMLSDHSIRSFDDLQRLLHGDPGEEDGAEL    60
          *     * *  * **  * **    ** * *    * *  ****** *            *
A-chain   MRTLACLLLLGCGYLAHVLAEEAEIPREVIERLARSQIHSIRDLQRLLEIDSVGSEDS L    59
                             ↑


                            ↓          ●
B-chain   DLNMTRSHSGGELESLARGRRSLGSLTIAEPAMIAECKTRTEVFEISRRLIDRTNANFLV   120
          *                          * *  * *****  ** *   * * ****
A-chain   DTSLRAHGVHATKHVPEKRPLPIRRKRSIEEAVPAVCKTRTVIYEIPRSQVDPTSANFLI   119
                                    ↑

             ●       ●    ●●         ●                              ● ●
B-chain   WPPCVEVQRCSGCCNNRNVQCRPTQVQLRPVQVRKIEIVRKKPIFKKATVTLEDHLACKC   180
          ******* ** ****    * *  *  * * * * ***** *   * ** ** * *
A-chain   WPPCVEVKRCTGCCNTSSVKCQPSRVHHRSVKVAKVEYVRKKPKLKEVQVRLEEHLECAC   179


B-chain   ETVAAARPVTRSPGGSQEQRAKTPQTRVTIRTVRVRRPPKGKHRKFKHTHDKTALKETLGA   241
             *        *       *       *
                                    GRPRESGKKRKRKRLKPT      long version    211
A-chain   ATTSLNPDYREEDT DVR                                  short version    196
```

Figure 1. Amino acid sequences of the human PDGF A- and B-chain precursors. Processing sites for removal of leader sequences and cleavage of precursor molecules are indicated by arrows and cysteine residues in the mature proteins by dots.

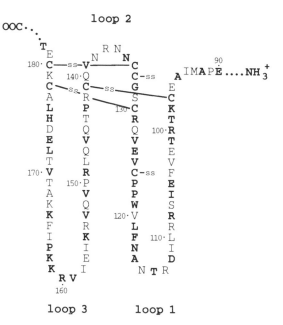

Figure 2. Schematic illustration of the three-dimensional structure of the mature PDGF B-chain. Amino acid identical between the A- and B-chains are bold.

PDGF genes and transcription

The human A-chain gene is localized on the tip of the p-arm of chromosome 7 (Stenman et al. 1992) and the B-chain gene on chromosome 22 q12.2–13.1 (Swan et al. 1982). The two genes are similarly organized with seven exons. The A-chain gene is subject to alternative splicing; the sixth exon can be present or not in the transcript, giving rise to the long or the short form of the protein, respectively. The B-chain gene generates a single transcript of about 3.5 kb, and a minor one of 2.5 kb, as the result of differential promoter usage (Fen and Daniel 1991). The A-chain gene gives rise to transcripts of 2.8, 2.3 and 1.8 kb as a result of differential promoter usage and use of different poly A addition sites (Rorsman et al. 1992).

Both PDGF A- and B-chain genes are transcribed in endothelial cells, activated macrophages, smooth muscle cells, placental cytotrophoblasts, and nerve cells. Moreover, the A-chain transcript and protein are induced in fibroblasts upon growth stimulation (Heldin and Westermark 1990; Raines et al. 1990).

The human B-chain gene contains both positive and negative regulatory elements in its 5' flanking region (Pech et al. 1989) as well as in the first intron (Franklin et al. 1991). The A-chain gene contains a consensus sequence for binding of the WT-1 tumour suppressor gene product (Gashler et al. 1992).

stretches of basic amino acid residues in their C-termini which mediate retention of the factors intracellularly and in the extracellular matrix (Raines and Ross 1992).

The second and fourth cysteine residues in PDGF form the interchain disulphide bonds; the three intrachain disulphide bonds are arranged such that the first cysteine residue bridges with the sixth, the third with the seventh, and the fifth with the eighth (Östman et al. 1993).

The three-dimensional structure of PDGF–BB has been solved at 3.0 Å resolution (Oefner et al. 1992). It shows an unusual fold also found in nerve growth factor and transforming growth factor-β2; the three intrachain disulphide bonds form a knot structure at one end of the molecule with one of them passing through a hole formed by the other two and intervening amino acid sequences. The other end of the molecule consists of two loops of twisted β-strands (Fig. 2). The two subunits in the dimer are arranged in an antiparallel manner.

PDGF is a stable molecule and resists extremes of pH, heat, and chaotropic agents. It is a highly basic protein (isoelectric point about 10) which also has hydrophobic epitopes. PDGF can be purified using conventional chromatography steps; because of its physicochemical features, ion exchange chromatography, and reverse-phase HPLC are the most efficient steps. The different isoforms can be separated from each other by immobilized metal ion chromatography since the A-chain contains three histidine residues but the B–chain only one (Hammacher et al. 1988). Alternatively, chain-specific monoclonal antibodies can be used (Hart et al. 1990).

Biological functions

PDGF A-chain binds to PDGF α-receptors which have intrinsic tyrosine kinase activity, whereas the B-chain binds both to the α-receptor and to the related β-receptor (Heldin and Westermark 1990; Raines et al. 1990). Ligand binding induces receptor dimerization. Thus, depending on the stimulating PDGF isoform and which receptor types the responder cell expresses, different responses will be obtained. Both receptors mediate stimulation of cell growth. In contrast, whereas the β-receptor mediates stimulation of chemotaxis, the α-receptor in certain cell types mediates inhibition of chemotaxis.

The specific expression of PDGF and PDGF receptors in various tissues during development suggests that PDGF regulates growth and differentiation of cells in the fetus and the placenta (Palmieri et al. 1993). In the adult organism PDGF stimulates wound healing (Robson et al. 1992).

Pathology

The findings that PDGF B-chain corresponds to the sis-oncogene product and that sis-transformation is exerted by autocrine stimulation by a PDGF-BB-like factor, have stimulated efforts to explore the possibility that anomalous PDGF production may be involved in the development of human neoplasia (Heldin and Westermark 1990; Raines *et al.* 1990). The accumulated data support the notion that PDGF overactivity is an important feature in the progression of certain malignancies, e.g. glioblastomas and sarcomas. Overproduction of PDGF may also be associated with non-malignant disorders characterized by excessive cell proliferation, e.g., atherosclerosis, rheumatoid arthritis, glomerulonephritis, and various fibrotic conditions (Heldin and Westermark 1990; Raines *et al.* 1990).

■ References

Betsholtz, C., Johnsson, A., Heldin, C.-H., Westermark, B., Lind, P., Urdea, M.S., Eddy, R., Shows, T.B., Philpott, K., Mellor, A.L., Knott, T.J., and Scott, J. (1986). cDNA sequence and chromosomal localization of human platelet-derived growth factor A-chain and its expression in tumour cell lines. *Nature*, **320**, 695–9.

Bonthron, D.T., Sultan, P., and Collins, T. (1991). Structure of the murine c-sis proto-oncogene (Sis, PDGFB) encoding the B chain of platelet-derived growth factor. *Genomics*, **10**, 287–92.

Fen, Z., and Daniel, T.O. (1991). 5′ untranslated sequences determine degradative pathway for alternate PDGF B/c-sis mRNA's. *Oncogene*, **6**, 953–9.

Franklin, G.C., Donovan, M., Adam, G.I., Holmgren, L., Pfeifer-Ohlsson, S., and Ohlsson, R. (1991). Expression of the human PDGF-B gene is regulated by both positively and negatively acting cell type specific regulatory elements located in the first intron. *EMBO J.*, **10**, 1365–73.

Gashler, A.L., Bonthron, D.T., Madden, S.L., Rauscher III, F.J., Collins, T., and Sukhatme, V.P. (1992). Human platelet-derived growth factor A chain is transcriptionally repressed by the Wilms tumour suppressor WT1. *Proc. Natl. Acad. Sci. (USA)*, **89**, 10984–8.

Hammacher, A., Hellman, U., Johnsson, A., Östman, A., Gunnarsson K., Westermark, B., Wasteson, Å., and Heldin, C.-H. (1988). A major part of platelet-derived growth factor purified from human platelets is a heterodimer of one A and one B chain. *J. Biol. Chem.*, **263**, 16493–8.

Hart, C.E., Bailey, M., Curtis, D.A., Osborn, S., Raines, E., Ross, R., and Forstrom, J.W. (1990). Purification of PDGF-AB and PDGF-BB from human platelet extracts and identification of all three PDGF dimers in human platelets. *Biochemistry*, **29**, 166–72.

Heldin, C.-H., and Westermark, B. (1990). Platelet-derived growth factor: mechanism of action and possible *in vivo* function. *Cell Regul.*, **1**, 555–66.

Heldin, C.-H., Johnsson, A., Wennergren, S., Wernstedt, C., Betsholtz, C., and Westermark, B. (1986). A human osteosarcoma cell line secretes a growth factor structurally related to a homodimer of PDGF A-chains. *Nature*, **319**, 511–4.

Josephs, S.F., Guo, C., Ratner, L., and Wong-Staal, F. (1984). Human-proto-oncogene nucleotide sequences corresponding to the transforming region of simian sarcoma virus. *Science*, **223**, 487–91.

Mercola, M., Melton, D.A., and Stiles, C.D. (1988). Platelet-derived growth factor A chain is maternally encoded in *Xenopus* embryos. *Science*, **241**, 1223–5.

Oefner, C., D'Arcy, A., Winkler, F.K., Eggimann, B., and Hosang, M. (1992). Crystal structure of human platelet-derived growth factor BB. *EMBO J.*, **11**, 3921–6.

Östman, A., Andersson, M., Bäckström, G., and Heldin, C.-H. (1993). Assignment of intrachain disulphide bonds in platelet-derived growth factor B-chain. *J. Biol. Chem.*, **268**, 13372–7.

Palmieri, S.L., Stiles, C.D., and Mercola, M. (1993). PDGF in the developing embryo. In *Biology of platelet-derived growth factor. Cytokines* (ed. B. Westermark and C. Sorg), pp. 115–28. Karger, Basel.

Pech, M., Rao, C.D., Robbins, K.C., and Aaronson, S.A. (1989). Functional identification of regulatory elements within the promoter region of platelet-derived growth factor 2. *Mol. Cell Biol.*, **9**, 396–405.

Raines, E.W., and Ross, R. (1992). Compartmentalizatian of PDGF on extracellular binding sites dependent on exon-6-encoded sequences. *J. Cell Biol.*, **116**, 533–43.

Raines, E.W., Bowen-Pope, D.F., and Ross, R. (1990). Platelet-derived growth factor. In *Handbook of experimental pharmacology. Peptide growth factors and their receptors* (ed. M.B. Sporn and A.B. Roberts), pp. 173–262. Springer, Heidelberg.

Robson, M.C., Phillips, L.G., Thomason, A., Robson, L.E., and Pierce, G.F. (1992). Platelet-derived growth factor BB for the treatment of chronic pressure ulcers. *Lancet*, **339**, 23–5.

Rorsman, F., Leveen, P., and Betsholtz, C. (1992). Platelet-derived growth factor (PDGF) A-chain mRNA heterogeneity generated by the use of alternative promoters and alternative polyadenylation sites. *Growth Factors*, **7**, 241–51.

Stenman, G., Rorsman, F., Huebner, K., and Betsholtz, C. (1992). The human platelet-derived growth factor alpha chain (PDGFA) gene maps to chromosome 7p22. *Cytogenet. Cell Genet.*, **60**, 206–7.

Stroobant, P., and Waterfield, M.D. (1984). Purification and properties of porcine platelet-derived growth factor. *EMBO J.*, **3**, 2963–7.

Swan, D.C., McBridge, O.W., Robbins, K.C., Keithley, D.A., Reddy, E.P., and Aaronson, S.A. (1982). Chromosomal mapping of the simian sarcoma virus onc gene analogue in human cells. *Proc. Natl. Acad. Sci. (USA)*, **79**, 4691–5.

Van den Ouweland, A.M., Van Groningen, J.J., Schalken, J.A., Van Neck, H.W., Bloemers, H.P., and Van de Ven, W.J. (1987). Genetic organization of the c-sis transcription unit. *Nucl. Acids Res.*, **15**, 959–70.

Carl-Henrik Heldin:
Ludwig Institute for Cancer Research,
Box 595, Biomedical Center,
S–751 24 Uppsala, Sweden

Bengt Westermark:
Department of Pathology,
University Hospital,
S–751 85 Uppsala, Sweden

Receptors for Platelet-derived Growth Factor (PDGF)

Two structurally similar tyrosine kinase receptors for PDGF have been identified. The α-receptor binds both PDGF A- and B-chains, whereas the β-receptor binds only B-chains. Ligand binding induces receptor dimerization and autophosphorylation. The autophosphorylated regions of the receptors then provide binding sites for down-stream components in the signal transduction pathways. Fibroblasts and smooth muscle cells express both α- and β-receptors, whereas oligodendrocyte progenitor cells, mesothelial cells, and liver endothelial cells express only α-receptors, and capillary endothelial cells, neurones, meningeal cells and Ito cells of the liver express only β-receptors (for reviews, see Raines et al. *1990; Claesson-Welsh 1993).*

■ PDGF α- and β-receptor proteins

The human α-receptor has 1089 amino acids including a 23-amino-acid leader sequence (GenBank accession number M22734 and M21574), and the β–receptor has 1106 amino acids including a 32-amino-acid leader (Gen-Bank accession number M21616 and JO3278) (Fig. 1). The mouse α- and β-receptors have also been cloned (GenBank accession numbers S40407 and X04367, respectively). The predicted molecular weight of the human α-receptor is 120 000 and of the β-receptor 121 000; however, the receptors run as components of apparent molecular weights 170 000 and 180 000 in SDS-gel electrophoresis, respectively, due to glycosylation and other post-translational modifications. Each receptor has five immunoglobulin-like domains in the extracellular part; the amino acid sequence similarity between the receptors is 30 per cent in this region. The intracellular part of each receptor contains a tyrosine kinase domain which is split into two parts by an intervening sequence of about 100 amino acids without homology to kinase domains. The sequence similarity between the α- and β-receptors is high in the juxtamembrane domain, and in the first and second part of the kinase domain (83, 87, and 74 per cent, respectively), but lower in the kinase insert and C-terminal tail (35 and 27 per cent, respectively). The PDGF receptors are structurally similar to the receptors for colony-stimulating factor-1 (c-fms) and stem cell factor (c-kit) (Yarden *et al.* 1986; Matsui *et al.* 1989; Claesson-Welsh *et al.* 1989).

■ PDGF receptor genes

The PDGF α- and β-receptors are localized on human chromosomes 4 and 5, respectively, the α-receptor gene being very close to the stem cell factor receptor gene and the β-receptor very close to the colony-stimulating factor-1 receptor gene (Yarden *et al.* 1986; Matsui *et al.* 1989).

The expression of PDGF α- and β-receptors is independently regulated. There are examples of cell types that express both receptors (fibroblasts, smooth muscle cells), only the α-receptor (oligodendrocyte progenitor cells, mesothelial cells, liver endothelial cells) and only the β-receptor (capillary endothelial cells, neurones, meningeal cells, Ito cells of the liver) (Raines *et al.* 1990; Claesson-Welsh 1993). The expression of the receptors in cultured cells is modulated by several growth regulatory molecules. *In vivo* an increased expression of PDGF β-receptors has been seen on connective tissue cells during wound healing and in conjunction with inflammation. Many tumour cells of different origins, particularly glioblastomas and fibrosarcomas, also express PDGF receptors (Raines *et al.* 1990; Claesson-Welsh 1993).

■ Binding characteristics of PDGF receptors

The α-receptor binds both A- and B-chains of PDGF, whereas the β-receptor binds only B-chains. Ligand binding induces receptor dimerization; thus, all PDGF isoforms induce αα receptor homodimers; PDGF-AB and PDGF-BB induce αβ receptor heterodimers; whereas only PDGF-BB induces ββ receptor homodimers (Heldin *et al.* 1989; Seifert *et al.* 1989) (Fig. 2).

PDGF-AA binds to α-receptors with a K_D of about 0.2 nM. PDGF-BB binds to α- and β-receptors with similar affinities (K_D about 0.5 nM) (Östman *et al.* 1989). It is likely that these high affinity interactions represent binding to dimeric receptor complexes. In support of this notion, PDGF-AB binds to the β-receptor with high affinity in the presence of α-receptors, but with low affinity in the absence of α-receptors (Seifert *et al.* 1993).

■ Signalling mechanisms

Ligand-induced receptor dimerization leads to auto-phosphorylation *in trans* between the receptors in the dimer (Kelly *et al.* 1991). The autophosphorylated tyrosine

Figure 1. Comparison of the amino acid sequences of PDGF α and β receptors (single letter code). Identical amino-acid residues are boxed. Arrowheads indicate the cleavage sites for signal peptidase. The cysteine residues in the extracellular parts of the mature proteins are indicated by filled circles. The transmembrane regions, and the first and second parts of the tyrosine kinase domains, are indicated by a solid line and grey bars, respectively.

residues then provide binding sites for SH2-domain-containing down-stream components in the signal transduction pathways (Fig. 3). In the β-receptor, eight autophosphorylation sites are currently known, several of which mediate interactions with the signal transduction molecules in a specific manner: phosphorylated Tyr579 and Tyr581 in the juxtamembrane part of the receptor bind members of the Src family of tyrosine kinases (Mori et al. 1993); phosphorylated Tyr740 and Tyr751 bind the p85 regulatory subunit of phosphatidylinositol-3'-kinase (Fantl et al. 1992; Kashishian et al. 1992), and Tyr751 also the adaptor molecule Nck (Nishimura et al. 1993); phosphorylated Tyr771 binds the GTPase activating protein of Ras (Fantl et al. 1992; Kashishian et al. 1992); phosphorylated

Tyr1009 and Tyr1021 in the C-terminal tail bind phospholipase Cγ (Rönnstrand et al. 1992), and Tyr1009 also binds phosphotyrosine phosphatase 1D (Kazlauskas et al. 1993).

Tyr857, located at a conserved position inside the kinase domain is also autophosphorylated; no signal transduction molecule has been found to bind to this autophosphorylation site and it may rather be involved in regulation of the catalytic activity of the kinase (Kazlauskas et al. 1991). Two other adapter molecules, Shc and Grb2, also bind to the β-receptor, but their exact docking sites have not been determined. Most of the signal transduction molecules mentioned above also bind to the α-receptor, but less information is available regarding their interaction with specific autophosphorylation sites.

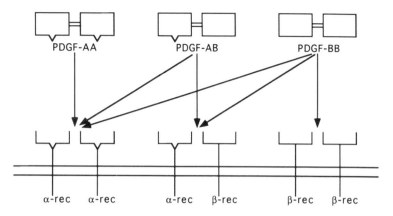

Figure 2. Binding of PDGF isoforms to homo- or heterodimeric complexes of PDGF a and β-receptors.

The binding of different signal transduction molecules to the PDGF receptors initiates different signal transduction pathways leading to cell growth, cell migration, and actin reorganization (Raines *et al.* 1990; Claesson-Welsh 1993). The β-receptor mediates stimulation of cell growth as well as motility. The PDGF a-receptor also mediates cell growth,

but its effect on migration is cell-type specific; the a-receptor in fact inhibits chemotaxis in certain cell types, like human fibroblasts (Siegbahn *et al.* 1990).

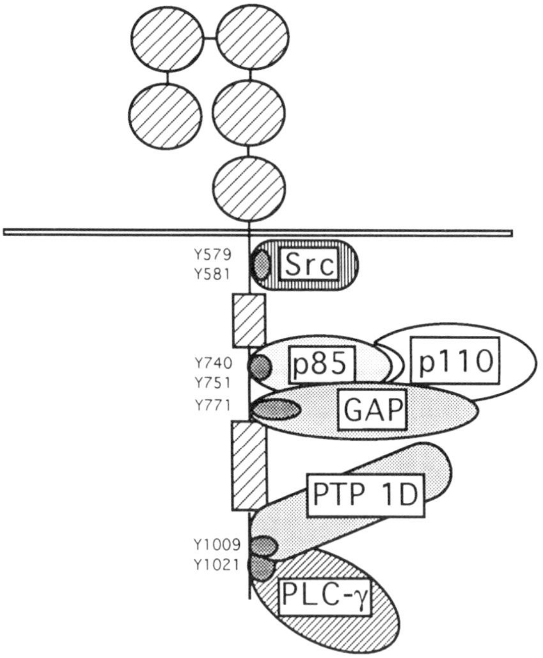

■ References

Claesson-Welsh, L. (1993). PDGF receptors: Structure and mechanisms of action. In *Biology of platelet-derived growth factor. Cytokines* (ed. B. Westermark and C. Sorg), pp. 31–43. Karger, Basel.

Claesson-Welsh, L., Eriksson, A., Westermark, B., and Heldin, C.-H. (1989). cDNA cloning and expression of the human A-type platelet-derived growth factor (PDGF) receptor establishes structural similarity to the B-type PDGF receptor. *Proc. Natl. Acad. Sci. (USA)*, **86**, 4917–21.

Fantl, W.J., Escobedo, J.A., Martin, G.A., Turck, C.W., del Rosario, M., McCormick, F., and Williams, L.T. (1992). Distinct phosphotyrosines on a growth factor receptor bind to specific molecules that mediate different signaling pathways. *Cell*, **69**, 413–23.

Heldin, C.-H., Ernlund, A., Rorsman, C., and Rönnstrand, L. (1989). Dimerization of B-type platelet-derived growth factor receptors occurs after ligand binding and is closely associated with receptor kinase activation. *J. Biol. Chem.*, **264**, 8905–12.

Kashishian, A., Kazlauskas, A., and Cooper, J.A. (1992). Phosphorylation sites in the PDGF receptor with different specificities for binding GAP and PI3 kinase *in vivo*. *EMBO J.*, **11**, 1373–82.

Kazlauskas, A., Durden, D.L., and Cooper, J.A. (1991). Functions of the major tyrosine phosphorylation site of the PDGF receptor beta subunit. *Cell Regul.*, **2**, 413–25.

Kazlauskas, A., Feng, G.-S., Pawson, T., and Valius, M. (1993). The 64-kDa protein that associates with the platelet-derived growth factor receptor β subunit *via* Tyr-1009 is the SH2-containing phosphotyrosine phosphatase Syp. *Proc. Natl. Acad. Sci. (USA)*, **90**, 6939–43.

Kelly, J.D., Haldeman, B.A., Grant, F.J., Murray, M.J., Seifert, R.A., Bowen-Pope, D.F., Cooper, J.A., and Kazlauskas, A. (1991). Platelet-derived growth factor (PDGF) stimulates PDGF receptor subunit dimerization and intersubunit *trans*-phosphorylation. *J. Biol. Chem.*, **266**, 8987–92.

Matsui, T., Heidaran, M., Miki, T., Popescu, N., La Rochelle, W., Kraus, M., Pierce, J., and Aaronson, S. (1989). Isolation of a novel receptor cDNA establishes the existence of two PDGF receptor genes. *Science*, **243**, 800–4.

Mori, S., Rönnstrand, L., Yokote, K., Engström, A., Courtneidge, S.A., Claesson-Welsh, L., and Heldin, C.-H. (1993). Identification of two

Figure 3. Binding of signal transduction molecules to different autophosphorylation sites of the PDGF β receptor. Src, members of the src family of tyrosine kinases; p85 and p110, regulatory and catalytic subunits of the phosphatidylinositol-3-kinase, respectively; GAP, GTPase activating protein of ras; PTP 1D, phosphotyrosine phosphatase 1D; PLC-γ, phospholipase Cγ.

juxtamembrane autophosphorylation sites in the PDGF beta-receptor; involvement in the interaction with src family tyrosine kinases. *EMBO J.*, **12**, 2257–64.

Nishimura, R., Li, W., Kashishian, A., Mondino, A., Zhou, M., Cooper, J., and Schlessinger, J. (1993). Two signaling molecules share a phosphotyrosine-containing binding site in the PDGF receptor. *Mol. Cell Biol.*, **13**, 6889–96.

Östman, A., Bäckström, G., Fong, N., Betsholtz, C., Wernstedt, C., Hellman, U., Westermark, B., Valenzuela, P., and Heldin, C.-H. (1989). Expression of three recombinant homodimeric isoforms of PDGF in *Saccharomyces cerevisiae*: evidence for difference in receptor binding and functional activities. *Growth Factors*, **1**, 271–81.

Raines, E.W., Bowen-Pope, D.F., and Ross, R. (1990). Platelet-derived growth factor. In *Handbook of experimental pharmacology. Peptide growth factors and their receptors* (ed. M.B. Sporn and A.B. Roberts), pp. 173–262. Springer, Heidelberg.

Rönnstrand, L., Mori, S., Arvidsson, A.K., Eriksson, A., Wernstedt, C., Hellman, U., Claesson-Welsh, L., and Heldin, C.-H. (1992). Identification of two C-terminal autophosphorylation sites in the PDGF beta-receptor: involvement in the interaction with phospholipase C-gamma. *EMBO J.*, **11**, 3911–9.

Seifert, R.A., Hart, C.E., Philips, P.E., Forstrom, J.W., Ross, R., Murray, M.J., and Bowen-Pope, D.F. (1989). Two different subunits

associate to create isoform-specific platelet-derived growth factor receptors. *J. Biol. Chem.*, **264**, 8771–8.

Seifert, R.A., van Koppen, A., and Bowen-Pope, D.F. (1993). PDGF-AB requires PDGF receptor alpha-subunits for high-affinity, but not for low-affinity, binding and signal transduction. *J. Biol. Chem.*, **268**, 4473–80.

Siegbahn, A., Hammacher, A., Westermark, B., and Heldin, C.-H. (1990). Differential effects of the various isoforms of platelet-derived growth factor on chemotaxis of fibroblasts, monocytes, and granulocytes. *J. Clin. Invest.*, **85**, 916–20.

Yarden, Y., Escobedo, J.A., Kuang, W.J., Yang-Feng, T.L., Daniel, T.O., Tremble, P.M., Chen, E.Y., Ando, M.E., Harkins, R.N., Francke, U., Friend, V.A., Ullrich, A., and Williams, L.T. (1986). Structure of the receptor for platelet-derived growth factor helps define a family of closely related growth factor receptors. *Nature*, **323**, 226–32.

Carl-Henrik Heldin and Lena Claesson-Welsh:
Ludwig Institute for Cancer Research,
Biomedical Center,
S–751 24 Uppsala,
Sweden

Insulin–like Growth Factors (IGFs)

IGF-I and IGF-II are growth factors with structures similar to insulin, but which are primarily involved in normal growth and development of vertebrates. The major source of circulating (endocrine) IGFs is the liver; most extrahepatic tissues, however, synthesize IGFs, where they function as paracrine/autocrine regulators of the differentiated function of these tissues. The IGFs are involved in numerous physiological and pathological states, and their potential clinical applications include wound healing, osteoporosis, nerve regeneration, reversal of catabolic states, and insulin-resistant diabetes.

■ Alternative name

Somatomedins.

■ IGF sequence

The nucleotide sequences encoding IGF-I and IGF-II have been characterized for most major vertebrate species. The mammalian IGF-I genes are large, spanning over 90 kb of chromosomal DNA [GenBank accession numbers X57025 (cDNA) and X03419 (gene)]. The human IGF-I gene is on the long arm of chromosome 12 and contains six exons (Jansen

et al. 1983; Tricoli *et al.* 1984). Exons 1 and 2 encode distinct, mutually exclusive 5′-untranslated regions (UTRs) as well as distinct N-termini of the IGF-I signal peptide, due to the presence of several in-frame translation initiation codons (Fig. 1) (de Pagter-Holthuizen *et al.* 1986). Expression of the alternative 5′ leader and E-peptide-encoding exons is regulated by both developmental and tissue-specific factors (Lowe *et al.* 1987, 1988). The common C-terminal amino acids of the signal peptide are encoded by the 5′ end of exon 3, and all of the mature peptide coding sequence is present in the remainder of exon 3 and part of exon 4. E-peptide coding sequences are contained within exons 4, 5, and 6 and, in the human gene, both exons 5 and 6 encode 3′-UTR sequences. Expression of the IGF-I gene gives rise to mRNA transcripts ranging from ∼1 kb to ∼7.5 kb, the size of the latter reflecting a long 3′-UTR encoded by exon 6 which arises due to alternative polyadenylation site usage

Figure 1.   Mammalian IGF genes.

(Roberts *et al.* 1987). There is a very high degree of conservation of IGF genes (especially the mature peptide coding region) throughout vertebrate species, including fish (Kavsan *et al.* 1993). The specific organization of the 5' and 3' ends of the genes, however, differs in most species.

The IGF-II gene spans about 30 kb of chromosomal DNA on the distal part of the short arm of chromosome 11 in the human genome [GenBank accession numbers X00910 (cDNA) and X03423 (gene)] (de Pagter-Holthuizen *et al.* 1986; Bell *et al.* 1985; Dull *et al.* 1984). This gene comprises nine exons, with the mature peptide being encoded by exons 7, 8 and 9, and the 3'-UTR encoded by 9 (Fig. 1). The promoters preceding exons 1, 4, 5,and 6 are differentially activated in a tissue- and development-specific manner, giving rise to multiple mRNA transcripts ranging from ~2 kb to ~6.0 kb. Promoters P2, P3, and P4 are active in fetal and most adult non-hepatic tissues. In adult liver, only promoter P1 is active (de Pagter-Holthuizen *et al.* 1988; Gray *et al.* 1987).

The IGF-II gene is imprinted by a mechanism involving methylation and only the paternally inherited gene is expressed (DeChiara *et al.* 1991). As with IGF-I, expression of the IGF-II gene is regulated at both translational and transcriptional levels.

■ IGF proteins

Mature, circulating IGF-I consists of B and A domains (homologous to the B and A chains of insulin). Unlike insulin, the B and A domains of the IGFs are connected by a C peptide and contain an eight-amino-acid extension at the C-terminus, termed the D domain. In all species studied, the gène sequence predicts an additional E peptide, which is cleaved during processing and whose function is still unknown. The 70-amino-acid mature polypeptide is highly conserved in vertebrates and exhibits 40 per cent amino

acid homology with insulin and 60 per cent homology with IGF-II. Native IGF-I contains three disulphide bonds involving the following residues: Cys B6–Cys A7, Cys A6–Cys A11, and Cys B 18–Cys A20. Specific binding of IGF-I to its homologous receptor involves residues 23–25 (B domain) and 28–37 (C peptide), whereas residues 3, 4, 15, and 16 (B domain) are important for binding to IGF-binding proteins (Cascieri *et al.* 1988; Bayne *et al.* 1988).

Mature circulating human IGF-II is a 67-amino acid molecule which is cleaved from a 180-amino acid precursor composed of a 24-amino-acid signal peptide and an 89-amino-acid E peptide. Like IGF-I, the mature peptide contains B, C, A, and D domains. Residues 6, 7, 18, and 19 (B domain) are important for interactions with IGF-binding proteins. Residues 55–56 (A domain) are critical for interactions with the IGF-II receptor, and residues 26, 27, and 43 are important for interactions with the IGF-I receptor (Sakano *et al.* 1991).

■ IGF gene expression and IGF function

In developing fetal tissues IGF is expressed primarily by mesenchymal-derived cells, which suggests that it plays an important role in tissue development (Table 1) (Bondy *et al.* 1990). Mice carrying null mutations of the IGF-I genes introduced by gene targeting exhibited decreased prenatal growth and immediate postnatal lethality (Liu *et al.* 1993; Baker *et al.* 1993). Postnatally, IGF-I expression by most tissues increases dramatically (Adamo *et al.* 1989). Liver expression becomes growth hormone(GH)-dependent at weaning and contributes to the circulating endocrine form involved in longitudinal bone growth. Extrahepatic expression is less GH-dependent and is tissue-specific. Examples include the reproductive system, where IGF-I expression is regulated by gonadotropins and sex steroids and where it, in turn, regulates the synthesis and function of these compounds. In bone, IGF-I expression is regulated by parathyroid hormone and sex steroids. Thus, the regulation of expression and function of IGF-I is specific for each tissue (Table 1).

IGF-II is highly expressed in fetal tissues where it plays a major role in overall growth and development. When the IGF-II gene is inactivated by homologous recombination in mice, the homozygous offspring are smaller but develop normally (DeChiara *et al.* 1990). Postnatally, the expression of IGF-II by certain tissues decreases and its function in adult tissues is as yet undefined.

■ Pathology

IGFs are mitogens and are therefore involved in certain malignant states, e.g. breast tumours, in which they

Table 1. Biological actions of IGFs (Pardee 1989; Ewton and Florini 1981; Daughaday and Rotwein 1989; Sara and Hall 1990)

1. Stimulation of DNA synthesis/cell proliferation in all cells studied (acting as a 'progression factor' in cell cycle control).
2. Differentiation of myoblasts into myocytes and fibroblasts into adipocytes.
3. Differentiated responses:
   a. Sex steroid synthesis by granulosa and Leydig cells.
   b. Early response gene expression (c-*myc*, c-*jun*, c-*fos*) in uterine tissue.
   c. Neurite outgrowth, synapse formation, and myelin synthesis by nervous tissue-derived cells.
   d. Production of collagen and proteoglycans and other matrix proteins by chondrocytes, osteoblasts, renal mesangial cells, and granulosa cells.
   e. Erythropoietin production and colony formation by haemopoietic cells.
4. Other *in vivo* responses:
   a. Reverses catabolism in states of starvation, severe illness, or injury.
   b. Enhances wound healing and nerve regeneration.
   c. Reduces insulin resistance in diabetics.

enhance proliferation *via* the IGF-I receptors. Vascular smooth muscle cells may also be subject to the mitogenic activity of the IGFs contributing to atherogenesis. Their effect on kidney enlargement and altered renal haemodynamics may play a role in early diabetic renal disease. On the other hand, chronic malnutrition and poorly controlled diabetes in the young are associated with lower circulating IGF-I levels and growth retardation.

■ References

Adamo, M., Lowe, W.L. Jr., LeRoith, D., and Roberts, C.T. Jr. (1989). Insulin-like growth factor I messenger ribonucleic acids with alternative 5'-untranslated regions are differentially expressed during development of the rat. *Endocrinology*, **124**, 2737–44.

Baker, J., Liu, J.-P., Robertson, E.J., and Efstratiadis, A. (1993). Role of insulin-like growth factors in embryonic and postnatal growth. *Cell*, **75**, 73–82.

Bayne, M.L., Applebaum, J., Chicchi, G.G., Hayes, N.S., Green, B.G., and Cascieri, M.A. (1988). Structural analogs of human insulin-like growth factor I with reduced affinity for serum binding proteins and the type 2 insulin-like growth factor receptor. *J. Biol. Chem.*, **263**, 6233–9.

Bell, G.I., Gerhard, D.S., Fong, N.M., Sanchez-Pescador, R., and Rall, L.B. (1985). Isolation of the human insulin-like growth factor genes: insulin-like growth factor II and insulin genes are contiguous. *Proc. Natl. Acad. Sci. (USA)*, **82**, 6450–4.

Bondy, C.A., Werner, H., Roberts, C.T. Jr., and LeRoith, D. (1990). Cellular pattern of insulin-like growth factor-I (IGF-I) and type I IGF receptor gene expression in early organogenesis: comparison with IGF-II gene expression. *Mol. Endocrinol.*, **4**, 1386–98.

Cascieri, M.A., Chicchi, G.G., Applebaum, S., Hayes, N.S., Green, B.G., and Bayne, M.L. (1988). Mutants of human insulin-like growth factor I with reduced affinity for the type 1 insulin-like growth factor receptor. *Biochemistry*, **27**, 3229–33.

Daughaday, W.H., and Rotwein, P. (1989). Insulin-like growth factors I and II. Peptide, messenger ribonucleic acid and gene structures, serum, and tissue concentrations. *Endocrin. Rev.*, **10**, 68–91.

de Pagter-Holthuizen, P., Van Schaik, F.M., Verduijn, G.M., Van Ommen, G.J.,Bouma, B.N., Jansen, M., and Sussenbach, J.S. (1986). Organization of the human genes for insulin-like growth factors I and II. *Febs Lett.*, **195**, 179–84.

de Pagter-Holthuizen, P., Jansen, M., van der Kammen, R.A., Van Schaik, F.M.A., and Sussenbach, J.S. (1988). Differential expression of the human insulin-like growth factor II gene. Characterization of the IGF-II mRNAs and an mRNA encoding a putative IGF-II-associated protein. *Biochcm. Biophys. Acta*, **950**, 282–95.

DeChiara, T.M., Efstratiadis, A., and Robertson, E.J. (1990). A growth-deficiency phenotype in heterozygous mice carrying an insulin-like growth factor II gene disrupted by targeting. *Nature*, **345**, 78–80.

DeChiara, T.M., Robertson, E.J., and Efstratiadis, A. (1991). Parental imprinting of the mouse insulin-like growth factor II gene. *Cell*, **64**, 849–59.

Dull, T.J., Gray, A., Hayflick, J.S., and Ullrich, A. (1984). Insulin-like growth factor II precursor gene organization in relation to insulin gene family. *Nature*, **310**, 777–81.

Ewton, D.Z., and Florini, J.R. (1981). Effects of the somatomedins and insulin on myoblast dfferentiation *in vitro*. *Develop. Biol.*, **86**, 31–6.

Gray, A., Tam, A.W., Dull, T.J., Hayflick, J., Pintar, J., Cavenee, W.K., Koufos, A., and Ullrich, A. (1987). Tissue-specific and developmentally regulated transcription of the insulin-like growth factor 2 gene. *DNA*, **6**, 283–95.

Jansen, M., Van Schaik, F.M., Ricker, A.T., Bullock, B., Woods, D.E., Gabbay, K.H., Nussbaum, A.L., Sussenbach, J.S., and Van Der Brande, J.L. (1983). Sequence of cDNA encoding human insulin-like growth factor I precursor. *Nature*, **306**, 609–11.

Kavsan, V.M., Koval, A.P., Grebenjuk, V.A., Chan, S.J., Steiner, D.F., Roberts, C.T. Jr., and LeRoith, D. (1993). *DNA and Cell Biology*, (In press.)

Liu, J.-P., Baker, J., Perkins, A.S., Robertson, E.J., and Efstratiadis, A. (1993). Mice carrying null mutations of the genes encoding insulin-like growth factor I (*Igf-1*) and type 1 IGF receptor (Igf1r). *Cell*, **75**, 59–72.

Lowe, W.L. Jr., Roberts, C.T. Jr., Lasky, S.R., and LeRoith, D. (1987). Differential expression of alternative 5' untranslated regions in mRNAs encoding rat insulin-like growth factor I. *Proc. Natl. Acad. Sci. (USA)*, **84**, 8946–50.

Lowe, W.L., Lasky, S.R, LeRoith, D., and Roberts, C.T. Jr. (1988). Distribution and regulation of rat insulin-like growth factor I messenger ribonucleic acids encoding alternative carboxyterminal E-peptides: evidence for differential processing and regulation in liver. *Mol. Endocrinol.*, **2**, 528–35.

Pardee, A.B. (1989). G1 events and regulation of cell proliferation. *Science*, **246**, 603–8.

Roberts, C.T. Jr., Lasky, S.R., Lowe, W.L. Jr., Seaman, W.T., and LeRoith, D. (1987). Molecular cloning of rat insulin-like growth factor I complementary deoxyribonucleic acids: differential messenger ribonucleic acid processing and regulation by growth hormone in extrahepatic tissues. *Mol. Endocrinol.*, **1**, 243–8.

Sakano, K., Enjoh, T., Numata, F., Fujiwara, H., Marumoto, Y., Higashihashi, N., Sato, Y., Perdue, J.F., and Fujita-Yamaguchi, Y. (1991). The design, expression, and characterization of human insulin-like growth factor II (IGF-II) mutants specific for either the IGF-II/cation-independent mannose 6-phosphate receptor or IGF-I receptor. *J. Biol. Chem.*, **266**, 20626–35.

Sara, V.R, and Hall, K. (1990). Insulin-like growth factors and their binding proteins. *Physiol. Rev.*, **70**, 591–614.

Tricoli, J.V., Rall, L.B., Scott, J., Bell, G.I., and Shows, T.B. (1984). Localization of insulin-like growth factor genes to human chromosomes 11 and 12. *Nature*, **310**, 784–6.

Derek LeRoith and Charles T. Roberts Jr.:
Diabetes Branch, NIDDK, NIH,
Bethesda, MD 20892, USA

Receptors and Binding Proteins for Insulin-like Growth Factors

The IGF-I receptor is a heterotetramic transmembrane glycoprotein containing a tyrosine kinase activity in its cytoplasmic domain. It initiates essentially all of the biological actions of the IGFs. The IGF-II/mannose-6-phosphate (IGF-II/M-6-P) receptor is involved in transporting hydrolases from the Golgi to lysosomal compartments and internalizing IGFII and hydrolases from the extracellular environment. Six IGF-binding proteins (IGFBPs 1–6) are present in the circulation or local tissue environment. Their functions include neutralization of circulating IGFs, transfer of circulating IGFs out of the vascular component, and modulation of the interactions of the IGFs with their receptors.

■ Receptor and binding protein structure

The IGF-I receptor is initially synthesized as a precursor composed of a signal peptide and contiguous α and β subunits. Processing includes cleavage of the α and β subunits at an Arg-Lys-Arg-Arg tetrapeptide sequence, and the formation of disulphide bonds to generate a mature heterotetramic ($\alpha_2\beta_2$) receptor (Ullrich *et al.* 1986). The two α subunits of the mature receptor lie entirely extracellularly, are N-glycosylated, and are involved in ligand binding. The two β subunits are inserted into the membrane by a hydrophobic transmembrane domain, and the cytoplasmic portions of the β subunit contain tyrosine kinase domains (Fig. 1). The IGF-II/M-6-P receptor is largely extracellular with the extracellular portion containing 15 contiguous repeats, each with a similar pattern of eight cysteine residues (Morgan *et al.* 1987). The cytoplasmic tail is short (163 amino acids) and does not contain a tyrosine kinase domain.

The IGFBPs are a family of closely related proteins (Table 1) (Clemmons 1992; Rechler 1993; Shimasaki *et al.* 1991; Baxter *et al.* 1989). There are 18 conserved cysteine residues, 12 in the N-terminal region and six in the C-terminal region, and the majority occur in disulphide bonds. IGFBP-1 and -2 each contain an Arg-Gly-Asp (RGD) sequence at their C-terminus which enables them to bind integrin receptors. The middle region of the IGFBP molecule is less conserved.

■ Structure and expression of receptor and binding protein genes

The mammalian IGF-I receptor is the product of a single-copy gene, consisting of 22 exons spanning 100 kb of chromosomal DNA, and, in the case of humans, is located at the distal end of the long arm of chromosome 15 (GenBank accession No. X04484) (Abbott *et al.* 1992). The promoter contains a single transcription site contained within an 'initiator' sequence and precedes a very long 5'-UTR of about 1 kb. Both the 5' flanking region and the 5'-untranslated region (UTR) are extremely GC-rich and there are no TATA or CCAAT boxes. In the proximal approximately 450 bp of 5' flanking region and the 5'-UTR there are numerous (>10) consensus binding sites for transcription factors SP1 and Egr-1/Wt-1. In transient expression assays, SP1 enhances IGF-I receptor promoter activity whereas Wt-1 inhibits. The IGF-I promoter also contains potential binding sites for AP-2 and ETF (Werner *et al.* 1992, 1993).

The IGF-I receptor is expressed in virtually every tissue and cell type. Fetal expression of the receptor is widespread and very high, supporting its essential role in organogenesis (Bondy *et al.* 1990). The continued widespread expression of the receptor postnatally is important for the specialized functions of IGFs in various tissues. The important role of the IGF-I receptor in growth and development has been

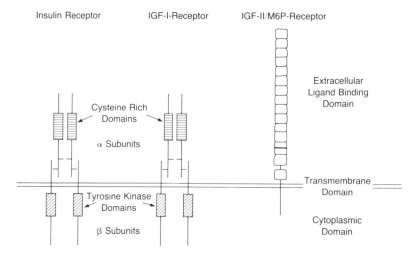

Figure 1. Insulin-growth factor receptors.

demonstrated by the impaired prenatal growth and post-natal death due to pulmonary failure seen in mice in which the IGF-I receptor gene has been inactivated by gene targeting (Liu *et al.* 1993; Baker *et al.* 1993).

■ IGF-II/M-6-P receptor

Cloning and sequencing of the IGF-II receptor revealed its identity to the cation-independent mannose-6-phosphate (M-6-P) receptor (Morgan *et al.* 1987). The gene is highly (80 per cent) conserved between human, bovine and rat species (GenBank accession No Y00285). This 270 kDa receptor is also widely expressed and particularly high levels of mRNA and protein are found in all fetal tissues, but these decrease significantly during postnatal development. Thus, the IGF-II/M-6-P receptor is also important in fetal development. The mouse *Tme* locus includes the IGF-II/M-6-P receptor and is genetically imprinted. Mice with this mutation exhibit defects in development that may be due to loss of expression of this receptor (Barlow *et al.* 1991).

■ IGFBPs

The IGFBP-1 gene is located on chromosome 7, contains four exons, TATA and CCAAT sequences in the promoter

region and a consensus binding site for hepatic nuclear factor-1 (HNF-1) (Brinkman *et al.* 1988). HNF-1 enhances basal promoter activity. cAMP positively regulates the gene, whereas insulin inhibits its expression. Expression of IGFBP-1 is highest in liver, especially during fetal and neo-natal development.

The IGFBP-2 gene is located on chromosome 2, contains four exons, and, in contrast to IGFBP-1, intron 1 is exception-ally large (>25 kb) (Binkert *et al.* 1992). The promoter does not contain a TATA box but is highly GC-rich, and SP1 enhances promoter activity. The IGFBP-2 gene is also ex-pressed in fetal tissues, especially liver.

The IGFBP-3 gene is located on chromosome 7, about 20 kb from the IGFBP-1 gene (Cubbage *et al.* 1989). Its promo-ter contains a TATA element and AP-2 and SP1 sites. Ex-pression of IGFBP-3 in liver is especially high in adults.

The IGFBP-4, -5, and -6 genes have not been characterized to the same extent. However, these genes are widely expressed.

■ Biological functions

The IGF-I receptor, like other tyrosine kinase receptors, autophosphorylates on specific tyrosine residues upon li-gand binding to the extracellular domain. This enhances its

Table 1. Insulin-like growth binding proteins

IGFBP	Human chromosome	rnRNA size (kb)	Molecular mass (kDa)	Special features
IGFBP-1	7p12–p13	1.6	26	RGD
IGFBP-2	2q33–q34	1.4	30	RGD
IGFBP-3	7	2.5	28	N-glycosylated
IGFBP-4	17	2.6	25	N-glycosylated
IGFBP-5	5	6.0	28	
IGFBP-6	12	1.3	21	O-glycosylated

inherent tyrosine kinase activity which then activates a cascade of events culminating in biological function. An immediate substrate, insulin receptor substrate-1 (IRS-1), acts as a 'docking protein' (Rothenberg *et al.* 1991; Sun *et al.* 1991). The phosphorylation of IRS-1 on tyrosine residues in a Tyr-Met-X-Met (YMXM) motif increases association with proteins containing src-homology-2 (SH2) domains and activates the activity of these enzymes. IRS-1 associates with the p85 domain of phosphatidylinositol-3'-kinase (PI-3' kinase), which then binds and activates the catalytic 110 kDa domain, resulting in increased PI-3' kinase activity. Other responses to IGF-I receptor activation include c-myc, c-jun, c-*fos*, and ornithine decarboxylase gene expression, increased thymidine incorporation, and MAP-2 kinase activation (Kato *et al.* 1993).

IGF-II/M-6-P receptors target recently synthesized lysosomal enzymes from the *trans*-Golgi network to lysosomes. Those enzymes that may have escaped the cell, and IGF-II, are internalized by receptors found on the cell surface. Despite its short cytoplasmic tail, the IGF-II/M-6-P receptor may be capable of binding and activating a G protein (G_i2) which could therefore mediate certain biological responses of IGF-II (Nishimoto *et al.* 1989).

■ IGFBPs

Circulating IGFBP-3 forms a ternary complex with an acid-labile subunit and the IGFs. This complex neutralizes the IGFs, prevents their degradation, and effectively transports them to distant sites. Once released from this complex, the IGFs are complexed with circulating IGFBP-1 and -2 which effectively transfer the IGFs out of the vascular compartment to the responsive tissues. Local tissue production of IGFBPs is involved in modulating the interaction of IGFs with cell-surface receptors, particularly the IGF-I receptor (Elgin *et al.* 1987). IGFBPs, while attached to the surrounding matrix bind the IGFs and may prevent IGFs interacting with their receptors; alternatively, slow release of IGFs to their receptors may enhance their biological actions, by preventing 'downregulation' of the receptors seen after massive doses of IGFs. Release of IGFs from these binding proteins may be facilitated by phosphorylation or proteolytic cleavage of the IGFBPs.

■ Pathology

IGF-I receptors are commonly over-expressed in different tumours, suggesting an important role in proliferation of these tumours. Synthesis and secretion of the IGFBPs may play an important role in regulating the interaction of IGFs with these receptors and represent a potential new mode of therapy.

■ References

Abbott, A.M., Bueno, R., Pedrini, M.T., Murray, J.M., and Smith, R.J. (1992). Insulin-like growth factor I receptor gene structure. *J. Biol. Chem.*, **267**, 10759–63.

Baker, J., Liu, J.-P., Robertson, E.J., and Efstratiadis, A. (1993). Role of insulin-like growth factors in embryonic and postnatal growth. *Cell*, **75**, 73–82.

Barlow, D., Stoger, R., Herrmann, B., Saito, K., and Schweifer, N. (1991). The mouse insulin-like growth factor type-2 receptor is imprinted and closely linked to the Tme locus. *Nature*, **349**, 84–7.

Baxter, R.C., Martin, J.L., and Beniac, V.A. (1989). High molecular weight insulin-like growth factor binding protein complex. Purification and properties of the acid-labile subunit from human serum. *J. Biol. Chem.*, **264**, 11843–8.

Binkert, C., Margot, J.B., Landwehr, H., Heinrich, G., and Schwander, J. (1992). Structure of the human insulin-like growth factor binding protein-2 gene. *Mol. Endocrinol.*, **6**, 826–36.

Bondy, C.A., Werner, H., Roberts, C.T. Jr., and LeRoith, D. (1990). Cellular pattern of insulin-like growth factor-I (IGF-I) and type I IGF receptor gene expression in early organogenesis: comparison with IGF-II gene expression. *Mol. Endocrinol.*, **4**, 1386–98.

Brinkman, A., Groffen, C.A., Kortleve, D.J., and Drop, S.L.S. (1988). Organization of the gene encoding the insulin-like growth factor binding protein IBP-1. *Biochem. Biophys. Res. Commun.*, **157**, 898–907.

Clemmons, D.R. (1992). IGF binding proteins: regulation of cellular actions. *Growth Regulation*, **2**, 80–7.

Cubbage, M.L., Suwanichkul, A., and Powell, D.R. (1989). Structure of the human chromosomal gene for the 25 kilodalton insulin-like growth factor binding protein. *Mol. Endocrinol.*, **3**, 846–51.

Elgin, G.R., Busby, W.H., and Clemmons, D.R. (1987). An insulin-like growth factor (IGF) binding protein enhances the biologic response to IGF-I. *Proc. Natl. Acad. Sci. (USA)*, **84**, 3254–8.

Kato, H., Faria, T.N., Stannard, B., Roberts, C.T. Jr., and LeRoith, D. (1993). Role of tyrosine kinase activity in signal transduction by the insulin-like growth factor-I (IGF-I) receptor. Characterization of kinase-deficient IGF-I receptors and the action of an IGF-I-mimetic antibody (alpha IR-3). *J. Biol. Chem.*, **268**, 2655–61.

Liu, J.-P., Baker, J., Perkins, A.S., Robertson, E.J., and Efstratiadis, A. (1993). Mice carrying null mutations of the genes encoding insulin-like growth factor I (*Igf-1*) and type 1 IGF receptor (Igf1r). *Cell*, **75**, 59–72.

Morgan, D.O., Edman, J.C., Standring, D.N., Fried, V.A., Smith, M.C., Roth, R.A., and Rutter, W.J. (1987). Insulin-like growth factor II receptor as a multifunctional binding protein. *Nature*, **329**, 301–7.

Nishimoto, L., Murayama, Y., Katada, T., Ui, M., and Ogata, E. (1989). Possible direct linkage of insulin-like growth factor-II receptor with guanine nucleotide-binding proteins. *J. Biol. Chem.*, **264**, 14029–38.

Rechler, M.M. (1993). Insulin-like growth factor binding proteins. *Vitamins and Hormones*, **47**, 1–114.

Rothenberg, P.L., Lane, W.S., Kasasik, A., Backer, J., White, M., and Kahn, C.R. (1991). Purification and partial sequence analysis of pp185, the major cellular substrate of the insulin receptor tyrosine kinase. *J. Biol. Chem.*, **266**, 8302–11.

Shimasaki, S., Shimonaka, M., Zhang, H-P., and Ling, N. (1991). Identification of five different insulin-like growth factor binding proteins (IGFBPs) from adult rat serum and molecular cloning of a novel IGFBP-5 in rat and human. *J. Biol. Chem.*, **266**, 10646–53.

Sun, X-J., Rothenberg, P., Kahn, C.R., Backer, J.M., Araki, E., Cahill, D., Goldstein, B.J., and White, M.F. (1991). Structure of the insulin receptor substrate IRS-1 defines a unique signal transduction protein. *Nature*, **352**, 73–7.

Ullrich, A., Gray, A., Tam, A.W., Yang-Feng, T., Tsubokawa, M., Collins, C., Henzel, W., LeBon, T., Kathuria, S., Chen, E., Jacobs, S.,

Franke, U., Ramachandran, J., and Fujita-Yamaguchi, Y. (1986). Insulin-like growth factor I receptor primary structure: comparison with insulin receptor suggests structural determinants that define functional specificity. *EMBO J.*, **5**, 2503–12.

Werner, H., Bach, M.A., Stannard, B., Roberts, C.T. Jr., and LeRoith, D. (1992). Structural and functional analysis of the insulin-like growth factor I receptor gene promoter. *Mol. Endocrinol.*, **6**, 1545–58.

Werner, H., Re, G.G., Drummond, L.A., Sukhatame, V.P., Rauscher, F.J., Sens, D.A., Garvin, A.J., LeRoith, D., and Roberts, C.T. Jr. (1993). Increased expression of the insulin-like growth factor I receptor gene, IGF1R, in Wilms tumour is correlated with modulation of IGF1R promoter activity by the WT1 Wilms tumour gene product. *Proc. Natl. Acad. Sci. (USA)*, **90**, 5828–32.

Derek LeRoith and Charles T. Roberts, Jr.:
Diabetes Branch, NIDDK, NIH,
Bethesda, MD 20892, USA

Fibroblast Growth Factors (FGFs)

Fibroblast growth factors (FGFs) are a family of 15–32 kDa heparin–binding, single-chain polypeptides having a role in a variety of biological processes such as cell growth, differentiation, angiogenesis, tissue repair, and transformation (reviewed by Burgess and Maciag 1989; Basilico and Moscatelli 1992; Partanen et al. 1993). At present the FGF family consists of nine members. These proteins share 30–70% per cent amino acid sequence identity with each other.

■ Nomenclature

Acidic FGF (aFGF, FGF-1), basic FGF (bFGF, FGF-2), Int-2 (FGF-3), K-FGF/hst-1 (FGF-4), FGF-5, FGF-6, keratinocyte growth factor (KGF, FGF-7), androgen-induced growth factor (AIGF, FGF-8), glia-activating factor (GAF, FGF-9).

■ Cloning of FGF cDNAs and purification of different FGFs

The first characterized members of the FGF family were aFGF and bFGF, which were purified as mitogens for fibroblasts from bovine pituitary. The Int-2 oncogene was identified as a gene activated by mouse mammary tumour virus insertion. K-FGF and FGF5 genes were identified by their ability to transform NIH 3T3 cells and FGF-6 was discovered because of its homology with the K-FGF/hst-1 gene. KGF was purified and cloned as a mitogen for cultured keratinocytes (see Burgess and Maciag 1989; Basilico and Moscatelli 1992), AIGF as an androgen-induced growth factor from the conditioned medium of a mouse mammary carcinoma cell line (Tanaka et al. 1992) and GAF from the culture supernatant of a human glioma cell line (Miyamoto et al. 1993). A great advantage in the purification of FGFs has been their affinity for heparin.

■ Protein structure

In contrast to other members of the family, aFGF, bFGF, and GAF lack a signal sequence and it is not yet understood how they are secreted from cells. bFGF can be expressed in four different forms, an 18 kDa form generated by initiation at an AUG codon (M1 in Fig. 1), and 22, 22.5, and 24 kDa forms initiated from upstream CUG codons. The Int-2 protein is also initiated from an upstream CUG codon, as well as an AUG codon. The N-terminally extended forms of bFGF and Int-2 are localized in the nucleus, suggesting that these higher molecular weight forms contain a nuclear translocation sequence. With the exception of AIGF, there are two conserved cysteine residues in all members of the FGF family and with the exception of bFGF, one to two N-linked glycosylation sites. However, the cysteine residues are not necessary for the mitogenic activity of bFGF.

The three-dimensional structure of bFGF has been described as a trigonal pyramid folded in a similar manner to interleukin-1α and -1β. The data obtained from bFGF structure suggests that the binding site for heparin is a cluster of basic residues including Lys-128, Arg-129, Lys-134 and Lys-138 (Eriksson et al. 1991; Zhang et al. 1991). Six-residue peptides capable of inhibiting the binding of bFGF to its receptor, as well as bFGF-induced proliferation of vascular endothelial cells, had the consensus sequence corresponding to amino acids 22–27 (Pro-Pro-Gly-His-Phe-Lys) and 129–134 (Arg-Thr-Gly-Gln-Tyr-Lys). Thus, these residues may constitute at least part of the bFGF receptor binding determinants (Yayon et al. 1993).

```
FGF 6   MSRGAGRLQG TLWALVFLGI LV........ .GMVVPSPAG TR.ANNTLLD  40
K-FGF   MS.GPGTAAV ALLPAVLLAL LA........ .PWAGRGGAA APTAPNGTLE  40
FGF5    .......MSL SFLLLLFFSH LILSAWAHGE KRLAPKGQPG PAATDRNPRG  43
aFGF    .......... .......... .......... .......... ..........
bFGF    .......... .......... .......... .......... ..........
GAF     .......... .......... .......... MAPLGEVGNY FGVQDAVPFG  20
KGF     .......... .......... ....MHKWI LTWILPTLLY RSCFHIICLV  25
INT-2   .......... .......... .......... .......... .......MG   2
AIGF    .......... .......... .......... .......... ...MGSPRS   6
                                        <       >
FGF 6   S...RGWGTL LSRSRAGLAG EI...AGVN WESG.YLVGI .KRQRRLY..  79
K-FGF   AELERRWESL VALSLARLPV AAQPKEAAVQ SGAGDYLLGI .KRLRRLY..  87
FGF5    SSSRQSSSSA MSSSSASSSP AASLGSQGSG LEQSSFQWSL GARTGSLY..  91
aFGF    ....MAEGEI TTFTALTEKF N...LPPG.. .....NYK.. .KP.KLLY..  30
bFGF    ....MAAGSI TTLPALPEDG GSGAFPPG.. .....HFK.. .DP.KRLY..  33
GAF     NVPVLPVDSP VLLSDHLGQS EAGGLPRGPA VTDLDHLKGI LRR.RQLY..  67
KGF     GTISLACNDM TPEQMATNVN CSSPERHTRS YDYMEG..GD IRV.RRLF..  70
INT-2   LIWLLLLSLL EPGWPAAGPG ARLRRDAGGR GGVYEHLGGA PRR.RKLY..  49
AIGF    ALSCLLLHLL VLCLQAQVTV QSSPNFTQHV REQSLVTDQL SRRLIRTYQL  56
        x
FGF 6   CNVGIGFHLQ VLPDGRISGT HEE.NPYSLL EISTVERGV. VSLFGVRSAL 127
K-FGF   CNVGIGFHLQ ALPDGRIGGA HAD.TRDSLL ELSPVERGV. VSIFGVASRF 135
FGF5    CRVGIGFHLQ IYPDGKVNGS HEA.NMLSVL EIFAVSQGI. VGIRGVFSNK 139
aFGF    CSNG.GHFLR ILPDGTVDGT RDRSDQHIQL QLSAESVGE. VYIKSTETGQ  78
bFGF    CKNG.GFFLR IHPDGRVDGV REKSDPHIKL QLQAEERGV. VSIKGVCANR  81
GAF     CR.T.GFHLE IFPNGTIQGT RKDHSRFGIL EFISIAVGL. VSIRGVDSGL 114
KGF     CR.T.QWYLR IDKRGKVKGT QEMKNNYNIM EIRTVAVGI. VAIKGVESEF 117
INT-2   C.AT.KYHLQ LHPSGRVNGS LENSA.YSIL EITAVEVGI. VAIRGLFSGR  95
AIGF    YSRTSGKHVQ VLANKRINAM AEDGDPFAKL IVETDTFGSR VRVRGAETGL 106
                                        x
FGF 6   FVAMNSKGRL YA.TPSFQEE CKFRETLLPN NYNAYESDLY QG....... 168
K-FGF   FVAMSSKGKL YG.SPFFTDE CTFKEILLPN NYNAYESYKY PG....... 176
FGF5    FLAMSKKGKL HA.SAKFTDD CKFRERFQEN SYNTYASAIH RT....... 180
aFGF    YLAMDTDGLL YG.SQTPNEE CLFLERLEEN HYNTYISKKH .......... 117
bFGF    YLAMKEDGRL LA.SKCVTDE CFFFERLESN NYNTYRSRKY .......... 120
GAF     YLGMNEKGEL YG.SEKLTQE CVFREQFEEN WYNTYSSNLY .......... 153
KGF     YLAMNKEGKL YA.KKECNED CNFKELILEN HYNTYASAKW .......... 156
INT-2   YLAMNKRGRL YA.SEHYSAE CEFVERIHEL GYNTYASRLY RTVSSTPGAR 144
AIGF    YICMNKKGKL IAKSNGKGKD CVFTEIVLEN NYTALQNAKY EG....... 148
                        <      >
                        oo     o        o
FGF 6   ..T......Y IALSKYGRVK RG..SKVSPI MTVTHFLPRI .......... 198
K-FGF   ..M......F IALSKNGKTK KG..NRVSPT MKVTHFLPRL .......... 206
FGF5    ..EKTGREWY VALNKRGKAK RGCSPRVKPQ HISTHFLPRF KQSEQPELSF 228
aFGF    ....AEKNWF VGLKKNGSCK RG..PRTHYG QKAILFLPLP VSSD...... 155
bFGF    ....T...SWY VALKRTGQYK LG..SKTGPG QKAILFLPMS AKS....... 155
GAF     KHVDTGRRYY VALNKDGTPR EG..TRTKRH QKFTHFLPRP VDPDKVPELY 201
KGF     THNGGEM..F VALNQKGIPV RG..KKTKKE QKTAHFLPMA IT........ 194
INT-2   RQPSAERLWY VSVNGKGRPR RG..FKTRRT QKSSLFLPRV LDHRDHEMVR 192
AIGF    .......WY MAFTRKGRPR KG..SKTRQH QREVHFMKRL PRGHHTTEQS 188

FGF 6   .......... .......... .......... ..........
K-FGF   .......... .......... .......... ..........
FGF5    TVTVPEKKNP PSPIKSKIPL SAPRKNTNSV KYRLKFRFG. ....... 267
aFGF    .......... .......... .......... ..........
bFGF    .......... .......... .......... ..........
GAF     KDILSQS... .......... .......... .......... 208
KGF     .......... .......... .......... ..........
INT-2   QLQSGLPRPP GKGVQPRRRR QKQSPDNLEP SHVQASRLGS QLEASAH 239
AIGF    LRFEFLNYPP FTRSLRGSQR TWAPEPR... .......... 215
```

Figure 1. Comparison of amino acid sequences of known FGFs. Marked above the sequences are: Conserved cysteines (**x**), regions corresponding to peptides that inhibit bFGF binding to FGFR-1 (<>) and amino acids involved in bFGF binding to heparan sulphate (°). The numbering of amino acids is according to the starting methionine. Sequences are for human FGFs, with the exception of AIGF, which is murine. Accession numbers: aFGF: P05230 (Swiss Prot); bFGF: X04431 (GenBank); Int-2: P11487 (Swiss Prot); K-FGF: M17446 (GenBank); FGF-5: P12034 (Swiss Prot); FGF-6: X63454 (GenBank); KGF: M60828 (GenBank); AIGF: D12482 (GenBank); GAF: D14839 (GenBank).

Interaction of FGFs with heparan sulphate

FGFs transduce their signals by binding to cell surface tyrosine kinase receptors (FGFRs) (for reviews, see Partanen *et al.* 1993; Johnson and Williams 1993). In addition, FGFs bind heparan sulphate proteoglycans (HSPGs, such as syndecan), present on the cell surface and in the extracellular matrix. Reported affinities for these two classes of receptors vary between $2–15 \times 10^{-11}$M and $2–600 \times 10^{-9}$M, respectively.

Interaction of bFGF, aFGF and K-FGF with heparan sulphate is necessary for their interaction with FGFRs, perhaps because of a conformational change induced by FGF binding to heparin (Yayon *et al.* 1991). Binding to heparin or heparan sulphate also protects bFGF from denaturation and proteolytic degradation, and many of the FGFs in tissues are apparently present as HSPG matrix-bound forms which can promote cell growth (Salmivirta *et al.* 1992). Distinct classes of HSPGs may regulate, for example, neural responses to aFGF and bFGF during development (Nurcombe *et al.* 1993).

Biological functions

FGFs are potent mitogens for a wide variety of cells of mesenchymal and neuroectodermal origin. Besides, at least aFGF and bFGF induce angiogenesis *in vivo*, possess chemotactic and mitogenic effects on endothelial cells *in vitro*, and stimulate cultured endothelial cells to invade a basement membrane matrix, a process dependent on the proteolytic enzymes collagenase and plasminogen activator. Transfected FGF constructs are capable of cellular transformation through autocrine or paracrine mechanisms and exogenous administration of bFGF has been found to promote angiogenesis and progression of tumours (reviewed by Basilico and Moscatelli 1992).

In addition, FGFs are involved in the differentiation of a variety of cells: aFGF and bFGF stimulate neurite outgrowth of PC12 rat phaeochromocytoma cells. At least aFGF and AIGF cause profound phenotypic changes in epithelial tumour cells in culture and aFGF induces neurite outgrowth from retinal ganglion cells. The expression of bFGF in quail embryos has been localized to neurones of the neural tube and crest, and at later stages to the spinal cord and dorsal root ganglia, suggesting a role in neural development. On the other hand, aFGF, bFGF, and FGF-5 are more abundant in adult brain than in other adult tissues and aFGF is highly expressed in motoneurones, primary sensory neurones, and retinal ganglion neurones, suggesting that FGFs are important in neural physiology in adults (Tanaka *et al.* 1992; Basilico and Moscatelli 1992; Elde *et al.* 1991) (Fig. 2).

K-FGF stimulates the proliferation of mouse embryo limb-bud mesenchyme and thus may have a function in limb growth (Niswander and Martin 1993). Furthermore, FGFs inhibit myoblast differentiation by causing phosphorylation and inactivation of myogenic basic helix–loop–helix (bHLH) transcription factors and repressing their transcription. bFGF also induces the expression of negatively acting partners of these factors (Li *et al.* 1992; Hardy *et al.* 1993). bFGF is involved in mesoderm induction in *Xenopus* embryos, while aFGF, K-FGF and Int-2 can also induce mesodermal structures. Expression of a dominant negative FGF receptor construct inhibits the formation of mesodermal tissues in *Xenopus* and causes specific defects in gastrulation and development (Amaya *et al.* 1991). Xenopus embryonic FGF (XeFGF) may be the actual secreted mesoderm inducing factor (Isaacs *et al.* 1992).

The tissue- and stage-specific expression of FGFs also supports the idea that FGFs have functions in development. High levels of K-FGF can be detected only in early embryos and Int-2 and FGF-5 are highly expressed in specific tissues during embryonic and fetal growth (reviewed by Burgess and Maciag 1989; Basilico and Moscatelli 1992). In addition, homozygous mice with a targeted disruption of *int*-2 have been shown to have defects in the development of the tail and inner ear. These defects correlate with abnormalities of mesoderm derived from the posterior primitive streak and failure of the otocyst to form an endolymphatic duct, respectively (Mansour *et al.* 1993).

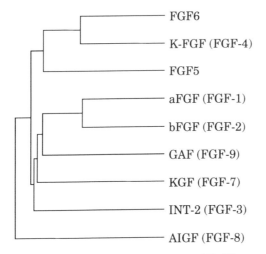

Figure 2. Evolutionary relationships of FGFs. The tree was constructed using the Pileup program of the software package of the Genetics Computer Group, Wisconsin University.

Pathology

FGFs may have a role in tumour vascularization. One of the first angiogenic factors isolated from tumours was bFGF. The ability of FGFs to stimulate the secretion of collagenase and plasminogen activator may be involved in tumour invasion and metastasis as well as angiogenesis. Neovascularization and tumourigenicity of fibrosarcomas in transgenic mice carrying the bovine papilloma virus genome are associated with enhanced bFGF secretion (Kandel *et al.* 1991).

A K-FGF recombinant retrovirus induces tumours with a high frequency and a short latency. Also, amplification of the *K-FGF/hst*-1 and *int*-2 genes has been detected in breast and squamous cell carcinomas and may correlate with poor prognosis. However, expression of neither of these genes has been detected in the tumours. KGF is expressed in the stromal cells of the human prostate, whereas prostatic adenocarcinoma cells express both KGF and its receptor, suggesting an autocrine mechanism for growth of these tumours.

■ References

Amaya, E., Musci, T.J., and Kirschner, M.W. (1991). Expression of a dominant negative mutant of the FGF receptor disrupts mesoderm formation in *Xenopus* embryos. *Cell*, **66**, 257–70.

Basilico, C., and Moscatelli, D. (1992). The FGF family of growth factors and oncogenes. *Adv. Cancer Res.*, **59**, 115–65.

Burgess, W.H., and Maciag, T. (1989). The heparin-binding (fibroblast) growth factor family of proteins. *Annu. Rev. Biochem.*, **58**, 575–606.

Elde, R., Cao, Y.H., Cintra, A., Brelje, T.C., Pelto-Huikko, M., Junttila, T., Fuxe, K., Pettersson, R.F., and Hökfelt, T. (1991). Prominent expression of acidic fibroblast growth factor in motor and sensory neurones. *Neuron*, **7**, 349–64.

Eriksson, A.E., Cousens, L.S., Weaver, L.H., and Matthews, B.W. (1991). Three-dimensional structure of human basic fibroblast growth factor. *Proc. Natl. Acad. Sci. (USA)*, **88**, 3441–5.

Hardy, S., Kong, Y., and Konieczny, S.F. (1993). Fibroblast growth factor inhibits MRF4 activity independently of the phosphorylation status of a conserved threonine residue within the DNA-binding domain. *Mol. Cell. Biol.*, **13**, 5943–56.

Isaacs, H.V., Tannahill, D., and Slack, J.M. (1992). Expression of a novel FGF in the Xenopus embryo. A new candidate inducing factor for mesoderm formation and anteroposterior specification. *Development*, **114**, 711–20.

Johnson, D.E., and Williams, L.T. (1993). Structural and functional diversity in the FGF receptor multigene family. *Adv. Cancer Res.*, **60**, 1–41.

Kandel, J., Bossy-Wetzel, E., Radvanyi, F., Klagsbrun, M., Folkman, J., and Hanahan, D. (1991). Neovascularization is associated with a switch to the export of bFGF in the multistep development of fibrosarcoma. *Cell*, **66**, 1095–104.

Li, L., Zhou, J., James, G., Heller-Harrison, R., Czech, M.P., and Olson, E.N. (1992). FGF inactivates myogenic helix–loop–helix proteins through phosphorylation of a conserved protein kinase C site in their DNA-binding domains. *Cell*, **71**, 1181–94.

Mansour, S.L., Goddard, J.M., and Capecchi, M.R. (1993). Mice homozygous for a targeted disruption of the proto-oncogene *int*-2 have developmental defects in the tail and inner ear. *Development*, **117**, 13–28.

Miyamoto, M., Naruo, K., Seko, C., Matsumoto, S., Kondo, T., and Kurokawa, T. (1993). Molecular cloning of a novel cytokine cDNA encoding the ninth member of the fibroblast growth factor family, which has a unique secretion property. *Mol. Cell. Biol.*, **13**, 4251–9.

Niswander, L., and Martin, G.R. (1993). FGF-4 and BMP-2 have opposite effects on limb growth. *Nature*, **361**, 68–71.

Nurcombe, V., Ford, M.D., Wildschut, J.A., and Bartlett, P.F. (1993). Developmental regulation of neural response to FGF-1 and FGF-2 by heparan sulfate proteoglycan. *Science*, **260**, 103–6.

Partanen, J., Vainikka, S., and Alitalo, K. (1993). Structural and functional specificity of FGF receptors. *Phil. Trans. R. Soc. Lond. -Series B: Biol. Sci.*, **340**, 297–303.

Salmivirta, M., Heino, J., and Jalkanen, M. (1992). Basic fibroblast growth factor–syndecan complex at cell surface or immobilized to matrix promotes cell growth. *J. Biol. Chem.*, **267**, 17606–10.

Tanaka, A., Miyamoto, K., Minamino, N., Takeda, M., Sato, B., Matsuo, H., and Matsumoto, K. (1992). Cloning and characterization of an androgen-induced growth factor essential for the androgen-dependent growth of mouse mammary carcinoma cells. *Proc. Natl. Acad. Sci. (USA)*, **89**, 8928–32.

Yayon, A., Klagsbrun, M., Esko, J.D., Leder, P., and Ornitz, D.M. (1991). Cell surface, heparin-like molecules are required for binding of basic fibroblast growth factor to its high affinity receptor. *Cell*, **64**, 841–8.

Yayon, A., Aviezer, D., Safran, M., Gross, J.L., Heldman, Y., Cabilly, S., Givol, D., and Katchalski-Katzir, E. (1993). Isolatioan of peptides that inhibit binding of basic fibroblast growth factor to its receptor from a random phage-epitope library. *Proc. Natl. Acd. Sci. (USA)*, **90**, 10643–7.

Zhang, J.D., Cousens, L.S., Barr, P.J., and Sprang, S.R. (1991). Three-dimensional structure of human basic fibroblast growth factor, a structural homolog of interleukin 1 beta [published erratum appears in *Proc. Natl. Acad. Sci. (USA)* 1991, **88** 5477]. *Proc. Natl. Acad. Sci. (USA)*, **88**, 3446–50.

Satu Vainikka, Tuija Mustonen, and Kari Alitalo:
Molecular/Cancer Biology Laboratory,
P.O. BOX 21 (Haartmaninkatu 3),
00014 University of Helsinki, Finland

The FGF receptors (FGFRs) form a multigene family of at least five members, all having a glycosylated extracellular ligand-binding domain consisting of either two or three immunoglobulin-like (Ig-like) domains, an acidic region of eight amino acids, a transmembrane region, and a cytoplasmic tyrosine kinase domain which is split by a short insertion of fourteen amino acids. Like other tyrosine kinase receptors, the FGFRs form dimers upon ligand binding which facilitates activation and intermolecular tyrosine phosphorylation of the cytoplasmic kinase domains. There is a high degree of redundancy among FGFRs as they can bind more than one ligand. FGFR-1, FGFR-2, FGFR-3 and FGFR-4 bind both acidic and basic FGF with high affinity. A splice variant of FGFR-2 also known as KGFR binds keratinocyte growth factor (KGF) and acidic FGF with equal high affinity. The FGFRs are expressed in a large variety of cells and ligand binding may induce either mitogenic or differentiation responses in a wide array of biological processes such as embryonal development, tissue repair, and survival and differentiation of neuronal and glial cells.

■ Nomenclature of the FGFR genes

The different FGFR genes and their spliced variants are described in the literature under various names (Table 1). The five genes are now known as FGFR-1 (Lee *et al.* 1989), FGFR-2 (Dionne *et al.* 1990), FGFR-3 (Keegan *et al.* 1991), FGFR-4 (Partanen *et al.* 1991) and FGFR-5 (Avivi *et al.* 1991), in the order in which they were first identified.

■ FGFR-1 protein

The nucleotide sequence of FGFR-1 (Dionne *et al.* 1990; Reid *et al.* 1990) cDNA predicts a protein of 822 amino acids and a molecular mass of 92 kDa in mouse and man. The protein contains an amino-terminal signal peptide, a single predicted transmembrane segment of 22 amino acids and an extracellular domain comprised of three immunoglobulin-like (Ig-like) domains. A unique region which is specific to FGFRs lies between the first and second Ig-like domains and consists of eight acidic amino acids and is known as the

'acidic box'. The intracellular domain contains a consensus tyrosine kinase sequence which is split by a short insertion of 14 amino acids. The region between the transmembrane and the kinase domain, known as the juxtamembrane domain, is unusually long in the FGFR-1 protein and consists of 79 amino acids (Fig. 1). A high degree of amino-acid identity (98 per cent) exists between mouse and human FGFR-1 proteins. The least conserved regions are the signal peptide (93 per cent), the first Ig-like domain (93 per cent), the transmembrane domain (90 per cent) and the kinase insert (93 per cent). Multiple forms of the FGF receptor were isolated for both FGFR-1 and FGFR-2 and have been demonstrated to result from alternative mRNA splicing (Fig. 2). cDNA clones lacking the first Ig-like domain were isolated from both mouse (Reid *et al.* 1990) and human (Johnson *et al.* 1990) cells. Binding studies demonstrated that the three Ig-like domain and the two Ig-like domain forms bind aFGF and bFGF with similar affinities of K_D 20–80 pM and K_D 50–150 pM, respectively (Dionne *et al.* 1990; Johnson *et al.* 1990). These findings suggest that Ig-like domain I is not necessary for high-affinity binding of acidic (aFGF) and basic (bFGF). Another splice variant of FGFR-1 is a secreted form which contains the extracellular Ig-like domain but

Table 1. Alternative nomenclature of the FGF receptor genes

FGFR-1	FGFR-2	FGFR-3	FGFR-4	FGFR-5
flg-1 (Dionne *et al.* 1990)	*bek* (Kornbluth *et al.* 1988)	Cek-2 (Pasquale 1990)	FGFR-4 (Partanen *et al.* 1991)	*flg-2* (Avivi *et al.* 1991)
bFGFR (Reid *et al.* 1990)	Cek-3 (Pasquale 1990)	FGFR-3 (Keegan *et al.* 1991)		FGFR-5 (Matthew *et al.* 1994)
Cek-1 (Pasquale 1990)	K-sam (Hattori *et al.* 1990)			
N-bGFGR (Reid *et al.* 1990)	KGFR (Miki *et al.* 1991)			
FGFR-1 (Lee *et al.* 1989)	FGFR-2 (Miki *et al.* 1991)			

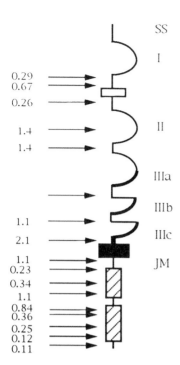

Figure 1. Schematic diagram of the human FGFR-1 gene. Arrows indicate the intronic positions next to the respective size (kb). Exon regions are indicated: the hydrophobic signal sequence (SS), three Ig-like domains (I, II, III), and three alternative exons for the 3′ half of Ig-like domain III (thick black line: a, b, and c), acidic region (open box), transmembrane domain (black box), juxtamembrane region (JM), and the split tyrosine kinase domain (hatched box).

lacks the transmembrane and kinase domains (Johnson et al. 1990). The function of the secreted receptor is not yet known. There are three alternative exons for the second half of the third Ig-like domain (III), designated IIIa, IIIb, and IIIc (Fig. 2). Exon IIIa was found only in the secreted form of FGFR-1 (Johnson et al. 1990) while the other two exons are present in the receptors containing the tyrosine kinase domain (Johnson et al. 1991). The presence of exon IIIb or IIIc confers distinct ligand binding specificities to both FGFR-1 (Werner et al. 1992) and FGFR-2 (Crumley et al. 1991; Miki et al. 1991). Kaposi sarcoma-FGF (K-FGF) can also bind to FGFR-1 and compete with bFGF in a competition assay (Mansukhani et al. 1990). FGFR-1, like all other FGFRs, is heavily glycosylated.

■ Characterization of FGFR-2, FGFR-3, FGFR-4, and FGFR-5

These additional FGFR genes are very similar in structure to FGFR-1 and are highly conserved at the amino acid level (Table 2). FGFR-2 also known as *bek* was demonstrated to

have 72 per cent amino acid identity with FGFR-1. As for FGFR-1, two splice variants of the second half of the third Ig-like domain also exist for FGFR-2, and they contain structures corresponding to either the IIIb or the IIIc exons of FGFR-1 (Crumley et al. 1991; Miki et al. 1991). The FGFR-2 cDNAs which contain the IIIb sequences are known as the keratinocyte growth factor receptors (KGFR) (Miki et al. 1991). They bind with high affinity both KGF (K_D 180–480 pM) as well as aFGF. The KGFR was isolated from 3T3 cells which normally express KGF, but not the KGFR, after they were transfected with cDNAs derived from KGFR-expressing cells. Cells from transformed foci were found to express high levels of KGFR. The KGFR isolated from these cells corresponds to FGFR-2 without the first Ig-like domain and the acidic box (Miki et al. 1991, 1992). The cDNAs containing the IIIc exon are known as *bek;* they do not bind KGF, but bind both aFGF (K_D 40–100pM) and bFGF (K_D 80–150pM) with high affinity (Dionne et al. 1990).

FGFR-3 cDNA was isolated from a human cDNA library derived from K562 leukaemia cells (Keegan et al. 1991) and bears significant homology with FGFR-1, -2, -4, and -5 (very high homology to FGFR-5: see Table 2). FGFR-3 is also stimulated by both aFGF and bFGF. Only one form of FGFR-3 has been identified. This cDNA contains the three Ig-like domains and the IIIc exon of the third Ig-like domain.

FGFR-4 cDNAs were isolated from human (Partanen et al. 1991; Ron et al. 1993) and mouse (Stark et al. 1991) by PCR using tyrosine kinase specific oligonucleotides. Comparison of amino acid sequences shows that FGFR-4 is the most divergent of all known FGFRs, exhibiting the lowest homology to other FGFRs (Table 2). The human FGFR-4 was reported to bind with high affinity to both aFGF (K_D of 10–15pM) and bFGF (K_D 120pM) (Stark et al. 1991).

FGFR-5 was first isolated from a human keratinocyte cDNA library and was initially named *flg-2* as it exhibited high homology to *flg-1* (Avivi et al. 1991) (Table 1). Recently a new mouse FGFR was identified with 94 per cent identity to human FGFR-3 and 99 per cent identity to *flg-2* (Table 2). This new mouse receptor was isolated from a brain cDNA library and was designated FGFR-5 (Matthew et al. 1994). All human and murine FGFR DNA database accession numbers are shown in Table 3.

■ Expression patterns of the FGFRs

FGFR-1 and FGFR-2 genes exhibit a broad but distinct pattern of expression during embryonal development and in adults, while the expression of FGFR-3 and FGFR-4 is more restricted. In the developing embryo, FGFR-1 is mainly expressed in the brain and mesenchymal tissue (Wanaka et al. 1991), while FGFR-2 is expressed in the brain and epithelium (Peters et al. 1992a). FGFR-3 is expressed mostly in brain, spinal cord and cartilage rudiments of the developing bone (Peters et al. 1993). FGFR-4 transcripts are found in developing endoderm and the myotomal components of the somite and the myotomally derived skeletal muscle and in most human embryonal tissues except the brain (Partanen et al. 1991; Stark et al. 1991). In adults, FGFR-1 is expressed in brain, bone, kidney, skin, lung, heart and muscle but not

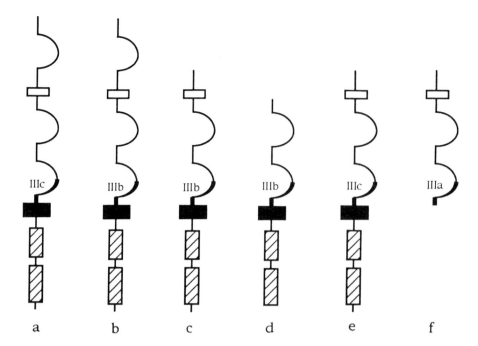

Figure 2. FGFR protein structure forms. Schematic diagram of FGFR structures indicating (a) full length FGFR-1, 2, 3, 4, and 5 forms containing three Ig-like domains and the IIIc 3′ exon of Ig-like domain III; (b) K-sam′ (Kornbluth *et al.* 1988), an FGFR-2 variant containing the IIIb exon of the third Ig-like domain; (c) K-sam, another FGFR-2 form similar to K-sam′ with only two Ig-like domains and a truncated C-terminus; (d) KGFR, another FGFR-2 form containing the IIIb exon of Ig-like domain three but missing the first Ig-like domain and the acidic region; (e) Bek′, an FGFR-2 variant with two Ig-like domains and the IIIc exon of the third Ig-like domain; (f) a secreted FGFR-1 variant which consists of the acidic box, Ig-like domains two and three, and contains exon IIIa of the third Ig-like domain.

in liver (Reid *et al.* 1990; Wanaka *et al.* 1991). FGFR-2 transcripts exhibit a similar pattern to FGFR-1 as they are expressed in brain, kidney, skin, and lung but are also detected in liver (Peters *et al.* 1992a). No transcripts of FGFR-2 were observed in heart, spleen or muscle, which contrasts with the FGFR-1 pattern. Expression of FGFR-1 in the brain was observed mainly in neurones while the presence of FGFR-2 is more consistent with glial expression (Wanaka *et al.* 1991; Peters *et al.* 1992a, 1993; Heuer *et al.* 1990). FGFR-3 transcripts were detected in brain, kidney, skin, and lung of adults (Johnson and Williams 1993). FGFR-4 expression was detected in lung, liver, and kidney but not in other tissues of the adult mouse (Stark *et al.* 1991). Limited studies on the expression of FGFR-5 indi-

cated that it is expressed in the brain, skin, lung and testes but not in seminal vesicles, heart, spleen, and kidney of adult mice (Avivi *et al.* 1991).

■ Signalling mechanisms

Binding of the FGFs to their receptors leads to either proliferation, differentiation, inhibition of differentiation, or maintenance of the differentiated phenotype depending on cell type (Burgess and Maciag 1989). For example, in fibroblasts both aFGF and bFGF are mitogenic, but they can also stimulate differentiation in hippocampal neurones as indicated by neurite outgrowth. FGFs and their receptors are also thought to be important in mesoderm formation, in embryogenesis, and in angiogenesis (Burgess and Maciag 1989). The signalling mechanisms that give rise to such a variety of responses are not yet understood. Heparan sulphate proteoglycan molecules bind FGFs with low affinity and play an important role in potentiating binding of FGFs to the high affinity receptors (Kiefer *et al.* 1990; Yayon *et al.* 1991). The presence of these molecules on the cell surface or in the media is essential for FGF binding. Upon the binding of ligand to its receptor, dimerization of the receptors takes place. Both homodimers and heterodimers

Table 2. Amino acid identities of FGF receptors

| | Sequence identity (per cent) | | | | |
	FGFR-1	FGFR-2	FGFR-3	FGFR-4	FGFR-5
FGFR-1	–	72	62	56	64
FGFR-2	72	–	66	57	68
FGFR-3	62	66	–	62	93
FGFR-4	56	57	62	–	61
FGFR-5	64	68	93	61	–

Table 3. Sequence database accession numbers for FGF receptors

	Murine	Human
FGFR-1	M 28998 GenBank	X 52883 EMBL
FGFR-2	M 97193 GenBank	X 52832 EMBL
FGFR-3	–	M 58051 GenBank
FGFR-4	X 59927 EMBL	X 57205 EMBL
FGFR-5	–	X 58255 EMBL

can be formed between FGFR-1, FGFR-2, and FGFR-3. The binding also results in tyrosine kinase activation and in phosphorylation of dimerized receptors by intramolecular transphosphorylation mechanisms (Bellot et al. 1991; Ueno et al. 1992). Activation of the kinase domain results in binding of phospholipase Cγ (PLC-γ) to the FGFR and its phosphorylation on a tyrosine (Tyr) residue (Mohammadi et al. 1991). This binding is mediated via the SH2 domain of PLC-γ and the phosphorylated Tyr (766) on the C-terminus of FGFR. This Tyr is conserved in all the FGF receptors and mutation of this Tyr to phenylalanine abolishes the capacity of the receptor to associate with or phosphorylate PLC-γ. This mutant, however, retained its kinase activity and can autophosphorylate and increase tyrosine phosphorylation of other cellular proteins. Moreover, cells expressing this mutated FGFR can proliferate in response to FGF, implying that PLC-γ may not be essential in the pathways involved in FGF-induced mitogenesis (Mohammadi et al. 1993; Peters et al. 1992b). The role of other molecules in FGFR signalling is not yet known.

■ Tumorigenicity of FGFRs

Members of the FGFR family have been demonstrated to be involved in tumourigenesis. Transfection of FGFR-1 into Rat-2 cells resulted in the formation of transformed foci. Nude mice injected with these focus-forming cells developed tumours (Bernard et al. 1991). FGFR-1 and FGFR-2 were found to be amplified in 12 per cent of breast cancer cases examined (Adnane et al. 1991). A variant of FGFR-2 named K-sam was isolated from a stomach tumour. K-sam cDNA has two Ig-like domains and a truncated carboxy-terminal tail (Fig. 2) and the gene was observed to be amplified and rearranged in the stomach tumour cells (Hattori et al. 1990). Tests of the oncogenic potential of other FGFR family members have not yet been reported.

■ Conclusion

Although five members of the FGFR family have been identified and their ligand specificities partially determined, the exact relationship between the seven known members of the FGF family and their receptors has not been elucidated.

Binding studies demonstrated that different FGFRs exhibit common binding specificities. However, differences in binding properties are also evident. Different ligand-binding specificities can also be conferred by alternative splicing of RNA as in the case of FGFR-2 and KGFR which differ in the second half of the Ig-like domain III. Therefore, alternative splicing and differential expression of FGFR genes may be mechanisms by which different tissues respond to various FGFs. To date, only four of the seven known FGFs have been shown to bind with high affinity or exert biological activity upon the FGFRs. These include: aFGF, bFGF, KGF, and possibly K-FGF. When the precise binding patterns and signalling pathways of all FGFs and their receptors are known it will be possible to elucidate which receptor forms and pathways are involved in mitogenesis, differentiation, inhibition of differentiation, or maintenance of specific cell types.

■ References

Adnane, J., Gaudray, P., Dionne, C.A., Crumley, G., Jaye, M., Schlessinger, J., Jeanteur, P., Birnbaum, D., and Theillet, C. (1991). BEK and FLG, two receptors to members of the FGF family, are amplified in subsets of human breast cancers. Oncogene, 6, 659–63.

Avivi, A., Zimmer, Y., Yayon, A., Yarden, Y., and Givol, D. (1991). Flg-2, a new member of the family of fibroblast growth factor receptors [published erratum appears in Oncogene 1992 Apr;7(4):823]. Oncogene, 6, 1089–92.

Bellot, F., Crumley, G., Kaplow, J.M., Schlessinger, J., Jaye, M., and Dionne, C.A. (1991). Ligand-induced transphosphorylation between different FGF receptors. EMBO J., 10, 2849–54.

Bernard, O., Li, M., and Reid, H.H. (1991). Expression of two different forms of fibroblast growth factor receptor 1 in different mouse tissues and cell lines. Proc. Natl. Acad. Sci. (USA), 88, 7625–9.

Burgess, W.H., and Maciag, T. (1989). The heparin-binding (fibroblast) growth factor family of proteins. Annu. Rev. Biochem., 58, 575–606.

Crumley, G., Bellot, F., Kaplow, J.M., Schlessinger, J., Jaye, M., and Dionne, C.A. (1991). High-affinity binding and activation of a truncated FGF receptor by both aFGF and bFGF. Oncogene, 6, 2255–62.

Dionne, C.A., Crumley, G., Bellot, F., Kaplow, J.M., Searfross, G., Ruta, M., Burgess, W.H., Jaye, M., and Schlessinger, J. (1990). Cloning and expression of two distinct high-affinity receptors cross-reacting with acidic and basic fibroblast growth factors. EMBO J., 9, 2685–92.

Hattori, Y., Odagiri, H., Nakatani, H., Miyagawa, K., Naito, K., Sakamoto, H., Katoh, O., Yoshida, T., Sugimura, T., and Terada, M. (1990). K-sam, an amplified gene in stomach cancer, is a member of the heparin-binding growth factor receptor genes. Proc. Natl. Acad. Sci. (USA), 87, 5983–7.

Heuer, J.G., von Bartheld, C.S., Kinoshita, Y., Evers, P.C., and Bothwell, M. (1990). Alternating phases of FGF receptor and NGF receptor expression in the developing chicken nervous system. Neuron, 5, 283–96.

Johnson, D.E., and Williams, L.T. (1993). Structural and functional diversity in the FGF receptor multigene family. Adv. Cancer Res., 60, 1–41.

Johnson, D.E., Lee, P.L., Lu, J., and Williams, L.T. (1990). Diverse forms of a receptor for acidic and basic fibroblast growth factors. Mol. Cell Biol., 10, 4728–36.

Johnson, D.E., Lu, J., Chen, H., Werner, S., and Williams, L.T. (1991). The human fibroblast growth factor receptor genes: a common structural arrangement underlies the mechanisms for generating receptor forms that differ in their third immunoglobulin domain. *Mol. Cell Biol.*, **11**, 4627–34.

Keegan, K., Johnson, D.E., Williams, L.T., and Hayman, M.J. (1991). Isolation of an additional member of the fibroblast growth factor receptor family, FGFR-3. *Proc. Natl. Acad. Sci. (USA)*, **88**, 1095–9.

Kiefer, M.C., Stephans, J.C., Crawford, K., Okino, K., and Barr, P.J. (1990). Ligand-affinity cloning and structure of a cell surface heparan sulfate proteoglycan that binds basic fibroblast growth factor. *Proc. Natl. Acad. Sci. (USA)*, **87**, 6985–9.

Kornbluth, S., Paulson, K.E., and Hanafusa, H. (1988). Novel tyrosine kinase identified by phosphotyrosine antibody screening of cDNA libraries. *Mol. Cell Biol.*, **8**, 5541–4.

Lee, P.L., Johnson, D.E., Cousens, L.S., Fried, V.A., and Williams, L.T. (1989). Purification and complementary DNA cloning of a receptor for basic fibroblast growth factor. *Science*, **245**, 57–60.

Mansukhani, A., Moscatelli, D., Talarico, D., Levytska, V., and Basilico, C. (1990). A murine fibroblast growth factor (FGF) receptor expressed in CHO cells is activated by basic FGF and Kaposi FGF. *Proc. Natl. Acad. Sci. (USA)*, **87**, 4378–82.

Matthew, P., Li, M., and Bernard, O. (1994). Characterization and genomic organization of a new murine fibroblast growth factor receptor FGFR-5. Manuscript in preparation.

Miki, T., Fleming, T.P., Bottaro, D.P., Rubin, J.S., Ron, D., and Aaronson, S.A. (1991). Expression cDNA cloning of the KGF receptor by creation of a transforming autocrine loop. *Science*, **251**, 72–5.

Miki, T., Bottaro, D.P., Fleming, T.P., Smith, C.L., Burgess, W.H., Chan, A.M., and Aaronson, S.A. (1992). Determination of ligand-binding specificity by alternative splicing: two distinct growth factor receptors encoded by a single gene. *Proc. Natl. Acad. Sci. (USA)*, **89**, 246–50.

Mohammadi, M., Honegger, A.M., Rotin, D., Fischer, R., Bellot, F., Li, W., Dionne, C.A., Jaye, M., Rubinstein, M., and Schlessinger, J. (1991). A tyrosine-phosphorylated carboxy-terminal peptide of the fibroblast growth factor receptor (Flg) is a binding site for the SH2 domain of phospholipase C-gamma 1. *Mol. Cell Biol.*, **11**, 5068–78.

Mohammadi, M., Dionne, C.A., Li, W., Li, N., Spivak, T., Honegger, A.M., Jaye, M., and Schlessinger, J. (1993). Point mutation in FGF receptor eliminates phosphatidylinositol hydrolysis without affecting mitogenesis. *Nature*, **358**, 681–4.

Partanen, J., Makela, T.P., Eerola, E., Korhonen, J., Hirvonen, H., Claesson-Welsh, L., and Alitalo, K. (1991). FGFR-4, a novel acidic fibroblast growth factor receptor with a distinct expression pattern. *EMBO J.*, **10**, 1347–54.

Pasquale, E.B. (1990). A distinctive family of embryonic protein-tyrosine kinase receptors. *Proc. Natl. Acad. Sci. (USA)*, **87**, 5812–6.

Peters, K.G., Werner, S., Chen, G., and Williams, L.T. (1992a). Two FGF receptor genes are differentially expressed in epithelial and mesenchymal tissues during limb formation and organogenesis in the mouse. *Development*, **114**, 233–43.

Peters, K.G., Marie, J., Wilson, E., Ives, H.E., Escobedo, J., Del Rosario, M., Mirda, D., and Williams, L.T. (1992b). Point mutation of an FGF receptor abolishes phosphatidylinositol turnover and Ca²⁺ flux but not mitogenesis. *Nature*, **358**, 678–81.

Peters, K.G., Ornitz, D., Werner, S., and Williams, L. (1993). Unique expression pattern of the FGF receptor 3 gene during mouse organogenesis. *Dev. Biol.*, **155**, 423–30.

Reid, H.H., Wilks, A.F., and Bernard, O. (1990). Two forms of the basic fibroblast growth factor receptor-like mRNA are expressed in the developing mouse brain. *Proc. Natl. Acad. Sci. (USA)*, **87**, 1596–600.

Ron, D., Reich, R., Chedid, M., Lengel, C., Cohen, O.E., Chan, A.M., Neufeld, G., Miki, T., and Tronick, S.R. (1993). Fibroblast growth factor receptor 4 is a high affinity receptor for both acidic and basic fibroblast growth factor but not for keratinocyte growth factor. *J. Biol. Chem.*, **268**, 5388–94.

Stark, K.L., McMahon, J.A., and McMahon, A.P. (1991). FGFR-4, a new member of the fibroblast growth factor receptor family, expressed in the definitive endoderm and skeletal muscle lineages of the mouse. *Development*, **113**, 641–51.

Ueno, H., Gunn, M., Dell, K., Tseng, A. Jr., and Williams, L.T. (1992). A truncated form of fibroblast growth factor receptor 1 inhibits signal transduction by multiple types of fibroblast growth factor receptor. *J. Biol. Chem.*, **267**, 1470–6.

Wanaka, A., Milbrandt, J., and Johnson, E.M. Jr. (1991). Expression of FGF receptor gene in rat development. *Development*, **111**, 455–68.

Werner, S., Duan, D.S., de Vries, C., Peters, K.G., Johnson, D.E., and Williams, L.T. (1992). Differential splicing in the extracellular region of fibroblast growth factor receptor 1 generates receptor variants with different ligand-binding specificities. *Mol. Cell Biol.*, **12**, 82–8.

Yayon, A., Klagsbrun, M., Esko, J.D., Leder, P., and Ornitz, D.M. (1991). Cell surface, heparin-like molecules are required for binding of basic fibroblast growth factor to its high affinity receptor. *Cell*, **64**, 841–8.

Ora Bernard* and Paul Matthew†:
* The Walter and Eliza Hall Institute of Medical Research,
† Centre for Animal Biotechnology, School of Veterinary Science,
The University of Melbourne, Australia

Transforming Growth Factor-β (TGFβ)

TGFβ was originally discovered as a secreted factor that induced malignant transformation in vitro. It is now recognized as a prototype member of a growing superfamily of secreted, disulphide-linked homodimeric polypeptides. These factors affect a variety of biological processes in both transformed and normal cells, including regulation of cellular proliferation and differentiation. Several extensive reviews are available on TGFβ, and should be consulted for a more detailed reference list than can be provided in this chapter (Derynck 1994; Massagué 1990; Roberts and Sporn 1990).

■ Superfamily

The TGFβ superfamily may be divided into subfamilies according to sequence homology. One group consists of the closely related TGFβ1, -2, and -3. cDNAs described previously as chicken TGFβ4 and *Xenopus* TGFβ5 are now thought to represent TGFβ1 cDNAs. The other subfamilies include the inhibins and activins, Müllerian inhibitory substance, and the decapentaplegic/Vg1/BMP group, which has been referred to as DVR (Lyons *et al.* 1991). This latter group includes Vg1 from *Xenopus laevis*, a maternally inherited transcript localized to the vegetal pole of embryos. The bone morphogenetic proteins (BMPs) were originally characterized as osteoinductive factors purified from demineralized bone, and there are now at least eight such related factors (Wozney 1992). BMPs-2 and -4 bear closest homology to the decapentaplegic complex protein, a *Drosophila* protein mediating dorsal/ventral axis specification. BMPs-5, -6 (also known as vgr-1), -7, and -8 most closely resemble *Drosophila* protein 60A. No other TGFβ-like factors have been identified in *Drosophila* to date. This superfamily, and especially the BMP-related family, continues to grow through additional cDNA cloning strategies (Jones *et al.* 1992; McPherron and Lee 1993; Ozkaynak *et al.* 1992; Basler *et al.* 1993).

It is beyond the scope of this review to consider the structural characteristics and biological functions attributed to all of these various factors, and we will concentrate on the TGFβ isoforms. A recurrent theme with this superfamily is that these factors are initially purified and characterized with a specific functional assay, but often have broader biological activities that may be especially relevant in development. For example, the BMPs are not only expressed in bone and cartilage, but have widespread distribution, and are present early in development. In fact, BMP-4 appears to function not only in osteogenesis but also as a ventralizing factor for mesoderm induction in *Xenopus* (Dale *et al.* 1992).

■ Structure

The structure of TGFβ1 is prototypical for all members of this superfamily. It is synthesized as a 390-amino-acid precursor containing an N-terminal secretory signal sequence, a long precursor segment, and C-terminal 112-amino-acid

sequence, which corresponds to the mature, bioactive TGFβ monomer. The C-terminal region is cleaved from the remaining precursor segment following a series of four basic amino acids. The protease mediating this cleavage likely belongs to the KEX/furin-like proteases (Barr 1991).

The biologically active form of TGFβ is a 25 kDa disulphide-linked homodimer of the mature segments, although heterodimers have also been found (Cheifetz *et al.* 1987). The monomeric fragment contains nine conserved cysteines, eight of which form intramolecular disulphide bonds. The crystalline structure of TGFβ2 (Daopin *et al.* 1992; Schlunegger and Grutter 1992) shows that each monomer contains two antiparallel pairs of β-strands and a separate long α-helix. One cysteine residue from each monomer forms an intermolecular disulphide bridge, and the remainder form an unusual clustering of intramolecular disulphide bridges termed a cystine knot (McDonald and Hendrickson 1993). Similar motifs have been delineated for nerve growth factor and platelet-derived growth factor, and suggests that these factors may form part of a primordial structural superfamily.

■ Sequence

TFGβ1 cDNAs from different animal species show an extremely high degree of conservation, with virtual amino acid identity over the C-terminal mature TGFβ domain (GenBank accession number for human TGFβ1 cDNA is X02812, J05114; mouse cDNA is M13177). This region has a 70–80 per cent amino acid identity among the three TGFβ isoforms and 25 per cent or greater among all members of the superfamily. Furthermore, at least seven of the nine cysteine residues found in this domain are conserved in all members of this superfamily. Considerably less homology is found among the pro-segments, both across species for a particular TGFβ isoform, and within a single species for different TGFβ isoforms. However, three cysteine residues and N-linked glycosylation sites are conserved in this region.

■ Gene and transcription

The human TGFβ1 gene encompasses seven exons, and splice site junctions are highly conserved among species

(GenBank accession number Y00112). Furthermore, the same structure is found in the TGFβ2, and -β3 genes, suggesting that these factors arose from a primordial gene through gene duplication. The various isoform mRNAs range in size from 1.7 to 6.5 kb. The size extensions originate from both 5′ and 3′ untranslated regions, and the former appear to play an important role in the translational regulation of the mRNAs.

The transcription of the three TGFβ isoforms is differentially regulated, which is largely due to divergence in the 5′ flanking sequence for each of these genes. Thus, transcription for each of these isoforms is mediated by a distinct collection of *cis* elements and transcription factors. The nature and complexity of this regulation has been illustrated elsewhere (Roberts and Sporn 1990; Roberts and Sporn 1992).

■ Localization

The three TGFβ isoforms have complex but well defined patterns of expression during development, which are partially overlapping (for example, see Pelton *et al.* 1991). Most if not all organs express one or more isoforms at defined stages of development, usually coinciding with tissue differentiation and morphogenesis. TGFβ is frequently associated with mesenchyme, especially during cartilage and bone formation, and at sites of mesenchymal–epithelial interactions. It is also expressed at high levels in epithelia, such as skin. Such widespread expression suggests that TGFβ plays a major role in directing embryogenesis. TGFβ1 expression is also upregulated in tumours, and may play an important role in oncogenesis.

■ Secretion and latent form

TGFβ is assembled and secreted in a four-protein complex, with the C-terminal homodimer interacting non-covalently with the two pro-segments (see Fig. 1). The N-terminal domain is required for the secretion of the C-terminal homodimer and may function as a chaperone that mediates folding of the mature TGFβ and its secretion. This complex is considered inactive or 'latent', because it prevents mature TGFβ from interacting with cell surface receptors. An additional protein termed latent TGFβ binding protein (LTBP) is often associated with this secreted complex, interacting through at least one disulphide bond from one of the pro-segments of the latent complex. LTBP is made by a wide variety of cell types, suggesting that this five protein latent complex is common (Miyazono *et al.* 1991).

■ Activation

Latent TGFβ can be activated by treatment with heat and acid, but under normal physiological conditions the mature bioactive homodimer is most likely released from the latent complex through degradation of the pro-segments by such proteases as plasmin and cathepsins (Harpel *et al.* 1992).

This activation process probably occurs at the cell surface, where the pro-segments are retained at least in part through binding to the mannose-6-phospahte receptor (Dennis and Rifkin 1991). The role of LTBP in this process remains unclear, but it may direct the latent complex to the cell surface.

An additional level of control for TGFβ activity is that the latent and active TGFβ complexes can interact with a variety of other proteins in the extracellular milieu. A protein with particularly high affinity is α2-macroglobulin, present in high concentrations in the circulation, which may act as an efficient scavenger of free active TGFβ (James 1990). TGFβ may also interact with a number of components of the extracellular matrix, including decorin, biglycan, thrombospondin, fibronectin, and various collagens. These interactions may serve to sequester active TGFβ, and thereby form a reservoir that is available only as the matrix is remodelled (Flaumenhaft and Rifkin 1992).

■ Protein purification

TGFβ was originally purified from its most abundant source, platelets (Assoian *et al.* 1983), and also from placenta, kidney, and bone. It was identified by its ability to induce normal rat kidney fibroblasts to grow and form colonies in soft agar in the presence of epidermal growth factor. Many other assays have since been devised to detect the purified

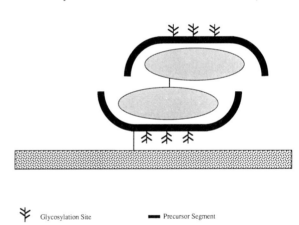

⅄ Glycosylation Site	▬ Precursor Segment
◯ Mature Segment	— Disulfide Bond
▦ Latent TGF-β Binding Protein	

Figure 1. Structure of secreted TGFβ latent complex. The mature C-terminal bioactive portion of the molecule is secreted as a homodimer, surrounded by precursor segments. This complex associates with the latent TGFβ-binding protein (LTBP). Note the disulphide bonds that exist between the mature fragments, and between the LTBP and precursor segment. Not represented are intramolecular disulphide bonds in mature domain, *nor* interactions between precursor segments and mature segments.

factor, with perhaps the most widely used being either the inhibition of growth of Mv1Lu mink lung epithelial cells (CCL-64) (Meager 1991), or antibody-based assays. Recombinant TGFβ1 has since been produced at high levels in transfected Chinese hamster ovary cells overexpressing the factor (Gentry et al. 1987). The secreted TGFβ is in its latent form, but following acid activation it exhibits appropriate biological activity. Such recombinant protein is now widely available on a commercial basis.

■ Antibodies

Great difficulty was initially encountered in raising high titre antibodies against TGFβ, and in obtaining antibodies specific for different isoforms. Polyclonal antibodies have been developed against synthetic peptides from various regions of the molecule (Flanders et al. 1988), and this approach has yielded polyclonal antisera against specific TGFβ isoforms (Pelton et al. 1991). Such antibodies are suitable for western blot analysis, radioimmunoassays, and immunohistochemistry. High titre blocking antibodies are also now available which react specifically with TGFβ1 or β2, or that cross-react with both molecules (Danielpour et al. 1989).

■ Biological functions

TGFβ has been implicated in various biological processes, affecting growth and differentiation (Derynck 1994; Massagué 1990; Roberts and Sporn 1990). The nature of these effects depends on cell type, degree of differentiation, and culture conditions. These biological effects may be divided into several categories. One set of such activities is related to those that affect cellular proliferation. TGFβ inhibits the proliferation of many cells in culture, including epithelial, endothelial and many haemopoietic cells. The TGFβ induced growth arrest in epithelial cells occurs at late G1 and may involve a physiological interaction with the retinoblastoma gene product pRB, since inactivation of pRB results in abolition of the antiproliferative effect, at least in some cells. In addition, TGFβ induces a mitogenic response in some cells, particularly in cells of mesenchymal origin, which, in some cases, is mediated through production of platelet-derived growth factor. Injection of TGFβ in vivo induces hypercellular lesions, but this may be due to a chemoattractant activity on fibroblasts and monocytes which undergo subsequent activation, rather than a direct proliferative effect (Moses et al. 1990).

Another set of TGFβ actions involves those that affect cellular interaction with the extracellular matrix. These effects are mediated at multiple levels. TGFβ augments synthesis and secretion of many extracellular matrix proteins. It increases production of integrins, cell surface receptors that bind to these extracellular matrix proteins. Finally, it decreases the synthesis of proteases that degrade extracellular matrix proteins, such as metalloproteases, and increases production of protease inhibitors, such as tissue inhibitor of metalloprotease (TIMP). These effects are highly variable and dependent on the cell line and type. However, the net effect of this TGFβ activity is expected to increase production of extracellular matrix, and enhance the cell's interaction with its substrate (Derynck 1994; Massagué 1990; Roberts and Sporn 1990).

TGFβ has an important role in the function and differentiation of the immune system, and as an immunosuppressive agent (Derynck 1994; Roberts and Sporn 1990; Kehrl 1991). It inhibits both the proliferation and function of B cells, cytotoxic T cells and NK cells, inhibits cytokine production in lymphocytes and often antagonizes the effects of tumour necrosis factor. Furthermore, TGFβ activates and is strongly chemotactic for monocytes, yet deactivates macrophage function. The immunosuppressive and anti-inflammatory activities in several in vivo studies are complemented by the finding that inactivation of the endogenous TGFβ1 gene results in multifocal inflammatory disease, massive lymphocytic infiltration, and early demise of the newborn mice (Shull et al. 1992).

TGFβ is also important for wound healing and tissue repair (reviewed by Amento and Beck 1991). It is released at sites of wound healing by platelets or activated macrophages and moncytes. The effects may be related to induced changes in extracellular matrix, but also to secondary effects on monocyte and macrophage influx, fibroblast influx, and angiogenesis. Exogenous TGFβ accelerates wound healing, but excessive endogenous secretion of TGFβ may cause an exuberant healing response, and lead to fibrosis.

The widespread distribution of TGFβ (and its receptors) during development suggests that it also plays an important role in embryogenesis (Pelton et al. 1991). Most studies to date have focused on its role in mesenchymal differentiation. It inhibits myoblast maturation and myotube formation in some cell lines, but under other culture conditions it may stimulate myotube formation. It also inhibits adipocytic differentiation, but stimulates chondrocytic and osteoblastic differentiation (although the latter effect is again dependent on the differentiation state of the cell). TGFβ may also play an important role in the development of epithelia and skin, and in the branching morphogenesis of complex epithelial structures such as the lung and mammary gland (Robinson et al. 1991).

Finally, TGFβ1 is overproduced in many tumours. Even though increased TGFβ synthesis may not provide a growth advantage in vitro, it may stimulate tumour development in vivo, presumably due to its stimulation of extracellular matrix formation and increased adhesiveness of the cells to the matrix, and its local immmunosuppressive effect on cytotoxic T cells and NK cells (Arteaga et al. 1993; Chang et al. 1993).

■ References

Amento, E.P., and Beck, L.S. (1991). TGF-beta and wound healing. *Ciba Foundation Symposium*, **157**, 115–23.

Arteaga, C.L., Carty-Dugger, T., Moses, H.L., Hurd, S.D., and Pietenpol, J.A. (1993). Transforming growth factor beta 1 can induce estrogen-independent tumourigenicity of human breast cancer cells in athymic mice. *Cell Growth Diff.*, **4**, 193–201.

Assoian, R.K., Komoriya, A., Meyers, C.A., Miller, D.M., and Sporn, M.B. (1983). Transforming growth factor beta in human platelets. *J. Biol. Chem.*, **258**, 7155–60.

Barr, P.J. (1991). Mammalian subtilisins: the long-sought dibasic processing endoproteases. *Cell*, **66**, 1–3 .

Basler, K., Edlund, T., Jessell, T.M., and Yamada, T. (1993). Control of cell pattern in the neural tube: regulation of cell differentiation by dorsalin-1, a novel TGF beta family member. *Cell*, **73**, 687–702 .

Chang, H.L., Gillett, N., Figari, I., Lopez, A.R., Palladino, M.A., and Derynck, R. (1993). Increased transforming growth factor-b expression inhibits cell proliferation *in vitro*, yet increases tumourigenicity and tumour growth of Meth A sarcoma cells. *Cancer Res.*, **53**, 4391–8.

Cheifetz, S., Weatherbee, J.A., Tsang, M.L., Andersen, J.K., Mole, J.E., Lucas, R., and Massagué, J. (1987). The transforming growth factor-beta system, a complex pattern of cross-reactive ligands and receptors. *Cell*, **48**, 409–15.

Dale, L., Howes, G., Price, B.M., and Smith, J.C. (1992). Bone morphogenetic protein 4: a ventralizing factor in early *Xenopus* development. *Development*, **115**, 573–85 .

Danielpour, D., Dart, L.L., Flanders, K.C., Roberts, A.B., and Sporn, M.B. (1989). Immunodetection and quantitation of the two forms of transforming growth factor-beta (TGF-beta 1 and TGF-beta 2) secreted by cells in culture. *J. Cell.Physiol.*, **138**, 79–86 .

Daopin, S., Piez, K.A., Ogawa, Y., and Davies, D.R. (1992). Crystal structure of transforming growth factor-beta 2: an unusual fold for the superfamily [see comments]. *Science*, **257**, 369–73 .

Dennis, P.A., and Rifkin, D.B. (1991). Cellular activation of latent transforming growth factor beta requires binding to the cation-independent mannose 6-phosphate/insulin-like growth factor type II receptor. *Proc. Natl. Acad. Sci. (USA)*, **88**, 580–4 .

Derynck, R. (1994) The biological complexity of transforming growth factor-β. In *The cytokine handbook* (ed. A. Thompson). Academic, (In press.)

Flanders, K.C., Roberts, A.B., Ling, N., Fleurdelys, B.E., and Sporn, M.B. (1988). Antibodies to peptide determinants in transforming growth factor beta and their applications. *Biochem.*, **27**, 739–46.

Flaumenhaft, R., and Rifkin, D.B. (1992). The extracellular regulation of growth factor action. *Mol. Biol. Cell*, **3**, 1057–65.

Gentry, L.E., Webb, N.R., Lim, G.J., Brunner, A.M., Ranchalis, J.E., Twardzik, D.R., Lioubin, M.N., Marquardt, H., and Purchio, A.F. (1987). Type 1 transforming growth factor beta: amplified expression and secretion of mature and precursor polypeptides in Chinese hamster ovary cells. *Mol. Cell. Biol.*, **7**, 3418–27.

Harpel, J.G., Metz, C.M., Kojima, S., and Rifkin, D.B. (1992). Control of transforming growth factor-beta activity: Latency vs. activation. *Prog. Growth Factor Res.*, **4**, 321–35.

James, K. (1990). Interactions between cytokines and alpha 2-macroglobulin. *Immunol. Today*, **11**, 163–6.

Jones, C.M., Simon-Chazottes, D., Guenet, J.L., and Hogan, B.L. (1992). Isolation of Vgr-2, a novel member of the transforming growth factor-beta-related gene family. *Molec. Endocrinol.*, **6**, 1961–8 .

Kehrl, J.H. (1991). Transforming growth factor-beta: an important mediator of immunoregulation. *Int. J. Cell Cloning*, **9**, 438–50.

Lyons, K.M., Jones, C.M., and Hogan, B.L. (1991). The DVR gene family in embryonic development. *Trends Genet.*, **7**, 408–12 .

Massagué, J. (1990). The transforming growth factor-beta family. *Annu. Rev. Cell Biol.*, **6**, 597–641.

McDonald, N.Q., and Hendrickson, W.A. (1993). A structural superfamily of growth factors containing a cystine knot motif. *Cell*, **73**, 421–4 .

McPherron, A.C., and Lee, S.J. (1993). GDF-3 and GDF-9: two new members of the transforming growth factor-beta superfamily containing a novel pattern of cysteines. *J. Biol. Chem.*, **268**, 3444–9.

Meager, A. (1991). Assays for transforming growth factor beta. *J. Immunol. Meth.*, **141**, 1–14.

Miyazono, K., Oloffson, A., Colosetti, P., and Heldin, C.H. (1991). A role of the latent TGF-beta 1-binding protein in the assembly and secretion of TGF-beta 1. *EMBO J.*, **10**, 1091–101.

Moses, H.L., Yang, E.Y., and Pietenpol, J.A. (1990). TGF-beta stimulation and inhibition of cell proliferation: new mechanistic insights. *Cell*, **63**, 245–7.

Ozkaynak, E., Schnegelsberg, P.N., Jin, D.F., Clifford, G.M., Warren, F.D., Drier, E.A., and Oppermann, H. (1992). Osteogenic protein-2. A new member of the transforming growth factor-beta superfamily expressed early in embryogenesis. *J. Biol. Chem.*, **267**, 25220–7.

Pelton, R.W., Saxena, B., Jones, M., Moses, H.L., and Gold, L.I. (1991). Immunohistochemical localization of TGF beta 1, TGF beta 2, and TGF beta 3 in the mouse embryo: expression patterns suggest multiple roles during embryonic development. *J. Cell Biol.*, **115**, 1091–105 .

Roberts, A.B., and Sporn, M.B. (1990). The transforming growth factor-βs. In *Peptide growth factors and their receptors* (ed. M.B. Sporn and A.B. Roberts), pp. 419–72. Springer, Heidelberg.

Roberts, A.B., and Sporn, M.B. (1992). Mechanistic interrelationships between two superfamilies: the steroid/retinoid receptors and transforming growth factor-beta. *Cancer Surveys*, **14**, 205–20 .

Robinson, S.D., Silberstein, G.B., Roberts, A.B., Flanders, K.C., and Daniel, C.W. (1991). Regulated expression and growth inhibitory effects of transforming growth factor-beta isoforms in mouse mammary gland development. *Development*, **113**, 867–78.

Schlunegger, M.P., and Grutter, M.G. (1992). An unusual feature revealed by the crystal structure at 2.2 Å resolution of human transforming growth factor-beta 2. *Nature*, **358**, 430–4 .

Shull, M.M., Ormsby, I., Kier, A.B., Pawlowski, S., Diebold, R.J., Yin, M., Allen, R., Sidman, C., Proetzel, G., Calvin, D., Annunziata, N., and Doetschman, T. (1992). Targeted disruption of the mouse transforming growth factor-beta 1 gene results in multifocal inflammatory disease. *Nature*, **359**, 693–9 .

Wozney, J.M. (1992). The bone morphogenetic protein family and osteogenesis. *Molec. Reprod. Dev.*, **32**, 160–7 .

Stephen E. Gitelman* and Rik Derynck†:
Departments of Pediatrics* and Growth and Development†, and Anatomy†,
Programs in Cell Biology and Developmental Biology†,
University of California at San Francisco,
San Francisco, CA, USA

Receptors for Transforming Growth Factor-β (TGFβ)

TGFβ, a multifunctional cytokine, binds specifically and with high affinity to a wide variety of cell types (Segarini 1991; Wakefield et al. 1987). Cross-linking to radiolabelled TGFβ has allowed the identification of a number of proteins that bind TGFβ with high affinity (Lin and Lodish 1993). These include soluble or ECM (extracellular matrix)-associated proteins and cell-surface receptors. The most widely distributed high affinity TGFβ-receptors on the cell surface are the types I, II and III receptors with apparent molecular weights of 55, 80, and 280 kDa, respectively. Receptors type I and II bind TGFβ with higher affinity (K_D = 5–25 pM) than does type III (K_D = 200 pM). The type III receptor is a transmembrane proteoglycan (Segarini and Seyedin 1988; Cheifetz et al. 1992) that binds TGFβ through its 100 kDa deglycosylated core protein and that has a short cytoplasmic tail with no obvious signalling motif (Wang et al. 1991). Cell surface expression of the type III receptor modulates the binding of TGFβ to the type II receptor (Wang et al. 1991; Lopez-Casillias et al. 1991). The TGFβ type II receptor is a member of a family of transmembrane serine/threonine kinases which include several type II activin receptors (Mathews and Vale 1991; Attisano et al. 1992) and the daf-4 gene product of Caenorhabditis elegans (Georgi et al. 1990; Estevez et al. 1993). Expression of the cloned TGFβ type II receptor can restore growth inhibition by TGFβ1 to cells lacking this receptor (Inagaki et al. 1993; Wrana et al. 1992). The TGFβ type II receptor forms a heteromeric complex with the type I receptor which seems to be essential for TGFβ signal transmission. The TGFβ type I receptor has distinctive biochemical properties (Cheifetz and Massagué 1991). Recent cloning of TGFβ type I receptor (Ebner et al. 1993; Franzén et al. 1993; Attisano et al. 1993) reveals a transmembrane serine/threonine kinase which shares conserved sequences in both the extracellular and intracellular domains with the type II TGFβ receptor. TGFβ receptors type II and I are components of the TGFβ receptor signalling complex and are involved in two distinct signalling pathways triggered by TGFβ, growth inhibition and gene regulation. Since both receptors are serine/threonine kinases they modulate different molecules that are distinct from substrates of tyrosine kinase receptors. The composition of the actual receptor complex (homo/heterooligomer) may also affect the multiple intracellular signals generated by TGFβ.

■ The TGFβ type III receptor

The cDNA for the TGFβ type III receptor was first isolated from A10 cells (rat vascular smooth muscle cell line) by expression cloning in COS 7 cells (Wang et al. 1991) (GenBank accession number M 80784) and from rat fetal tissue by a peptide-sequencing-PCR approach (Lopez-Casillias et al. 1991) (GenBank accession number M 77809). The human and pig cDNAs encoding the type III receptor have since been isolated (Moren et al. 1992). The encoded receptor is an 853-amino-acid protein (Fig. 1) with a single hydrophobic transmembrane domain, a 43-amino-acid-long cytoplasmic tail with no obvious signalling motif and a large N-terminal extracellular domain containing at least two sites for glycosaminoglycan addition and six consensus sites for N-linked glycosylation (Wang et al. 1991; Lopez-Casillias et al. 1991). There is 63 per cent amino acid identity between the transmembrane and cytoplasmic domains of the type III receptor and endoglin, a glycoprotein of 95 kDa found on endothelial cells as well as some haemopoietic precursor cells (Lin and Lodish 1993; Wang et al. 1991; Lopez-Casillias et al. 1991). Overall, the proteins are only 27 per cent identical. The biological role of endoglin is not known, but it contains an 'RGD' sequence in its ectodomain which might promote interaction with other adhesion molecules. Endoglin also binds TGFβ (Cheifetz et al. 1992). Furthermore, the extracellular domain of the type III receptor has homology to the sperm receptors Zp2 and Zp3, the urinary protein uromodulin, and the zymogen granule membrane protein GP-2 (Bork and Sander 1992). The type III receptor is heavily modified by glycosaminoglycan (GAG) groups, which cause it to migrate heterogenously upon SDS-polyacrylamide gels with apparent molecular masses in the range of 280–330 kDa (Segarini and Seyedin 1988; Cheifetz et al. 1988a). Deglycosylation experiments have shown that the covalently linked heparan sulphate and chondroitin sulphate chains are not required for TGFβ binding (Cheifetz and Massagué 1989).

In addition to the membrane anchored form of the type III receptor, a soluble form is secreted by some cell types. This soluble secreted ectodomain may act as a 'reservoir' for ligand retention or for presentation of TGF-β to the signalling TGFβ type II and I receptors (Lin and Lodish 1993; Massagué 1990; Andres et al. 1991).

Expression of recombinant type III receptor in rat L6 myoblasts, which lack this receptor, resulted in an increase in

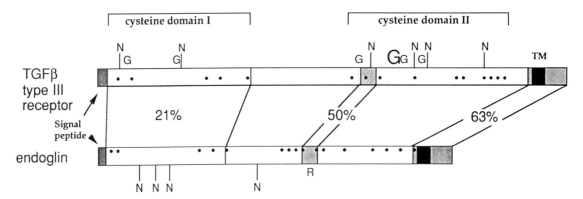

Figure 1. Schematic representation of TGFβ receptor type III and endoglin: a comparison between the rat TGFβ type III receptor and the human endoglin molecule. Amino acid identity between boxed regions is given as a percentage. TM, transmembrane region; N, N-glycosylation site; G, potential GAG-attachment site (large G denotes consensus site); R, RGD sequence in endoglin; filled circles, position of cysteines.

binding of iodinated TGFβ1 to the TGFβ type II receptor, without affecting binding to the type I receptor (Wang et al. 1991). Further studies showed that binding of TGFβ2 to the type II receptor can only occur after interaction with the TGFβ2-type III receptor complex (Lopez-Casillias et al. 1993; Lin et al. 1994; Moustakas et al. 1993). Since TGFβ2 is equally as potent as TGFβ1 and β3 in its ability to arrest the growth of cells (ED50 ~ 5–20 pM) (Cheifetz et al. 1987; Zugmaier et al. 1989), it can only cause growth inhibition of cells which express the TGFβ type III receptor (Lopez-Casillias et al. 1993). In the presence of ligand the type III receptor forms a heteromeric complex with the signalling type II receptor. However, only a minor fraction of the type III receptor interacts with type II receptor, which reflects either the relative amounts of the receptors on the cell surface (about ten times more TGFβ type III receptor) or the instability of the heteromeric complex (Moustakas et al. 1993, Henis et al. 1994). Even though the type III receptor has a lower affinity for any TGFβ isoform than the types I and II, because it is much more abundant (~ 200 000 receptors per cell)

than the others it might act as an important determinant of binding of TGFβ to the types I and II and possibly other signalling receptors.

■ Chromosomal location of the TGFβ type III receptor gene

Human chromosome 1 (J. Lawrence, University of Massachusetts, Worcester, MA, USA, personal communication).

■ The TGFβ type II receptor

The cDNA for the TGFβ type II receptor was isolated by expression cloning in COS 7 cells (Lin et al. 1992) (GenBank accession number M 85079). The human TGFβ type II receptor has a short 127-amino-acid, cysteine-rich extracellular domain, a single hydrophobic transmembrane domain, and an intracellular region dominated by a serine/threonine

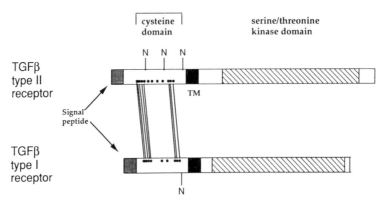

Figure 2. Schematic representation of TGFβ receptor type II and I: a comparison between the TGFβ type II and type I receptor. The conserved cysteine clusters in the extracellular domain are indicated by lines. TM, transmembrane region; N, N-glycosylation site; filled circles, position of cysteines.

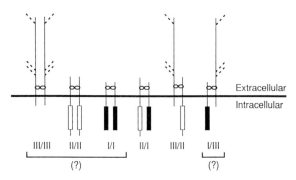

III/III	II/II	I/I	II/I	III/II	I/III	Extracellular Intracellular
(?)					(?)	

Figure 3. Schematic representation of the three high-affinity TGFβ receptors on the cell surface. The types I, II, and III TGFβ receptors are depicted as homo- or heterodimers bound to a dimeric TGFβ (open double ellipse). The three glycosaminoglycan side chains of TGFβ type III receptor are shown as dotted lines in the extracellular domain of the receptor. The Ser/Thr kinase segments in the cytoplasmic domain of the type I and II receptors are drawn as filled and open boxes, respectively. Question marks indicate receptor complexes for which there is no direct evidence (We thank A. Moustakas for preparing this figure.)

kinase motif (Fig. 2). The full-length receptor is 567 amino acids in length and has a predicted molecular mass of 65 kDa. The mature, complex-glycosylated form of the receptor has a heterogeneous molecular mass of ∼ 80–90 kDa. The receptor has been cloned from human, pig (Lin et al. 1992), mink (Wrana et al. 1992), and rat (Tsuchida et al. 1993).

The TGFβ Type II receptor is a member of a family of transmembrane serine/threonine kinases which includes daf-4, a presumed receptor important in C. elegans development, and two subtypes of activin receptors, ActR-II and ActR-IIB (Attisano et al. 1992; Mathews et al. 1992; Kondo et al. 1991; Legerski 1992). It has recently been shown that the daf-4 gene encodes a bone morphogenic protein receptor (Estevez et al. 1993). In the kinase region the TGFβ type II receptor shares 45 per cent amino acid identity with the activin receptor and 34 per cent with the daf-4 gene product. Expression in E. coli of a chimeric protein containing nearly the entirety of the intracellular domain of the TGFβ type II receptor fused to a glutathione S-transferase (GST) gene showed that it can be autophosphorylated on serine and threonine residues in vitro (Lin et al. 1992). Neither ligand-induced activation of the receptor kinase nor ligand-dependent or independent kinase activity towards cytoplasmic substrates has yet been described. It is also unknown whether homo- or heterodimerization is necessary for kinase function. Interestingly a TGFβ type II receptor with a truncated cytoplasmic domain acts as a dominant negative inhibitor specific for growth inhibition mediated via the type II receptor (Chen et al. 1993).

The cell-surface TGFβ type II receptor binds TGFβ1 and β3 with greater affinity ($K_D \sim$ 25–50 pM) than TGFβ2 ($K_D \sim$ 500 pM) (Massagué 1990). No binding of TGFβ2 can be detected to the secreted, recombinant ectodomain of the TGFβ type II receptor (K_D higher than 10 nM), though this soluble receptor does bind efficiently to TGFβ1 and β3 ($K_D \sim$ 100pM) (Lin et al. 1994). However, specific binding of TGFβ2 to the TGFβ type II receptor occurs when cells also express the type III receptor on their surface (Lopez-Casillas et al. 1993; Moustakas et al. 1993). There is a physical interaction between type II and type III receptor, since antibodies to one can immunoprecipitate the other (Lopez-Casillias et al. 1993).

The extracellular domains of daf-4, ActRII and ActRIIB and the TGFβ type II receptor have very little similarity in sequence (∼10 per cent). All of the cloned TGFβ type II receptors, however, have short cysteine-rich extracellular domains that are relatively homologous (Lin et al. 1992). Recently cloned TGFβ type I receptor shares with the type II receptor a conserved structural motif, containing the five cysteines just following the signal peptide (Fig. 3). This motif is absent in activin receptors and may be important in determining ligand-binding specificity (Ebner et al. 1993; Franzén et al. 1993; Attisano et al. 1993).

The functionality of the cloned TGFβ type II receptor was established by two distinct assay systems, expression of the receptor in early Xenopus embryos which naturally lack the receptor, and in cultured cells lacking a functional gene for the type II receptor. Expression of the TGFβ type II receptor in Xenopus embryos allows TGFβ1 to induce mesoderm formation in the same way as does activin via the endogenous or exogenous activin receptors. In addition, the response of animal caps to TGFβ1, like activin, is blocked by co-expression of a dominant negative ras protein, suggesting that the activin receptor and TGFβ type II receptor share a common downstream effector, ras (Bhushan et al. 1994).

Expression of the recombinant type II receptor restores growth inhibition by TGFβ1 to cells lacking the type II receptor gene (Hep3B-TR, [Inagaki et al. 1993]) or to cells expressing a mutated type II receptor (MvLu-cells DR, [Wrana et al. 1992]).

No cell line has been described which expresses only the TGFβ type II receptor without the types I and III. While it is clear that the TGFβ type II receptor has a biologically active serine/threonine kinase, signalling through the type II receptor occurs via a heteromeric complex with the TGFβ type I receptor (Wrana et al. 1992). A physical association between type II and type I receptor was shown by co-immunoprecipitation of the type I receptor by antibodies against the type II receptor (Inagaki et al. 1993; Wrana et al. 1992; Lopez-Casillas et al. 1993). Growth inhibition by TGFβ1 is mediated by the heteromeric complex of receptor types II and I. However, cells exist that have no detectable type II receptor (e.g. SW480, 293 cells) and which are not responsive to the anti-proliferative activity of TGFβ, but which exhibit TGFβ-induced gene regulation. These cells express a type I receptor on their cell surface, in the absence of detectable type II. Thus, there is evidence that the two TGFβ-triggered signalling pathways are mediated by different signalling receptor complexes: growth inhibition by the complex of types II and I, and gene regulation by the type I alone. The functions of these receptor complexes may differ from cell type to cell type (Chen *et al.* 1993; Brand *et al.* 1993; Wieser *et al.* 1993).

■ Chromosomal location of the TGFβ type II gene

Human chromosome 3 (J. Lawrence, University of Massachusetts, Worcester, MA, USA, personal communication).

■ The TGFβ type I receptor

The type I receptor forms together with the type II receptor a heteromeric signalling receptor complex that, as noted above, seems essential for mediating the anti-proliferation effects of TGFβ. As judged by binding and cross-linking of radioiodinated TGFβ, the type I receptor is a 53 kDa transmembrane glycoprotein with the same binding properties as the type II receptor: it binds TGFβ1 and β3 with higher affinity than TGFβ2.

Recently type I receptors have been cloned by a PCR approach (Ebner *et al.* 1993; He *et al.* 1993; Franzén *et al.* 1993; Attisano *et al.* 1993, ten Dijke *et al.* 1993). Type I receptors for TGFβ, activin, and BMP classify a second family of transmembrane serine/theonine kinases, which can only bind their ligand when coexpressed with their corresponding type II receptors. The TGFβ type I receptor (Franzén *et al.* 1993) encodes a 503-amino-acid polypeptide with one N-glycosylation site and a large number of cysteines in the extracellular domain. These cysteines are thought to be important in determining the correct three-dimensional structure necessary for ligand binding. The cytoplasmic domain of the receptor is dominated by a serine/threonine kinase motif. In many cell lines, the type I and type II receptor associate as interdependent components of a heteromeric complex: type I receptor requires type II receptor to bind TGFβ1 and type II requires type I to signal (Inagaki *et al.* 1993; Wrana *et al.* 1992).

This complex has been detected on the cell surface in the presence of ligand. However, we do not know if formation of the complex occurs in the absence of ligand, nor whether such a complex forms on the plasma membrane or during biosynthesis in the endoplasmic reticulum. However, as noted, some haemopoietic cell lines express a type I receptor that can bind TGFβ in the absence of any detectable type II receptor, and some of these cells show transcriptional responses to TGFβ. Thus, there may be multiple signalling receptor complexes.

■ Other TGFβ receptors

- Glycosyl-phospatidyl-inositol (GPI)-linked proteins in fetal bovine heart endothelial and other cells: a 180 kDa protein that binds TGFβ1 preferentially, not TGFβ2; a 140 kDa species that binds TGFβ2 preferentially and a 60 kDa protein (Cheifetz and Massagué 1991).
- Type IV receptor in GH3 rat pituitary cells, which binds activin, inhibin, and TGFβ1 and TGFβ2 (Cheifetz *et al.* 1988*b*).
- Type V receptor, a 400 kDa protein that exhibits ser/thr autophosphorylation activity (O'Grady *et al.* 1991*a,b*, 1992).

- A 38 kDa protein in Be Wo choriocarcinoma that binds TGFβ2 (Mitchell *et al.* 1992).

■ TGFβ receptor complexes and signalling

TGFβ is a multifunctional regulatory peptide that controls cell growth, differentiation and composition of the extracellular matrix (Massagué 1990; Roberts and Sporn 1990). The cellular mechanism of action of TGFβ is unknown; none of the known pathways for intracellular signalling, including phosphatidylinositol turnover, modulation of intracellular cAMP levels and tyrosine phosphorylation have been directly linked to TGFβ action (Massagué 1990; Roberts and Sporn 1990; Sporn and Roberts 1992). The most direct effect of TGFβ in many cells is the inhibition of cell cycle progression by arresting the cell in late G1 phase and preventing DNA synthesis. Significant among these events is that the retinoblastoma gene product RB remains unphosphorylated after addition of TGFβ1 (Laiho *et al.* 1990). Cell cycle-dependent phosphorylation and dephosphorylation of RB regulate its activity in normal cells. TGFβ-treated cells fail to accumulate cyclin E-cdk2 complexes or cdk2 associated kinase activities, which may explain the absence of RB phosphorylation (Laiho *et al.* 1990).

The expression of many genes related to growth control, differentiation and cell adhesion are regulated by TGFβ, such as transcription factors (c-jun, junB, fos, myc), growth factors (PDGF-A, PDGF-B) (Leof *et al.* 1986) or their receptors (PDGFα receptor, IL-1 receptor, GM-CSF receptor, IL-3 receptor, G-CSF receptor) (Yamakage *et al.* 1992; Dubois *et al.* 1990; Jacobsen *et al.* 1991). TGFβ elicits these responses through its interaction with the high-affinity cell-surface receptors type III, II and I. As noted, these receptors fall into two categories: the type I and II show cytoplasmic signalling structures (Ser/Thr kinase), while the type III receptor shows no obvious signalling motif. In the presence of ligand, heteromeric complexes of type III/type II and type II/type I receptors are formed.

■ Acknowledgements

We thank colleagues for providing unpublished results and A. Moustakas, Y. Henis and R. Wells for critical reading and comments on the manuscript.

■ References

Andres, J.L., Ronnstrand, L., Cheifetz, S., and Massague, J. (1991). Purification of the transforming growth factor-beta (TGF-beta) binding proteoglycan betaglycan. *J. Biol. Chem.*, **266**, 23282–7.

Attisano, L., Wrana, J.L., Cheifetz, S., and Massague, J. (1992). Novel activin receptors: distinct genes and alternative mRNA splicing generate a repertoire of serine/threonine kinase receptors. *Cell*, **68**, 97–108.

Attisano, L., Carcamo, J., Ventura, F., Weis, F.M.B., Massagué, J., and Wrana, J.L. (1993). Identification of human activin and TGFb type I receptors that form heteromeric kinase complexes with type II receptors. *Cell*, **75**, 671–80.

Bhushan, A., Lin, H.Y., Lodish, H.F., and Kintner, C.I. (1994). *Development* (Submitted).

Bork, P., and Sander, C. (1992). A large domain common to sperm receptors (Zp2 and Zp3) and TGF-beta type III receptor. *FEBS Lett.*, **300**, 237–40.

Brand, T., MacLellan, W.R., and Schneider, M.D. (1993). A dominant negative receptor for type beta transforming growth factors created by deletion of the kinase domain. *J. Biol. Chem. (USA)*, **268**, 11500–3.

Cheifetz, S., and Massagué, J. (1989). Transforming growth factor-beta (TGF-beta) receptor proteoglycan. Cell surface expression and ligand binding in the absence of glycosaminoglycan chains. *J. Biol. Chem.*, **264**, 12025–8.

Cheifetz, S., and Massagué, J. (1991). Isoform-specific transforming growth factor-beta binding proteins with membrane attachments sensitive to phosphatidylinositol-specific phospholipase C. *J. Biol. Chem.*, **266**, 20767–72.

Cheifetz, S., Weatherbee, J.A., Tsang, M.L., Anderson, J.K., Mole, J.E., Lucas, R., and Massagué, J. (1987). The transforming growth factor-beta system, a complex pattern of cross-reactive ligands and receptors. *Cell*, **48**, 409–15.

Cheifetz, S., Andres, J.L., and Massagué, J. (1988a). The transforming growth factor-beta receptor type III is a membrane proteoglycan. Domain structure of the receptor. *J. Biol. Chem.*, **263**, 16984–91.

Cheifetz, S., Ling, N., Guillemin, R., and Massagué, J. (1988b). A surface component on GH3 pituitary cells that recognizes transforming growth factor-beta, activin, and inhibin. *J. Biol. Chem.*, **263**, 17225–8.

Cheifetz, S., Bellon, T., Cales, C., Vera, S., Bernabeu, C., Massagué, J., and Letarte, M. (1992). Endoglin is a component of the transforming growth factor-beta receptor system in human endothelial cells. *J. Biol. Chem.*, **267**, 19027–30.

Chen, R.-H., Ebner, R., and Derynck, R. (1993). Inactivation of the type II receptor reveals two receptor pathways for the diverse TGF-beta activities. *Science*, **260**, 1335–8.

ten Dijke, P., Ichijo, H., Franzén, P., Schulz, P., Saras, J., Toyoshima, H., Heldin, C.-H., and Miyazono, K. (1993). Activin receptor-like kinases: a novel subclass of cell-surface receptors with predicted serine/threonine kinase activity. *Oncogene*, **8**, 2879–87.

Dubois, C.M., Ruscetti, F.W., Palaszynski, E.W., Falk, L.A., Oppenheim, J.J., and Keller, J.R. (1990). Transforming growth factor beta is a potent inhibitor of interleukin 1 (IL-1) receptor expression: proposed mechanism of inhibition of IL-1 action. *J. Exp. Med.*, **172**, 737–44.

Ebner, R., Chen, R.-H., Shum, L., Lawler, S., Zioncheck, T. F., Lee, A., Lopez, A.R., and Derynck, R. (1993), Cloning of a type I TGF-beta receptor and its effect on TGF-beta binding to the type II receptor. *Science*, **260**, 1344–8.

Estevez, M., Attisano, L., Wrana, J., Albert, P.S., Massagué, J., and Riddle, D.L. (1993). The daf-4 gene encodes a bone morphogenic protein receptor controlling *C. elegans* dauer larva development. *Nature*, **365**, 644–9.

Franzén, P., ten Dijke, P., Ichijo, H., Yamashita, H., Schulz, P., Heldin, C.-H., and Miyazono, K. (1993). Cloning of a TGFb type I receptor that forms a heteromeric complex with the TGFb type II receptor. *Cell*, **75**, 681–92.

Georgi, L.L., Albert, P.S., and Riddle, D.L. (1990). daf-1, a *C. elegans* gene controlling dauer larva development, encodes a novel receptor protein kinase. *Cell*, **61**, 635–45.

He, W.W., Gustafson, M.L., Hirobe, S., and Donahoe, P.K. (1993). Developmental expression of four novel serine/threonine kinase receptors homologous to the activin/transforming factor-β type II receptor family. *Dev. Dyn.*, **196**, 133–42.

Henis, Y., Moustakas, A., Lin, H.Y., and Lodish, H.F. (1994). The types II and III transforming growth factor-β receptors form homo-oligomers. *J. Cell Biology*, **126**, 139–54.

Inagaki, M., Moustakas, A., Lin, H.Y., Lodish, H.F., and Carr, B.I. (1993). Growth inhibition by transforming growth factor

(TGF-beta) type I is restored in TGF-beta-resistant hepatoma cells after expression of TGF-beta receptor type II cDNA. *Proc. Natl. Acad. Sci. (USA)*, **90**, 5359–63.

Jacobsen, S.E., Ruscetti, F.W., Dubois, C.M., Lee, J., Boone, T.C., and Keller, J.R. (1991). Transforming growth factor-beta trans-modulates the expression of colony stimulating factor receptors on murine hematopoietic progenitor cell lines. *Blood*, **77**, 1706–16.

Kondo, M., Tashiro, K., Fujii, G., Asano, M., Miyoshi, R., Yamada, R., Muramatsu, M., and Shiokawa, K. (1991). Activin receptor mRNA is expressed early in *Xenopus* embryogenesis and the level of the expression affects the body axis formation. *Biochem. Biophys. Res. Commun.*, **181**, 684–90.

Laiho, M., DeCaprio, J.A., Ludlow, J.W., Livingston, D.M., and Massagué, J. (1990). Growth inhibition by TGF-beta linked to suppression of retinoblastoma protein phosphorylation. *Cell*, **62**, 175–85.

Legerski, R., Zhou, X., Dresback, J., Eberspaecher, H., McKinney, S., Segarini, P., and de Crombrugghe, B. (1992). Molecular cloning and characterization of a novel rat activin receptor. *Biochem. Biophys. Res. Commun.*, **183**, 672–9.

Leof, E.B., Proper, J.A., Goustin, A.S., Shipley, G.D., DiCorleto, P.E., and Moses, H.L. (1986). Induction of c-sis mRNA and activity similar to platelet-derived growth factor by transforming growth factor b: a proposed model for indirect mitogenesis involving autocrine activity. *Proc. Natl. Acad. Sci. (USA)*, **83**, 2453–7.

Lin, H.Y., and Lodish, H.F. (1993). Receptors for the TGF-β superfamily: multiple polypeptides and serine/threonine kinases. *Trends Cell Biol.*, **3**, 14–19.

Lin, H.Y., Wang, X.-F., Ng-Eaton, E., Weinberg, R.A., and Lodish, H.F. (1992). Expression cloning of the TGF-beta type II receptor, a functional transmembrane serine/threonine kinase. *Cell*, **68**, 775–85 [Erratum *Cell* **70**, following 1068.]

Lin, H.Y., Moustakas, A., Knaus, P., Henis, Y.I., Lodish, H.F. (1994). The soluble exoplasmic domain of the type II TGF-β receptor binds TGF-β1 but not TGF-β2. *J. Biol. Chem.* (Submitted.)

Lopez-Casillias, F., Cheifetz, S., Doody, J., Andres, J.L., Lane, W.S., and Massague, J. (1991). Structure and expression of the membrane proteoglycan betaglycan, a component of the TGF-beta receptor system. *Cell*, **67**, 785–95.

Lopez-Casillias, F., Wrana, J.L., and Massague, J. (1993). Betaglycan presents ligand to the TGFb signalling receptor. *Cell*, **73**, 1435–44.

Massagué, J. (1990). The transforming growth factor-beta family. *Annu. Rev. Cell. Biol.*, **6**, 597–641.

Mathews, L.S., and Vale, W.W. (1991). Expression cloning of an activin receptor, a predicted transmembrane serine kinase. *Cell*, **65**, 973–82.

Mathews, L.S., Vale, W.W., and Kintner, C.R. (1992). Cloning of a second type of activin receptor and functional characterization in *Xenopus* embryos. *Science*, **255**, 1702–5.

Mitchell, E.J., Fitz-Gibbon, L., and OConnor-McCourt, M.D. (1992). Subtypes of betaglycan and of type I and type II transforming growth factor-beta (TGF-beta) receptors with different affinities for TGF-beta1 and TGF-beta 2 are exhibited by human placental trophoblast cells. *J. Cell Physiol.*, **150**, 334–43.

Moren, A., Ichijo, H., and Miyazono, K. (1992). Molecular cloning and characterization of the human and porcine transforming growth factor-beta type III receptors. *Biochem. Biophys. Res. Commun.*, **189**, 356–62.

Moustakas, A., Lin, H.Y., Henis, Y., Plamondon, J., OConnor-McCourt, M.D., and Lodish, H. (1993). The transforming growth factor β receptors types I, II, and III form hetero-oligomeric complexes in the presence of the ligand. *J. Biol. Chem.*, **268**, 22215–8.

OGrady, P., Huang, S.S., and Huang, J.S. (1991a). Expresssion of a new type high molecular weight receptor (type V receptor) of transforming growth factor beta in normal and transformed cells. *Biochem. Biophys. Res. Commun.*, **179**, 378–85.

OGrady, P., Kuo, M.D., Baldassare, J.J., Huang, S.S., and Huang, J.S. (1991b). Purification of a new type high molecular weight receptor (type V receptor) of transforming growth factor beta

(TGF-beta) from bovine liver. Identification of the type V TGF-beta receptor in cultured cells. *J. Biol. Chem.*, **266**, 8583–9.

OGrady, P., Liu, Q., Huang, S.S., Huang, J.S. (1992). Transforming growth factor beta (TGF-beta) type V receptor has a TGF-beta-stimulated serine/threonine-specific autophosphorylation activity. *J. Biol. Chem.*, **267**, 21033–7.

Roberts, A.B., and Sporn, M.B. (1990). The transforming growth factor-bs. In *Peptide growth factors and their receptors*, Part I (ed. M.B. Sporn and A.B. Roberts), pp. 419–72. Springer, Berlin.

Segarini, P. (1991). Clinical application of TGFβ. In *Ciba Foundation Symposium*, pp. 29–50. Wiley, Chichester.

Segarini, P.R., and Seyedin, S.M. (1988). The high molecular weight receptor to transforming growth factor-beta contains glycosaminoglycan chains. *J. Biol. Chem.*, **263**, 8366–70.

Sporn, M.B., and Roberts, A.B. (1992). Transforming growth factor-beta: recent progress and new challenges. *J. Cell Biol.*, **119**, 1017–21.

Tsuchida, K., Lewis, K.A., Mathews, L.S., and Vale, W.W. (1993). Molecular characterization of rat transforming growth factor-beta type II receptor. *Biochem. Biophys. Res. Commun.*, **191**, 790–5.

Wakefield, L.M., Smith, D.M., Masui, T., Harris, C.C., and Sporn, M.B. (1987). Distribution and modulation of the cellular receptor for transforming growth factor-beta. *J. Cell. Biol.*, **105**, 965–75.

Wang, X.-F., Lin, H.Y., Ng-Eaton, E., Downward, J., Lodish, H.F., and Weinberg, R.A. (1991). Expression cloning and characterization of the TGF-beta type III receptor. *Cell*, **67**, 797–805.

Wieser, R., Attisano, L., Wrana, J., and Massagué, J. (1993). Signaling activity of transforming growth factor β type II receptors lacking specific domains in the cytoplasmic region. *Mol. Cell. Biol.*, **13**, 7239–47.

Wrana, J.L., Attisano, L., Carcamo, J., Zentella, A., Doody, J., Laiho, M., Wang, X.F., and Massague, J. (1992). TGF beta signals through a heteromeric protein kinase receptor complex. *Cell*, **71**, 1003–14.

Yamakage, A., Kikuchi, K., Smith, E.A., LeRoy, E.C., and Trojanowska, M. (1992). Selective upregulation of platelet-derived growth factor alpha receptors by transforming growth factor beta in scleroderma fibroblasts. *J. Exp. Med.*, **175**, 1227–34.

Zugmaier, G., Ennis, B.W., Deschauer, B., Katz, D., Knabbe, C., Wilding, G., Daly, P., Lippman, M.E., and Dickson, R.B. (1989). Transforming growth factors type beta 1 and beta 2 are equipotent growth inhibitors of human breast cancer cell lines. *J. Cell. Physiol.*, **141**, 353–61.

Petra I. Knaus:
Whitehead Institute for Biomedical Research,
Cambridge, MA 02142, USA

Harvey F. Lodish:
Department of Biology,
Massachusetts Institute of Technology,
Cambridge, MA 02139, USA

Vascular Endothelial Growth Factor (VEGF)

VEGF is an endothelial cell-specific mitogen that has the ability to promote angiogenesis in several in vivo models. By alternative splicing of mRNA, VEGF may exist in four different isoforms that differ markedly in their secretion pattern but have similar biological activities. Expression of VEGF mRNA is temporally and spatially related to the proliferation of blood vessels in a variety of physiological circumstances, such as embryonic development, formation of the ovarian corpus luteum, or wound healing. Recent studies suggest that VEGF is also an important mediator of angiogenesis in solid tumours and in rheumatoid arthritis.

■ The VEGF isoforms

VEGF purified from conditioned media of normal and neoplastic cells is a glycoprotein of $\sim$45 000 kDa (Ferrara and Henzel 1989). It is inactivated by reducing agents but it is heat- and acid-stable. By alternative splicing of RNA, VEGF may exist in four different homodimeric isoforms (Leung *et al.* 1989; Houck *et al.* 1991; Tisher *et al.* 1991). Following signal sequence cleavage, human VEGF monomers are expected to have 121, 165, 189, and 206 amino acids, respectively ($VEGF_{121}$, $VEGF_{165}$, $VEGF_{189}$, $VEGF_{206}$). Rodent or bovine VEGF monomers are shorter by one amino acid. The VEGF monomers have a single glycosylation site. The amino-acid sequence of VEGF has limited homology (15–18 per cent) to the A and B chains of platelet-derived growth factor (Leung *et al.* 1989). $VEGF_{165}$ appears to be the most abundant product of the VEGF gene in a variety of normal and transformed cells (Ferrara *et al.* 1992). The chromatographic properties of native VEGF purified from a variety of sources correspond closely to those of recombinant $VEGF_{165}$. Compared with $VEGF_{165}$, $VEGF_{121}$ lacks 44 amino acids; $VEGF_{189}$ has an insertion of 24 amino acids enriched in basic residues and $VEGF_{206}$ has an additional insertion of 17 amino acids (Fig. 1). $VEGF_{121}$ is a weakly acidic polypeptide that does not bind to heparin. In contrast, $VEGF_{165}$ is basic and binds to heparin. $VEGF_{189}$ and $VEGF_{206}$ are more basic and bind to heparin with even greater affinity (Houck *et al.* 1992). Recent studies have shown that such differences in affinity for heparin profoundly affect the fate of the VEGF isoforms (Houck *et al.* 1992; Park *et al.* 1993). $VEGF_{121}$ is secreted and is freely diffusible in the conditioned medium of transfected cells. $VEGF_{165}$ is also secreted but a significant fraction remains bound to heparin-containing proteoglycans. The longer forms are almost completely bound

NH$_2$– MNFLLSWVHWSLALLLYLHHAKWSQAAPMAEGGGQNHHE

    VVKFMDVYQRSYCHPIETLVDIFQEYPDEIEYIFKPSCV

    PLMRCGGCCNDEGLECVPTEESNITMQIMRIKPHQGQHI

    GEMSFLQHNKCECRPKKDRARQEK KSVRGKGKGQKRKRK

    KSRYKSWSV YVGARCCLMPWSLPGPH PCGPCSERRKHLF

    VQDPQTCKCSCKNTDSRCKARQLELNERTCRCDKPRR –COOH

Figure 1. Amino acid sequence of the VEGF isoforms. Boxed residues indicate the 17-amino-acid insertion unique to VEGF$_{206}$. Full line, amino acids contained in both VEGF$_{189}$ and VEGF$_{206}$. Broken line, amino acids shared by VEGF$_{165}$, VEGF$_{189}$, and VEGF$_{206}$. In VEGF$_{169}$, the lysine residue marked by an asterisk is replaced by asparagine. All remaining amino acids are shared by the four isoforms. The arrow marks the signal peptide cleavage site.

to the extracellular matrix (ECM) but they may be released in a diffusible form by heparin or suramin (Houck *et al.* 1992; Park *et al.* 1993). Furthermore, plasmin is able to cleave ECM-bound VEGF$_{189}$ or VEGF$_{206}$ at the COOH terminus, releasing a diffusible and bioactive proteolytic fragment that, by a variety of criteria, including affinity for anion-exchange matrices, behaves similarly to VEGF$_{121}$ (Houck *et al.* 1992).

■ Purification of VEGF

Several schemes have been proposed for the purification of native VEGF from conditioned medium of normal or transformed cells (Ferrara and Henzel 1989; Connolly *et al.* 1989; Plöuet *et al.* 1989; Ferrara *et al.* 1991). These schemes are primarily based on the affinity of the protein for heparin or cation-exchange matrices. According to one protocol (Ferrara *et al.* 1991), serum-free conditioned medium from normal bovine pituitary folliculo-stellate cells is concentrated 10–20 fold by Amicon stir cells or by ammonium sulphate precipitation. The concentrated conditioned medium is equilibrated in 10 mM Tris/Cl, 50 mM NaCl, pH 7.2 and then applied to a heparin–Sepharose column pre-equilibrated with the same buffer. Following washes at low salt concentrations, the column is eluted in the presence of 0.9 M NaCl. 80–90 per cent of the activity is recovered in such fractions. VEGF is further purified by sequential reversed phase HPLC steps on C4 columns. Alternatively, following heparin–Sepharose, the protein is subjected to gel filtration, cation-exchange chromatography on a Mono-S column and, finally, reversed phase HPLC (Plöuet *et al.* 1989). VEGF has also been purified to homogeneity as a vascular permeability factor (VPF) from conditioned medium of U937 cells (Connolly *et al.* 1989).

■ Biological properties of VEGF

Purified VEGF is able to promote the growth of vascular endothelial cells derived from small and large vessels (Ferrara and Henzel 1989; Plöuet *et al.* 1989). Half-maximal stimulation of bovine capillary endothelial cell growth is obtained at 100–300 pg/ml (2–6 pM) and a maximal stimulation at 1–5 ng/ml (22–110 pM). However, VEGF has no appreciable mitogenic activity on non-endothelial cell types. The VEGF binding sites are localized to endothelial cells but not other cell types in fetal or adult tissues (see chapter on VEGF receptors, p. 235). VEGF also has been shown to promote angiogenesis in several *in vivo* models, including the chick chorioallantoic membrane (Leung *et al.* 1989; Plöuet *et al.* 1989). According to recent studies, intramuscular or intra-arterial administration of VEGF has the ability to augment perfusion and collateral vessel formation in a rabbit model of hind-limb ischemia (Takeshita *et al.* 1994). VEGF is able to promote chemotaxis of endothelial cells (Koch *et al.* 1994).

VEGF has been reported also to induce monocyte chemotaxis, although at concentrations higher than those required to stimulate endothelial cell growth (ED$_{50}$ = 300–500 pM). VEGF induces synthesis of the serine proteases, urokinase-type and tissue type plasminogen activators (PAs) and also of PA inhibitor 1 in cultured microvascular endothelial cells. Also, VEGF induces synthesis of interstitial collagenase in human umbilical vein endothelial cells but not in dermal fibroblasts (Ferrara *et al.* 1992). The co-induction of PAs and collagenase by VEGF is expected to promote a prodegradative environment that facilitates migration of endothelial cells, a key step in the angiogenesis cascade (Folkman and Klagsbrun 1987).

Expression of VEGF$_{165}$ or VEGF$_{121}$ confers on a non-tumourigenic clone of Chinese hamster ovary cells the ability to form vascularized tumours in nude mice. However, VEGF has no mitogenic affect on such cells, suggesting that their ability to grow *in vivo* is due to paracrine rather than autocrine mechanisms. Anti-VEGF monoclonal antibodies are able to suppress the growth of several human tumours in nude mice, suggesting that VEGF is an important mediator of tumour angiogenesis (Kim *et al.* 1993). Recent studies also suggest that VEGF is a mediator of angiogenesis in rheumatoid arthritis (Koch *et al.* 1994).

As mentioned above, VEGF has been isolated also as a tumour-derived VPF based on its ability to induce protein-bound dye extravasation when applied intradermally in the guinea pig skin (Connolly *et al.* 1989). On the basis of this activity, the protein was proposed to be a mediator of hyperpermeability of tumour vessels.

■ Antibodies to VEGF

A series of high-affinity anti-VEGF murine monoclonal antibodies (mAbs) have been described recently (Kim *et al.* 1992). They were generated using rhVEGF$_{165}$ as an immunogen and belong to the IgG$_1$ isotype. Two different epitopes were detected with these mAbs. One of these mAbs (4.6.1) immunoprecipitates all four molecular species of rhVEGF and has been shown to inhibit VEGF-induced mitogenesis *in vitro*, as well as angiogenesis in the chick chorioallantoic membrane. This mAb is able to suppress the growth of several human tumour cell lines in nude mice (Kim *et al.*

1993). A characteristic of these mAbs is a high degree of species-specificity. Rabbit polyclonal antisera raised against rhVEGF$_{165}$ have the ability to neutralize VEGF action across species.

■ The VEGF gene and expression of VEGF mRNA

The VEGF gene is organized in eight exons and the size of its coding region has been estimated to be ~14 kb (Tisher *et al.* 1991). VEGF$_{165}$ is missing the residues encoded by exon 6, whereas VEGF$_{121}$ is missing the residues encoded by exons 6 and 7. Interestingly, there is no intron between the coding sequences of the 24-amino-acid insertion found in VEGF$_{189}$ and VEGF$_{206}$ and the additional 17-amino-acid insertion unique to VEGF$_{206}$ (Houck *et al.* 1991). The 5' end of the 51 base-pair insertion unique to VEGF$_{206}$ begins with GT, the consensus sequence for the 5' splice donor necessary for RNA processing. Therefore, the definition of the 5' splice donor for removal of a 1 kb intron appears to be variable.

VEGF mRNA expression is temporally and spatially related to the proliferation of blood vessels in the developing embryo (Breier *et al.* 1992), in the ovary (Phillips *et al.* 1990) or in a healing wound (Brown *et al.* 1992), suggesting that VEGF is a regulator of physiological angiogenesis. Hypoxia can potently induce expression of VEGF mRNA (Shweiki *et al.* 1992). A variety of normal cell types have been shown to express the VEGF mRNA, including podocytes in the Bowman's capsule, pituitary cells, luteal cells, and neurones (Ferrara *et al.* 1992). However, there is no evidence that endothelial cells express VEGF *in vivo*, suggesting that VEGF is a purely paracrine mediator. Additionally, VEGF mRNA is expressed by a variety of human tumours, including renal cell carcinoma, colon carcinoma and several intracranial tumours such as glioblastoma multiforme and capillary haemangioblastoma (Berse *et al.* 1992; Shweiki *et al.* 1992; Berkman *et al.* 1993). A strong correlation exists between the degree of vascularization of the malignancy and VEGF mRNA expression.

■ References

Berkman, R.A., Merril,I M.J., Reinhold, W.C., Monacci, W.T., Saxena, A., Clark, W.C., Robertson, J.T., Ali, I.U., and Oldfield, E.H. (1993). Expression of vascular permeability/vascular endothelial growth factor gene in human intracranial neoplasms. *J. Clin. Invest.*, **91**, 153–9.

Berse, B., Brown, L.F., Van de Vater, L., Dvorak, H.F., and Senger, D.R. (1992). The vascular permeability factor (vascular endothelial growth factor) gene is differentially expressed in normal tissues, macrophages and tumours. *Molec. Biol. Cell*, **3**, 211–20.

Breier, G., Albrecht, U., Sterrer, S., and Risau, W. (1992). Expression of vascular endothelial growth factor during embryonic angiogenesis and endothelial cell differentiation. *Development*, **114**, 521–32.

Brown, L.F., Yeo, K.T., Berse, B., Yeo, K.T., Senger, D.R., Dvorak, H.F., and Van de Water, L. (1992). Expression of vascular permeability factor (vascular endothelial growth factor) by epidermal keratinocytes during wound healing. *J. Exp. Med.*, **176**, 1375–9.

Connolly, D.T., Olander, J.V., Heuvelman, D., Nelson, R., Monsell, R., Siegel, N., Haymore, B.L., Leimgruber, R., and Feder, J. (1989). Purification of human vascular permeability factor from U937 cells. *J. Biol. Chem.*, **264**, 20017–24.

Ferrara, N., and Henzel, W.J. (1989). Pituitary follicular cells secrete a novel heparin-binding growth factor specific for vascular endothelial cells. *Biochem. Biophys. Res. Commun.*, **161**, 851–859.

Ferrara, N., Leung, D.W., Cachianes, G., Winer, J., and Henzel, W.J. (1991). Purification and cloning of vascular endothelial growth factor secreted by pituitary folliculo-stellate cells. *Meth. Enzymol.*, **198**, 391–405.

Ferrara, N., Houck, K., Jakeman, L., and Leung, D.W. (1992). Molecular and biological properties of the vascular endothelial growth factor family of proteins. *Endocrinol Rev.*, **13**, 18–32.

Folkman, J., and Klagsbrun, M. (1987). Angiogenic Factors. *Science*, **235**, 442–7.

Houck, K.A., Ferrara, N., Winer, J., Cachianes, G., Li, B., and Leung, D.W. (1991). The vascular endothelial growth factor family: Identification of a fourth molecular species and characterization of alternative splicing of RNA. *Molec. Endocrinol.*, **5**, 1806–14.

Houck, K.A., Leung, D.W., Rowland, A.M., Winer, J., and Ferrara, N. (1992). Dual regulation of vascular endothelial growth factor bioavailability by genetic and proteolytic mechanisms. *J. Biol. Chem.*, **267**, 26031–7.

Kim, K.J., Li, B., Houck, K.A., Winer, J., and Ferrara, N. (1992). The vascular endothelial growth factor proteins: Identification of biologically relevant regions by neutralizing monoclonal antibodies. *Growth Factors*, **7**, 53–62.

Kim, K.J., Li, B., Winer, J., Armanini, M., Gillett, N., Phillips, H.S., and Ferrara, N. (1993). Inhibition of vascular endothelial growth factor-induced angiogenesis suppresses tumour growth *in vivo*. *Nature*, **362**, 841–4.

Koch, A.E., Harlow, L.A., Haines, G.K., Amento, E.P., Unemori, E.N., Wong, W.L., Pope, R.M., and Ferrara, N. (1994). Vascular endothelial growth factor: a cytokine modulating endothelial function in rheumatoid arthritis. *J. Immunol.*, **152**, 4149–55.

Leung, D.W., Cachianes, G., Kuang, W.-J., Goeddel, D.V., and Ferrara, N. (1989). Vascular endothelial growth factor is a secreted angiogenic mitogen. *Science*, **246**, 1306–9.

Park, J.E., Keller, G.-A. and Ferrara, N. (1993). The vascular endothelial growth factor (VEGF) isoforms: Differential deposition in the subepithelial extracellular matrix and bioactivity of extracellular matrix-bound VEGF. *Mol. Biol. Cell*, **4**, 1317–26.

Phillips, H.S., Hains, J., Leung, D.W., and Ferrara, N. (1990). Vascular endothelial growth factor is expressed in rat corpus luteum. *Endocrinol.*, **127**, 965–7.

Plöuet, J., Schilling, J., and Gospodarowicz, D. (1989). Isolation and characterization of a newly identified endothelial cell mitogen from AtT 20 cells. *EMBO J.*, **8**, 3801–8.

Schweiki, D., Itin, A., Soffer, D., and Keshet, E. (1992). Vascular endothelial growth factor induced by hypoxia may mediate hypoxia-induced angiogenesis. *Nature*, **359**, 843–5.

Takeshita, S., Zhung, L., Brogi, E., Kearney, M., Asahara, T., Pu, L.-Q., Bunting, S., Ferrara, N., Synes, J. F., and Isner, J.M. (1994). Therapeutic angiogenesis: a single intraarterial bolus of vascular endothelial growth factor augments collateral vessel formation in a rabbit ischemic hindlimb model. *J. Clin. Invest.*, **93**, 662–70.

Tisher, E., Mitchell, R., Hartman, M., Silva, M., Gospodarowicz, D., Fiddes, J.C., and Abraham, J.A. (1991). The human gene for vascular endothelial growth factor. *J. Biol. Chem.*, **266**, 11947–54.

Napoleone Ferrara and John E. Park:
Department of Cardiovascular Research,
Genentech Inc.,
So. San Francisco, CA 94080 USA

Vascular Endothelial Growth Factor (VEGF) Receptors

Two tyrosine kinases have been identified as potential VEGF receptors, flt-1 and KDR. Both proteins bind VEGF with high affinity and share several structural features with the M-CSF and PDGF receptors. Expression of flt-1 and KDR mRNAs appears to be endothelial cell-specific. Binding of VEGF to endothelial cells stimulates tyrosine phosphorylation events and calcium fluxes. It is unknown which receptor(s) mediates these processes or which intracellular components are involved in signalling. Ligand binding studies in tissue sections from developing or adult rats demonstrate that high affinity binding sites for VEGF are present in differentiated endothelial cells as well as in their progenitors in the blood islands.

■ The Flt-1 and KDR/Flk-1 proteins

Two tyrosine kinases have been recently identified as putative VEGF receptors (deVries *et al.* 1992; Terman *et al.* 1992a). The flt-1 (fms-like-tyrosine kinase; Shibuya *et al.* 1990) and KDR (kinase domain region; Terman *et al.* 1991) proteins have been shown to bind VEGF with high affinity. The overall amino acid sequence identity between the two proteins is 44 per cent. The murine homologue of KDR has been named flk-1 (fetal liver kinase; Matthews *et al.* 1991) and shares 85 per cent sequence identity with human KDR. For simplicity, the two receptors will be referred to as KDR/flk-1. Flt-1 and KDR/flk-1 each have a single hydrophobic leader peptide, a single transmembrane domain, seven immunoglobulin-like domains in its extracellular domain, and a consensus tyrosine kinase sequence which is interrupted by a kinase-insert domain. Flt-1 and KDR share 30 per cent sequence identity with c-fms, a known tyrosine kinase receptor for M-CSF (CSF-1).

Flt-1 (GenBank accession number X51602) is produced as a 1338-amino-acid precursor with a predicted 22-amino-acid signal peptide. The extracellular domain of flt-1 is composed of 736 amino acids; its transmembrane spanning domain is 22 amino acids and its cytoplasmic domain is 558 amino acids long. Flt-1 contains 13 potential N-linked glycosylation sites, several of which appear to be utilized since the observed molecular weight of 190 000–200 000 on SDS-PAGE gels is about 25 per cent greater than predicted (150 000). Recently, a cDNA encoding a truncated form of flt-1 missing the seventh immunoglobulin-like domain, the cytoplasmic domain, and transmembrane sequence has been identified in human umbilical vein endothelial cells (Kendell and Thomas 1993). The encoded protein is expected to have 687 amino acids and is soluble following secretion. This soluble flt-1 protein is able to inhibit VEGF-induced mitogenesis and has been proposed to be a physiological negative regulator of VEGF action. KDR (GenBank accession number L04947) is expressed as a 1356-amino-acid precursor with a predicted 19-amino-acid signal peptide. Its extracellular domain is 745 amino acids long after signal removal, its transmembrane domain is 25 amino

acids long and its cytoplasmic domain is 567 amino acids long. KDR contains 18 potential N-linked glycosylation sites. Its observed molecular weight is 190 000–200 000 by SDS-PAGE gel, whereas its predicted molecular weight is only 150 000. This suggests that, similar to flt-1, some of the glycosylation sites in KDR are actually utilized. The physical characteristics of flk-1 are very similar to those of KDR.

■ Characteristics of VEGF binding to endothelial cells

Two classes of high affinity VEGF binding sites have been identified on the cell surface of cultured endothelial cells, having K_D values of 10^{-12} and 10^{-11}M, respectively (Vaisman *et al.* 1990; Plöuet and Moukadiri 1990). Cross-linking studies using ^{125}I-VEGF revealed bands in the range of 180–230 kDa, as well as a lower molecular mass (110 kDa) band in endothelial cells (Vaisman *et al.* 1990; Plóet and Moukadiri 1990; Olander *et al.* 1991). Scatchard analysis (Bikfalvi *et al.* 1989; Myoken *et al.* 1991; Olander *et al.* 1991) indicates that cultured endothelial cells express more moderate-affinity binding sites (3000–60 000 per cell) than high affinity binding sites (500–3000 per cell). The binding of VEGF to endothelial cells is enhanced by low concentrations of heparin and is inhibited by removal of cell-surface heparan sulphate by heparinases (Gitay-Goren *et al.* 1992). This presumably reflects the heparin-binding nature of VEGF and suggests also that a cell-surface proteoglycan may be required for binding. Lower affinity binding sites on mononuclear phagocytes (K_D~300–500 pM) have recently been described (Shen *et al.* 1993). The nature of these sites remains to be elucidated.

■ Characteristics of VEGF binding to flt-1 and KDR/flk-1

Flt-1 has the highest affinity for rhVEGF$_{165}$, with a K_D of approximately 10–20 pM (deVries *et al.* 1992). KDR has a somewhat lower affinity for VEGF: the K_D has been estimated to be approximately 75 pM (Terman et al. 1992a). The K_D for binding of rhVEGF$_{165}$ to Flk-1 is 500–600 pM (Millauer et al. 1993). Therefore, it is likely that the binding of KDR/flk-1 to VEGF is partially species-specific. VEGF binding to these receptors was not competed for by structurally related peptides such as PDGF (deVries et al. 1992; Quinn et al. 1993). Affinity cross-linking of ^{125}I-rhVEGF$_{165}$ to transfected COS cells expressing KDR/flk-1 revealed bands of 190–230 kDa (Terman et al. 1992a; Millauer et al. 1993).

Distribution of binding sites for VEGF

Ligand autoradiography studies on tissue sections of adult rats revealed that high affinity ^{125}I-VEGF binding sites were localized to the vascular endothelium of large or small vessels but not to other cell types (Jakeman et al. 1992). Scatchard analysis of saturation isotherms in sections from a variety of tissues revealed a single class of binding sites with high affinity (K_D = 16–35 pM). Specific binding colocalized with factor VIII immunoreactivity and was apparent on both proliferating and quiescent endothelial cells. Binding of ^{125}I-VEGF during development of rat embryos is first detectable in the blood islands of the yolk sac, which contain the earliest progenitors of haemopoietic and endothelial cells (Jakeman et al. 1993). As the vascular system develops, VEGF binding sites continue to colocalize with the endothelium of blood vessels.

VEGF receptor genes and expression

The flt-1 gene is localized to chromosome 13q12 in humans near the gene for flt-3, as assessed by in situ hybridization (Shibuya et al. 1990). The flt-1/flt-3 linkage is also repeated in the mouse on chromosome 5 (Rosnet et al. 1993). The flt-1 gene lies near a chromosomal breakpoint characteristic of alveolar rhabdomyosarcomas. However, the flt-1 gene is located proximal to the breakpoint and does not appear to be disrupted in rhabdomyosarcoma. The flk-1 gene is closely linked to c-kit and PDGF receptor A genes on mouse chromosome 5 (Matthews et al. 1991). This linkage is not conserved in humans, as kit and PDGF receptor A map to chromosome 4q11–4q13, whereas the gene for KDR lies on chromosome 4q31.2–4q32, in a region not known to contain any other tyrosine kinase growth factor receptors (Terman et al. 1992b).

In situ hybridization studies demonstrate that flk-1 mRNA expression is restricted to endothelial cells in the mouse embryo at all stages of development (Millauer et al. 1993; Quinn et al. 1993). Expression was first detected in the blood islands and is confined to the mesoderm which gives rise to angioblasts. The stroma of the umbilical cord is the only non-endothelial cell type which expresses flk-1 mRNA. Expression of flt-1 mRNA correlates with the degree of vascularity of a variety of rat tissues (Shibuya et al. 1990).

Recent studies (Peters et al. 1993) indicate that flk-1 mRNA is expressed in endothelial cells in fetal and adult mice. The expression pattern of flt-1 appears identical to the distribution of high efficiency VEGF binding sites (Jakeman et al. 1992, 1993).

Signal transduction mechanisms

Little is known at the present time regarding signal transduction mechanisms that follow binding of VEGF to its receptors. VEGF binding to the cell surface of human umbilical vein endothelial cells stimulates rapid tyrosine phosphorylation of an approximately 200 000 Da protein, presumably one (or both) of the known VEGF receptors (Myoken et al. 1991). Cytosolic calcium fluxes in response to VEGF have been observed in endothelial cells derived from human umbilical vein, bovine adrenal cortex, and bovine pulmonary artery (Bikfalvi et al. 1991; Brock et al. 1991).

Neither flt-1 nor KDR/flk-1 confers on non-endothelial cells the ability to proliferate in response to VEGF. Therefore, it is unknown whether both receptors are necessary or whether additional endothelial cell-specific cytoplasmic factors are required for mitogenic activity. Experiments where flt-1 or KDR/flk-1 cDNAs were transfected in COS cells so far have demonstrated VEGF-dependent tyrosine phosphorylation only on cells expressing KDR/flk-1 (deVries et al. 1992; Millauer et al. 1993; Quinn et al. 1993). This is surprising since both flt-1 and KDR/flk-1 proteins contain consensus tyrosine kinase domains in their cytoplasmic regions (Shibuya et al. 1990; Terman et al. 1991; Matthews et al. 1991). However, VEGF can induce a Ca^{2+} efflux in Xenopus laevis oocytes expressing either flt-1 or KDR/flk-1 (deVries et al. 1992; Quinn et al. 1993). The relevance of these effects to the regulation of vascular endothelial cell growth remains to be determined.

References

Bikfalvi, A., Sauzeau, C., Moukadiri, H., Maclouf, J., Busso, N., Bryckaert, M., Plouet, J., and Tobelem, G. (1991). Interaction of vasculotropin/vascular endothelial cell growth factor with human umbilical vein endothelial cells: binding, internalization, degradation, and biological effects. J. Cell. Physiol., **149**, 50–9.

Brock, T.A., Dvorak, H.F., and Senger, D.R. (1991). Tumor-secreted vascular permeability factor increases cytosolic Ca2+ and von Willebrand factor release in human endothelial cells. Am. J. Pathol., **138**, 213–21.

deVries, C., Escobedo, J.A., Ueno, H., Houck, K., Ferrara, F., and Williams, L.T. (1992). The fms-like tyrosine kinase, a receptor for vascular endothelial growth factor. Science, **255**, 989–91.

Gitay-Goren, H., Soker, S., Vlodasky, I., and Neufeld, G. (1992). The binding of vascular endothelial growth factor to its receptors is dependent on cell surface-associated heparin-like molecules. J. Biol. Chem., **267**, 6093–8.

Jakeman, L.B., Winer, J., Bennett, G.L., Altar, C.A., and Ferrara, N. (1992). Binding sites for vascular endothelial growth factor are localized on endothelial cells in adult rat tissues. J. Clin. Invest., **89**, 244–53.

Jakeman, L.B., Armanini, M., Philips, H.S., and Ferrara, N. (1993). Developmental expression of binding sites and messenger ribonucleic acid for vascular endothelial growth factor suggests a role for this protein in vasculogenesis and angiogenesis. Endocrinol., **133**, 848–59.

Kendell, R.L., and Thomas, K.A. (1993). Inhibition of vascular endo-thelial growth factor by an endogenously encoded soluble recep-tor. *Proc. Natl. Acad. Sci. (USA)*, **90**, 10705–9.

Matthews, W., Jordan, C.T., Gavin, M., Jenkins, N.A., Copeland, N.G., and Lemischka, I.R. (1991). A receptor tyrosine kinase cDNA iso-lated from a population of enriched primitive hematopoetic cells and exhibiting close genetic linkage to c-kit. *Proc. Natl. Acad. Sci. (USA)*, **88**, 9026–30.

Millauer, B., Wizigmann-Voos, S., Schnurch, H., Martinez, R., Moller, N.P.H., Risau, W., and Ullrich, A. (1993). High affinity VEGF bind-ing and developmental expression suggest flk-1 as a major regu-lator of vasculogenesis and angiogenesis. *Cell*, **72**, 835–46.

Myoken, Y., Kayada, Y., Okamoto, T., Kan, M., Sato, G.H., and Sato, J.D. (1991). Vascular endothelial cell growth factor (VEGF) pro-duced by A-431 human epidermoid carcinoma cells and identifi-cation of VEGF membrane binding sites. *Proc. Natl. Acad. Sci. (USA)*, **88**, 5819–23.

Olander, J.V., Connolly, D.T., and DeLarco, J.E. (1991). Specific bind-ing of vascular permeability factor to endothelial cells. *Biochem. Biophys. Res. Commun.*, **175**, 68–76.

Peters, K.G., De Vries, C., and Williams, L.T. (1993). Expression of vascular endothelial growth factor receptor during embryogen-esis and tissue repair suggests a role in endothelial cell differen-tiation and blood vessel growth. *Proc. Natl. Acad. Sci.(USA)*, **90**, 8915–19.

Ploüet, J. and Moukadiri, H.J. (1990). Characterization of the recep-tors to vasculotropin on bovine adrenal cortex-derived capillary endothelial cells. *J. Biol. Chem.*, **265**, 22071–7.

Quinn, T.P., Peters, K.G., De Vries, C., Ferrara, N., and Williams, L.T. (1993). Fetal liver kinase 1 is a receptor for vascular endothelial growth factor and is selectively expressed in vascular endothe-lium. *Proc. Natl. Acad. Sci. (USA)*, **90**, 7533–7.

Rosnet, O., Stephenson, D., Mattei, M.G., Marchetto, S., Shibuya, M., Chapman, V.M., and Birnbaum, D. (1993). Close physical linkage of the FLT1 and FLT3 genes on chromosome 13 in man and chro-mosome 5 in mouse. *Oncogene*, **8**, 173–9.

Shen, H., Clauss, M., Ryan, J., Schmidt, A.M., Tijburg, P., Borden, L., Connolly, D., Stern, D., and Kao, J. (1993). Characterization of vascular permeability factor/vascular endothelial growth factor receptors on mononuclear phagocytes. *Blood*, **81**, 2767–73.

Shibuya, M., Yamaguchi, S., Yamane, A., Ikeda, T., Tojo, A., Mat-sushime, H., and Sato, M. (1990). Nucleotide sequence and ex-pression of a novel human receptor-type tyrosine kinase gene (flt) closely related to the fms family. *Oncogene*, **5**, 519–24.

Terman, B.I., Carrion, M.E., Kovacs, E., Rasmussen, B.A., Eddy, R.L., and Shows, T.B. (1991). Identification of a new endothelial cell growth factor receptor tyrosine kinase. *Oncogene*, **6**, 1677–83.

Terman, B.I., Dougher-Vermazen, M., Carrion, M.E., Dimitrov, D., Armellino, D.C., Gospodarowicz, D., and Bohlen, P. (1992a). Identification of the KDR tyrosine kinase as a receptor for vascular endothelial growth factor. *Biochem. Biophys. Res. Commun.*, **187**, 1579–86.

Terman, B.I., Jani-Sait, S., Carrion, M.E., and Shows, T.B. (1992b). The KDR gene maps to human chromosome 4q31.2—-q32, a locus which is distinct from locations for other type III growth factor receptor tyrosine kinases. *Cytogenet. Cell. Genet.*, **60**, 214–5.

Vaisman, N., Gospodarowicz, D., and Neufeld, G. (1990). Character-ization of the receptors for vascular endothelial growth factor. *J. Biol. Chem.*, **265**, 19461–9.

John E. Park and Napoleone Ferrara:
Department of Cardiovascular Research,
Genentech Inc.,
San Francisco, Ca 94080, USA

Endothelin and Its Receptors

Endothelins belong to a family of peptides with 21 amino acids which differ in two to five amino acids; all isoforms, however, contain two disulphide bridges. Endothelin-1 is a product of endothelial cells; in addition the human genome processes the genes for endothelin-2 and endothelin-3. Endothelins have potent biological effects, in particular vasoconstriction, vasodilation, proliferative properties as well as renal and endocrine effects. Production of endothelin is rather low under physiological conditions, but increased in most forms of vascular disease as well as in heart and kidney disease.

■ Endothelin genes

Endothelins are products of three distinct genes [GenBank accession numbers (AC): Y00749 (Endothelin-1), M65199 and X55177 (Endothelin-2), and X52001 (Endothelin-3)]. 2.2 kb human preproendothelin messenger RNA is encoded on chromosome 6 in five exons which are distributed over nearly 7 kb of the genome. The 5′ flanking region contains several response elements which confer inducibility of tran-scription by phorbol esters (e.g. fos/jun complex inducibil-ity), nuclear factor-1 (responsible for transforming growth factor beta-induced expression) and hexanucleotide se-quence elements, which may be involved in induction by acute physical stress. The 3′ untranslated region of the messenger-RNA contains 'suicide-motives' (AUUUA) poss-ibly involved in messenger-RNA instability. The life span of endothelin-1 messenger-RNA has been shown to be ex-tremely short (approximately 50 minutes) and is super in-duced by cycloheximide treatment of cells.

■ Endothelin peptides

Endothelins belong to a family of peptides with 21 amino acids and two disulphide bridges (Fig. 1) of molecular weight 2492 Da. Three isoforms are known: endothelin-1, endothelin-2, and endothelin-3. Safaratoxins (i.e. snake

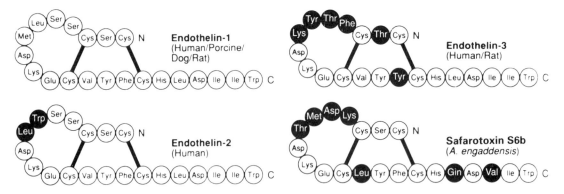

Figure 1. Structure and amino acid sequence of endothelins as well as of the snake-venom safaratoxin.

venoms) also have a high degree of homology with the endothelins.

Endothelin is produced by endothelial cells, neuronal cells and possibly numerous other cells. Endothelial cells produce only endothelin-1. Its production requires *de novo* protein synthesis and the protein does not appear to be stored in the cells. Endothelin is formed from preproendothelin (203 amino acids) to big endothelin or proendothelin (38 amino acids; molecular weight 4283 Da) to its final product endothelin. The last step in the synthesis involves an endothelin converting enzyme, most likely a metalloprotease which has not been identified yet. Plasma levels of endothelin can be measured most conveniently with radioimmunoassay techniques. A number of anti-

bodies are commercially available with different cross reactivity patterns for the different isoforms and the precursors (i.e. big endothelin) of the peptide. For the measurement of plasma levels, the extraction procedure seems to be crucial and different techniques may explain the wide spread of the range of normal values reported in the literature.

The production of endothelin is increased by physical factors (i.e. hypoxia), or receptor-operating agonists such as thrombin, angiotensin II, arginine vasopressin, epinephrine, platelet-derived products, as well as cytokines such as interleukin-1. The fact that the calcium ionophore A23187 is a potent stimulator for endothelin production demonstrates that increases in intracellular calcium in endothelial cells are crucial for its production. In addition,

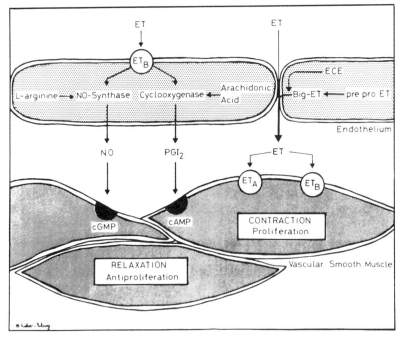

Figure 2. Putative effects of endothelin-1 (ET-1) binding to receptors present on either vascular smooth muscle or endothelial cells. NO, nitric oxide; PGI$_2$, ECE, endothelin converting enzyme; cAMP, cyclic 3',5'-adenosine monophosphate; cGMP, 3',5'-guanosine monophosphate.

endothelin production is inhibited by endothelium-derived nitric oxide via the formation of cyclic GMP, possibly also via prostaglandins activating cAMP, as well as by a putative vascular smooth muscle cell-derived inhibitory factor (Fig. 2). These events may explain why the vascular production and circulating levels of endothelin in plasma are very low under normal conditions. Furthermore, most of the endothelin produced in the vessel wall is released abluminally towards smooth muscle and hence is not detected in the supernatant or plasma respectively.

■ Endothelin function

The biological role of endothelin is uncertain, but clearly the peptide does have potent biological properties such as vasoconstriction, vasodilation, stimulation of proliferation of different cell lines, in particular vascular smooth muscle cells, as well as profound renal and endocrine effects.

■ Endothelin receptors

Vasoconstriction induced by endothelin-1 involves activation of endothelin receptors on vascular smooth muscle. At low concentrations of endothelin both ET_A and ET_B receptors contribute, while at higher concentrations of the peptide, ET_A receptors appear to be more dominant. Genes encoding the ET_A and ET_B receptors in mammals have been cloned [GenBank accession numbers L06622, S63938, D90348, and X61950 (ET_A receptor); L06623 and S57283 (ET_B receptor)] and a gene encoding the ET_C receptor has been cloned in frogs (GenBank accession number L20299). The receptors are linked to phospholipase C which in turn leads to the formation of inositol triphosphate (IP_3) as well as diacyl-glycerol (DAG). IP_3 stimulates the release of intracellular calcium which is important for the initiation of endothelin-induced contraction. In addition, the receptors are linked (via a G_i-protein) to voltage-operated calcium channels. If this pathway is expressed, calcium antagonists are potent inhibitors of the response to endothelin. Endothelins decrease membrane potential of vascular smooth muscle cells, in particular those obtained from veins, and thereby further increase the influx of extracellular calcium. As compared to other vasoconstrictor hormones, endothelin is extremely potent with half maximally active concentration 1–2 orders of magnitude lower than those of other agonists.

The vasodilation induced by endothelins is mediated by ET_B receptors on endothelial cells linked to the formation of nitric oxide and prostacyclin. This endothelium-dependent inhibition of its own effects is particularly prominent with intraluminal infusion of endothelin.

■ Endothelin receptor agonists

Recently, a number of endothelin receptor antagonists have been developed which either interfere with ET_A or both ET_A and ET_B receptors. Depending on the vascular bed studied, and whether ET_A or both ET_A and ET_B receptors are expressed, selective ET_A antagonists are able to interfere with the effects of endothelin or combined endothelin receptor antagonists are required. Selective agonists for the ET_B receptor are safratoxins as well as some newly developed synthetic compounds. These tools now allow the delineation of the physiological and pathophysiological roles of endothelins as well as the receptors involved.

■ Endothelin and disease

Endothelin plasma levels have been shown to be increased in various disease states, in particular vascular disease such as atherosclerosis, Raynaud's disease, and Takayashu's disease. Myocardial infarction, congestive heart failure and cardiogenic shock are all associated with high levels of endothelin. Furthermore, pulmonary hypertension (that occurring at high altitude as well as other forms of the disease) is characteristically associated with increased plasma endothelin levels. Other disease states which are associated with increased endothelin levels are all forms of renal disease, hepato-renal syndrome and subarachnoid haemorrhage. The currently available endothelin antagonists will allow the determination of whether endothelin is an epiphenomenon and/or an important pathophysiological mediator under these conditions.

Transgenic animals expressing the endothelin-2 gene have been developed. In spite of high circulating levels of endothelin-2, the animals do not appear to suffer from any cardiovascular abnormality nor do they have hypertension. On the other hand, animals in which the endothelin gene has been knocked out exhibit developmental abnormalities of the oro-pharynx and they are hypertensive.

■ References

Boulanger, C., and Lüscher, T.F. (1990). Release of endothelin from the porcine aorta. Inhibition by endothelium-derived nitric oxide. *J. Clin. Invest.*, **85**, 587–90.

Inoue, A., Yanagisawa, M., Kimura, S., Kasuya, Y., Miyauchi, T., Goto, K., and Masaki, T. (1989). The human endothelin family: three structurally and pharmacologically distinct isopeptides predicted by three separate genes. *Proc. Natl. Acad. Sci. (USA)*, **86**, 2863–7.

Inoue, Y., Yanagisawa, M., Takuwa, Y., Mitsui, Y., Kobayashi, M., and Masaki, T. (1990). The human prepro-endothelin-1 gene. *J. Biol. Chem.*, **264**, 14954–9,

Lin, H.Y., Kaji, E.H., Winkel, G.K., Ives, H.E., and Lodish, H.F. (1991). Cloning and functional expression of a vascular smooth muscle endothelin 1 receptor. *Proc. Natl. Acad. Sci. (USA)*, **88**, 3185–9.

Lüscher, T.F., Oemar, B.S., Boulanger, C.M., and Hahn, A.W.A. (1993). Molecular and cellular biology of endothelin and its receptors–Part I. *J. Hypertension*, **11**, 7–11.

Lüscher, T.F., Oemar, B.S., Boulanger, C.M., and Hahn, A.W.A. (1993). Molecular and cellular biology of endothelin and its receptors–Part II. *J. Hypertension*, **11**, 121–6.

Sakurai, T., Yanagisawa, M., Takuwa, Y., Miyazaki, H., Kimura, S., Goto, K., and Masaki, T. (1990). Cloning of a cDNA encoding a non-isopeptide-selective subtype of the endothelin receptor. *Nature*, **348**, 732–5.

Yanagisawa, M., Kurihara, H., Kimura, S., Tomobe, Y., Kobayashi, M., Mitsui, Y., Yazaki, Y., Goto, K., and Masaki, T. (1988). A novel potent vasoconstrictor peptide produced by vascular endothelial cells. *Nature*, **332**, 411–5.

Yanagisawa, M., Inoue, A., Ishikawa, T., Kasuya, Y., Kimura, S., Kumagaye, S., Nakajima, K., Watanabe, T.X., Sakakibara, S., Goto, K., and Masaki, T. (1988). Primary structure, synthesis, and biological activity of rat endothelin, an endothelium-derived vasoconstrictor peptide. *Proc. Natl. Acad. Sci. (USA)*, **85**, 6964–7.

Yanagisawa, M. ,and Masaki, T. (1989). Biochemistry and molecular biology of the endothelins. *Tr. Pharmacol. Sci.*, **10**, 374–8.

Thomas F. Lüscher:
Cardiology,
University Hospital,
Inselspital,
CH-3010 Bern, Switzerland

Appendix: Chromosomal and Genetic Localization of Genes for Murine and Human Cytokines and Their Receptors

Over the past few years, the genes encoding a substantial number of murine and human cytokines and their cognate cellular receptors have been cloned. In most cases, the chromosomal or genetic localizations of these genes have also been determined. While in many cases little immediate information, beyond the actual localization, has been gleaned from these studies, in several cases tantalizing locations and linkages have emerged. This chapter contains an extensive compilation of currently available information regarding the localization of the murine and human genes for the cytokines and their cognate receptors discussed in this Guidebook. It should be noted that, since in many instances issues beyond the strict localization of a particular gene have emerged (for example, tight linkage to other, perhaps related, genes), the literature cited also includes references to these issues.

With the notable exception of the IFNγ receptor, which was initially localized by monitoring the display of the human IFNγ receptor on the surface of a panel of human: mouse somatic cell hybrids which retain various human chromosomes, the chromosomal localizations given were determined using cloned gene probes in conjunction with either *in situ* hybridization to metaphase chromosome spreads, genetic linkage analysis or using panels of interspecies somatic cell hybrids. Moreover, in certain instances, close physical linkage has been established by pulsed-field gel electrophoresis, and by analysis of overlapping genomic clones.

The currently available chromosomal and genetic localizations of genes for various cytokines and their cognate receptors discussed in this Guidebook are summarized in Tables 1 and 2, respectively. In several instances, correspondence between a gene defined physically and a marker defined phenotypically has been established, in which case the genetic name is given in parenthesis after the chromosomal/genetic localization, with correspondences that have not been formally proven indicated by a question mark.

Table 1. Chromosal and genetic location of cytokine loci

Cytokine	Mouse gene	Reference	Human gene	Reference
IL-1α/β	2F	1, 2	2q13-21	3, 4
IL-1 receptor antagonist	2 proximal	5	2q14-21	6, 7
IL-1β convertase	9 proximal	8	11q23	9
IL-2	3B-C	10, 11, 12	4q26-28	13
IL-3 (Multi-CSF)	11A5-B1	10, 14, 15, 16, 17, 18, 19, 20	5q23-31	21, 22, 23, 24, 25, 26, 27
IL-4	11A5-B1	10, 28, 29	5q31	23, 25, 26, 27, 30, 31, 32
IL-5	11A5-B1	18, 28, 33	5q31	23, 25, 26, 27, 29, 32, 34
IL-6	5 proximal	35	7p15-p21	30, 36, 37, 38
IL-7	3	39, 40	8q12-13	39
MIP1/IL-8 family (MCP-1/JE; MIP-1α; MIP-1β; TCA3)	11	33, 41, 42	17q11.2-12	43, 44, 45
MIP2 family (MGSA; PF4; γIP-10; βTG)			4q13-21	47, 48, 49, 50
IL-9	13	51	5q22-35	26, 27, 51
IL-10	1	52	1	52
IL-11			19q13.3-13.4	53
IL-12A (p35)			3p12-q13.2	54
IL-12B (p40)			5q31-33	26, 54
IL-13	11	55	5q23-31	55, 56, 57
BCGF (12kD)			16q11.2-12.1	271
CD40L			Xq26.3-27.1 (HIGM1)	103, 104, 105, 106, 107, 108
TNFα/β	17	100	6p	101, 102
IFNα	4C3-C7	88, 89, 90, 91	9p	92
IFNβ1	4	91, 93	9p21	94
IFNγ	10	82, 83, 95, 96, 97	12q24.1	98, 99
LIF	11A1-A2	75, 76 (See Footnote 2)	22q12.1-12.2	77, 78, 79
Oncostatin M (OSM)			22q12	79, 80
CNTF	19	142	11q12	143, 144, 145
NGFβ	3	123, 124, 133, 134	1p22	132, 135
NGFγ	7	116, 136		
BDNF	2	137	11p13-14	137, 138, 139
NT-3	6	137	12p13	137, 139
NT-4			19q13.38	140
NT-5	7	141	19	141
Neu Differentiation Factor (NDF)			8p12-21	132
Erythropoietin	5G	70, 71	7q11-22	72, 73, 74

Table 1. (*continued*)

G-CSF	11D-E1	16, 19	17q11.2-21	61, 62, 63, 64
M-CSF (CSF-1)	3F3 (*op*)	65, 66, 67, 68	1p13-p21	69 (See Footnote 1)
GM-CSF	11A5-B1	14, 16, 17, 19, 20, 58	5q23-31	23, 24, 25, 59, 60
Stem cell factor (SCF)	10 (*Sl*)	81, 82, 83, 84, 85, 86	12q14.3-qter	87
HGF (Hepatocyte growth factor; Scatter factor)			7q11-22	177, 178, 179, 180, 181
Growth hormone	11 distal	158, 159	17q22-24	160
Chorionic somatomammotropin			17	160
GH-like gene			17	160
Prolactin	13	161	6p	102, 162
EGF	3	40, 122, 123, 124	4q25-27	124, 125
TGFα	6 (*wa-1*)	109, 110, 111	2p13	112, 113
Amphiregulin			4q13-21	272
HB-EGF			5	273
PDGF A	5	71	7p21-22	126, 127
PDGF B (c-sis)	15E	128, 129	22q12.3-13.1	130, 131
Insulin-1	6	146		
Insulin-2	7	147	11p15	148, 149, 150, 151
IGF-1	10	82, 83, 96, 152	12q22-24.1	125, 153, 154, 155
IGF-II	7	146, 156	11p15	125, 148, 154, 157
FGF1 (aFGF)	18	129, 163, 164	5q23-33	27, 165
FGF2 (bFGF)	3	71, 166	4q26-27	167
FGF3			11q13	168
FGF4			11q13	169
FGF5	5	166, 170	4q21	171
FGF6	6F3-G1	172, 173, 174, 175	12p13	176
TGFβ1	7	114, 115, 116	19q13.1-13.3	115, 117
TGFβ2	1	114, 119	1q41	119, 120
TGFβ3	12	114, 119	14q23-24	119, 121
PlGF			14	278
Endothelin-1			6p23-24	274, 275
Endothelin-2			1p34	275
Endothelin-3			20q13.2-13.3	275, 276, 277

Footnotes to Table 1:

1. This gene was originally assigned to 5q33.1 (see ref. 182).

2. The murine LIF gene, under the synonym HILDA, was erroneously assigned to chromosome 13 (ref. 183) presumably due to hybridisation to repeated sequences in the 3' untranslated region (ref. 184).

Table 2. Chromosal and genetic location of cytokine receptor loci

Receptor	Mouse gene	Reference	Human gene	Reference
IL-1-R type I	1 centro	185	2q12	185
IL-2-Rα	2 A2-A3	12	10p14-15	186, 187
IL-2-Rβ	15E	128, 188	22q11.2-12	189, 190
IL-2-Rγ			Xq13	279
IL-3-Rα	14 proximal	191	X/Y pseudoautosomal region	192
IL-3/GM-CSF/IL-5Rβ (Aic2b)	15	193	22q12.3-13.1	194
Aic2a	15	193		
IL-4-R	7 distal	71, 195	16p11.2-12.1	195
IL-5-Rα	6 distal	196	3p24-26	197
IL-6-Rα			1	198
IL-6-R-like gene			9	198
IL-6/LIF-Rβ (gp130)			5 and 17	199
IL-7-R	15 proximal	200, 201	5p13	202
IL-8-RA (type I)			2q35	203, 204, 205
IL-8-RB (type II)	1	206	2q35	203, 205
IL-8-R pseudogene			2q35	203, 205
CD40	2 distal	235		
TNF-R type I (p80)	4 distal	232	1p36	233
TNF-R type II (p60)	6 distal	175, 232	12p13	233, 234
IFNα-R	16	92	21	228
IFNβ-R	16	92	21	228
IFNγ-R	10	92, 97, 229	6q16-22	230, 231
LIF-Rα	15 proximal	224	5p12-13	224
CNTF-R			9p13	249
NGF-R	11	19	17q12-22	248
TRKA			1q23-24	280
neu (c-erbB2)	11 distal	16, 19, 245	17q11.2-12	246, 247
EPO-R	9	222	19pter-q12	222, 223
G-CSF-Ra	4 distal	211	1p34.2-35.1	212, 213
G-CSF-Rb	19	211		
M-CSF-R (c-fms)	18D (*Fim-2*)	19, 129, 163, 214, 215, 216, 217	5q33.2-33.3	26, 27, 60, 218, 219, 220, 221
GM-CSF-Rα	19D2	207	X/Y pseudoautosomal region	208, 209, 210
SCF-R (c-kit)	5 (*W*)	81, 225, 226	4q11-12	225, 227
HGF-R (c-met)	6	281	7q21-31	282
Growth hormone-R	15	201, 263, 264	5p13.1-12	263, 265

Table 2. (*continued*)

Prolactin-R	15	264	5p13-14	265
EGF-R (c-erbB1)	11 proximal (*wa-2*)	16, 19, 20, 236, 237	7p12-13	238
PDGF-RA	5 (*Ph?*)	239, 240, 241	4q11-12	239, 242
PDGF-RB	18D	163, 243, 244	5q33.2-33.3	219
Insulin-R	8	216	19p13.2-13.3	250
IGF-I-R	7	116, 136, 251	15q25-26	252
IGF-II-R (CIM6PR)	17 (*Tme?*)	253, 254, 255	6q25-27	254
IGFBP1			7p12-14	253, 256, 257, 258, 259
IGFBP2	1	260	2q33-34	253, 256, 260, 261
IGFBP3			7p12-14	253, 256, 257
IGFBP4			17q12-21.1	253, 256, 262
IGFBP5			5	253
IGFBP6			12	253
FGFR1 (flg)			8p11.2-12	266, 267
FGFR2			10q26	268
FGFR3			4p16.3	269
FGFR4			5q33-qter	270
TGF-βR (Endoglin)			9q34-qter	283
TGF-βR Type II			3	284
TGF-βR Type III			1	285
VEGF-R FLT-1	5	286	13q12	286
VEGF-R KDR/FLK-1	5	287	4q31.2-32	288
Endothelin-R A			4	289, 290
Endothelin-R B			13	291

References

1. D'Eustachio, P., Jadidi, S., Fuhlbrigge, F.C., Gray, P.W., and Chaplin, D.D. (1987). Interleukin-1 α and β genes: linkage on chromosome 2 in the mouse. *Immunogenetics*, **26**, 339–43.
2. Boultwood, J., Breckon, G., Birch, D., and Cox, R. (1989). Chromosomal localization of murine interleukin-1 α and β genes. *Genomics*, **5**, 481–5.
3. Webb, A.C., Collins, K.L., Auron, P.E., Eddy, R.L., Nakai, H., Byers, M.G., Haley, L.L., Henry, W.M., and Shows, T.B. (1986). Interleukin-1 gene [IL-1] assigned to long arm of human chromosome 2. *Lymphokine Res.*, **5**, 77–85.
4. LaFage, M., Maroc, N., Dubrevil, P., de Waal Malefijt, R., Pebusque, M.-J., Carcassone, Y., and Mannoni, P. (1989). The human interleukin-1α gene is located on the long arm of chromosome 2 at band q13. *Blood*, **73**, 104–7.
5. Zahedi, K., Seldin, M.F., Rits, M., Ezekowitz, R.A.B., and Whitehead, A.S. (1991). Mouse IL-1 receptor antagonist protein: molecular characterization, gene mapping, and expression of mRNA *in vitro* and *in vivo*. *J. Immunol.*, **146**, 4228–33.
6. Steinkasserer, A., Spurr, N.K., Cox, S., Jegga, P., and Sim, R.B. (1992). The human IL-1 receptor antagonist gene (IL1RN) maps to chromosome 2q14-q21, in the region of the IL-1α and IL-1β loci. *Genomics*, **13**, 654–7.
7. Patterson, D., Jones, C., Hart, I., Bleska, J., Berger, R., Geyer, D., Eisenberg, S.P., Smith, M.F., and Arend, W.P. (1993). The human interleukin-1 receptor antagonist (IL1RN) gene is located in the chromosome 2q14 region. *Genomics*, **15**, 173–6.
8. Nett, M.A., Cerretti, D.P., Berson, D.R., Seavitt, J., Gilbert, D.J., Jenkins, N.A., Copeland, N.G., Black, R.A., and Chaplin, D.D. (1992). Molecular cloning of the murine IL-1β converting enzyme cDNA. *J. Immunol.*, **149**, 3254–9.
9. Cerretti, D.P., Kozlosky, C.J., Mosley, B., Nelson, N., Van Ness, K., Greenstreet, T.A., March, C.J., Kronheim, S.R., Druck, T., Cannizzaro, L.A., Huebner, K., and Black, R.A. (1992). Molecular cloning of the interleukin-1β converting enzyme. *Science*, **256**, 97–100.
10. D'Eustachio, P., Brown, M., Watson, C., and Paul, W.E. (1988). The IL-4 gene maps to chromosome 11, near the gene encoding IL-3. *J. Immunol.*, **141**, 3067–71.
11. Fiorentino, L., Austen, D., Pravtcheva, D., Ruddle, F.H., and Brownell, E. (1989). Assignment of the interleukin-2 locus to mouse chromosome 3. *Genomics*, **5**, 651–3.
12. Webb, G.C., Campbell, H.D., Lee, J.S., and Young, I.G. (1990). Mapping the gene for murine T-cell growth factor, IL-2, to bands B–C on chromosome 3 and for the α chain of the IL-2 receptor, IL-2ra, to bands A2-A3 on chromosome 2. *Cytogenet. Cell Genet.*, **54**, 164–8.
13. Seigel, L.J., Harper, M.E., Wong-Staal, F.W., Gallo, R.C., Nash, W.G., and O'Brien, S.J. (1984). Gene for T-cell growth factor: location on human chromosome 4q and feline chromosome B1. *Science*, **223**, 175–8.
14. Barlow, D.P., Bucan, M., Lehrach, H., Hogan, B.L.M., and Gough, N.M. (1987). Close genetic and physical linkage between the murine haemopoietic growth factor genes GM-CSF and Multi-CSF [IL3]. *EMBO J.*, **6**, 617–23.
15. Ihle, J.N., Silver, J., and Kozak, C.A. (1987). Genetic mapping of the mouse interleukin-3 gene to chromosome 11. *J. Immunol.*, **138**, 3051–4.
16. Buchberg, A.M., Bedigan, H.G., Taylor, B.A., Brownell, E., Ihle, J.N., Nagata, S., Jenkins, N.A., and Copeland, N.G. (1988). Localization of Evi-2 to chromosome 11: linkage to other, proto-oncogene and growth factor loci using interspecific backcross mice. *Oncogene Res.*, **2**, 149–65.
17. Lee, J.S., and Young, I.G. (1989). Fine-structure mapping of the murine IL3 and GM-CSF genes by pulsed-field gel electrophoresis and molecular cloning. *Genomics*, **5**, 359–62.
18. Webb, G.C., Lee, J.S., Campbell, H.D., and Young, I.G. (1989). The genes for interleukins 3 and 5 map to the same locus on mouse chromosome 11. *Cytogenet. Cell. Genet.*, **50**, 107–10.
19. Buchberg, A.M., Brownell, E., Nagata, S., Jenkins, N.A, and Copeland, N.G. (1990). A comprehensive genetic map of murine chromosome 11 reveals extensive linkage conservation between mouse and human. *Genetics*, **122**, 153–61.
20. Buckwalter, M.S., Katz, R.W., and Camper, S.A. (1991). Localization of the panhypopituitary dwarf mutation (df) on mouse chromosome 11 in an intersubspecific backcross. *Genomics*, **10**, 515–26.
21. Le Beau, M.M., Epstein, N.D., O'Brien, S.J., Nienhuis, A.W., Yang, Y-C., Clark, S.C., and Rowley, J.D. (1987). The interleukin 3 gene is located on human chromosome 5 and is deleted in myeloid leukaemias with a deletion of 5q. *Proc. Natl. Acad. Sci. (USA)*, **84**, 5913–7.
22. Yang, Y-C., Kovacis, S., Kriz, R., Wolf, S., Clark, S.C., Wellems, T.E., Nienhuis, A., and Epstein, N. (1988). The human genes for GM-CSF and IL-3 are closely linked in tandem on chromosome 5. *Blood*, **71**, 958–61.
23. van Leeuwen, B.H., Martinson, M.E., Webb, G.C., and Young, I.G. (1988). Molecular organization of the cytokine gene cluster, involving the human IL-3, IL-4, IL-5 and GM-CSF genes, on chromosome 5. *Blood*, **73**, 1142–8.
24. Frolova, E.I., Dolganov, G.M., Mazo, I.A., Smitnov, D.V., Copeland, P., Stewart, C., O'Brien, S.J., and Dean, M. (1991). Linkage mapping of the human *CSF2* and *IL3* genes. *Proc. Natl. Acad. Sci. (USA)*, **88**, 4821–4.
25. Huebner, K., Nagarajan, L., Besa, E., Angert, E., Lange, B.J., Cannizzaro, L.A., van den Berghe, H., Santoli, D.,

Finan, J., Croce, C.M., and Nowell, P.C. (1990). Order of genes on human chromosome 5q with respect to 5q interstitial deletions. *Am. J. Hum. Genet.*, **46**, 26–36.

26. Warrington, J.A., Bailey, S.K., Armstrong, E., Aprelikova, O., Alitalo, K., Dolganov, G.M., Wilcox, A.S., Sikela, J.M., Wolfe, S.F., Lovett, M., and Wasmuth, J.J. (1992). A radiation hybrid map of 18 growth factor, growth factor receptor, hormone receptor, or neurotransmitter receptor genes on the distal region of the long arm of chromosome 5. *Genomics*, **13**, 803–8.

27. Saltman, D.L., Dolganov, G.M., Warrington, J.A., Wasmuth, J.J., and Lovett, M. (1993). A physical map of 15 loci on human chromosome 5q23-q33 by two-color fluorescence *in situ* hybridization. *Genomics*, **16**, 726–32.

28. Lee, J.S., Campbell, H.D., Kozak, C.A., and Young, I.G. (1989). The IL-4 and IL-5 genes are closely linked and are part of a cytokine gene cluster on mouse chromosome 11. *Somatic Cell and Mol.Genet*, **15**, 143–52.

29. Takahashi, M., Yoshida, M.C., Satoh, H., Hilgers, J., Yaoita, Y., and Honjo, T. (1989). Chromosomal mapping of the mouse IL-4 and human IL-5 genes. *Genomics*, **4**, 47–52.

30. Sutherland, G.R., Baker, E., Callen, D.F., Hyland, V.J., Wong, G., Clark, S., Jones, S.S., Eglington, L.K., Shannon, M.F., Lopez, A.F., and Vadas, M.A. (1988). Interleukin 4 is at 5q31 and interleukin 6 is at 7p15. *Hum. Genet.*, **79**, 335–7.

31. Arai, N., Nomura, D., Villaret, D., De Waal Malefijt, R., Seiki, M., Yoshida, M., Minoshima, S., Fukuyama, R., Maekawa, M., Kudoch, J., Shimizu, N., Yokota, K., Abe, E., Yokota, T., Takebe, Y., and Arai, K. (1989). Complete nucleotide sequence of the chromosomal gene for human IL-4 and its expression. *J. Immunol.*, **142**, 274–82.

32. Chanrasekharappa, S.C., Rebelsky, M.S., Firak, T.A., Lebeau, M.M., and Westbrook, C.A. (1990). A long-range restriction map of the interleukin-4 and interleukin-5 linkage group on chromosome 5. *Genomics*, **6**, 94–9.

33. Wilson, S.D., Billings, P.R., D'Eustachio, P., Fournier, R.E.K., Geissler, E., Lalley, P.A., Burd, P.R., Housman, D.E., Taylor, B.A., and Dorf, M.E. (1990). Clustering of cytokine genes on mouse chromosome 11. *J. Exp. Med.*, **171**, 1301–14.

34. Sutherland, G.R., Baker, E., Callen, D.F., Campbell, H.D., Young, I.G., Sanderson, C.J., Garson, O.M., Lopez, A., and Vadas, M.A. (1988). Interleukin 5 is at 5q31 and is deleted in the 5q⁻ syndrome. *Blood*, **71**, 1150–2.

35. Mock, B.A., Nordan, R.P., Justice, M.J., Kozak, C., Jenkins, N.A., Copeland, N.G., Clark, S.C., Wong, G.G., and Rudikoff, S. (1989). The murine IL-6 gene maps to the proximal region of chromosome 5. *J. Immunol.*, **141**, 1372–6.

36. Bowcock, A.M., Kidd, J.R., Lathrop, G.M., Daneshvar, L., May, L.T., Ray, A., Sehgal, P.B., Kidd, K.K., and Cavalli-Sforza, L.L. (1988). The human 'Interferon-β2/hepatocyte stimulating factor/Interleukin-6' gene: DNA polymorphism studies and localization to chromosome 7p21. *Genomics*, **3**, 8–16.

37. Sehgal, P.B., Zilberstein, A., Ruggieri, R-M., May, L.M.,

Fersuson-Smith, A., Slate, D.L., Revel, M., and Ruddle, F.H. (1986). Human chromosome 7 carries the β_2-interferon gene. *Proc. Natl. Acad. Sci. (USA)*, **83**, 5219–22.

38. Ferguson-Smith, A.C., Chen, Y.F., Newmans, M.S., May, L.T., Sehgal, P.B., and Ruddle, F.H. (1988). Regional localization of the interferon-beta 2/B-cell stimulatory factor 2/hepatocyte stimulating factor gene to human chromosome 7p15-p21. *Genomics*, **2**, 203–8.

39. Goodwin, R. (1991). *Cancer Cells*, **3**, 72.

40. Siracusa, L.D., Rosner, M.H., Vigano, M.A., Gilbert, D.J., Staudt, L.M., Copeland, N.G., and Jenkins, N.A. (1991). Chromosomal location of the octamer transcription factors, *Otf-1*, *Otf-2*, and *Otf-3*, defines multiple *Otf-3*-related sequences dispersed in the mouse genome. *Genomics*, **10**, 313–26.

41. Wilson, S., and Dorf, M. quoted in: Wolpe, S.D., and Cerami, A. (1989). Macrophage inflammatory proteins 1 and 2: members of a novel superfamily of cytokines. *FASEB J.*, **3**, 2565–73.

42. Smith, A., Lalley, P.A., Killary, A.M., Ghosh-Choudhury, G., Wang, L.M., Han, E.S., Martinez, L., Naylor, S.L., and Sakaguchi, A.Y. (1989). Sigje, a member of the small inducible gene family that includes platelet factor 4 and melanoma growth stimulatory activity, is on mouse chromosome 11. *Cytogenet. Cell Genet.*, **52**, 194–6.

43. Irving, S.G., Zipfel, P.F., Balke, J., McBride, O.W., Morton, C.C., Burd, P.R., Siebenlist, U., and Kelly, K. (1990). Two inflammatory mediator cytokine genes are closely linked and variably amplified on chromosome 17q. *Nucleic Acids Res.*, **18**, 3261–70.

44. Mehrabian, M., Sparkes, R.S., Mohandas, T., Fogelman, A.M., and Lusis, A.J. (1991). Localization of monocyte chemotactic protein-1 gene (SCYA2) to human chromosome 17q11.2-q21.1. *Genomics*, **9**, 200–3.

45. Rollins, B.J., Morton, C.C., Ledbetter, D.H., Eddy, R.L., and Shows, T.B. (1991). Assignment of the human small inducible cytokine A2 gene, SYCA2 (encoding JE or MCP-1), to 17q11.2–12: Evolutionary relatedness of cytokines clustered at the same gene locus. *Genomics*, **10**, 489–92.

46. Reference 46 deleted from reference list.

47. Richmond, A., Balentien, E., Thomas, H.G., Flaggs, G., Barton, D.E., Spiess, J., Borodoni, R., Francke, U., and Derynck, R. (1988). Molecular characterization and chromosomal mapping of melanoma growth stimulatory activity, a growth factor structurally related to beta-thromboglobulin. *EMBO J.*, **7**, 2025–33.

48. Griffin, C.A., Emanuel, B.S., La Rocco, P., Schwartz, E., and Poncz, M. (1987). Human platelet factor 4 gene is mapped to 4q12-q21. *Cytogenet. Cell Genet.*, **45**, 67–9.

49. Luster, A.D., Jhanwar, S.C., Chaganti, R.S.K., Kersey, J.H., and Ravetch, J.V. (1987). Interferon-inducible gene maps to a chromosomal band associated with a (4;11) translocation in acute leukaemia cells. *Proc. Natl. Acad. Sci. (USA)*, **84**, 2868–71.

50. Tunnacliffe, A., Majumdar, S., Yan, B., and Poncz, M. (1992). Genes for β-thromboglobulin and platelet factor 4 are closely linked and form part of a cluster of related genes on chromosome 4. *Blood*, **79**, 2896–900.

51. Mock, B.A., Krall, M., Kozak, C.A., Nesbitt, M.N., McBride, O.W., Renauld, J-C., and Van Snick, J. (1990). IL9 maps to mouse chromosome 13 and human chromosome 5. *Immunogenetics*, **31**, 265–70.

52. Kim, J.M., Brannan, C.I., Copeland, N.G., Jenkins, N.A., Khan, T.A., and Moore, K.W. (1992). Structure of the mouse IL-10 gene and chromosomal localization of the mouse and human genes. *J. Immunol.*, **148**, 3618–23.

53. McKinley, D., Wu, Q., Yang-Feng, T., and Yang, Y-C. (1992). Genomic sequence and chromosomal location of human interleukin-11 gene (IL11). *Genomics*, **13**, 814–9.

54. Sieburth, D., Jabs, E.W., Warrington, J.A., Li, X., Lasota, J., LaForgia, S., Kelleher, K., Huebner, K., Wasmuth, J.J., and Wolf, S.F. (1992). Assignment of genes encoding a unique cytokine (IL12) composed of two unrelated subunits to chromosomes 3 and 5. *Genomics*, **14**, 59–62.

55. McKenzie, A.N.J., Li, X., Largaespada, D.A., Sato, A., Kaneda, A., Zurawski, S.M., Doyle, E.L., Milatovich, A., Francke, U., Copeland, N.G., Jenkins, N.A., and Zurawski, G. (1993). Structural comparison and chromosomal localization of the human and mouse IL-13 genes. *J. Immunol.*, **150**, 5436–44.

56. Minty, A., Chalon, P., Derocq, J-M., Dumont, X., Guillemot, J-C., Kaghad, M., Labit, C., Leplatois, P., Liauzun, P., Miloux, B., Minty, C., Casellas, P., Loison, G., Lupker, J., Shire, D., Ferrara, P., and Caput, D. (1993). Interleukin-13 is a new human lymphokine regulating inflammatory and immune responses. *Nature*, **362**, 248–50.

57. Morgan, J.G., Dolganov, G.M., Robbins, S.E., Hinton, L.M., and Lovett, M. (1992). The selective isolation of novel cDNAs encoded by the regions surrounding the human interleukin 4 and 5 genes. *Nucleic Acids Res.*, **20**, 5173–9.

58. Gough, N.M., Gough, J., Metcalf, D., Kelso, A., Grail, D., Nicola, N.A., Burgess, A.W., and Dunn, A.R. (1984). Molecular cloning of cDNA encoding a murine haemopoietic growth regulator, granulocyte-macrophage colony stimulating factor. *Nature*, **309**, 763–7.

59. Huebner, K., Isobe, M., Croce, C.M., Golde, D.W., Kaufman, S.E., and Gasson, J.C. (1985). The human gene encoding GM-CSF is at 5q21-q32, the chromosome region deleted in the 5q– anomaly. *Science*, **230**, 1282–5.

60. Le Beau, M.M., Westbrook, C.A., Diaz, M.O., Larson, R.A., Rowley, J.D., Gasson, J.C., Golde, D.W., and Sherr, C.J. (1986). Evidence for the involvement of *GM-CSF* and *FMS* in the deletion [5q] in myeloid disorders. *Science*, **231**, 984–7.

61. Simmers, R.N., Webber, L.M., Shannon, M.F., Garson, O.M., Wong, G., Vadas, M.A., and Sutherland, G.R. (1987). Localization of the G-CSF gene on chromosome 17 proximal to the breakpoint in the t[15;17] in acute promyelocytic leukaemia. *Blood*, **70**, 330–2.

62. Simmers, R.N., Smith, J., Shannon, M.F., Wong, G., Lopez, A.F., Baker, E., Sutherland, G.R., and Vadas, M.A. (1988). Localization of the human G-CSF gene to the region of a breakpoint in the translocation typical of acute promyelocytic leukaemia. *Hum. Genet.*, **78**, 134–6.

63. Kanda, N., Fukushige, S., Murotsu, T., Yoshida, M.C., Tsuchiya, M., Asano, S., Kaziro, Y., and Nagata, S. (1987). Human gene coding for granulocyte-colony stimulating factor is assigned to the q21-q22 region of chromosome 17. *Somat. Cell. Mol. Genet.*, **13**, 679–84.

64. Xu, W., Gorman, P.A., Rider, S.H., Hedge, P.J., Moore, G., Prichard, C., Sheer, D., and Solomon, E. (1988). Construction of a genetic map of human chromosome 17 by use of chromosome-mediated gene transfer. *Proc. Natl. Acad. Sci. (USA)*, **85**, 8563–7.

65. Gisselbrecht, S., Sola, B., Fichelson, S., Bordereaux, D., Tambourin, P., Mattei, M.G., Simon, D., and Guenet, J.L. (1989). The murine M-CSF gene is localized on chromosome 3. *Blood*, **73**, 1742–6.

66. Buchberg, A.M., Jenkins, N.A., and Copeland, N.G. (1989). Localization of the murine macrophage colony stimulating factor gene to chromosome 3 using interspecific backcross analysis. *Genomics*, **5**, 363–7.

67. Yoshida, H., Hayashi, S., Kunisada, T., Ogawa, M., Nishikawa, S., Okamura, H., Sudo, T., Shultz, L.D., and Nishikawa, S. (1990). The murine mutation osteopetrosis is in the coding region of the macrophage colony stimulating factor gene. *Nature*, **345**, 442–4.

68. Jedrzejczak, W.W., Bartocci, A., Ferrante, A.W. Jr., Ahmed-Ansari, A., Sell, K.W., Pollard, J.W., and Stanley, E.R. (1990). Total absence of colony-stimulating factor 1 in the macrophage-deficient osteopetrotic (*op/op*) mouse. *Proc. Natl. Acad. Sci. (USA)*, **87**, 4828–32. [Published erratum appears in *Proc. Natl. Acad. Sci. (USA)*, (1991) **88**, 5937.]

69. Morris, S.W., Valentine, M.B., Shapiro, D.N., Sublett, J.E., Deaven, L.L., Foust, J.T., Roberts, M.W., Cerretti, D.G., and Look, A.T. (1991). Reassignment of the human CSF1 gene to chromosome 1p13-p21. *Blood*, **78**, 2013–20.

70. LaCombe, C., Tambourine, P., Mattei, M.G., Simon, D., and Guenet, J.L. (1988). The murine erythropoietin gene is localized on chromosome 5. *Blood*, **72**, 1440–2.

71. Lock, L.F., Pines, J., Hunter, T., Gilbert, D.J., Gopalan, G., Jenkins, N.A., Copeland, N.G., and Donovan, P.J. (1992). A single cyclin A gene and multiple cyclin B1-related sequences are dispersed in the mouse genome. *Genomics*, **13**, 415–24.

72. Law, M.L., Cai, G.Y., Lin, F.K., Wei, Q., Huang, S.Z., Hartz, J.H., Morse, H., Lin, C.H., Jones, C., and Kao, F.T. (1986). Chromosomal assignment of the human erythropoietin gene and its DNA polymorphism. *Proc. Natl. Acad. Sci. (USA)*, **83**, 6920–4.

73. Watkins, P.C., Eddy, R., Hoffman, N., Stanislovitis, P., Beck, A.K., Galli, J., Vellucci, V., Gusella, J.F., and Shows, T.B. (1986). Regional assignment of the erythropoietin gene to human chromosome region 7pter-q22. *Cytogenet. Cell. Genet.*, **42**, 214–8.

74. Powell, J.S., Berkner, K.L., Lebo, R.V., and Adamson, J.W. (1986). Human erythropoietin gene: high level expression in stably transfected mammalian cells and chromosome localization. *Proc. Natl. Acad. Sci. (USA)*, **83**, 6465–9.

75. Kola, I., Davey, A., and Gough, N.M. (1989). Localization of the murine Leukaemia inhibitory factor gene

near the centremere on chromosome 11. *Growth Factors*, **2**, 235–40.

76. Bottoroff, D., and Stone, J.C. (1992). The murine leukaemia inhibitory factor gene (LIF) is located on proximal chromosome 11, not chromosome 13. *Mamm. Genome*, **3**, 681–4.

77. Sutherland, G.R., Baker, E., Hyland, V.J., Callen, D.F., Stahl, J., and Gough, N.M. (1989). The gene for human leukaemia inhibitory factor [LIF] maps to 22q12. *Leukaemia*, **3**, 9–13.

78. Budorf, M., Emanuel, B.S., Mohandes, T., Goeddel, D.V., and Lowe, D.G. (1989). Human differentiation-stimulating factor (leukaemia inhibitory factor, human interleukin DA) gene maps distal to the Ewing sarcoma breakpoint on 22q. *Cytogenet. Cell. Genet.*, **52**, 19–22.

79. Rose, T.M., Lagrou, M.J., Fransson, I., Werelius, B., Delattre, O., Thomas, G., de Jong, P.J., Todaro, G.J., and Kumanski, J.P. (1993). The genes for oncostatin M (OSM) and leukaemia inhibitory factor (LIF) are tightly linked on human chromosome 22. *Genomics*, **17**, 136–40.

80. Giovannini, M., Selleri, L., Hermanson, G.G., and Evans, G.A. (1993). Localization of the human oncostatin M gene (OSM) to chromosome 22q12, distal to the Ewing sarcoma breakpoint. *Cytogenet. Cell. Genet.*, **62**, 32–4.

81. Russell, E.S. (1979). Hereditary anemias of the mouse: a review for geneticists. *Adv. Genet.*, **20**, 357–459.

82. Taylor, R.G., Grieco, D., Clarke, G.A., McInnes, R.R., and Taylor, B.A. (1993). Identification of the mutation in murine histidinemia (*his*) and genetic mapping of the murine histidase locus (*Hal*) on chromosome 10. *Genomics*, **16**, 231–40.

83. Kwon, B.S., Chintamaneni, C., Kozak, C.A., Copeland, N.G., Gilbert, D.J., Jenkins, N., Barton, D., Francke, U., Kobayashi, Y., and Kim, K.K. (1991). A melanocyte-specific gene, Pmel 17, maps near the silver coat color locus on mouse chromosome 10 and is in a syntenic region on human chromosome 12. *Proc. Natl. Acad. Sci. (USA)*, **88**, 9228–32.

84. Copeland, N.G., Gilbert, D.J., Cho, B.C., Donovan, P.J., Jenkins, N.A., Cosman, D., Anderson, D., Lyman, S.D., and Williams, D.E. (1990). Mast cell growth factor maps near the steel locus on mouse chromosome 10 and is deleted in a number of steel alleles. *Cell*, **63**, 175–83.

85. Zsebo, K.M., Williams, D.A., Geissler, E.N., Broudy, V.C., Martin, F.H., Atkins, H.L., Hsu, R.Y., Birkett, N.C., Okino, K.H., Murdock, D.C., Jacobsen, F.W., Langley, K.E., Smith, K.A., Takeishi, T., Cattanach, B.M., Galli, S.J., and Suggs, S.V. (1990). Stem cell factor is encoded at the *Sl* locus of the mouse and is the ligand for the *c-kit* tyrosine kinase receptor. *Cell*, **63**, 213–24.

86. Huang, E., Nocka, K., Beier, D.R., Chu, T.Y., Buck, J., Lahm, H.W., Wellner, D., Leder, P., and Besmer, P. (1990). The hematopoietic growth factor KL is encoded by the *Sl* locus and is the ligand of the *c-kit* receptor, the gene product of the *W* locus. *Cell*, **63**, 225–33.

87. Geissler, E.N., Liao, M., Brook, J.D., Martin, F.H., Zsesbo, K.M., Housman, D.E., and Calli, S.J. (1991). Stem cell factor (*SCF*), a novel hematopoietic growth factor and ligand for c-kit tyrosine kinase receptor, maps on human chromosome 12 between 12q14.3 and 12qter. *Som. Cell and Mol. Genet.*, **17**, 207–14.

88. Dandoy, F., Kelley, K.A., DeMaeyer-Guignard, J., DeMaeyer, E., and Pitha, P.M. (1984). Linkage analysis of the murine interferon-α locus on chromosome 4. *J. Exp. Med.*, **160**, 294–302.

89. Lovett, M., Cox, D.R., Yee, D., Boll, W., Weissmann, C., Epstein, C.J., and Epstein, L.B. (1984). The chromosomal location of mouse interferon α genes. *EMBO J.*, **3**, 1643–6.

90. Cahilly, L.A., George, D., Daugherty, B.L., and Pestka, S. (1985). Subchromosomal localization of mouse IFN-α genes by *in situ* hybridization. *J. Interferon Res.*, **5**, 391–5.

91. vander Korput, J.A.G.M., Hilkens, J., Kroezen, V., Zwarthoff, E.C., and Trapman, J. (1985). Mouse interferon alpha and beta genes are linked at the centromere proximal region of chromosome 4. *J. Gen. Virol.*, **66**, 493–502.

92. Langer, J.A., and Pestka, S. (1988). Interferon receptors. *Immunol. Today*, **9**, 393–400.

93. Kelley, K.A., Kozak, C.A., and Pitha, P.M. (1985). Localization of the mouse interferon-β1 gene to chromosome 4. *J. Interferon Res.*, **5**, 409–13.

94. Henry, L., Sizun, J., Turleau, C., Boue, J., Azoulay, M., and Junien, C. (1984). The gene for human fibroblast interferon (IFB) maps to 9p21. *Hum. Genet.*, **68**, 67–9.

95. Naylor, S.L., Gray, P.W., and Lalley, P.A. (1984). Mouse immune interferon [IFN-gamma] gene is on chromosome 10. *Somat. Cell Mol. Genet.*, **10**, 531–4.

96. Justice, M.J., Siracusa, L.D., Gilbert, D.J., Heisterkamp, N., Groffen, J., Chada, K., Silan, C.M., Copeland, N.G., and Jenkins, N.A. (1990). A genetic linkage map of mouse chromosome 10: localization of eighteen molecular markers using a single interspecific backcross. *Genetics*, **125**, 855–66.

97. Kozak, C.A., Peyser, M., Krall, M., Mariano, T.M., Kumar, C.S., Pestka, S., and Mock, B.A. (1990). Molecular genetic markers spanning mouse chromosome 10. *Genomics*, **8**, 519–24.

98. Naylor, S.L., Sakaguchi, A.Y., Shows, T.B., Law, M.L., Goeddel, D.V., and Gray, P.W. (1983). Human immune interferon gene is located on chromosome 12. *J. Exp. Med.*, **57**, 1020–7.

99. Trent, J.M., Olson, S., and Lawn, R.M. (1982). Chromosomal localization of human leukocyte, fibroblast and immune interferon genes by means of *in situ* hybridization. *Proc. Natl. Acad. Sci. (USA)*, **79**, 7809–13.

100. Nedospasov, S.A., Hirt, B., Shakhov, A.N., Dobrynin, V.N., Kawashima, E., Accolla, R.S., and Jongeneel, C.V. (1986). The genes for tumour necrosis factor [TNFα] and lymphotoxin [TNFβ] are tandemly arranged on chromosome 17 of the mouse. *Nucleic Acids Res.*, **14**, 7713–25.

101. Nedwin, G.E., Naylor, S.L., Sakaguchi, A.Y., Smith, D., Jarrett-Nedwin, J., Pennica, D., Goeddel, D.V., and Gray, P.W. (1985). Human lymphotoxin and tumour necrosis factor genes: structure, homology and chromosomal localization. *Nucleic Acids Res.*, **13**, 6361–73.

102. Evans, A.M., Petersen, J.W., Sekhon, G.S., and DeMars,

R. (1989). Mapping of prolactin and tumour necrosis factor—beta genes on human chromosome 6p using lymphoblastoid cell deletion mutants. *Somat. Cell Mol. Genet.*, **15**, 203–13.

103. Allen, R.C., Armitage, R.J., Conley, M.E., Rosenblatt, H., Jenkins, N.A., and Copeland, N.G. (1993). CD40 ligand gene defects responsible for X-linked hyper-IgM syndrome. *Science*, **259**, 990–3.

104. Aruffo, A., Farrington, M., Hollenbaugh, D., Li, X., Milatovich, A., Nonoyama, S., Bajorath, J., Grosmaire, L.S., Stenkamp, R., Neubauer, M., Roberts, R.L., Noelle, R.J., Ledbetter, J.A., Francke, U., and Ocks, H.D. (1993). The CD40 ligand gp39, is defective in activated T cells from patients with X-linked hyper-IgM syndrome. *Cell*, **72**, 291–300.

105. Fuleihan, R., Ramesh, N., Loh, R., Jabara, H., Rosen, R.S., Chatila, T., Fu, S.M., Stamenkovic, I., and Geha, R.S. (1993). Defective expression of the CD40 ligand in X chromosome-linked immunoglobulin deficiency with normal or elevated IgM. *Proc. Natl. Acad. Sci. (USA)*, **90**, 2170–3.

106. Korthauer, U., Graf, D., Mages, H.W., Briere, F., Padayachee, M., Malcolm, S., Ugazio, A.G., Notarangelo, L.D., Levinsky, R.J., and Kroczek, R.A. (1992). Defective expression of T-cell CD40 ligand causes X-linked immunodeficiency with hyper-IgM. *Nature*, **361**, 539–41.

107. Di Santo, J.P., Bonnefoy, J.Y., Gauchat, J.F., Fischer, A., and de Saint Basile, G. (1993). CD40 ligand mutations in X-linked immunodeficiency with hyper-IgM. *Nature*, **361**, 541–3.

108. Graf, D., Korthauer, U., Mages, H.W., Senger, G., and Kroczek, R.A. (1992). Cloning of TRAP, a ligand for CD40 on human T cells. *Eur. J. Immunol.*, **12**, 3191–4.

109. Fowler, K.J., Mann, G.B., and Dunn, A.R. (1993). Linkage of the murine transforming growth factor α gene with *Igk*, *Ly-2*, and *Fabp1* on chromosome 6. *Genomics*, **16**, 782–4.

110. Mann, G.B., Fowler, J., Gabriel, A., Nice, E.C., Williams, L., and Dunn, A.R. (1993). Mice with a null mutation of the TGFα gene have abnormal skin architecture, wavy hair, and curly whiskers and often develop corneal inflammation. *Cell*, **73**, 249–61.

111. Luetteke, N.C., Qui, T.H., Peiffer, R.L., Oliver, P., Smithies, O., and Lee, D.C. (1993). TGFα deficiency results in hair follicle and eye abnormalities in targeted and waved-1 mice. *Cell*, **73**, 263–78.

112. Brissenden, J.E., Derynck, R., and Francke, U. (1985). Mapping of transforming growth factor α gene on human chromosome 2 close to the breakpoint of the Burkitt's lymphoma t(2;8) variant translocation. *Cancer Res.*, **45**, 5593–7.

113. Tricoli, J.V., Nakai, H., Byers, M.G., Rall, L.B., Bell, G.I., and Shows, T.B. (1986). The gene for human transforming growth factor α is on the short arm of chromosome 2. *Cytogenet. Cell Genet.*, **42**, 94–8.

114. Dickinson, M.E., Kobrin, M.S., Silan, C.M., Kingsley, D.M., Justice, M.J., Miller, D.A., Ceci, J.D., Lock, L.F., Lee, A., Buchberg, A.M., Siracusa, L.D., Lyons, K.M., Derynck, R., Hogan, B.L.M., Copeland, N.G., and Jenkins, N.A. (1990). Chromosomal localization of seven

115. Fujii, D., Brissenden, J.E., Derynk, R., and Francke, U. (1986). Transforming growth factor β gene maps to human chromosome 19 long arm and to mouse chromosome 7. *Somat. Cell Mol. Genet.*, **12**, 281–8.

116. Saunders, A., and Seldin, M.F. (1990). A molecular genetic linkage map of mouse chromosome 7. *Genomics*, **8**, 525–35.

117. Ardinger, H.H., Ardinger, R.H., Bell, G.I., and Murray, J.C. (1988). RFLP for the human transforming growth factor beta-1 gene (TGFβ) on chromosome 19. *Nucleic Acids Res.*, **16**, 8202.

118. Reference 118 deleted from reference list.

119. Barton, D.E., Foellmer, B.E., Du, J., Tamm, J., Derynck, R., and Francke, U. (1988). Chromosomal mapping of genes for transforming growth factors $\beta2$ and $\beta3$ in man and mouse: dispersion of TGF-β gene family. *Oncogene Res.*, **3**, 323–31.

120. Nishimura, D.Y., Purchio, A.F., and Murray, J.C. (1993). Linkage localization of TGFB2 and the human homeobox gene HLX1 to chromosome 1q. *Genomics*, **15**, 357–64.

121. Dijke, P.T., Geurts van Kessel, A.H.M., Foulkes, J.G., and Le Beau, M.M. (1988). Transforming growth factor type $\beta3$ maps to human chromosome 14, region q23–24. *Oncogene*, **3**, 721–4.

122. Moseley, W.S., and Seldin, M.F. (1989). Definition of mouse chromosome 1 and 3 gene linkage groups that are conserved on human chromosome 1: evidence that a conserved linkage group spans the centromere of human chromosome 1. *Genomics*, **5**, 899–905.

123. Mucenski, M.L., Taylor, B.A., Copeland, N.G., and Jenkins, N.A. (1988). Chromosomal location of *Evi-1*, a common site of ecotropic viral integration in AKXD murine myeloid tumours. *Oncogene Res.*, **2**, 219–33.

124. Zabel, B.U., Eddy, R.L., Lalley, P.A., Scott, J., Bell, G.I., and Shows, T.B. (1985). Chromosomal locations of the human and mouse genes for precursors of epidermal growth factor and the beta subunit of nerve growth factor. *Proc. Natl. Acad. Sci. (USA)*, **82**, 469–73.

125. Brissenden, J.E., Ullrich, A., and Francke, U. (1984). Human chromosomal mapping of genes for insulin-like growth factors I and II and epidermal growth factor. *Nature*, **310**, 781–4.

126. Bonthron, D.T., Morton, C.C., Orkin, S.H., and Collins, T. (1988). Platelet-derived growth factor A chain: Gene structure, chromosomal location, and basis for alternative mRNA splicing. *Proc. Natl. Acad. Sci. (USA)*, **85**, 1492–6.

127. Betsholtz, C., Johnsson, A., Heldin, C.-H., Westermark, B., Lind, P., Urdea, M.S., Eddy, R., Shows, T.B., Philpott, K., Mellor, A.L., Knott, T.J., and Scott, J. (1986). cDNA sequence and chromosomal localization of human platelet-derived growth factor A-chain and its expression in tumour cell lines. *Nature*, **320**, 695–9.

128. Malek, T.R., Vincek, V., Gatalica, B., and Bucan, M. (1993). The IL-2 receptor β chain gene (*Il-2rb*) is closely

linked to the *Pdgfb* locus on mouse chromosome 15. *Immunogenetics*, **38**, 154–6.

129. Johnson, K.R., and Davisson, M.T. (1992). A multipoint genetic linkage map of mouse chromosome 18. *Genomics*, **13**, 1143–9.

130. Swan, D.C., McBride, O.W., Robbins, K.C., Keithley, D.A., Reddy, E.P., and Aaronson, S.A. (1982). Chromosomal mapping of the simian sarcoma virus *onc* gene analogue in human cells. *Proc. Natl. Acad. Sci. (USA)*, **79**, 4691–5.

131. Bartram, C.R., de Klein, A., Hagemeijer, A., Grosveld, G., Heisterkamp, N., and Groffen, J. (1984). Localization of the human c-*sis* oncogene in Ph'-positive and Ph'-negative chronic myelocytic leukaemia by *in situ* hybridization. *Blood*, **63**, 223–5.

132. Orr-Urtreger, A., Trakhtenbrot, L., Ben-Levy, R., Wen, D., Rechavi, G., Lonai, P., and Yarden, Y. (1993). Neural expression and chromosomal mapping of Neu differentiation factor to 8p12-p21. *Proc. Natl. Acad. Sci. (USA)*, **90**, 1867–71.

133. Kingsmore, S.F., Moseley, W.S., Watson, M.L., Sabina, R.L., Holmes, E.W., and Seldin, M-F. (1990). Long-range restriction site mapping of a syntenic segment conserved between human chromosome 1 and mouse chromosome 3. *Genomics*, **7**, 75–83.

134. Dracopoli, N.C., Rose, E., Whitfield, G.K., Guidon, P.T. Jr., Bale, S.J., Chance, P.A., Kourides, I.A., and Housman, D.E. (1988). Two thyroid hormone regulated genes, the beta-subunits of nerve growth factor (NGFB) and thyroid stimulating hormone (TSHB), are located less than 310 kb apart in both human and mouse genomes. *Genomics*, **3**, 161–7.

135. Francke, U., De Martinville, B., Coussens, L., and Ullrich, A. (1983). The human gene for the β subunit of nerve growth factor is located on the proximal short arm of chromosome 1. *Science*, **222**, 1248–51.

136. Saunders, A.M., and Seldin, M.F. (1990). The syntenic relationship of proximal mouse chromosome 7 and the myotonic dystrophy gene region on human chromosome 19q. *Genomics*, **6**, 324–32.

137. Ozcelik, T., Rosenthal, A., and Francke, U. (1991). Chromosomal mapping of brain-derived neurotrophic factor and neurotrophin-3 genes in man and mouse. *Genomics*, **10**, 569–75.

138. Hanson, I.M., Seawright, A., and Van Heyningen, V. (1992). The human BDNF gene maps between FSHB and HVBS1 at the boundary of 11p13-p14. *Genomics*, **13**, 1331–3.

139. Maisonpierre, P.C., Le Beau, M.M., Espinosa, R. III, Ip, N.Y., Belluscio, L., De La Monte, S.M., Squinto, S., Furth, M.E., and Yancopoulos, G.D. (1991). Human and rat brain-derived neurotrophic factor and neurotrophin-3: gene structures, distributions, and chromosomal localizations. *Genomics*, **10**, 558–68.

140. Ip, N.Y., Ibanez, C.F., Nye, S.H., McClain, J., Jones, P.F., Gies, D.R., Belluscio, L., Le Beau, M.M., Espinosa, R. III, Squinto, S.P., Persson, H., and Yancopoulos, G.D. (1992). Mammalian neurotrophin-4: Structure, chromosomal localization, tissue distribution, and receptor specificity. *Proc. Natl. Acad. Sci. (USA)*, **89**, 3060–4.

141. Berkemeier, L.R., Ozcelik, T., Francke, U., and Rosenthal, A. (1992). Human chromosome 19 contains the neurotrophin-5 gene locus and three related genes that may encode novel acidic neurotrophins. *Somat. Cell and Mol. Genet.*, **18**, 233–45.

142. Kaupmann, K., Sendtner, M., and Jockush, H. (1991). The gene for ciliary neurotrophic factor (CNTF) maps to chromosome 19 of the mouse. *Mouse Genome*, **89**, 246.

143. Giovannini, M., Romo, A.J., and Evans, G.A. (1993). Chromosomal localization of the human ciliary neurotrophic factor gene (CNTF) to 11q12 by fluorescence *in situ* hybridization. *Cytogenet. Cell Genet.*, **63**, 62–3.

144. Lam, A., Fuller, F., Miller, J., Kloss, J., Manthorpe, M., Varon, S., and Cordell, B. (1991). Sequence and structural organization of the human gene encoding ciliary neurotrophic factor. *Gene*, **102**, 271–6.

145. Lev, A.A., Rosen, D.R., Kos, C., Clifford, E., Landes, G., Hauser, S.L., and Brown, R.H. Jr. (1993). Human ciliary neurotrophic factor: localization to the proximal region of the long arm of chromosome 11 and association with CA/GT dinucleotide repeat. *Genomics*, **16**, 539–41.

146. Zemel, S., Bartolomei, M.S., and Tilghman, S.M. (1992). Physical linkage of two mammalian imprinted genes, H19 and insulin-like growth factor 2. *Nature Genetics*, **2**, 61–5.

147. Jones, J.M., Meisler, M.H., Seldin, M.F., Lee, B.K., and Eicher, E.M. (1992). Localization of insulin-2 (*Ins-2*) and the obesity mutant tubby (*tub*) to distinct regions of mouse chromosome 7. *Genomics*, **14**, 197–9.

148. Bell, G.I., Gerhard, D.S., Fong, N.M., Sanchez-Pescado, R., and Rall, L.B. (1985). Isolation of the human insulin-like growth factor genes: insulin-like growth factor II and insulin genes are contiguous. *Proc. Natl. Acad. Sci. (USA)*, **82**, 6450–4.

149. Glaser, T., Housman, D., Lewis, W.H., Gerhard, D., and Jones, C. (1989). A fine-structure deletion map of human chromosome 11p: analysis of J1 series hybrids. *Somat. Cell Mol. Genet.*, **15**, 477–501.

150. Owerbach, D., Bell, G.I., Rutter, W.J., and Shows, T.B. (1980). The insulin gene is located on chromosome 11 in humans. *Nature*, **286**, 82–4.

151. Zabel, B.U., Kronenberg, H.M., Bell, G.I., and Shows, T.B. (1985). Chromosome mapping of genes on the short arm of human chromosome 11: Parathyroid hormone gene is at 11p15 together with the genes for insulin, c-Harvey-ras 1, and beta-hemoglobin. *Cytogenet. Cell Genet.*, **39**, 200–5.

152. Taylor, B.A., and Grieco, D. (1991). Localization of the gene encoding insulin-like growth factor I on mouse chromosome 10. *Cytogenet. Cell Genet.*, **56**, 57–8.

153. Francke, U., Yang-Feng, T.L., Brissenden, J.E., Ullrich, A. (1986). Chromosomal mapping of genes involved in growth control. *Cold Spring Harbor Symp. Quant. Biol.*, **51**, 855–66.

154. Tricoli, J.V., Rall, L.B., Scott, J., Bell, G.I., and Shows, T.B. (1984). Localization of insulin-like growth factor genes to human chromosomes 11 and 12. *Nature*, **310**, 784–6.

155. Hoppener, J.W.M., de Pagter-Holthuizen, P., Geurts van Kessel, A.H.M., Jansen, M., Kittur, S.D., Anton-arakis, S.E., Lips, C.J.M., and Sussenbach, J.S. (1985). The human gene encoding insulin-like growth factor I is located on chromosome 12. *Hum. Genet.*, **69**, 157–60.

156. Lalley, P.A., and Chirgwin, J.M. (1984). Mapping of the mouse insulin genes. *Cytogenet. Cell Genet.*, **37**, 515.

157. de Pagter-Holthuizen, P., Hoppener, J.W.M., Jansen, M., Geurts van Kessel, A.H.M., van Ommen, G.J.B., and Sussenbach, J.S. (1985). Chromosomal localization and preliminary characterization of the human gene encoding insulin-like growth factor II. *Hum. Genet.*, **69**, 170–3.

158. Elliott, R.W., Lee, B.K., and Eicher, E.M. (1990). Local-ization of the growth hormone gene to the distal half of mouse chromosome 11. *Genomics*, **8**, 591–4.

159. Jackson-Grusby, L.L., Pravtcheva, D., Ruddle, F.H., and Linzer, D.I.H. (1988). Chromosomal mapping of the prolactin/growth hormone gene family in the mouse. *Endocrinology*, **122**, 2462–6.

160. Owerbach, D., Rutter, W.J., Martial, J.A., Baxter, J.D., and Shows, T.B. (1980). Genes for growth hormone, chorionic somatomammotropin and growth hor-mone-like gene on chromosome 17 in humans. *Science*, **209**, 289–92.

161. Holcombe, R.F., Stephenson, D.A., Zweidler, A., Stew-art, R.M., Chapman, V.M., and Sedman, J.G. (1991). Linkage of loci associated with two pigment mu-tations on mouse chromosome 13. *Genet. Res.*, **58**, 41–50.

162. Owerbach, D., Rutter, W.J., Cooke, N.E., Martial, J.A., and Shows, T.B. (1981). The prolactin gene is located on chromosome 6 in humans. *Science*, **212**, 815–6.

163. Justice, M.J., Gilbert, D.J., Kinzler, K.W., Vogelstein, B., Buchberg, A.M., Ceci, J.D., Matsuda, Y., Chapman, V.M., Patriotis, C., Makris, A., Tsichlis, P.N., Jenkins, N.A., and Copeland, N.G. (1992). A molecular genetic linkage map of mouse chromosome 18 reveals exten-sive linkage conservation with human chromosomes 5 and 18. *Genomics*, **13**, 1281–8.

164. Cox, R.D., Copeland, N.G., Jenkins, N.A., and Lehrach, H. (1991). Interspersed repetitive element polymerase chain reaction product mapping using a mouse inter-specific backcross. *Genomics*, **10**, 375–84.

165. Jaye, M., Howk, R., Burgess, W., Ricca, G.A., Chiu, I.M., Ravera, M.W., O'Brien, S.J., Modi, W.S., Maciag, T., and Drohan, W.N. (1986). Human endothelial cell growth factor: cloning, nucleotide sequence, and chromosome localization. *Science*, **233**, 541–5.

166. Mattei, M.G., Pebusque, M.J., and Birnbaum, D. (1992). Chromosomal localizations of mouse Fgf2 and Fgf5 genes. *Mammal. Genome*, **2**, 135–7.

167. Lafage, M., Galland, F., Simonetti, J., Prats, H., Mattei, M.G., and Birnbaum, D. (1990). The human basic fibro-blast growth factor gene is located on the long arm of chromosome 4 at bands q26-q27. *Oncogene Res.*, **5**, 241–4.

168. Casey, G., Smith, R., McGillivray, D., Peters, G., and Dickson, C. (1986). Characterization and chromosome assignment of the human homolog of *int-2*, a poten-tial proto-oncogene. *Mol. Cell Biol.*, **6**, 502–10.

169. Adelaide, J., Mattei, M.G., Marics, I., Raybaud, F., Planche, J., de Lapeyriere, O., and Birnbaum, D. (1988). Chromosomal localization of the *hst* onco-gene and its co-amplification with the *int-2* oncogene in a human melanoma. *Oncogene*, **2**, 413–6.

170. Benovic, J.L., Onorato, J.J., Arriza, J.L., Stone, W.C., Lohse, M., Jenkins, N.A., Gilbert, D.J., Copeland, N.G., Caron, M.G., and Lefkowitz, R.J. (1991). Cloning, ex-pression, and chromosomal localization of β-adrener-gic receptor kinase 2. *J. Biol. Chem.*, **266**, 14939–46.

171. Nguyen, C., Roux, D., Mattei, M.G., de Lapeyriere, O., Goldfarb, M., Birnbaum, D., and Jordan, B. (1988). The FGF-related oncogenes *hst* and *int2*, and the *bcl1* lo-cus are contained within one megabase in band q13 of chromosome 11, while the *fgf5* oncogene maps to 4q21. *Oncogene*, **3**, 703–8.

172. de Lapeyriere, O., Rosnet, O., Benharroch, D., Ray-baud, F., Marchetto, S., Planche, J., Galland, F., Matti, M.G., Copeland, N.G., Jenkins, N.A., Coulier, F., and Birnbaum, D. (1990). Structure, chromosome map-ping and expression of the murine *Fgf-6* gene. *Onco-gene*, **5**, 823–31.

173. Ludwig, T., Ruther, U., Metzger, R., Copeland, N.G., Jenkins, N.A., Lobel, P., and Hoflack, B. (1992). Gene and pseudogene of the mouse cation-dependent mannose 6-phosphate receptor. *J. Biol. Chem.*, **267**, 12211–9.

174. Goodwin, R.G., Anderson, D., Jerzy, R., Davis, T., Bran-nan, C.I., Copeland, N.G., Jenkins, N.A., and Smith, C.A. (1991). Molecular cloning and expression of the type 1 and type 2 murine receptors for tumour necro-sis factor. *Mol. Cell Biol.*, **11**, 3020–6.

175. Yi, T., Gilbert, D.J., Jenkins, N.A., Copeland, N.G., and Ihle, J.N. (1992). Assignment of a novel protein tyro-sine phosphatase gene (Hcph) to mouse chromosome 6. *Genomics*, **14**, 793–5.

176. Marics, I., Adelaide, J., Raybaud, F., Mattei, M.G., Cou-lier, F., Planche, J., de Lapeyriere, O., and Birnbaum, D. (1989). Characterization of the *HST*-related *FGF6* gene, a new member of the fibroblast growth factor gene family. *Oncogene*, **4**, 335–40.

177. Szpirer, C., Riviere, M., Cortese, R., Nakamura, T., Is-lam, M.Q., Levan, G., and Szpirer, J. (1992). Chromo-somal localization in man and rat of the genes encod-ing the liver-enriched transcription factors C/EBP, DBP, and HNF1/LFB-1 (CEBP, DBP, and transcription factor 1, TCF1, respectively) and of the hepatocyte growth factor/scatter factor gene (HGF). *Genomics*, **13**, 293–300.

178. Weidner, K.M., Arakaki, N., Hartman, G., Vandekerck-hove, J., Weingart, S., Rieder, H., Fonasch, C., Tsu-bouchi, H. T., Daikuhara, Y., and Birchmeier, W. (1991). Evidence for the identity of human scatter factor and human hepatocyte growth factor. *Proc. Natl. Acad. Sci. (USA)*, **88**, 7001–5.

179. Fukuyama, R., Ichijoh, Y., Minoshima, S., Kitamura, N., and Shimizu, N. (1991). Regional localization of the

179. hepatocyte growth factor (HGF) gene to human chromosome 7 band q21.1. *Genomics*, **11**, 410–5.

180. Zarnegar, R., Petersen, B., DeFrances, M.C., and Michalopoulos, G. (1992). Localization of hepatocyte growth factor (HGF) gene on human chromosome 7. *Genomics*, **12**, 147–50.

181. Laguda, B., Selden, C., Jones, M., Hodgson, H., and Spurr, N.K. (1991). Assignment of the hepatocyte growth factor (HGF) to chromosome 7q22-qter. *Ann. Hum. Genet.*, **55**, 213–6.

182. Pettenati, M.J., Le Beau, M.M., Lemons, R.S., Shima, E.A., Kawasaki, E.S., Larson, R.A., Sherr, C.J., Diaz, M.O., and Rowley, J.D. (1987). Assignment of *CSF-1* to 5q33.1: Evidence for clustering of genes regulating hematopoiesis and their involvement in the deletion of the long arm of chromosome 5 in myeloid disorders. *Proc. Natl. Acad. Sci. (USA)*, **84**, 2970–4.

183. Justice, M.J., Silan, C.M., Ceci, J.D., Buchberg, A.M., Copeland, N.G., and Jenkins, N.A. (1990). A molecular genetic linkage map of mouse chromosome 13 anchored by the Beige (bg) and Satin (Sa) loci. *Genomics*, **6**, 341–51.

184. Stahl, J., Gearing, D.P., Willson, T.A., Brown, M.A., King, J.A., and Gough, N.M. (1990). Structural organization of the genes for murine and human leukaemia inhibitory factor (LIF): evolutionary conservation of coding and non-coding regions. *J. Biol. Chem.*, **265**, 8833–41.

185. Copeland, N.G., Silan, C.M., Kingsley, D.M., Jenkins, N.A., Cannizzaro, L.A., Croce, C.M., Huebner, K., and Sims, J.E. (1991). Chromosomal location of murine and human IL-1 receptor genes. *Genomics*, **9**, 44–50.

186. Leonard, W.J., Donlon, T.A., Lebo, R.V., and Green, W.C. (1985). Localization of the gene encoding the human interleukin-2 receptor on chromosome 10. *Science*, **228**, 1547–9.

187. Ardinger, R.H., and Murray, J.C. (1988). A Bgl I polymorphism for the interleukin-2 receptor gene (IL2R) on chromosome 10. *Nucleic Acids Res.*, **16**, 8201.

188. Campbell, H.D., Webb, G.C., Kono, T., Taniguchi, T., Ford, J.H., and Young, I.G. (1992). Assignment of the interleukin-2 receptor β chain gene (*Il-2rb*) to band E on mouse chromosome 15. *Genomics*, **12**, 179–80.

189. Gnarra, J.R., Otani, H., Wang, M.G., McBride, O.W., Sharon, M., and Leonard, W.J. (1990). Human interleukin 2 receptor β-chain gene: chromosomal localization and identification of 5′ regulatory sequences. *Proc. Natl. Acad. Sci. (USA)*, **87**, 3440–4.

190. Shibuya, H., Yoneyama, M., Nakamura, Y., Harada, H., Hatakeyama, M., Minamoto, S., Kono, T., Doi, T., White, R., and Taniguchi, T. (1990). The human interleukin 2 receptor β-chain gene: genomic organization, promoter analysis and chromosomal assignment. *Nucleic Acids Res.*, **18**, 3697–703.

191. Rakar, S.J., Goodall, G.J., Maretti, P.A.B., D'Andrea, R.J., and Gough, N.M. (1993). Localization of the murine IL-3 receptor a chain gene (*Il3ra*) to the proximal end of chromosome 14 (submitted).

192. Kremer, E., Baker, E., D'Andrea, R.J., Phillips, H., Moretti, P.A.B., Lopez, A.F., Vadas, M.A., Sutherland, G.R., and Goodall, G.J. (1993). A cytokine receptor gene cluster in the X-Y pseudoautosomal region? *Blood*, **82**, 22–8.

193. Gorman, D.M., Itoh, N., Jenkins, N.A., Gilbert, D.J., Copeland, N.G., and Miyajima, A. (1992). Chromosomal localization and organization of the murine genes encoding the β subunits (AIC2A and AIC2B) of the IL-3, GM-CSF and IL-5 receptors. *J. Biol. Chem.*, **267**, 15842–8.

194. Shen, Y., Baker, E., Callen, D.F., Sutherland, G.R., Willson, T.A., Rakar, S., and Gough, N.M. (1992). Localization of the human GM-CSF receptor β chain gene to chromosome 22q12.2–22q13.1 *Cytogenet. Cell Genet.*, **61**, 175–7.

195. Pritchard, M.A., Baker, E., Whitmore, S.A., Sutherland, G.R., Idzerda, R.L., Park, L.S., Cosman, D., Jenkins, N.A., Gilbert, D.J., Copeland, N.G., and Beckmann, M.P. (1991). The interleukin-4 receptor gene (IL4R) maps to 16p11.2–16p12.1 in human and to the distal region of mouse chromosome 7. *Genomics*, **10**, 801–6.

196. Gough, N.M., and Rakar, S. (1992). Localization of the IL5 receptor gene to distal half of murine chromosome 6 using recombinant inbred strains of mice. *Genomics*, **12**, 855–6.

197. Isobe, M., Kumura, Y., Murata, Y., Takaki, S., Tominaga, A., Takatsu, K., and Ogita, Z. (1992). Localization of the gene encoding the a subunit of the human interleukin-5 receptor (IL5RA) to chromosome region 3p24–3p26. *Genomics*, **14**, 755–8.

198. Szpirer, J., Szpirer, C., Riviere, M., Houart, C., Baumann, M., Fey, G.H., Poli, V., Cortese, R., Islam, M.Q., and Levan, G. (1991). The interleukin-6-dependent DNA-binding protein gene (transcription factor 5: TCF5) maps to human chromosome 20 and rat chromosome 3, the IL6 receptor locus (IL6R) to human chromosome 1 and rat chromosome 2, and the rat IL6 gene to rat chromosome 4. *Genomics*, **10**, 539–46.

199. Kidd, V.J., Nesbitt, J.E., and Fuller, G.M. (1992). Chromosomal localization of the IL-6 receptor signal transducing subunit, gp130 (IL6ST). *Somat. Cell Mol. Genet.*, **18**, 477–83.

200. Goodwin, R. (1991). IL-7 receptor. *Cancer Cells*, **3**, 73.

201. Brannan, C.I., Gilbert, D.J., Ceci, J.D., Matsuda, Y., Chapman, V.M., Mercer, J.A., Eisen, H., Johnston, L.A., Copeland, N.G., and Jenkins, N.A. (1992). An interspecific linkage map of mouse chromosome 15 positioned with respect to the centromere. *Genomics*, **13**, 1075–81.

202. Lynch, M., Baker, E., Park, L.S., Sutherland, G.R., and Goodwin, R.G. (1992). The interleukin-7 receptor gene is at 5p13. *Hum. Genet.*, **89**, 566–8.

203. Morris, S.W., Nelson, N., Valentine, M.B., Shapiro, D.N., Look, A.T., Kozlosky, C.J., Beckmann, M.P., and Cerretti, D.P. (1992). Assignment of the genes encoding human interleukin-8 receptor Types 1 and 2 and an interleukin-8 receptor pseudogene to chromosome 2q35. *Genomics*, **14**, 685–91.

204. Mollereau, C., Muscatelli, F., Mattei, M.G., Vassart, G., and Parmentier, M. (1993). The high-affinity interleukin 8 receptor gene (IL8RA) maps to the 2q33-q36

region of the human genome: cloning of a pseudo-gene (IL8RBP) for the low-affinity receptor. *Genomics*, **16**, 248–51.

205. Ahuja, S.K., Özçelik, T., Milatovitch, A., Francke, U., and Murphy, P.M. (1992). Molecular evolution of the human interleukin-8 gene cluster. *Nature Genet.*, **2**, 31–6.

206. Cerretti, D.P., Nelson, N., Kozlosky, C.J., Morrissey, P.J., Copeland, N.G., Gilbert, D.J., Jenkins, N.A., Dosik, J.K., and Mock, B.A. (1993). The murine homologue of the interleukin-8 receptor type B maps near the *Ity-Lsh-Bcg* disease resistance locus. *Genomics*, **18**, 410–3.

207. Disteche, C.M., Brannan, C.I., Larsen, A., Adler, D.A., Schorderet, D.F., Gearing, D., Copeland, N.G., Jenkins, N.A., and Park, L.S. (1992). The human pseudoautosomal GM-CSF receptor α subunit gene is autosomal in mouse. *Nature Genet.*, **1**, 333–6.

208. Gough, N.M., Gearing, D.P., Nicola, N.A., Baker, E., Pritchard, M., Callen, D.F., and Sutherland, G.R. (1990). Localization of the human GM-CSF receptor gene to the X–Y pseudoautosomal region. *Nature*, **345**, 734–6.

209. Rappold, G., Willson, T.A., Henke, A., and Gough, N.M. (1992). Arrangement and localization of the human GM-CSF receptor α chain gene CSF2RA within the X–Y pseudoautosomal region. *Genomics*, **14**, 455–61.

210. Slim, R., Levilliers, J., Horsthenke, B., Nguyen, V.C., Gough, N.M., and Petit, C. (1992). A human pseudoautosomal gene encodes the T2 ADP/ATP translocase and escapes X-inactivation. *Genomics*, **16**, 26–33.

211. Gough, N.M., Rakar, S., and Willson, T.A. (1991). Localization on chromosome 4 and 19 of two murine G-CSF receptor loci. (in preparation.)

212. Inazawa, J., Fukunaga, R., Seto, Y., Nakagawa, H., Misawa, S., Abe, T., and Nagata, S. (1991). Assignment of the human granulocyte colony stimulating factor receptor gene (CSF3R) to chromosome 1 at region p35-p34.3. *Genomics*, **10**, 1075–8.

213. Tweardy, D.J., Anderson, K., Cannizzaro, L.A., Steinman, R.A., Croce, C.M., and Huebner, K. (1992). Molecular cloning of cDNAs for the human granulocyte colony-stimulating factor receptor from HL-60 and mapping of the gene to chromosome region 1p32–34. *Blood*, **79**, 1148–54.

214. Hoggan, M.D., Halden, N.F., Buckler, C.B.E., and Kozak, C.A. (1988). Genetic mapping of the mouse c-fms proto-oncogene to chromosome 18. *J. Virol.*, **62**, 1055–6.

215. Sola, B., Simon, D., Mattei, M.G., Fichelson, S., Bordereaux, D., Tambourin, P.E., Guenet, J.L., and Gisselbrecht, S. (1988). Fim-1, Fim-2/c-fms, and Fim-3, three common integration sites of Friend murine leukaemia virus in myeloblastic leukaemias, map to mouse chromosome 13, 18 and 3, respectively. *J. Virol.*, **62**, 3973–7.

216. Wang, L.M., Killary, A.M., Fang, X.E., Parriott, S.K., Lalley, P.A., Bell, G.I., and Sakaguchi, A.Y. (1988). Chromosome assignment of mouse insulin, colony stimulating factor 1 and low-density lipoprotein receptors. *Genomics*, **3**, 172–5.

217. Sakai, Y., Miyawaki, S., Shimizu, A., Ohno, K., and Watanabe, T. (1991). A molecular genetic linkage map of mouse chromosome 18, including *spm*, *Grl-1*, *Fim-2/c-fms*, and *Mbp. Biochem. Genet.*, **29**, 103–13.

218. Bartram, C.R., Böhlke, J.V., Adolph, S., Haemister, H., Ganser, A., Anger, B., Heisterkamp, N., and Groffen, J. (1987). Deletion of the *c-fms* sequence in the 5q- syndrome. *Leukemia*, **1**, 146–9.

219. Roberts, W.M., Look, A.T., Roussel, M.F., and Sherr, C.J. (1988). Tandem linkage of human CSF-1 receptor [c-fms] and PDGF receptor genes. *Cell*, **55**, 655–61.

220. Angert, E., Nagarajan, L., and Huebner, K. (1989). Previously unreported NcoI RFLP for human CSF1R. *Nucleic Acids Res.*, **17**, 2153.

221. de Parseval, N., Fichelson, S., Mayeux, P., Gisselbrecht, S., and Sola, B. (1993). Expression of functional β-platelet-derived growth factor receptors on hematopoietic cell lines. *Cytokine*, **5**, 8–15.

222. Budorf, M., Huebner, K., Emanuel, B., Croce, C.M., Copeland, N.G., Jenkins, N.A., and D'Andrea, A.D. (1990). Assignment of the erythropoietin receptor (EPOR) gene to mouse chromosome 9 and human chromosome 19. *Genomics*, **8**, 575–8.

223. Winkelmann, J.C., Penny, L.A., Deaven, L.L., Forget, B.G., and Jenkins, R.B. (1990). The gene for the human erythropoietin receptor: analysis of the coding sequence and assignment to chromosome 19p. *Blood*, **76**, 24–30.

224. Gearing, D.P., Druck, T., Huebner, K., Overhauser, J., Gilbert, D.J., Copeland, N.G., and Jenkins, N.A. (1993). The leukaemia inhibitory factor receptor (LIFR) gene is located within a cluster of cytokine receptor loci on mouse chromosome 15 and human chromosome 5p12-p13. *Genomics*, **18**, 148–50.

225. Yarden, Y., Kuang, W-J., Yang-Feng, T., Coussens, L., Munemitsu, S., Dull, T.J., Chen, E., Schlessinger, J., Francke, U., and Ullrich, A. (1987). Human proto-oncogene c-kit: a new cell surface receptor tyrosine kinase for an unidentified ligand. *EMBO J.*, **6**, 3341–51.

226. Chabot, B., Stephenson, D.A., Chapman, V.M., Besmer, P., and Bernstein, A. (1988). The proto-oncogene c-kit encoding a transmembrane tyrosine kinase receptor maps to the mouse *w* locus. *Nature*, **335**, 88–9.

227. d'Auriol, L., Mattei, M.-G., André, C., and Galibert, F. (1988). Localization of the human *c-kit* proto-oncogene on the q11-q12 region of chromosome 4. *Hum. Genet.*, **78**, 374–6.

228. Fournier, A., Zhang, Z.Q., and Tan, Y.H. (1985). Human β, α but not γ interferon binding site is a product of the chromosome 21 interferon action gene. *Somat. Cell Mol. Genet.*, **11**, 291–5.

229. Mariano, T.M., Kozak, C.A., Langer, J.A., and Pestka, S. (1987). The mouse immune interferon receptor gene is located on chromosome 10. *J. Biol. Chem.*, **262**, 5812–4.

230. Pfizenmaier, K., Wiegmann, K., Scheurich, P., Krönke, M., Merlin, G., Aguet, M., Knowles, B.B., and Ucer, U. (1988). High affinity human IFNg-binding capacity is encoded by a single receptor gene located in

proximity to c-ros on human chromosome region 6q16 to 6q22. *J. Immunol.*, **141**, 856–60.

231. Aguet, M., Dembic, Z., and Merlin, G. (1988). Molecular cloning and expression of the human interferon-γ receptor. *Cell*, **55**, 273–80.

232. Goodwin, R.G., Anderson, D., Jerzy, R., Davis, T., Brannan, C.I., Copeland, N.G., Jenkins, N.A., and Smith, C.A. (1991). Molecular cloning and expression of the type 1 and type 2 murine receptors for tumour necrosis factor. *Mol. Cell. Biol.*, **11**, 3020–6.

233. Baker, E., Chen, L.Z., Smith, C.A., Callen, D.F., Goodwin, R., and Sutherland, G.R. (1991). Chromosomal location of the human tumour necrosis factor receptor genes. *Cytogenet. Cell. Genet.*, **57**, 117–8.

234. Fuchs, P., Strehl, S., Dworzak, M., Himmler, A., and Ambros, P.F. (1992). Structure of the human TNF receptor 1 (p60) gene (TNFR1) and localization to chromosome 12p13. *Genomics*, **13**, 219–24.

235. Grimaldi, J.C., Torres, R., Kozak, C.A., Chang, R., Clark, E.A., Howard, M., and Cockayne, D.A. (1992). Genomic structure and chromosomal mapping of the murine CD40 gene. *J. Immunol.*, **149**, 3921–6.

236. Silver, J., Whitney, J.B. III, Kozak, C., Hollis, G., and Kirsch, I. (1985). *Erbb* is linked to the alpha-globin locus on mouse chromosome 11. *Mol. Cell. Biol.*, **5**, 1784–6.

237. Fowler, K.J., Walker, F., Alexander, W., Hibbs, M.L., Chetty, R., Thumwood, C., Nice, E.C., Mann, G.B., Burgess, A.W., and Dunn, A.R. (1993). The wavy hair, curly whiskers and occasional eye defects that characterize waved-2 mice is a consequence of a loss-of-function mutation in the epidermal growth factor receptor gene. (Submitted).

238. Spurr, N.K., Solomon, E., Jansson, M., Sheer, D., Goodfellow, P.N., Bodmer, W.F., and Vennstrom, B. (1984). Chromosomal localization of the human homologues to the oncogenes erbA and B. *EMBO J.*, **3**, 159–63.

239. Hsieh, C.L., Navankasattusas, S., Escobedo, J.A., Williams, L.T., and Francke, U. (1991). Chromosomal localization of the gene for AA-type platelet-derived growth factor receptor (PDGFRA) in humans and mice. *Cytogenet. Cell Genet.*, **56**, 160–3.

240. Stephenson, D.A., Mercola, M., Anderson, E., Wang, C., Stiles, C.D., Bowen-Pope, D.F., and Chapman, V.M. (1991). Platelet-derived growth factor receptor α-subunit gene (*Pdgfra*) is deleted in the mouse patch (*Ph*) mutation. *Proc. Natl. Acad. Sci. (USA)*, **88**, 6–10.

241. Smith, E.A., Seldin, M.F., Martinez, L., Watson, M.L., Choudhury, G.G., Lalley, P. A., Pierce, J., Aaronson, S., Barker, J., Naylor, S.L., and Sakaguchi, A.Y. (1991). Mouse platelet-derived growth factor receptor α gene is deleted in W[19H] and patch mutations on chromosome 5. *Proc. Natl. Acad. Sci. (USA)*, **88**, 4811–5.

242. Stenman, G., Eriksson, A., and Claesson-Welsh, L. (1989). Human PDGFA receptor gene maps to the same region on chromosome 4 as the kit oncogene. *Genes, Chromosomes & Cancer*, **1**, 155–8.

243. Sundaresan, S., and Francke, U. (1989). Genes for β2-adrenergic receptor and platelet-derived growth factor receptor map to mouse chromosome 18. *Somat. Cell Mol. Genet.*, **15**, 367–71.

244. Oakey, R.J., Caron, M.G., Lefkowitz, R.J., and Seldin, M.F. (1991). Genomic organization of adrenergic and serotonin receptors in the mouse: Linkage mapping of sequence-related genes provides a method for examining mammalian chromosome evolution. *Genomics*, **10**, 338–44.

245. Stetler-Stevenson, W.G., Liotta, L.A., and Seldin, M.F. (1992). Linkage analysis demonstrates that the Timp-2 locus is on mouse chromosome 11. *Genomics*, **14**, 828–9.

246. Fukushige, S.I., Matsubara, K.I., Yoshida, M., Sasaki, M., Suzuki, T., Semba, K., Toyoshima, K., and Yamamoto, T. (1986). Localization of a novel v-*erb*B-related gene, c-*erb*B-2 on human chromosome 17 and its amplification in a gastric cancer cell line. *Mol. Cell. Biol.*, **6**, 955–8.

247. van de Vijuer, M., van de Bersselaar, R., Devillee, P., Cornelisse, C., Petersen, J., and Nusse, R. (1987). Amplification of the *neu* (c-*erb*B-2) oncogene in human mammary tumours is relatively frequent and is often accompanied by amplification of the linked c-*erb*A oncogene. *Mol. Cell. Biol.*, **7**, 2019–23.

248. Huebner, K., Isobe, M., Chao, M., Bothwell, M., Ross, A.H., Finan, J., Hoxie, J.A., Sehgal, A., Buck, C.R., Lanahan, A., Nowell, P.C., Koprowski, H., and Croce, C.M. (1986). The nerve growth factor receptor gene is at human chromosome region 17q12–17q22, distal to the chromosome 17 breakpoint in acute leukaemias. *Proc. Natl. Acad. Sci. (USA)*, **83**, 1403–7.

249. Donaldson, D.H., Britt, D.E., Jones, C., Jackson, C.L., and Patterson, D. (1993). Localization of the gene for the ciliary neurotrophic factor receptor (CNTFR) to human chromosome 9. *Genomics*, **17**, 782–4.

250. Yang-Feng, T.L., Francke, U., and Ullrich, A. (1985). Gene for human insulin receptor: localization to site on chromosome 19 involved in pre-B-cell leukaemia. *Science*, **228**, 728–30.

251. Sundaresan, S., and Francke, U. (1989). Insulin-like growth factor I receptor gene is concordant with c-Fes protooncogene and mouse chromosome 7 in somatic cell hybrids. *Somat. Cell Mol. Genet.*, **15**, 373–7.

252. Ullrich, A., Gray, A., Tam, A.W., Yang-Feng, T., Tsubokawa, M., Collins, C., Henzel, W., Le Bon, T., Kathuria, S., Chen, E., Jacobs, S., Francke, U., Ramachandran, J., and Fujita-Yamaguchi, Y. (1986). Insulin-like growth factor I receptor primary structure: comparison with insulin receptor suggests structural determinants that define functional specificity. *EMBO J.*, **5**, 2503–12.

253. LeRoith, D., and Roberts, C.T. Jr. (1994). Receptors and binding proteins for insulin-like growth factors. In this Volume, p. 211.

254. Laureys, G., Barton, D.E., Ullrich, A., and Francke, U. (1988). Chromosomal mapping of the gene for the type II insulin-like growth factor receptor/cation-independent mannose 6-phosphate receptor in man and mouse. *Genomics*, **3**, 224–9.

255. Barlow, D.P., Stoger, R., Herrmann, B.G., Saito, K., and Schweifer, N. (1991). The mouse insulin-like growth

factor type-2 receptor is imprinted and closely linked to the *Tme* locus. *Nature*, **349**, 84–7.

256. Allander, S.V., Bajalica, S., Larsson, C., Luthman, H., Powell, D.R., Stern, L., Weber, G., Zazzi, H., and Ehrenborg, E. (1993). Structure and chromosomal localization of human insulin-like growth factor binding genes. *Growth Regul.*, **3**, 3–5.

257. Ehrenborg, E., Larsson, C., Stern, I., Janson, M., Powell, D.R., and Luthman, H. (1992). Contiguous localization of the genes encoding human insulin-like growth factor binding proteins 1 (IGBP1) and 3 (IGBP3) on chromosome 7. *Genomics*, **12**, 497–502.

258. Alitalo, T., Kontula, K., Koistinen, R., Aalto-Setala, K., Julkunen, M., Janne, O.A., Seppala, M., and del la Chapelle, A. (1989). The gene encoding human low-molecular weight insulin-like growth-factor binding protein (IGF-BP25): Regional localization to 7p12-p13 and description of a DNA polymorphism. *Hum. Genet.*, **83**, 335–8.

259. Ekstrand, J., Ehrenborg, E., Stern, I., Stellan, B., Zech, L., and Luthman, H. (1990). The gene for insulin-like growth-factor-binding protein-1 is localized to human chromosomal region 7p14–12. *Genomics*, **6**, 413–8.

260. Agarwal, N., Hsieh, C.L., Sills, D., Swaroop, M., Desai, B., Francke, U., and Swaroop, A. (1991). Sequence analysis, expression and chromosomal localization of a gene, isolated from a subtracted human retina cDNA library, that encodes an insulin-like growth factor binding protein (IGFBP2) *Exp. Eye Res.*, **52**, 549–61.

261. Ehrenborg, E., Vilhelmsdotter, S., Bajalica, S., Larsson, C., Stern, L., Koch, J., Brondum-Nilsen, K., and Luthman, H. (1991). Structure and localization of the human insulin-like growth factor-binding protein 2 gene. *Biochem. Biophys. Res. Commun.*, **176**, 1250–5.

262. Tonin, P., Ehrenborg, E., Lenoir, G., Feunteun, J., Lynch, H., Morgan, K., Zazzi, H., Vivier, A., Pollak, M., Huynh, H., Luthman, H., Larsson, C., and Narod, S. (1993). The human insulin-like growth factor-binding protein 4 gene maps to chromosome region 17q12-q21.1 and is close to the gene for hereditary breast-ovarian cancer. *Genomics*, **18**, 414–7.

263. Barton, D.E., Foellmer, B.E., Wood, W.I., and Francke, U. (1989). Chromosome mapping of the growth hormone receptor gene in man and mouse. *Cytogenet. Cell Genet.*, **50**, 137–41.

264. Barker, C.S., Bear, S.E., Keler, T., Copeland, N.G., Gilbert, D.J., Jenkins, N.A., Yeung, R.S., and Tsichlis, P.N. (1992). Activation of the prolactin receptor gene by promoter insertion in a moloney murine leukaemia virus-induced rat thymoma. *J. Virol.*, **66**, 6763–8.

265. Arden, K.C., Boutin, J.M., Djiane, J., Kelly, P.A., and Cavenee, W.K. (1990). The receptors for prolactin and growth hormone are localized in the same region of human chromosome 5. *Cytogenet. Cell Genet.*, **53**, 161–5.

266. Ruta, M., Howk, R., Ricca, G., Drohan, W., Zabelshansky, M., Laureys, G., Barton, D.E., Francke, U., Schlessinger, J., and Givol, D. (1988). A novel protein tyrosine kinase gene whose expression is modulated during endothelial cell differentiation. *Oncogene*, **3**, 9–15.

267. Lafage, M., Pedeutour, F., Marchetto, S., Simonetti, J., Prosperi, M.T., Gaudray, P., and Birnbaum, D. (1992). Fusion and amplification of two originally non-syntenic chromosomal regions in a mammary carcinoma cell line. *Genes Chromosomes & Cancer*, **5**, 40–9.

268. Mattei, M.G., Moreau, A., Gesnel, M.C., Houssaint, E., and Brethnach, R. (1991). Assignment by *in situ* by-bridization of a fibroblast growth factor receptor gene to human chromosome band 10q26. *Hum. Genet.*, **87**, 84–6.

269. Thompson, L., Plummer, S., Shalling, M., Altherr, M., Gusella, J., Housman, D., and Wasmuth, J. (1991). A gene encoding a fibroblast growth factor receptor isolated from the Huntington disease gene region of human chromosome 4. *Genomics*, **11**, 1133–42.

270. Armstrong, E., Portanen, J., Cannizzaro, L., Huebner, K., and Alitalo, K. (1992). Localization of the fibroblast growth factor receptor-4 gene to chromosome region 5q33-qter. *Genes Chromosomes & Cancer*, **4**, 94–8.

271. Sharma, S. (1994). Human B cell growth factor-12kD (BCGF-12kD). In this Volume, p. 97.

272. Plowman, G.D., Green, J.M., McDonald, V.L., Neubauer, M.G., Disteche, C.M., Todaro, G.J., and Shoyab, M. (1990). The amphiregulin gene encodes a novel epidermal growth factor-related protein with tumour-inhibiting activity. *Mol. Cell. Biol.*, **10**, 1969–81.

273. Fen, Z., Dhadly, M., Yoshizumi, M., Hilkert, R.J., Quertermous, T., Eddy, R.L., Shows, T.B., and Lee, M.E. (1993). Structural organization and chromosomal assignment of the gene encoding the human heparin-binding epidermal growth factor-like diptheria toxin receptor. *Biochemistry*, **32**, 7932–8.

274. Hoehe, M.R., Ehrenreich, H., Otterud, B., Caenazzo, L., Plaetke, R., Zander, H., and Leppert, M. (1993). The human endothelin-1 gene (EDN1) encoding a peptide with potent vasoactive properties maps distal to HLA on chromosome are 6p in close linage to D6S89. *Cytogenet. Cell Genet.*, **62**, 131–5.

275. Arinami, T., Ishikawa, M., Inoue, A., Yanagisawa, M., Masaki, T., Yoshida, M.C., and Hamaguchi, H. (1991). Chromosomal assignments of the human endothelin family genes: the endothelin-1 gene (EDN1) to 6p23-p24, the endothelin-2 gene (EDN2) to 1p34, and the endothelin-3 gene (EDN3) to 20q13.2-q13.3. *Am. J. Hum. Genet.*, **48**, 990–6.

276. Bloch, K.D., Eddy, R.L., Shows, T.B., and Quertermous, T. (1989). cDNA cloning and chromosomal assignment of the gene encoding endothelin 3. *J. Biol. Chem.*, **264**, 18156–61.

277. Rao, V.V., Loffler, C., and Hansmann, I. (1991). The gene for the novel vasoactive peptide endothelin 3 (EDN3) is localized to human chromosome 20q13.2-qter. *Genomics*, **10**, 840–1.

278. Maglione, D., Guerriero, V., Viglietto, G., Ferraro, M.G., Aprelikova, O., Alitalo, K., Del-Vecchio, S., Lei, K.J., Chou, J.Y., and Persico, M.G. (1993). Two alternative mRNAs coding for the angiogenic factor,

placental growth factor (PIGF), are transcribed from a single gene of chromosome 14. *Oncogene*, **8**, 925–31.

279. Noguchi, M., Yi, H., Rosenblatt, H.M., Filipovich, A.H., Adelsfern, S., Modi, W.S., McBride, O.W., and Leonard, W.J. (1993). Interleukin-2 receptor gamma chain mutation results in X-linked severe combined immunodeficiency in humans. *Cell*, **73**, 147–57.

280. Morris, C.M., Hao, Q.L., Heisterkamp, N., Fitzgerald, P.H., and Groffen, J. (1991). Localization of the TRK proto-oncogene to human chromosome bands 1q23–1q24. *Oncogene*, **6**, 1093–5.

281. Dean, M., Kozak, C., Robbins, J., Callahan, R., O'Brien, S., and Vande-Woude, G.F. (1987). Chromosomal localizaton of the met proto-oncogene in the mouse and cat genome. *Genomics*, **1**, 167–73.

282. Dean, M., Park, M., Le-Beau, M.M., Robins, T.S., Diaz, M.O., Rowley, J.D., Blair, D.G., and Vande-Woude, G.F. (1985). The human met oncogene is related to the tyrosine kinase oncogenes. *Nature*, **318**, 385–8.

283. Fernandez-Ruiz, E., St-Jacques, S., Bellon, T., Letarte, M., and Bernabeu, C. (1993). Assignment of the human endoglin gene (END) to 9q34–>qter. *Cytogenet. Cell Genet.*, **64**, 204–7.

284. Knaus, P.I., and Lodish, H.F. (1994). Receptors for transforming growth factor-β (TGF-β). In this Volume, p. 227.

285. Reference 285 deleted from reference list.

286. Rosnet, O., Stephenson, D., Mattei, M.G., Marchetto, S., Shibuya, M., Chapman, V.M., and Birnbaum, D. (1993). Close physical linkage of the FLT1 and FLT3 genes on chromosome 13 in man and chromosome 5 in mouse. *Oncogene*, **8**, 173–9.

287. Mathews, W., Jordan, C.T., Galvin, M., Jenkins, N.A., Copeland, N.G., and Lemischka, I.R. (1991). A receptor tyrosine kinase cDNA isolated from a population of enriched primitive hematopoietic cells and exhibiting close genetic linkage to c-kit. *Proc. Natl. Acad. Sci. (USA)*, **88**, 9026–30.

288. Terman, B.I., Jani-Sait, S., Carrion, M.E., and Shows, T.B. (1992). The KDR gene maps to human chromosome 4q31.2-q32, a locus which is distinct from locations for other type III growth factor receptor tyrosine kinases. *Cytogenet. Cell Genet.*, **60**, 214–5.

289. Hosoda, K., Nakao, K., Tamura, N., Arai, H., Ogawa, Y., Suga, S., Nakanishi, S., and Imura, H. (1992). Organization, structure, chromosomal assignment, and expression of the gene encoding the human endothelin-A receptor. *J. Biol. Chem.*, **267**, 18797–804.

290. Cyr, C., Huebner, K., Druck, T., and Kris, R. (1991). Cloning and chromosomal localization of a human endothelin ETA receptor. *Biochem. Biophys. Res. Commun.*, **181**, 184–90.

291. Arai, H., Nakao, K., Takaya, K., Hosoda, K., Ogawa, Y., Nakanishi, S., and Imura, H. (1993). The human endothelin-B receptor gene. Structural organization and chromosomal assignment. *J. Biol. Chem.*, **268**, 3463–70.

Nicholas M. Gough* and Steven J. Rakar*,
The Walter and Eliza Hall Institute of Medical Research,
Post Office, Royal Melbourne Hospital,
Parkville, Victoria 3050,
Phone: 61–3–345 2500,
Fax: 61–3–347 0852

* Present address: AMRAD Operations Pty. Ltd.,
17–27 Cotham Road,
Kew, Victoria 3101,
Phone: 61–3–8530022,
Fax: 61–3–8530202

Index

A list of alternative names and abbreviations is provided on p. xiii.